T0282486

Book of Abstracts of the 58th Annual Meeting of the European Association for Animal Production

The EAAP Book of Abstracts is published under the direction of Ynze van der Honing

EAAP - European Federation of Animal Science

The European Association for Animal Production wishes to express its appreciation to the
Ministero delle Politiche Agricole e Forestali (Italy) and the
Associazione Italiana Allevatori (Italy)
for their valuable support of its activities.

Book of Abstracts of the 58th Annual Meeting of the European Association for Animal Production

Dublin, Ireland, August 26th - 29th, 2007

Ynze van der Honing, Editor-in-chief

E. Strandberg, E. Cenkvari, C. Fourichon, M. Vestergard, J. Hermansen, C. Lazzaroni, G. Nieuwhof, C. Wenk and W. Martin-Rosset

ISBN 978-90-8686-045-6
ISSN 1382-6077

First published, 2007

Wageningen Academic Publishers
The Netherlands, 2007

Preface

The 58th annual meeting of the European Association for Animal Production (EAAP) is held in Dublin, Ireland, 26 to 29 August 2007. In Ireland the main theme of the meeting is "**Sustainable animal production - meeting the challenges for quality food**".

In total 39 sessions are planned.The start will be with a plenary paper:"**Meeting the challenges for quality food - Societal and industry aspects"** by B. Hubert (FR).

The annual EAAP meeting gives the opportunity to present new scientific results and discuss their potential applicability in animal production practices.This year's meeting is of particular interest for participants from a wide range of animal production organisations and institutions. Discussions stimulate developments in animal production and encourage research on relevant topics.

The book of abstracts is the main publication of the scientific contributions to this meeting; it covers a wide range of disciplines and livestock species. It contains the full programme and abstracts of the invited as well as the contributing speakers, including posters, of all 39 sessions.The number of abstracts submitted for presentation at this meeting is a true challenge for the different study commissions and chairpersons to put together a scientific programme. Some abstracts were not acceptable for the meeting due to poor quality and have been rejected. In addition to the theatre presentations, there will be a large number of poster presentations during the conference.

Several persons have been involved in the development of the book of abstracts. Wageningen Academic Publishers has been responsible for organising the administrative and editing work and the production of the book.The contact persons of the study commissions have been responsible for organising the scientific programme and communicating with the chairpersons and invited speakers.Their help in the programme is highly appreciated.

The programme is very interesting and I trust we will have a good meeting in Dublin. I hope that you will find this book a useful reference source as well as a reminder of a good meeting during which a large number of people actively involved in livestock science and production will meet and exchange ideas.

Ynze van der Honing
Editor-in Chief

EAAP Program Foundation

Aims

EAAP aims to bring to our annual meetings, speakers who can present the latest findings and views on developments in the various fields of science relevant to animal production and its allied industries. In order to sustain the quality of the scientific program that will continue to entice the broad interest in EAAP meetings we have created the "EAAP Program Foundation". This Foundation aims to support

- Invited speakers with a high international profile by funding part or all of registration and travel costs.
- Delegates from less favoured areas by offering scholarships to attend EAAP meetings
- Young scientists by providing prizes for best presentations

The "**EAAP Program Foundation**" is an initiative of the Scientific committee (SC) of EAAP. The Foundation aims to stimulate the quality of the scientific program of the EAAP meetings and to ensure that the science meets societal needs. The Foundation Board of Trustees oversees these aims and seeks to recruit sponsors to support its activities.

Sponsorships

1. **Meeting sponsor – From 3000 euro up**

- Acknowledgements in the book of abstracts with contact address and logo (on page X of abstract book of abstracts);
- one page allowance in the final programme booklet of Dublin;
- advertising/information material inserted in the bags of delegates;
- advertising/information material on a stand display;
- aknowledgement in the EAAP Newsletter with possibility of a page of publicity;
- possibility to add session and speaker support (at additional cost to be negotiated).

2. **Session sponsor – from 2000 euro up**

- Acknowledgements in the book of abstracts with contact address and logo (on page X of abstract book of abstracts);
- one page allowance in the final programme booklet of Dublin;
- advertising/ information material in the delegate bag;
- ppt at beginning of session to acknowledge support and recognition by session chair;
- acknowledgement in the EAAP Newsletter.

3. **Speaker sponsor - from 1000 euro up**

- half page allowance in the final programme booklet of Dublin;
- advertising/ information material in the delegate bag;
- recognition by speaker of the support at session;
- acknowledgement in the EAAP Newsletter.

4. **Registration Sponsor - (equivalent to a full registration fee of the Annual Meeting)**

- acknowledgements in the book of abstracts with contact address and logo (on page X of abstract book of abstracts);
- advertising/information material in the delegate bag.
-

The association

EAAP (The European Federation of Animal Science) organises every year an international meeting which attracts between 800 and 1000 people. The main aims of EAAP are to promote, by means of active co-operation between its members and other relevant international and national organisations, the advancement of scientific research, sustainable development and systems of production; experimentation, application and extension; to improve the technical and economic conditions of the livestock sector; to promote the welfare of farm animals and the conservation of the rural environment; to control and optimise the use of natural resources in general and animal genetic resources in particular; to encourage the involvement of young scientists and technicians. More information on the organisation and its activities can be found at www.eaap.org

Contact and further information

If you are interested to become a sponsor of the "EAAP Program Foundation" or want to have further information, please contact the secretary (rosati@eaap.org, Phone +39 06 44202639)

Acknowledgements

www.teagasc.ie

www.geno.no

Healthy cows ... join the future with us!
www.dansire.dk

IP Publishing Ltd, UK - www.ippublishing.com

www.dsm.com

European Association for Animal Production (EAAP)

President: Jim Flanagan
Secretary General: Andrea Rosati
Address: Via G.Tomassetti 3, A/I
I-00161 Rome, Italy
Phone +39 06 4420 2639
Fax: +39 06 8632 9263
E-mail: eaap@eaap.org
Web: www.eaap.org

Organising Committee

President: Dr.Tom Teehan, Department of Agriculture, Fisheries and Food
Secretary: Mr John Byrne, Department of Agriculture, Fisheries and Food
Members: Dr Dave Beehan, Department of Agriculture, Fisheries and Food
Dr Ignatius Byrne, Department of Agriculture, Fisheries and Food
Ms Louise Byrne, Department of Agriculture, Fisheries and Food
Mr Jim Flanagan, Teagasc
Dr Seamus Crosse, Teagasc
Dr Frank O'Mara, Teagasc
Prof Maurice Boland, University College Dublin
Ms Grainne Dwyer, Irish Grassland Association
Mr Cormac Healy, Irish Business and Employers Confederation
Dr Brian Wickham, Irish Cattle Breeding Federation

Organising Secretariat

Mr John Byrne
Department of Agriculture, Fisheries and Food
Agriculture House, Kildare Street, Dublin 2, Ireland
Tel : + 353 1 6072530
Fax : + 353 1 6072385
Email: John.Byrne@agriculture.gov.ie

Scientific Committee

President: Prof Maurice Boland, University College Dublin, Belfield, Dublin 4
Members: Dr Frank Buckley, Teagasc, Moorepark Dairy Production Research Centre, Co Cork
Dr John Murphy, Teagasc, Moorepark Dairy Production Research Centre, Co Cork
Dr Mark Crowe, University College Dublin, Belfield, Dublin 4
Dr Gerry Keane, Teagasc, Grange Beef Research Centre, Co Meath
Dr Seamus Hanrahan, Teagasc, Animal Production Research Centre, Athenry, Co Galway
Dr John O'Doherty, University College Dublin, Belfield, Dublin 4
Mr Nicholas Finnerty, Irish Horse Board, Maynooth, Co Kildare
Dr Frank O' Mara, Teagasc, Oakpark, Carlow

EAAP 2007 Conference Organisers

Ovation Group
1 Clarinda Park North
Dun Laoghaire, Co Dublin, Ireland
Phone: +353 1 2802641
Fax: +353 1 2802665
Email: eaap2007@ovation.ie

59th Annual Meeting of the European Association for Animal Production

Vilnius, Lithuania, 24 - 27 August 2008

Local Organisers
The Ministry of Agriculture of the Republic of Lithuania
Institute of Animal Sciences of Lithuanian Veterinary Academy
Lithuanian Veterinary Academy

Organising Committee
Prof. Kazimira Danutė Prunskienė, President

Scientific Committee
Prof. Henrikas Žilinskas
Dr. Violeta Juškienė

Organising Secretariat
Dalia Laureckaitė, Secretary
Head of unit of Integration to EU
State Animal Breeding Supervision Service under MoA
Gediminas ave. 19
LT-01103, Vilnius
Lithuania
Tel. Mob.+370-616-06345
Tel. +370-5-2391293, fax 94
E-mail: Dalia_L@zum.lt
Website: www.eaap2008.org

Abstract submission
www.wageningenacademic.com/eaap

36th ICAR Session and INTERBULL Meeting

Niagara Falls, New York, USA, 16-24 June, 2008

Contact: ICAR Secretariat, Rome, Italy
Email: icar@eaap.org

Incorporating the 42nd South African Society for Animal Science Congress and the 5th All-Africa Conference on Animal Production

23rd to 28th November 2008, Cape Town, South Africa

wcap2008@icon.co.za
www.wcap2008co.za
Tel. +27 12 420 3270
Fax. +27 12 420 3290

Scientific Programme EAAP-2007

Plenary session, Sunday 26th August, 8.15-9.15:
Meeting the challenges for quality food – Societal and industry aspects: B. Hubert (Fr)
Next EU strategy to support research on animal science: C. Patermann (B)

Sunday 26 August 9.30 – 13.00	Sunday 26 August 14.00 – 18.00	Monday 27 August 8.30 – 12.30
Session 1 (C*, G, Interbull) Breeding for robustness in cattle Chair: M. Klopcic (SI) and R. Reents (DE)	**Session 8 (Ph*, C, G, N, S) Sustainable animal production - Biological aspects related to milk and meat quality Chair: M. Vestergaard (DK) and T. Kristensen (DK)**	**Session 15 (C*, S, L, N) Sustainable animal production - Productivity aspects related to milk and meat quality Chair: A. Kuipers (NL) and M. Schneeberger (CH)**
Session 2 (Ph) Regulation of milk synthesis Chair: K. Sejrsen (DK)	Session 9 (P) Open session -Uniformity in pigs Chair: P. Knap (DE)	Session 16 (M*, P, Ph) Genetics and physiology of behaviour in relation to housing and transport Chair: B. Earley (IE)
Session 3 (M*, OIE) Disease transmission and epidemiology Chair: D. Chaisemartin (FR) and J.T. Sorensen (DK)	Session 10 (G) Free communications on Animal Genetics Chair: E. Carlen (SE)	Session 17 (H) Breeding evaluation in horses Chair: D. Lewczuk (PL)
Session 4 (N) Impact of feed processing on nutritive value Chair: M. Crovetto (IT) and D. Sauvant (FR)	Session 11 (C) Use of crosses and dairy calves for beef production Chair: G. Keane (IE)	Session 18 (G) Statistical analysis of genomics data Chair: T. Meuwissen (NO)
Session 5 (P) Environmental pollution through pig production Chair: D. Peskovicova (SK)	Session 12 (H*, M) Herd and stable management: health and performance issues Chair: D. Burger (CH)	
Session 6 (H) Human-horse relationships Chair: M. Hausberger (FR)	Session 13 (L) Understanding and assessing farmers' decision making Chair: A. Bernués (ES)	
Session 7 (L*, S) Changes in land use due to the CAP reform Chair: A. Pflimlin (FR)	Session 14 (S) Artificial Insemination Chair: W. Holtz (DE)	
	Session 39 (= S1 Continuation) (C*, G, Interbull) Breeding for robustness in cattle Chair: M. Klopcic (SI) and R. Reents (DE)	

Key – C, Cattle Production; G, Genetics; H, Horse production; L, Livestock Farming Systems; M, Management and Health; N, Nutrition; P, Pig production; Ph, Physiology; S, Sheep and Goat Production; (*) Denotes organising commission.

Bold - Sessions contributing to the theme of the meeting: **Sustainable animal production - meeting the challenges for quality food**

Monday 27 August 14.00 – 18.00	Tuesday 28 August 8.30 – 12.30	Wednesday 29 August 8.30 – 12.30

Monday 27 August, 14.00 – 18.00

Programme and elections meetings followed by free communications on:

Session 19
Animal Genetics
Chair: V. Ducrocq (FR)

Session 20
Animal Nutrition
Chair: J.E. Lindberg (SE)

Session 21
Animal Management & Health
Chair: S. Edwards (UK)

Session 22
Animal Physiology
Chair: M. Vestergaard (DK)

Session 23
Livestock Farming Systems
Chair: E. Matlova (CZ)

Session 24
Cattle Production
Chair: J.F. Hocquette (FR)

Session 25
Sheep and Goat Production
Chair: M. Schneeberger (CH)

Session 26
Pig Production
Chair: C. Wenk (CH)

Session 27
Horse Production
Chair: B. Younge (IE)

Poster session – Monday 27 August 18.00 – 20.00

Tuesday 28 August, 8.30 – 12.30

Session 28
(M*, C, H, L, N, S)
Maximizing forage and pasture use in the diet of herbivores
Chair: P. O'Kiely (IE)

Session 29 (H)
Applications of molecular genetics to breeding programmes
Chair: G. Guerin (FR)

Session 30 (G*, C, S)
Crossbreeding in ruminants
Chair: F. Buckley (IE)

Session 31 (L)
Approaches to livestock farm multifunctionality
Chair: S. Oosting (NL)

Session 32 (P*, N, Ph)
Nutrition and management of lactating sows
Chair: S. Chadd (UK)

Plenary session – Tuesday 28 August 13.30 – 19.00

Wednesday 29 August, 8.30 – 12.30

Session 33 (N*, C, S)
Free communications on Ruminant Nutrition
Chair: M. Crovetto (IT)

Session 34 (G*, Ph)
Biology and genetics of udder health
Chair: C. Kühn (DE)

Session 35 (P*, N, Ph, M)
Feed for Pig Health Workshop
Chair: D. Torrallardona (ES)

Session 36 (H)
(Full day: 07.45 -18.00)
Horse production in Ireland + Horse tour
Chair: N. Finnerty (IE)

Session 37 (C)
Free communications on Cattle Production
Chair: C. Lazzaroni (IT)

Session 38 (G)
Free communications on Animal Genetics
Chair: J. Szyda (PL)

Key – C, Cattle Production; G, Genetics; H, Horse production; L, Livestock Farming Systems; M, Management and Health; N, Nutrition; P, Pig production; Ph, Physiology; S, Sheep and Goat Production; (*) Denotes organising commission.

Bold - Sessions contributing to the theme of the meeting: **Sustainable animal production - meeting the challenges for quality food**

Commission on Animal Genetics

Dr Ducrocq	President	INRA
	France	Vincent.ducrocq@dga.jouy.inra.fr
Prof. Dr Simianer	Vice-President	University of Goettingen
	Germany	simianer@genetics-network.de
Dr Gandini	Vice-President	University of Milan
	Italy	gustavo.gandini@unimi.it
Dr Strandberg	Secretary	SLU
	Sweden	Erling.Strandberg@hgen.slu.se
Dr Szyda	Secretary	Agricultural University of Wroclaw
	Poland	szyda@karnet.ar.wroc.pl

Commission on Animal Nutrition

Prof. Crovetto	President	University of Milano
	Italy	matteo.crovetto@unimi.it
Dr Ortigues-Marty	Vice-President	INRA
	France	ortigues@sancy.clermont.inra.fr
Dr Moreira	Secretary	University of the Azores
	Portugal	ocmoreira@netcabo.pt
Dr Cenkvàri	Secretary	Szent Istvan University
	Hungary	Czenkvari.Eva@aotk.szie.hu

Commission on Animal Management & Health

Prof. von Borell	President	University Halle-Wittenberg
	Germany	borell@landw.uni-halle.de
Dr Sorensen	Vice-President	DIAS
	Denmark	jantind.sorensen@agrsci.dk
Dr Geers	Vice-President	Zootechnical Centre - K.U.Leuven
	Belgium	Rony.geers@agr.kuleuven.ac.be
Dr Edwards	Secretary	University of Newcastle Upon Tyne
	United Kingdom	Sandra.edwards@ncl.ac.uk
Prof. Fourichon	Secretary	INRA
	France	fourichon@vet-nantes.fr

Commission on Animal Physiology

Dr Sejrsen	President Denmark	DIAS Kr.Sejrsen@agrsci.dk
Dr Knight	Vice-President UK	Hannah Institute Knightc@hri.sari.ac.uk
Dr Royal	Vice-President UK	University Liverpool mdroyal@liverpool.ac.uk
Dr Ratky	Vice-President Hungary	Research Institute jozsef.ratky@atk.hu
Dr Chilliard	Secretary France	INRA chilliar@clermont.inra.fr
Dr Vestergaard	Secretary Denmark	DIAS Mogens.Vestergaard@agrsci.dk

Commission on Livestock Farming Systems

Dr Zervas	President Greece	Agricultural University of Athens gzervas@aua.gr
Dr Hermansen	Vice-President Denmark	DIAS john.hermansen@agrsci.dk
Dr Matlova	Vice-President Czech Republic	Res. Institute for Animal Production matlova.vera@vuzv.cz
Dr Peters	Vice-President Germany	Humboldt-University Berlin k.peters@agrar.hu-berlin.de
Dr Bernués Jal	Secretary Spain	C.I.T.A. abernues@aragon.es
Dr Ingrand	Secretary France	INRA/SAD ingrand@clermond.inra.fr

Commission on Cattle Production

Dr Kuipers	President Netherlands	Wageningen UR abele.kuipers@wur.nl
Dr Gigli	Vice-President Italy	ISZ sergio.gigli@isz.it
Dr Hocquette	Vice-President France	INRA hocquet@clermont.inra.fr
Dr Keane	Vice-President Ireland	TEAGASC gkeane@grange.teagasc.ie
Dr Klopcic	Secretary Slovenia	University of Ljublijana Marija.Klopcic@bfro.uni-lj.si
Dr Lazzaroni	Secretary Italy	Università di Torino carla.lazzaroni@unito.it

Commission on Sheep and Goat Production

Dr Schneeberger	President	ETH Zentrum
	Switzerland	markus.schneeberger@inw.agrl.ethz.ch
Dr Rihani	Vice-President	DPA
	Morocco	n.rihani@iav.ac.ma
Dr Dýrmundsson	Vice-President	Farmers Assoc.
	Iceland	ord@bondi.is
Dr C. Papachristoforou	Vice President	Agricultural Research Institute
	Cyprus	Chr.Papachristoforou@arinet.ari.gov.cy
Prof. Gauly	Secretary	University Göttingen
	Germany	mgauly@gwdg.de
Dr Nieuwhof	Secretary	Meat and Livestock Commission
	United Kingdom	gert_nieuwhof@mlc.org.uk

Commission on Pig Production

Prof. Dr Wenk	President	ETH Zentrum
	Switzerland	caspar.wenk@inw.agrl.ethz.ch
Dr Chadd	Vice-President	Royal Agric. College
	UK	steve.chadd@royagcol.ac.uk
Dr Knap	Vice-President	PIC Deutschland
	Germany	KnaP@de.pic.co.uk
Dr Pescovicova	Secretary	Research Institute of Animal Production
	Slovak Republic	peskovic@vuzv.sk
Dr Torrallardona	Secretary	IRTA
	Spain	David.Torrallardona@irta.es

Commission on Horse Production

Dr Martin-Rosset	President	INRA
	France	wrosset@clermont.inra.fr
Dr Kennedy	Vice-President	Writtle College
	UK	mjk@writtle.ac.uk
Dr Miraglia	Vice president	Molise University
	Italy	miraglia@unimol.it
Dr Koenen	Vice-President	NRS BV
	Netherlands	Koenen.E@cr-delta.nl
Dr Saastamoinen	Secretary	MTT Equines
	Finland	markku.saastamoinen@mtt.fi
Dr Sondergaard	Secretary	DIAS
	Denmark	eva.sondergaard@agrsci.dk

Session 01. Breeding for robustness in cattle - part I

Date: 26 August '07; 09:30 - 13:00 hours
Chairperson: M. Klopcic (SI) and R. Reents (DE)

Session 02. Regulation of milk synthesis

Date: 26 August '07; 09:30 - 13:00 hours
Chairperson: K. Sejrsen (DK)

Theatre **Session 02 no. Page**

Session 03. Disease transmission and epidemiology

Date: 26 August '07; 09:30 - 13:00 hours
Chairperson: D. Chaisemartin (FR)

Session 04. Impact of feed processing on nutritive value

Date: 26 August '07; 09:30 - 13:00 hours
Chairperson: M. Crovetto (IT) and D. Sauvant (FR)

Session 05. Environmental pollution through pig production

Date: 26 August '07; 09:30 - 13:00 hours
Chairperson: D. Peskovicova (SK)

Theatre **Session 05 no. Page**

Poster **Session 05 no. Page**

Session 06. Human-horse relationships

Date: 26 August '07; 09:30 - 13:00 hours
Chairperson: M. Hausberger (FR)

Session 07. Changes in land use due to the CAP reform

Date: 26 August '07; 09:30 - 13:00 hours
Chairperson: A. Pflimlin (FR)

Session 08. Sustainable animal production - Biological aspects related to milk and meat quality

Date: 26 August '07; 14:00 - 18:00 hours
Chairperson: M. Vestergaard (DK) and T. Kristensen (DK)

Session 09. Open session - Uniformity in pigs

Date: 26 August '07; 14:00 - 18:00 hours
Chairperson: P. Knap (DE)

Session 10. Free communications on Animal Genetics

Date: 26 August '07; 14:00 - 18:00 hours
Chairperson: E. Carlen (SE)

Poster **Session 10 no. Page**

Session 11. Use of crosses and dairy calves for beef production

Date: 26 August '07; 14:00 - 18:00 hours
Chairperson: G. Keane (IE)

Theatre **Session 11 no. Page**

Poster **Session 11 no. Page**

Session 12. Herd and stable management: health and performance issues

Date: 26 August '07; 14:00 - 18:00 hours
Chairperson: D. Burger (CH)

Theatre **Session 12 no. Page**

Session 13. Understanding and assessing farmers' decision making

Date: 26 August '07; 14:00 - 18:00 hours
Chairperson: A. Bernués (ES)

Session 14. Artificial Insemination

Date: 26 August '07; 14:00 - 18:00 hours
Chairperson: W. Holtz (DE)

Theatre **Session 14 no. Page**

Poster **Session 14 no. Page**

Session 15. Sustainable animal production - Productivity aspects related to milk and meat quality

Date: 27 August '07; 08:30 - 12:30 hours
Chairperson: A. Kuipers (NL) and M. Schneeberger (CH)

Session 16. Genetics and physiology of behaviour in relation to housing and transport

Date: 27 August '07; 08:30 - 12:30 hours
Chairperson: B. Earley (IE)

Session 17. Breeding evaluation in horses

Date: 27 August '07; 08:30 - 12:30 hours
Chairperson: D. Lewczuk (PL)

Theatre **Session 17 no. Page**

Poster **Session 17 no. Page**

Session 18. Statistical analysis of genomics data

Date: 27 August '07; 08:30 - 12:30 hours
Chairperson: T. Meuwissen (NO)

Session 20. Programme and elections meeting followed by Free communications on Animal Nutrition

Date: 27 August '07; 14:00 - 18:00 hours
Chairperson: J.E. Lindberg (SE)

Theatre **Session 20 no. Page**

Poster **Session 20 no. Page**

Session 21. Programme and elections meeting followed by Free communications on Animal Management and Health

Date: 27 August '07; 14:00 - 18:00 hours
Chairperson: S. Edwards (UK)

Session 22. Programme and elections meeting followed by Free communications on Animal Physiology

Date: 27 August '07; 14:00 - 18:00 hours
Chairperson: M. Vestergaard (DK)

Theatre **Session 22 no. Page**

Session 23. Programme and elections meeting followed by Free communications on Livestock Farming Systems

Date: 27 August '07; 14:00 - 18:00 hours
Chairperson: E. Matlova (CZ)

Theatre **Session 23 no. Page**

Poster **Session 23 no. Page**

Session 24. Programme and elections meeting followed by Free communications on Cattle Production

Date: 27 August '07; 14:00 - 18:00 hours
Chairperson: J.F. Hocquette (FR)

Theatre **Session 24 no. Page**

Session 25. Programme and elections meeting followed by Free communications on Sheep and Goat Production

Date: 27 August '07; 14:00 - 18:00 hours
Chairperson: M. Schneeberger (CH)

Theatre **Session 25 no. Page**

Poster **Session 25 no. Page**

Session 26. Programme and elections meeting followed by Free communications on Pig Production

Date: 27 August '07; 14:00 - 18:00 hours
Chairperson: C. Wenk (CH)

Theatre **Session 26 no. Page**

Poster **Session 26 no. Page**

Session 27. Programme and elections meeting followed by Free communications in Equine nutrition and physiology

Date: 27 August '07; 14:00 - 18:00 hours
Chairperson: B.Younge (IE)

Theatre **Session 27 no. Page**

Session 28. Maximizing forage and pasture use in the diet of herbivores

Date: 28 August '07; 08:30 - 12:30 hours
Chairperson: P. O'Kiely (IE)

Session 29. Applications of molecular genetics to breeding programmes

Date: 28 August '07; 08:30 - 12:30 hours
Chairperson: G. Guerin (FR)

Theatre **Session 29 no. Page**

Poster **Session 29 no. Page**

Session 30. Crossbreeding in ruminants

Date: 28 August '07; 08:30 - 12:30 hours
Chairperson: F. Buckley (IE)

Theatre **Session 30 no. Page**

Poster **Session 30 no. Page**

Session 31. Approaches to livestock farm multifunctionality

Date: 28 August '07; 08:30 - 12:30 hours
Chairperson: S. Oosting (NL)

Theatre **Session 31 no. Page**

Session 32. Nutrition and management of lactating sows

Date: 28 August '07; 08:30 - 12:30 hours
Chairperson: S. Chadd (UK)

Session 33. Open session - Ruminant Nutrition

Date: 29 August '07; 08:30 - 12:30 hours
Chairperson: M. Crovetto (IT)

Session 34. Biology and genetics of udder health

Date: 28 August '07; 08:30 - 12:30 hours
Chairperson: C. Kühn (DE)

Session 35. Feed for Pig Health Workshop

Date: 29 August '07; 08:30 - 12:30 hours
Chairperson: D. Torrallardona (ES)

Theatre **Session 35 no. Page**

Poster **Session 35 no. Page**

Session 36. Horse production in Ireland / tour

Date: 29 August '07; 07:45 - 18:00 hours
Chairperson: N. Finnerty (IE)

Theatre **Session 36 no. Page**

Session 37. Free Communications on Cattle Production

Date: 29 August '07; 08:30 - 12:30 hours
Chairperson: C. Lazzaroni (IT)

Theatre **Session 37 no. Page**

Poster **Session 37 no. Page**

Session 38. Free communications on Animal Genetics

Date: 29 August '07; 08:30 - 12:30 hours
Chairperson: J. Szyda (PL)

Theatre **Session 38 no. Page**

Session 39. Breeding for robustness in cattle - part 2

Date: 26 August '07; 14:00 - 18:00 hours
Chairperson: M. Klopcic (SI)

Theatre **Session 39 no. Page**

Session 01 Theatre 1

Genetic concepts to improve robustness of dairy cows

R.F. Veerkamp, H.A. Mulder, P. Bijma and M.P.L. Calus, Animal Breeding and Genomics Centre, Animal Science Group, Wageningen UR, P.O. Box 65, 8200AB Lelystad, Netherlands

For some years, breeding indices for dairy cattle typically account for health, fertility and longevity. It can be questioned whether selection for these traits in an average environment addresses robustness of dairy cows sufficiently. Especially when robustness is defined as "the capacity to handle disturbances in common, sustainable and economical farming systems". Genetic concepts exists that may allow selection for robustness more directly. A first concept is to select against (or for) environmental sensitivity using reaction norms. Some studies indicate that in a continuously improving environment, selection for increased performance leads to increased sensitivity. Calus (2006) found GxE for several health and fertility traits, and concluded that environmental parameters describing nutrition and energy balance caused the strongest G×E, mainly heterogeneous variances. The second concept that may enable direct selection for robustness is genetic heterogeneity of environmental variance, that can be considered as sensitivity to random environmental fluctuations (Mulder, 2007). A third concept that might enhance selection for robustness might be to take into account the social effect that an animal has on its herd mates (Bijma *et al*., 2007), and might be of interest in dairy cattle, for example, to avoid spreading of mastitis within a herd or improving feeding and grazing behaviour. These concepts will help to anticipate on future demands for more robust.

Session 01 Theatre 2

Robustness in dairy cows: experimental approaches

A.B. Lawrence, G.E. Pollott, M. Haskell and M.P. Coffey, SAC Sustainable Livestock Systems Research Group, Sir Stephen Watson Building, Bush Estate, Penicuik, Midlothian, EH26 0PH, United Kingdom

A robust animal is one that is long-lived, healthy and happy in a range of environments. SAC has just completed a 3-year project on robustness in dairy cattle which combined analyses of national data with experimental approaches on both commercial farms and a research station. The ultimate objective of the project was to investigate the efficacy of incorporating indicators of robustness into national breeding programmes. Traits underlying robustness considered by the project were energy balance, health characteristics, fertility, temperament and behaviour. Reliable and repeatable tests of temperament were developed on an experimental farm and applied in a large-scale farm study. These tests were able to pick up differences between low and high robust animals at the commercial farm level. Detailed studies of fertility in relation to energy balance in early lactation using divergent genotypes and production systems outlined the linkage between these factors. These studies highlight the fact that it is possible to identify factors underlying robustness in dairy cows using experimental approaches and that some of these factors are readily available from on-farm records.

Session 01 Theatre 3

Breeding for improved robustness: the role of environmental sensitivity

E. Strandberg, Swedish University of Agricultural Sciences, Department of Animal Breeding and Genetics, PO Box 7023, 75007 Uppsala, Sweden

A dictionary definition of robust is: having or exhibiting vigorous health, strength or stamina; resilient; sturdy; capable of performing without failure under a wide range of conditions. This definition highlights two perspectives of robustness: 1) the ability of an individual to function well in the environment she lives in, being resilient to the changes in the "microenvironment" that she encounters during her life; and 2) the ability of individuals (genotypes) to function well over a wide range of macroenvironments, e.g., production systems or herds. Both of these perspectives are important and environmental sensitivity plays an important role for both. Given that the breeding goal is correctly and broadly defined, it seems reasonable that animals with high and even performance over environments in the total breeding goal, i.e. flat reaction norms, are desirable. However, it could also be of importance to have a high and even performance within their environment over their lifetime. It has been seen from studies of genotype by environment that the genetic variation varies across environment. This could be modeled as genetic variation in residual variance, where a high variation would be undesirable for the producer. Having a flat reaction norm for the breeding goal, does not necessarily mean that all component traits have a high and even performance. In general terms, one can envision several ways that an individual can achieve a high and flat reaction norm.

Session 01 Theatre 4

Breeding for functional longevity of dairy cow

F. Miglior[1,2] and A. Sewalem[1,2], [1]Agriculture and Agri-Food Canada, 2000, College Street, PO Box 90, J1M1Z3 Sherbrooke QC, Canada, [2]Canadian Dairy Network, 150 Research Lane, N1G 4T2 Guelph ON, Canada

Dairy cattle research breeding programs have led to substantial genetic changes in production traits which represent only one component contributing to overall efficiency and profitability of the dairy industry. Functional traits, such as reproduction, longevity, and health traits, are of increased interest to producers in order to improve herd profitability. However, despite their economic significance only recently some attention has been given in genetic selection programs. This has resulted in a shift towards a balanced breeding approach and demands dairy cow robustness. A robust cow can be defined as one that adapts well to a wide range of environmental conditions or in genetic terms expresses a reduced genotype by environment interaction. Genetic improvement of longevity involves breeding of dairy cows that can produce a live calf, cycle normally, show observable heat, conceive when inseminated, sustain adequate body condition, avoid udder injuries, resist to infection diseases, walk and stand comfortably and produce milk of desirable composition. Selection for longevity is achieved by combining past information of productive life with early indicators of future longevity, namely conformation, reproduction, health and management traits. As countries expand genetic evaluations for reproduction and health traits, enhanced indicators will be available to better predict and improve functional longevity of our dairy herd.

Session 01 Theatre 5

Do "robust" dairy cows already exist? The New Zealand experience

J.E. Pryce, B.L. Harris and W.A. Montgomerie, LIC, Strategy and Growth, Private Bag 3016, Hamilton, 3240, New Zealand

The introduction of multi-breed genetic evaluation in New Zealand encouraged the adoption of an economic index known as Breeding Worth (BW) that allows breeds and crosses to be compared directly. The BW index is fundamentally composed of income minus costs, expressed per kg DM. The economic value calculated for liveweight in BW has a negative weighting as heavier animals have higher energy demands for maintenance and outweighs the income from cull cow sales. The genetic trend for liveweight has marginally increased since the introduction of BW in 1996, implying that selection has improved gross efficiency. Good fertility is a fundamental part of the way in which farm systems operate, as most farmers aim to have a single concentrated seasonal calving pattern so that feed usage is optimised. The production environment also influences body tissue mobilisation patterns. Genetic correlations between body condition score and milk production traits change from being positive in early lactation to negative in late lactation, which could be because there is an advantage to cows having body tissue still available to mobilise in the later stages of lactation, which is late summer, when pasture availability is sometimes limited. The New Zealand environment has indirectly selected for robust cows, as only these animals perform well in the systems used. In practice, Holstein-Friesian cross Jersey cows are particularly well suited, this cross now forms the largest proportion of replacements being born per year.

Session 01 Theatre 6

Derivation of direct economic values for body tissue mobilisation in dairy cows

E. Wall[1], M.P. Coffey[1] and P.R. Amer[2], [1]SAC, Sustainable Livestock Systems, Sir Stephen Watson Building, Bush Estate, Pebicuik, Midlothian, EH26 0PH, United Kingdom, [2]Abacus Biotech Limited, PO Box 5585, Dunedin, New Zealand

This study presents a simplified schema for body energy mobilisation defining three traits to describe how body energy lost/gained in dairy cows. A theoretical framework was developed to derive economic weights for these traits accounting for changing feed costs during lactation. Results show that the economic values for body tissue mobilisation is dependent on the calving system employed. For example, the economic value for early lactation body mobilisation is positive (+11p) in an autumn calving system and negative (-14p) in a spring calving system. Any loss in early lactation in a spring calving system will need to be repaid towards the end of lactation when feed is more expensive. The opposite is true in an autumn calving system when it is economically sensible for a cow to lose body energy when feed is expensive in the winter and regain it when turned out to grass. This suggests that the economic cost of body tissue mobilisation is different dependent on the system of production. If it is, it may be necessary to consider customised indices allowing farmers to choose bulls on an index that is suitable for their system.

Session 01 Theatre 7

Potential to genetically alter intake and energy balance in grass fed dairy cows

D.P. Berry, B. Horan, M. O'Donovan, F. Buckley, E. Kennedy, M. Mc Evoy and P. Dillon, Moorepark Dairy Production Research Center, Fermoy, Co. Cork, Ireland

The objective of this study was to quantify the genetic variation in grass dry matter intake (DMI) and energy balance (EB) in pluriparous Irish Holstein-Friesian dairy cows fed predominantly grazed grass. Grass DMI and EB were estimated up to four times per lactation on 1,588 lactations from 755 cows on two research farms. Random regressions were used to model the additive genetic and permanent environmental variance across days in milk (DIM). Heritability for DMI and EB across lactation varied from 0.10 (8 DIM) to 0.30 (169 DIM) and from 0.06 (29 DIM) to 0.29 (305 DIM), respectively. Genetic correlations within each trait tended to decrease as the interval between time periods compared increased. The lowest genetic correlation between any two time periods was 0.10 and -0.36 for DMI and EB, respectively suggesting the impact of different genes at different stages of lactations. The eigenfunction associated with the main eigenvalue of the genetic covariance matrix for DMI changed sign during lactation indicating that genetic selection for differently shaped DMI lactation profiles may be fruitful. However, the sign of the main eigenfunction for EB was constant across all DIM although the second largest eigenvalue accounted for 30% of the genetic variation and its associated eigenfunction had different signs at both ends of the lactation. Genetic parameters presented are the first estimates from dairy cows fed predominantly grazed grass

---

Session 01 Theatre 8

An international perspective on breeding for robustness in dairy cattle

H. Jorjani, J.H. Jakobsen, F. Forabosco, E. Hjerpe and W.F. Fikse, Interbull Centre, Dept. of Animal Breeding & Genetics, Swedish University of Agricultural Sciences, Box 7023, S-75007 Uppsala, Sweden

Evidence on antagonistic correlated effects of selection for milk production traits in dairy cattle, and the resulting deterioration of functional traits, has been accumulating during the past decade. Consequently, it can be concluded that "robustness", which is the well-being and lifelong functionality of the cow in terms of superior health, fertility and welfare, should be given a more prominent place in the breeding objectives than before. Unfavorable genetic correlations between production and functional traits are about -0.25 to -030. Therefore, combining functional and production traits into a total merit index (TMI) to guarantee breeding of robust cows is necessary. Accordingly, number of countries combining traits in a TMI is increasing. There is, however, much variation among countries in the weights given to different traits. Therefore, international genetic evaluation of TMI is unrealistic. Fortunately, Interbull's service portfolio has been expanded steadily since 1994 and now, in addition to production and conformation traits, includes routine genetic evaluation for udder health, longevity, calving, fertility and evaluation for workability traits is under research. These traits constitute more than 95% of the total weight in any country's TMI and international breeding values for individual traits can be combined nationally to an international robust bull index on each country's scale.

Session 01 Theatre 9

Principal components analysis for conformation traits in international sire evaluations

M.P. Schneider[1] and W.F. Fikse[2], [1]INRA, UR337, 78352 Jouy-en-Josas, France, [2]Interbull Centre, Dept. Animal Breeding and Genetics, SLU, Box 7023, SE-75007 Uppsala, Sweden

The objective of the study was to detect patterns in the data of conformation traits. Genetic correlation matrices from the May 2006 Interbull routine international genetic evaluation were reparameterized using principal component analysis. Sixteen linear traits evaluated in 20 participating countries were studied. A clear pattern of variation was observed for body (*e.g.* stature) and mobility traits (*e.g.* foot angle). Australia, New Zealand and Switzerland were the most distinct countries compared to the rest. This variation could reflect differences in management production system (grazing *vs* intensive feeding system), trait definition, models, and genotype by environment interactions (*i.e.* tied *vs* free stalls for feet and legs traits). However, no clear pattern was found for udder traits (*e.g.* fore udder), indicating more agreement in trait definition among the countries. Three to 10 principal components for teat length and foot angle explained 98% of the total variation, respectively. Thus, countries contributing to the largest variation could be used in the estimation of genetic correlations with methods which summarize the data without much loss of information.

Session 01 Poster 10

How farmers think about characteristics of robustness

M. Klopčič[1], J. Osterc[1] and A. Kuipers[2], [1]University of Ljubljana, Biotechnical Faculty, Zootechnical Department, Groblje 3, 1230 Domžale, Slovenia, [2]Expertisecentre for Farm Management and Knowledge Transfer, Wageningen University and Research Centre, De Leeuwenborch, Postbox 35, 6700 AA Wageningen, Netherlands

In Slovenia an analysis is made of the plans and communication aspects of farm development under quota and CAP. As tool a questionnaire was used. 1.114 questionnaires (about 20 % of the distributed) have been anonymously returned. Strategies of the farmers and their interest in information are analysed in relation to base parameters, such as age of farmer, size of farm, breed, milk production level and less favoured area or not. One of the strategies asked for were the breeding goals including longevity, fertility, animal health traits, etc. Preliminary results show that about 25% of the farmers agreed with the existing breeding program. A high percentage of farmers desired more emphasis on health traits (62%), fertility (55%) and longevity (39%), but also 56% of farmers wanted to increase emphasis on protein content in milk. The reaction of farmers was influenced by age, education, type of farm, breed, etc. For instance, the interest in the mentioned traits was significant higher for farmers with HF breed than for farmers with Simmental and Brown breed. Those farmers showed a relatively higher interest in progress in milk yield and in beef characteristics. Too, a comparison was made with breeding goals as indicated in a similar study by suckler cow farmers.

Session 01 Poster 11

Genetics of tuberculosis in Irish dairy cows

M.L. Bermingham[1], S.J. More[2], M. Good[3], A.R. Cromie[4] and D.P. Berry[1], [1]Moorepark Production Research Centre, Fermoy, Co. Cork, Ireland, [2]Centre for Veterinary Epidemiology and Risk Analysis, University College Dublin, Belfield, Dublin 4, Ireland, [3]Department of Agriculture and Food, Kildare St, Dublin 2, Ireland, [4]The Irish Cattle Breeding Federation, Bandon, Co. Cork, Ireland

There is a lack of information on genetic parameters for tuberculosis (TB) susceptibility in dairy cows. Therefore, the objective of this study was to estimate the genetic variation in TB susceptibility in Irish dairy cows. A total of 24,286 animals (including 933 TB positive animals) from 1,176 herd test dates during the period of 1st November 2002 and 31st October 2005 were available for inclusion in the analysis. All animals were Holstein-Friesian and had an identified sire. TB positive animals were defined as animals with a skin change where the Mycobacterium bovis purified protein derivative response was >4 mm more than that of the M. avium purified protein derivative response. An animal linear mixed model was used to estimate the additive genetic and residual variance for TB susceptibility. Fixed effects in the model included herd test date, parity, age nested within parity and Holstein breed fraction. Heritability of TB susceptibility was 0.013 (SE=0.0062). This study demonstrates that exploitable genetic variation for tuberculosis susceptibility exists among Irish dairy cows.

Session 01 Poster 12

Genetic correlation between persistency and calving interval of Holstein in Japan

K. Hagiya[1], K. Togashi[2], H. Takeda[2], T. Yamasaki[2], T. Shirai[1], J. Saburi[1], Y. Masuda[3] and M. Suzuki[3], [1]National Livestock Breeding Center, Nishigo, Fukushima 961-8511, Japan, [2]Hokkaido National Agricultural Station, Sapporo, Hokkaido 062-8555, Japan, [3]Obihiro University of Agriculture and Veterinary Medicine, Obihiro, Hokkaido 080-8555, Japan

Effects of calving intervals (CI) on lactation curves were examined. In this study, genetic correlations among lactation yields, persistency and CI were estimated. Data included 8451 cows with 80,880 test-day records obtained from the DHI program in 2001. The persistency was defined as the difference between test day milk yields at 60 DIM and 150 DIM. A random regression model (RRM) was applied to this analysis, and the GIBBS3F90 program was used to estimate the effects of CI classes on lactation curves. The RRM included the fixed effects of herd-test-day, age, calving season and CI classes as well as random effects of animal, permanent environment and heterogeneous residuals with 10 intervals by days in milk (DIM). The genetic correlations were estimated using the AIREMLF90 program with a model including fixed effects of herd, age and calving season as well as random effects of animal and residual. The length of CI was affected by lactation curves when the DIM was longer than 150 d. The persistency had a negative correlation with the lactation yield (-0.17), but a positive correlation was estimated with the length of CI (0.60). The results indicated that selection for persistency might improve the lactation yields and the length of CI in a preferable direction.

Session 01 — Poster 13

Heritabilities of length of productive life of Holstein cows in Japan

Y. Terawaki[1] and V. Ducrocq[2], [1]Rakuno Gakuen University, Dairy Science Institute, 582 Midori-machi Bunkyo-dai Ebetsu, 069-8501, Japan, [2]INRA, Station de Génétique Quantitative et Appliquée, Jouy en Josas, 78352, France

Length of productive life (LPL) of Holstein cows in Japan was studied. The dataset was provided by the Hokkaido Dairy Milk Recording and Testing Association. It included 117,182 records from cows having their first calving between 1984 and 1999 in 860 herds in which more than 60 % of cows had type score. All analyses were carried out using the "Survival Kit V. 5.0" software. A previous study (Terawaki *et al.*, 2006) had shown that the origin of the cow's sire (SC) and the presence of a type score for the cows (TS) had a significant effect on LPL of Holstein cows in Japan. Four models were examined, including none, one or both factors SC and TS. Results indicated that the effect of TS on LPL was much larger than for SC. Herd-year variance was the largest (0.2766) with the model with factors SC and TS and smallest (0.2365) when both factors were excluded. Conversely, the effect of SC on sire variances was larger than for TS. The highest heritability of LPL (0.163) was obtained with the model without TS and SC, and the lowest one (0.134) with the model with both factors. These results indicate that in Japan, cows LPL is influenced by the country of origin of their sire and by whether they have been type scored in ways that may be confounded with sire effects. So, it was concluded that a model without correction should be adopted for genetic evaluation of LPL of Holstein cows in Japan.

Session 01 — Poster 14

A comparison between sire model and animal model for some economic traits on Friesian cattle in Egypt

A. Khattab, A. Abou Zeid, A.S. Omar and A. Nowier, Tanta university, Animal Production Department, Animal Production Department, 31527 Tanta, Egypt

A total of 2181 normal lactation records of Friesian cows kept at Sakha Farm during the period from 1996 to 2002 were used. Milk traits were analyzed by (1) Sire Model (SM) included the random effects of sire and cow within sires and the fixed effects of month and year of calving and parity and age at calving as a covariate and (2) Animal Model (AM) included the random effects of animals, permanent and errors and fixed effects of month and year of calving and parity and age at calving as a covariate. Estimates of heritability by using SM were 0.17, 0.15, 0.22, 0.11, 0.15, 0.13 and 0.14, for 305 day milk yield, 305 day fat yield, 305 day protein yield, lactation period, dry period, days open and calving interval, respectively. While the corresponding values from AM were 0.11, 0.12, 0.17, 0.10, 0.11, 0.04 and 0.05, respectively. Estimates of genetic and phenotypic correlations among all traits studied are near similar for the two models. Product moment correlations between sire breeding values (EBV' s) among all traits studied by using SM and AM are the same trend and there are a little differences between them and near similar to genetic correlations. Although both methods of analysis are succeeded in estimation EBV' s, the SM is the cheapest in terms of computing costs.

Session 01 Poster 15

DNA-polymorphism of BoLA-DRB3, kappa-casein, prolactin and growth hormone genes in Russian Yaroslavl cattle breed

G.E. Sulimova[1], S.R. Khatami[2] and M.R. Mohammadabadi[3], [1]Vavilov Institute of General Genetics, 3 Gubkin Str., 119991 Moscow, Russian Federation, [2]Akhvaz University, 2 Karun Blvd., Akhvaz, Iran, [3]3Kerman Shahid Bahonar University, 22 Bahman Blvd., Kerman, Iran

Yaroslavl cattle breed was formed in 19-th century in the Yaroslavl povince of Russia as the result of native selection. This breed has a series of valuable traits that discriminate it as a unique breed but its genofond did not else analyze with DNA-markers. We analyzed 120 cows in two farms of Yaroslavl region. It was identified 35 *BoLA-DRB3* alleles by RFLP. About 52% of the cumulative frequencies corresponded to five alleles (*DRB3.2*12, *13, *15, *24* and*28*). A high content (35%) of *BoLA-DRB3* alleles of resistance to leukemia was shown in Yaroslavl breed. High frequencies of *B*-allele (47.5%) and *BB*-genotypes (21.7%) of *CSN3* were found. The frequency of the allele MspI(-) of the *bGH* gene in the Yaroslavl breed was extremely low (2%) comparable only with that of the Holstein cattle. Homozygotes *BB* (RsaI+) of the *bPRL* gene (13.4%) were found in the Yaroslavl cattle. Since genotype *BB* of the *bPRL* has not been observed in European breeds previously. A new type of microsatellite repeat was desribed in the 5'-untranslated region of the gene *bPRL* at the Yaroslavl cattle. Thus Yaroslavl cattle has a great value for practical selection because it has a high content of economically important alleles of the gene studied.

Session 01 Poster 16

The characteristics of Busha cattle in the Republic of Serbia

P. Perisic[1], Z. Skalicki[1], S. Stojanovic[2], V. Bogdanovic[1] and R. Djedovic[1], [1]Faculty of Agriculture, Animal Husbandry, Nemanjina 6, 11081 Belgrade, Serbia, [2]Ministry of agriculture, forestry and water management, Nemanjina 22-26, 11000 Belgrade, Serbia

The size of Busha population as a direct descendant of *Bos brachyceros Adametza* in the Republic of Serbia is reduced to a minimum and according to FAO criteria belongs to a seriously endangered cattle species. The main goal of these activities was to preserve Busha cattle as a sustainable resource in a genetic, economic, cultural and every other sense. A total number of 50 female and 5 male animals of different types of Busha cattle (Grey, Red Metohian and Black) was included in the programme of *in situ* conservation, and used in measuring. The average established measures in cows were for withers height 113 cm (firstcalved 106 cm), sacrum height 116 cm (firstcalved 110 cm), body length 131 cm (firstcalved 129 cm), chest width 31 cm, chest depth 56cm, chest volume 158 cm, shank volume 16cm, and body mass ranging from 250 to 350 kg. The bulls used for breeding when approximately 3 years old, had withers height of 112 cm, sacrum height 115 cm, body length 139 cm, chest width 33 cm, chest depth 57 cm, chest volume 169 cm, and shank volume 19 cm. During lactation, which lasts for about 240 days on average, they can produce about 900-1000 kg milk with 4,4% milk fat and 3,1 % protein on average.

Session 02 — Theatre 1

Mammary growth and alveolar secretory cell differentiation: keys to milk production

R.M. Akers[1] and A.V. Capuco[2], [1]Virginia Polytechnic Institute and State University, Dairy Science, 2470 Litton Reaves Hall, Blacksburg, VA 24061, USA, [2]USDS-ARS, Bovine Functional Genomics Laboratory, Building 200, Beltsville, MD 20705, USA

Mammary growth and development is regulated by local and systemic endocrine and growth factor signaling. But signal strength is impacted by external stimuli. Effects may be immediate and short-lived, or long-term, depending upon the time when the stimulus is applied. Modulating events likely produce long term effects when applied during critical periods of mammary gland development. Subsequent responses may depend on epigenetic effects to change proliferation and differentiation of mammary stem cells. Thus lactation success depends on more than simply producing mammary parenchymal mass containing mature lobulo-alveolar structures. The secretory alveolar cells must achieve both biochemical and structural differentiation. Poor milk production corresponds with a relative differentiation failure in low milk production cattle. Thus there are two keys to maximization of lactation performance. Impairment of either mammary growth or differentiation of the secretory epithelium reduces lactation performance. An exciting, bewildering, universe of transcription factors, receptors, intracellular signaling intermediates, and extracellular molecules must ultimately interact to determine the mass of the mature mammary gland and the functional capacity.

Session 02 — Theatre 2

Variations in milk yield and regulation of lactose and casein synthesis

J. Guinard-Flament, H. Rulquin, M. Boutinaud and S. Lemosquet, INRA, Agrocampus Rennes, UMR1080 Production du Lait, Domaine de la Prise, 35590 St-Gilles, France

The present paper is a review of the main factors reducing high milk yield, to highlight the mechanisms involved in the regulation of milk lactose and protein production. These mechanisms determine the amount of glucose and amino acids taken up by the udder and metabolised either in lactose or proteins. The mammary blood flow that supplies nutrients to the udder plays an important role when milk yield is depressed by 20-30%. However, the reduced arterial availability of nutrients to the udder is not the only mechanism involved. The ability of the udder to extract nutrient could be affected either by altered transmembrane transport or by lower intra-cellular metabolic activity. Indeed, feed restriction reduces the mRNA levels of transmembrane transporters of glucose. The amount and activity of the enzymes responsible for the synthesis of milk components are also regulated. In addition to the lowered mRNA levels of caseins, once daily milking induces a decline in the activity of galactosyltransferase, the enzyme of the lactose synthase complex and in the mRNA levels of its co-factor, the a-lactalbumin.

Session 02 Theatre 3

Recent advances in the regulation of milk fat synthesis

K.J. Harvatine, J.L. Capper, Y.R. Boisclair and D.E. Bauman, Cornell University, 262 Morrison Hall, Ithaca, NY 14853, USA

In addition to its economic value, milk fat is responsible for many of milk's characteristics and can be markedly affected by diet. Diet-induced milk fat depression (MFD) was first described over a century ago. The biohydrogenation theory established that MFD is caused by an inhibition of mammary synthesis of milk fat by specific fatty acid intermediates of ruminal biohydrogenation. During MFD transcription of key mammary lipogenic genes are coordinately down-regulated. Our investigations have established that expression of SREBP1 and SREBP-activation proteins are down-regulated during MFD. Importantly, key lipogenic enzymes are transcriptionally regulated via SREBP1. Collectively, these results provide strong evidence for SREBP1 as a central signalling pathway in the regulation of mammary FA synthesis. Spot 14 is also down-regulated during MFD, consistent with a role for this novel nuclear protein, possibly as a lipogenic factor. In addition, knockouts of SREBP1 or Spot 14 in the mouse exhibit a milk fat reduction of similar magnitude and pattern to MFD. Knockout of lipogenic enzymes blocks milk fat production, but this appears to be related to the essentiality of the enzyme rather than a specific regulatory step. Genetic analysis has identified QTLs and SNPs that are able to explain a small portion of the variation in milk fat, but additional functional analysis will be valuable. Overall, genomic approaches continue to provide exciting insight into the regulation of milk fat synthesis

---

Session 02 Theatre 4

Effect of chronic inhibition of prolactin release on milk production of dairy cows

P. Lacasse[1], R.M. Bruckmaier[2], V. Lollivier[3], P.G. Marnet[3] and M. Boutinaud[3], [1]Dairy and Swine R&D Centre, PO Box 90, Sherbrooke J1M 1Z3, Canada, [2]University of Bern, Bremgartenstr. 109a, CH-3001, Switzerland, [3]INRA, Agrocampus Rennes, UMR1080, St-Gilles, 35590, France

In most mammals, suppression of prolactin (PRL) strongly inhibits lactation. Nevertheless, short term suppression of PRL by bromocriptine has produced inconsistent effects on milk yield in cows and goats. A preliminary experiment was carried out to evaluate the ability of a newer dopamine agonist, quinagolide, to suppress PRL release in cattle. The results indicated that daily injections of this molecule were able to suppress both basal PRL and milking-induced PRL release. An experiment was then carried out to evaluate the effect of long term inhibition of PRL release on milk production of dairy cows. Five Holstein dairy cows in early lactation received daily intramuscular injection of 1 mg of quinagolide during 8 weeks. Four control cows received the vehicle (water). Quinagolide reduced milk production ($P < 0.05$). At the end of the period, milk production averaged 36.4 ± 1.4 and 31.8 ± 1.2 kg/d for control and quinagolide treated cows, respectively. Milk composition was not affected by the treatments. Yield of fat ($P < 0.01$), protein ($P < 0.05$) and lactose ($P < 0.05$) were reduced by quinagolide. Feed intake was slightly lower ($P < 0.05$) in quinagolide-treated cows but, body weight gain were similar for both groups. In conclusion, chronic administration of an inhibitor of prolactin-release reduces milk production in cattle.

Session 02 Theatre 5

The 3'-UTR is crucial for differential allelic expression of bovine CSN3 gene

P. Frajman, M. Debeljak, T. Kunej and P. Dovc, University of Ljubljana, Deaprtment of Animal Science, Groblje 3, 1230 Domzale, Slovenia

Allelic variants of the bovine CSN3 gene have a major impact on technological properties of milk: CSN3 allele B has positive effect on micelle size, coagulation time, curd firmness and cheese yield. We analysed promoter elements of the bovine CSN3 gene and the 3'-UTR region in order to find molecular basis for differential allelic expression. Our study confirmed functionality of the bovine CSN3 gene promoter using bovine mammary gland derived cell line BME UV and identified important positive cis elements in the distal part of the promoter. However, the higher expression of CSN3B at RNA and protein level can not be explained by promoter polymorphisms. However, important regulatory role of allelic variants in the 3'-UTR region of the bovine CSN3 gene was found, where eight allele specific polymorphisms characterize CSN3 alleles A and B. One of them abolishes destabilizing ARE element in the CSN3B 3'-UTR region. Using reporter gene constructs containing CSN3 3'-UTR sequences A and B we demonstrated significant difference in regulatory potential between both variants *in vitro*. The expression of the Luc-CSN3B construct was higher than that of the Luc-CSN3A construct, supporting results obtained *in vivo* where CSN3 allele variants A and B represent 40 and 60% of the total CSN3, respectively. Bioinformatics analysis of the CSN3-3'-UTR revealed several potential miRNA targets which might play an important role in allele specific degradation of CSN3 mRNA.

Session 02 Theatre 6

Polymorphism of kappa casein gene (CSN3) in horse and comparative genomics approach to study conserved regions

S. Hobor[1], T. Kunej[1], T. Lenasi[1], G. Majdic[2] and P. Dovc[1], [1]University of Ljubljana, Biotechnical Faculty, Department of Animal Science, Groblje 3, SI-1230 Domzale, Slovenia, [2]University of Ljubljana, Veterinary Faculty, Gerbičeva 69, 1000 Ljubljana, Slovenia

Milk protein genes in farm animals belong to the best characterized genes in farm animals. The profound knowledge about allelic variants regulatory transcriptional mechanisms is available. In this study we sequenced promoter of the kappa casein gene (*CSN3*) in horse and identified 15 single nucleotide polymorphisms (SNPs), which were investigated for potential involvement in putative transcription factor binding sites (TFBSs). The 12 SNPs were involved in gain/loss of potential TFBSs. Using comparative genomics approach we obtained 1482 bp of the promoter. Phylogenetic footprinting revealed highly conserved blocks of promoter sequence among nine species. Transcriptional regulators STAT5, C/EBP, NF1 and STAT6 were located within conserved regions. Mutations in codogenic region of a gene are important, because of consequential allelic variants and arrangement into nomenclature of casein genes. We identified two SNPs in exon 1 and two in exon 4 in horse *CSN3* gene and genotyped them in six Slovenian horse breeds. Exon 1 is not coding, but could be involved in the regulation of transcription. The highest variation in genotype frequencies was present in Slovenian coldblood breed. Using horse primers we obtained 400 bp of exon 4 sequence in donkey and zebra. Two SNPs within zebra's exon 4 sequence were discovered, both causing aa substitutions.

Variation of the lactoferrin concentration in cow milk

L. Giordano, O. Dotreppe, J.L. Hornick, L. Istasse and I. Dufrasne, Liege University, Faculty of Veterinary Medicine, Animal Production - Nutrition Unit, Bd of Colonster, B43, 4000 Liege, Belgium

The objective of the present study was to determine some of the factors influencing the concentration of lactoferrin (Lf) in cow milk. The experiment was divided into two parts: milk samples from tanks of 99 farms were used in the first part of the experiment and the second part was based on 244 individual milk samples obtained from different farms. It appeared that the Lf concentration was influenced by cell counts ($p < 0.001$), protein concentration ($p < 0.001$), milk production ($p < 0.05$), stage ($p < 0.05$) and number of lactation ($p < 0.01$). The Lf concentration in milk also varied according to the herd ($p < 0.01$) but the herd effect had to be related to other factors such as genetics and management. By contrast, the fat and urea concentrations in milk did not have any effect on the production of Lf. Similarly there were no effects of agronomic areas or of the seasons. However the Lf concentration seemed to be higher in summer (796 mg/L) than in winter (487mg/L). The difference could be ascribed to the calving season, most calvings occurring in winter. So the cows were in a more advanced stage of lactation in summer producing more Lf during this period.

Milk enzymes and minerals related to mammary epithelial integrity in cows milked in an automated milking system or twice daily in a conventional milking parlor

F. Abeni, M.G. Terzano, M. Speroni, L. Migliorati, F. Calza and G. Pirlo, CRA - Istituto Sperimentale per la Zootecnia, Via Porcellasco 7, 26100 CREMONA, Italy

The aim of this paper was to evaluate the effects of automatic milking (AM) on milk enzymes and minerals related to mammary epithelial integrity in comparison to a twice daily conventional milking (CM). One cow from each of six pairs of twins (Italian Friesian) was assigned to be milked with AM or with CM throughout first lactation. Milk production was recorded and milk samples were collected at 4, 11, 18, 25, 32, and 39 wk of lactation, to determine fat and protein content, somatic cells, pH, plasminogen and plasmin activities, Na, K, and Cl. Statistical analysis was performed by a randomized block design, with milking system (AM *vs* CM), and time as main factors, with cow repeated in time. Milk production, fat and protein contents, and somatic cell count did not differ between milking systems. Both plasminogen and plasminogen+plasmin activities were lower ($P < 0.05$) in AM. Milk pH was higher ($P < 0.05$) in AM. Milk Na, K, Na/K ratio, and Cl did not differ across whole lactation. Milking system did not seem to affect mammary epithelial permeability. The differences in enzymatic (proteolytic) activity between milking systems, probably related to daily milking frequency, suggest a better preserved quality of protein fraction for the cheese-making process with AM, even if differences in pH might interfere negatively.

Session 02 Poster 9

Candidate genes affecting milk yield and quality in cattle

B. Moioli[1], L. Orrù[1], M. Savarese[1], M.C. Scatà[1], A. Crisà[2] and C. Marchitelli[2], [1]CRA-Istituto Sperimentale per la Zootecnia, via Salaria 31, 00016 Monterotondo, Italy, [2]Università della Tuscia, Dipartimento Produzioni Animali, via S. C. De Lellis, 01100 Viterbo, Italy

Nine candidate genes, that play different physiological roles, were selected beceuse they are expected to affect milk yield and quality: Growth Hormone Receptor (GHR) and Insulin-like Growth Factor (IGF1), for their involvement in the regulation of growth, skeletal and muscle development; Fatty Acid Synthase (FASN), Diacylglycerol O-acyltransferase 1 (DGAT1) and Stearoyl-CoA Desaturase (SCD), for their role in mediating and regulating fatty acid synthesis; Fatty Acid-Binding Protein 4 (FABP4), Phospholipid Transfer Protein (PLTP), Lipoprotein Lipase (LPL) and Plasminogen Activator Inhibitor 1 (PAI1), for their role in regulating adipogenesis, feeding intake and obesity. In each gene, the presence of at least one Single Nucleotide Polymorphism (SNP) had been identified either by direct sequencing or DHPLC analysis of PCR products. A sample of 27 Holstein sires was selected in the database of the International Genetic Evaluation among those who had either the highest or lowest genetic merit for milk fat percentage. We tested the allelic effect at each SNP on the genetic merit of milk yield and quality. We found that the SNP in DGAT1 influenced milk yield, fat and protein percentage; the SNP in SCD influenced milk yield and fat percentage; the SNP in PAI1 and GHR influenced fat and protein percentage.

Session 02 Poster 10

Synthetic activity and cell apoptosis in mammary epithelial cells in once daily milked goat

M.H. Ben Chedly[1], P.G. Marnet[1], J. Guinard-Flament[1], P. Lacasse[2] and M. Boutinaud[1], [1]INRA-Agrocampus Rennes, UMR Production du Lait, 35590 St-Gilles, France, [2]AAFC-Dairy and Swine R&D Centre, PO Box 90, Sherbrooke J1M 1Z3, Canada

The aim of this work was to investigate the regulations that take place in the mammary epithelial cells (MEC) in once daily milked goats. Two groups of 4 Alpines goats in mid-lactation were subjected alternately to ODM or twice daily milking (TDM) for a period of 5 wks according to a crossover design. After 1 and 5 weeks, cells were isolated from fresh milk and MEC were purified using magnetic beads for real-time rt-PCR determination of mRNA levels of k-casein, a-Lactalbuminand bax (pro-apoptotic protein) and Bcl2 (anti-apoptotic protein). A significant decrease in productions of milk (-22%, $P < 0.001$; -13%, $P < 0.01$), caseins (-24%, $P < 0.001$; -13%, $P < 0.01$) and lactose (-23%, $P < 0.001$; -12%, P=0.015) was observed after 1 and 5 wks of ODM. Accordingly, mRNA levels of a-Lactalbumin (-70%, P=0.02; -9%, P=0.04)and k-casein (-58%, $P < 0.01$; -1%, NS) were decreased after 1 and 5 wks of ODM. A surprising decrease in Bax /bcl2 mRNA ratio (-38%, P=0.07; -58%, P=0.08) as a result of a non significant decrease in Bax (-30%; -53%) and a non significant increase in Bcl2 (+5%; +27%) mRNA levels is observed after 1 and 5 wks of ODM. This study suggests that in the fist wk of ODM, the reduction of milk production is due to a down regulation of the synthetic activity of the MEC. Bax to Bcl2 ratio provides no evidence an apoptosis induction after 1 or 5 wks of ODM.

Session 02 Poster 11

Daily feed intake pattern can affect milk and blood composition

G. Bertoni, E. Trevisi, M.G. Maianti and F. Piccioli Cappelli, Università Cattolica S. Cuore, Institute of Zootechnics, via E. Parmense, 84, 29100 - Piacenza, Italy

To clarify the changes in milk and blood parameters due to a different daily pattern of forage intake, which nighttime appears affected by milking stimulation, 6 dairy cows with high or average milk yield were used. Cows were maintained tied and individually fed in a climatized stall. Forages, automatically fed in 2 meals at 12 h interval, were offered in a *latin* square design (14 day) after milking end (5.30 a.m. and p.m.; ME group), 2 h later (AM) or 5 h before (BM). Concentrate was fed by individual autofeeder in 8 meals/day, starting 0.5 h before the morning forage meal. Feed intake, milk yield and composition were frequently checked. Blood samples were collected at 7 a.m. (pre feeding for AM or 1.5 or 6.5 h after forage meal for ME or BM) and before forage meal for ME and BM, at the end of each period. Plasma samples were analysed for metabolic profile, NEFA, BHA, insulin and T3 content. Milk yield was slightly higher in AM, while milk protein and clot firmness were slightly higher in ME and BM. Some differences were observed comparing the data of blood taken before meals or later (1.5 h for ME and 6.5 h for BM). Shortly after meal, NEFA were reduced while BHA, urea and insulin were increased; later NEFA remained lower, while urea, BHA and insulin returned to basal values. To conclude, different feeding patterns could modify the milk yield and composition, but also some hormone and metabolite levels during the day, to be considered for a proper interpretation of their levels.

Session 03 Theatre 1

World animal health situation for animal movements

D.C. Chaisemartin, World Organisation for Animal Health, 12, rue de Prony, 75017 Paris, France

It is important that the movements of animals and products will be done with safety in order to facilitate animal and food production and allows its development. The good management for the early detection and control of diseases requires the maximum of transparency on the animal health situation. One of the historic mission of the OIE is to ensure transparency in the world animal disease situation including zoonosis. The member countries have the obligation to report diseases, emerging diseases or pahogenic detected on its territory. The OIE then disseminates the information to other countries which can take the necessary preventive action. The efficiency of this transparency is improved by an active search and verification of unofficial information based on various sources. Another step of improvement was reached with the establishment of a unique OIE list of diseases together with the change of notification procedures that started in January 2005. The modernisation of the OIE's World Animal Health Information is accomplished with the birth of WAHIS web application. The communication of information has also been improved. An Electronic distribution lists have been created for sending the alert messages and the weekly information. The modernisation of the Web is done through the development of WAHID Interface, which is the new interface of the World Animal Health Database that will allow for end-users a wide range of queries for a given country and a given disease.

Session 03 Theatre 2

How can veterinary epidemiology contribute to sustainable animal production?

U. Emanuelson, Swedish Univ of Agric Sci, Dept Clinical Sciences, POB 7054, SE-75007 Uppsala, Sweden

Sustainable animal production is characterised by conservation of resources, environmental care, good animal health and welfare, and should agree with consumer demand and be profitable for the producer. The relative importance of most of these characteristics, and how they are realised in practise, depends strongly on the situation at hand, due, for instance, to social and political conditions. However, irrespective of differences in local conditions with respect to available resources, waste management, etc, good animal health should always be an integrated part of a sustainable animal production, both from the animal perspective and from the human health perspective. Epidemiology is the study of the distribution of diseases in populations and of risk factors for diseases, and is well suited to address the oftentimes multifactorial nature of diseases in production animals. Epidemiological information can therefore be used to plan and evaluate strategies to prevent diseases, but also to monitor trends in order to identify emerging threats to animal health. As such, epidemiology as a discipline provides fundamental concepts and tools useful for basic research (from generating hypotheses to establishing causal relationships) as well as practical applications such as herd health services. This presentation outlines the value of epidemiology in promoting animal health, and highlights some applications of epidemiological methods with special reference to sustainable animal production.

Session 03 Theatre 3

Modelling individual patterns of somatic cell scores to derive cows infection status

J. Detilleux, University of Liege - Vet. Faculty, Quantitative Genetics, Bd de Colonster n° 20, 4000 Liege, Belgium

The absence of on farm recording systems in most countries makes direct identification of mastitis cases impossible. Therefore, in many countries high somatic cell scores (SCS) in milk are used as indicator for mastitis. However, individual SCS are not very accurate in identifying infected cows. Mathematical models may improve the accuracy of SCS by making better use of the information contained in SCS data. Here, a hidden Markov model (HMM) and a finite mixture model (FMM) were applied on simulated monthly SCS to evaluate their accuracy in estimating individual cows' infection status. Under the FMM, each SCS is assigned to one of two components hopefully representingSCS from cows with and without intra-mammary infection (IMI),respectively. Then, identification of animals at risk is computed as the posterior probability of putative IMI, given SCS, rather than on crude SCS. The HMM is a generalization of the FMM as it allows the estimation of a finite set of unknown probabilities of putative IMI given the observable set of SCS. The prevalence, incidence and recovery rates may be obtained given these probabilities. To simulate the SCS patterns, monthly SCS from cows with known IMI states were used. The maximum likelihood estimates (MLE) of the parameters under both HMM and FMM were obtained. Biases between MLE and simulated values were computed to evaluate the accuracy of MLE under both models. Results are shown and the pro and cons of the proposed methods are discussed.

Session 03 Theatre 4

Quantification of risk factors for the spread of foot and mouth disease

I. Witte and J. Krieter, Institute of Animal Breeding and Husbandry, Christian-Albrechts-University, Olshausenstrasse 40, 24098 Kiel, Germany

A spatial and temporal Monte-Carlo simulation model was developed to describe the spread of foot and mouth disease virus between cattle, pig and sheep farms. Starting from a defined index farm the virus could spread by direct and indirect contacts, local and airborne spread. Following the EU Directive 2003/85/EC contact tracing, establishment of protection and surveillance zones (Basic) as well as pre-emptive culling (P) within various circles were implemented. The input region consisted of 10,700 farms with a density of 1.45 farms/m^2. In general starting an epidemic on a farrowing farm (FF) resulted in significant more infected farms comparing to a dairy farm (DF). Moreover adding pre-emptive culling to the basic control measures reduced the number of infected farms (Basic, FF *vs* DF: 152.5 *vs* 311.9 farms, Basic+P, FF *vs* DF: 32.5 *vs* 63.5 farms) and the epidemic duration. Increasing the time period from the onset of clinical disease in the primary farm until its diagnosis (2, 4, 8 days) the mean number of infected farms increased significantly (18.3, 30.3, 65.5 farms). The same trend could be determined for the mean epidemic duration and mean number of culled and preventively culled farms. The control measures, type of the primary farm as well as time from infection until diagnosis were identified as risk factors for the course of foot and mouth epidemics.

Session 03 Theatre 5

Risk factors for dystocia in Irish dairy herds

J.F. Mee[1], A.R. Cromie[2] and D.P. Berry[1], [1]Teagasc, Moorepark Dairy Production Research Centre, Fermoy, Co. Cork, Ireland, [2]Irish Cattle Breeding Federation, Bandon, Co. Cork, Ireland

Published studies indicate a rise in dystocia (DYS) rate amongst Holstein-Friesian (HF) cows internationally. There are no recent studies on DYS rate in Irish dairy herds. The objectives of this study were to establish the DYS rate and associated risk factors in Irish HF cows. Data on DYS score (1=no assistance/unobserved; 2=slight assistance; 3=severe assistance; 4=veterinary assistance) and associated risk factors extracted from the Irish Cattle Breeding Federation database (2002 to 2005; 251,169 records) were analysed using generalised estimating equations. The average prevalence of no, slight, severe and veterinary calving assistance was 70%, 24%, 4% and 2%, respectively. After adjusting for year and calving month, the odds of DYS (score $\geq$ 3) were greater for twins (OR 2.4; 2.22-2.66) and cows with DYS in the preceding calving (OR 3.2; 2.83-3.57) and lesser for smaller (<20 cows) herds (OR 0.9: 0.76-0.95). Significant two-way interactions between parity and calf sex and between estimated breeding value (EBV) for direct calving ease (CE) and calf sex were detected. There were greater odds of DYS for male than female calves though the difference reduced with increasing parity. Irrespective of calf sex, DYS risk decreased with increasing parity. There was a greater effect of EBV for CE in male compared to female calves. In conclusion, dystocia rate and associated risk factors in Irish HF cows were similar to those in other EU countries.

Session 03 Theatre 6

Lying duration and hoof health of cows outwintered on forage crops

P. Gazzola[1], L. Boyle[1], P. French[1], A. Hanlon[2] and F. Mulligan[2], [1]Teagasc, Moorepark Research Centre, Fermoy, Co. Cork, Ireland, [2]UCD, School of Agriculture, Food Science and Veterinary Science, Veterinary Centre, Belfield, Dublin 4, Ireland

Brassica crops *in-situ* and pastured grass are cost effective options for out-wintering dairy cows. However, alternatives to indoor housing must not compromise animal welfare. The objective of this experiment was to evaluate the lying behaviour and hoof health of dry dairy cows out-wintered on (1) KALE, (2) SWEDE, (3) pasture GRASS and (4) indoors on grass SILAGE. Spring calving cows (n=88) were assigned to one of the four treatments in December 2005. Lying behaviour of 9 cows per treatment was recorded over two 24 hour periods using TinyTag Dataloggers. All cows were examined for hoof lesions prior to the experiment and again at one week and seven weeks post-calving. Sole haemorrhages, heel erosion, white line disease and digital dermatitis were scored according to severity. Lying times were 12, 9, 10, and 8 hours per day for KALE, SWEDE, GRASS and SILAGE respectively. At one week post-calving, total sole haemorrhaging was greatest in SWEDE compared to GRASS ($p < 0.01$) and SILAGE ($p < 0.05$). Total heel erosion was higher in GRASS and SILAGE compared to KALE and SWEDE ($p < 0.05$). SILAGE also had a greater severity of digital dermatitis ($p < 0.05$).The low lying times indicate inadequate lying surfaces. Hoof health is dependant upon ground surface and conditions. Further studies are required to determine how to improve out-wintering conditions for dry cows on forage crops.

Session 03 Poster 7

Tracing of the Hungarian poultry sector

P. Marlok, S.Z. Simai, T. Cserhidy and D. Mezőszentgyörgyi, Central Agricultural Office, Keleti Károly u. 24., 1024 Budapest, Hungary

In the last years the importance of integrated information systems has been increased in the field of animal breeding. These systems provide not only dates for the breeders or in connection with different subsidies, but for today they have become an important tool for veterinary authorities and food safety. On behalf of the Ministry the Hungarian Breeding Authorities elaborated a special information system, wich is able to identify the inland and import poultry livestocks, register the data of origin, the number and trace their movements. The persons taking part in the poutry sector are obliged to join the system by an order. The first step of developing the system is the registration of hatcheries, slaughterhouses and stock farms. Tracing the livestocks is based on special bills of delivery and the strict and regular supply of data by the members to the central database. The computer programme compares the data of bills of delivery automatically with the official register of parent stocks and with te the database of hatchery. The developed system fits to other official registers in Hungary (TIR, OÁIR) and in EU (TRACES). The introduction of this system means a qualitative change in the Hungarian poultry breeding since it spots the illegal livestock or livestock of doubtful origin and significantly improves the transparency of the sector. At the same time, stored data allow suitable base for elaborating a system of different certificates of origin.

Session 03 Poster 8

Risk factors for stillbirth in Irish dairy herds

J.F. Mee[1], A.R. Cromie[2] and D.P. Berry[1], [1]Teagasc, Moorepark Dairy Production Research Centre, Fermoy, Co. Cork, Ireland, [2]Irish Cattle Breeding Federation, Bandon, Co. Cork, Ireland

Stillbirth (SB) is the most important category of bovine mortality and SB rates are increasing in Holstein-Friesian (HF) cows internationally. There are no recent studies on SB rate in Irish dairy herds. The objectives of this study were to establish the SB rate and associated risk factors in Irish HF cows. The logit of the probability of a SB (dead at birth after full-term) and associated risk factors extracted from the Irish Cattle Breeding Federation database (2002 to 2005; 251,169 records) were modelled using generalised estimating equations. The average prevalence of SB was 5%. After adjusting for year and month of calving, the odds of SB were greater for assisted calvings (OR 4.0; 3.74-4.19) and for twins (OR 10.4; 9.67-11.26). Stillbirth in the previous calving and predicted transmitting ability for SB significantly increased the odds of SB. Significant interactions between parity and calf gender and between parity and a quadratic regression on age at calving, relative to the median within parity, were detected. There were greater odds of SB for male calves than female calves though the difference reduced with increasing parity. Irrespective of calf gender, SB risk decreased with increasing parity and age at first calving (median=761days). In conclusion, current stillbirth rate and associated risk factors in Irish HF cows were similar to those in other studies internationally.

Session 03 Poster 9

Breeding for resistance to footrot in sheep

J. Conington[1], G.J. Nieuwhof[2], B. Hosie[3], S.C. Bishop[4] and L. Bünger[1], [1]SAC, Sustainable Livestock Systems, W. Mains Rd., Edinburgh EH9 3JG, United Kingdom, [2]Meat & Livestock Commission, Milton Keynes, MK6 1AX, United Kingdom, [3]SAC, Vet Services, W.Mains Rd., Edinburgh EH9 3JG, United Kingdom, [4]Roslin Institute, Roslin, Midlothian EH25 9PS, United Kingdom

Footrot in sheep is a major welfare problem and is the main cause of lameness in UK sheep. It is possible that breeding sheep to be more resistant to footrot is a viable strategy contributing to its control. Using 9,103 Blackface, Texel and Mule sheep, research to investigate both the quantitative and molecular genetic properties of host resistance to footrot commenced in 2005. The aim is to combine new and existing knowledge of QTL for footrot resistance along with genetic parameters estimated from phenotypic scores. Hooves were graded according to the severity of footrot (0-4). Pedigree flocks that are linked genetically via their respective sire reference schemes were used (27 flocks). The prevalence of footrot for Blackface sheep differed both across years (2005=25%, 2006=18%) and across flocks (5-36%). Footrot levels were consistently high in Mule flocks, (2005=56%, 2006=43%). Texel flocks varied from <1% to 59% affected (average 29%). Phenotypic correlations between successive scores were generally low (0.0 to 0.3) so repeat footrot scoring on the same animals is beneficial. Heritability estimates using repeated measures differed according to the way in which footrot was defined, with the highest estimate being for the number of feet affected with score >2 (h^2=0.23).

Session 04 Theatre 1

Evaluation of technological treatments of starchy feeds for ruminants

D. Sauvant[1], A. Offner[2] and P. Chapoutot[1], [1]AgroParisTech, SVS, 16 rue Claude Bernard, 75231 Paris Cedex 05, France, [2]CYBELIA, ZAC - Lice Blossac, 35170 Bruz, France

Ruminant diets contain up to 50% of starch (ST) in dry matter (DM). Variations of ST digestion in the rumen are the major factor of digestible organic matter partition between rumen and intestines. These variations alter the profile of absorbed nutrients (volatile fatty acids, glucose) and impact animal responses (DM intake, energy digestibility, milk fat content). ST digestion in the rumen can be predicted by the in sacco method. Recent feed tables (INRA-AFZ, 2004) provided mean values of the parameters of ruminal ST disappearance. It is also possible to assess these criteria from *in vitro* hydrolysis with amylases. There are variations among feeds in the ST degradation rate. For maize silage, the genotype and the maturity are known to alter this parameter. Several processes can be applied to control the level of ruminal ST digestion. It is possible to quicken ST degradation rate by a physicochemical treatment (steam flaking, pelleting, expanding...) or to reduce it by formaldehyde or other treatments in order to prevent rumen acidosis. For treated maize and sorghum, feed density is a fairly good predictor of ST degradation rate. Endly, there is an inverse relationship between ruminal ST degradation rate and the mean particle size of feeds and rations altered by grinding.

Session 04 Theatre 2

Corn grain processing and its effect on ruminal degradability

F.J. Schwarz and H. Kurtz, Technical University of Munich, Section of Animal Nutrition, Department of Animal Science, Hochfeldweg 4+6, 85350 Freising-Weihenstephan, Germany

In ruminants the major site of cereal starch digestion is usually the rumen. Corn grain varies in ruminal degradability considerably and feed processing might be of high importance. This trial was conducted to determine the effect of corn grain processing - drying, ensiling, fresh - on the rate and extent of ruminal dry matter (starch) degradability especially depending on plant maturity. The trial used 37 different varieties of corn planted in the years 2000 to 2003. The plants were harvested each at four different harvest dates (HD1 - HD4) between August and October. Corn grain processing was varied in three different ways: The kernels were either dried as under practical farm conditions (cabinet dryer), ensiled or freeze-dried ("fresh"). The ruminal degradability of DM was measured by the nylon bag technique (in situ) using cannulated Holstein cows. The effective ruminal DM degradability (EDMD) was calculated with a ruminal passage rate of 8 %/h. Corn processing had a marked effect on the extent of the ruminal degradability which was further influenced by the harvest date. Dried corn significantly reduced EDMD8 to mean values of 55 % (HD1) to 44 % (HD4) in comparison to average data of fresh corn (79 % (HD1) to 55 % (HD4)) and to mean values of ensiled corn (87 % (HD1) to 81 % (HD4)). While fresh material showed the greatest differences in EDMD8 between HD1 and HD4, drying and especially ensiling reduced the differences between HD.

Session 04 Theatre 3

In vitro gas production profiles and formation of fermentation end-product in processed peas, lupins and faba beans

A. Azarfar[1,2], S. Tamminga[2], W.F. Pellikaan[2] and A.F.B. van der Poel[2], [1]University of Lorestan, Faculty of Agriculture, University of Lorestan PO Box 465, Khorramabad, Iran, [2]Wageningen University, Animal Nutrition Group, Marijkeweg 40, 6700 AA, Netherlands

The effects of using a pre-compacting device (expander) on the degradative behaviour of peas, lupins and faba beans and their different fractions were studied using an *in vitro* gas production technique. The entire samples (WHO) were fractionated into a soluble washable (SWF), insoluble washable (ISWF), and non-washable (NWF) fractions. Samples of the entire concentrate ingredients (WHO) and their different fractions (NWF, ISWF and SWF) were subjected to three processes (R, Retsch mill ground samples; E, expander treated samples; EP, expander-pelleted samples) and their fermentation characteristics were evaluated using an *in vitro* incubation technique for 72h. In peas and faba beans, both the E and EP process significantly ($P < 0.05$) increased the size of NWF compared to process R. The SWF in the processed samples E and EP was lower than in the R samples. In the legume seeds compared to the R samples; both the E and EP samples had a significantly higher maximum fractional rate of gas production in the first phase of fermentation ($P < 0.05$). In lupins and faba beans, E and EP shifted the pattern of fermentation towards a more glucogenic fermentation as represented by a lower non-glucogenic to glucogenic ratio (NGR). The produced ammonia for E and EP samples were significantly ($P < 0.05$) lower than in R samples.

Session 04 Theatre 4

Evaluation of four soybean meal products as protein sources for dairy cows

M. Awawdeh[1], E. Titgemeyer[2], J. Drouillard[2], R. Beyer[2] and J. Shirley[2], [1]Jordan University of Science & Technology, Pathology & Animal Health, Irbid, 22110, Jordan, [2]Kansas State University, Animal Sciences & Industry, Manhattan, KS, 66506, USA

In 3 experiments, 4 soybean meal (SBM) products were evaluated: solvent SBM (SSBM), expeller SBM (ESBM), lignosulfonate-treated SBM (LSBM), and SSBM treated with Baker's yeast and toasted at 100°C (YSBM). In Exp. 1, 32 multiparous Holstein cows (initially 152 ± 63 days in milk, BW = 708 ± 77 kg, 41 ± 7 kg/d milk) were used in a 4 × 4 Latin square to investigate cow responses to changes in SBM source. Diets were isonitrogenous and formulated to provide adequate ruminally degradable protein (RDP), but deficient ruminally undegradable protein (RUP) and metabolizable protein (MP) supplies. Diets had no effects ($P > 0.10$) on dry matter intake, BW gain, milk and component yields, or efficiency of milk production. Lack of response was likely due to adequate RUP and MP supply by all diets, when actual intakes and production levels were considered. In Exp. 2, in situ ruminal degradations were slower ($P < 0.05$) and RUP contents were greater ($P < 0.05$) for YSBM and LSBM than for SSBM or ESBM. The RUP of all SBM products had similar ($P > 0.20$) small intestinal digestibility. In Exp. 3, available lysine contents estimated chemically or using standard chick growth assay were less ($P < 0.05$) for YSBM and LSBM than for SSBM or ESBM. Although heating YSBM and LSBM decreased ruminal degradability, lysine bioavailability was also reduced.

Composition and nutrition value of heated rape seed cake

J. Dulbinskis[1], I.H. Konosonoka[1], V. Sterna[1], D. Ikauniece[1] and D. Lagzdins[2], [1]Latvia University of Agriculture, Research Institute of Biotechnology and Veterinary Medicine Sigra, Instituta 1, LV 2150, Sigulda, Latvia, [2]Ltd Iecavnieks, LS Kaltes, LV 3913, Iecava, Latvia

Rapeseed cake is one of richest source of protein for animals including dairy cows. From publications is known that heating increased nutrition value of rape seed cakes (Kamprda *et al*., 2002; Urbsiene, 2002). Rate of microbial contamination of rape seed cakes produced with cold pressure method and heated in different temperatures from 110° C up to 140° C is depended from used temperature. Extrusion temperature influenced also isolated microorganisms. During heating in mentioned temperatures yeast and lactic acid bacteria are killed, significantly decreased amount of fungi. Treatment in 140° C was reason why microorganisms produced butyric acid were not detected. More often was isolated fungi from the genus Mucor, Penicillum and Aspergillus. Content of fatty acids after heating in different temperatures did not change significantly ($p > 0.05$). Content of dry matter in heated rape seed cakes was by 2.5 – 3.5% higher than in unheated and degradable protein increased by 4.68 – 5.3 g per kg. In the groups where in cows diet was included rape seed cakes 2 kg per day, heated at 110° C or 140° C, content of saturated fatty acids decreased and content of unsaturated oleinic and miristicoleicacid increased. Amount of produced milk increased by 1.25 – 2.77 kg per day. Fat content of milk decreased in these groups by 0.2 % ($p < 0.05$).

Influence of mechanic-hydrothermic treatment of feed, on the rumen degradability and fermentescibility in buffalo

F. Sarubbi, R. Baculo, D. Balzarano, I. Pepe and L. Ferrara, CNR, ISPAAM, Via Argine 1085, 80147 Naples, Italy

Several treatments were used to increase the nutritive value of feed, the most of them were physical. The most used are milling, flaking and crushing. The use of flaking and crushing tratments showed weak homogeneous results on the rumen degradability. Moreover there is a lack of studies on buffalo that suggested to study, with *in situ* technique, the modifications induced by the treatments on cereals and leguminosae seeds on rumen degradability. Four fistulated buffaloes were used for this study. According to *curve peeling* method, potential degradability of crude proteins was determined by model: $p = a + b(1 - e^{-ct})$. The effective degradability evaluation was determined according to the equation: $Ed = a + b (c/c+k)$. PDI values were calculated from the effective degradabilities of each feedstuff in according to ASPA (1994). In buffalo, the effects of treatments on kinetic of degradation *in situ* and on the effective degradability of proteins and non-protein dry matter, were not readly determined. In the buffalo, like in bovine, flaking increases the degradability of proteins and the fermentescibility for corn, while crushing increases the degradability and the fermentescibility for oats but not for barley.

Session 04 Theatre 7

Effect of ruminally protected methionine and lysine on milk nitrogen fractions of Holstein dairy cows fed processed cottonseed

A.R. Foroughi[1], A.A. Naserian[2], R. Valizadeh[2] and M. Danesh Mesgaran[2], [1]High Educational centre of hashemi nejad, Jame university, Animal scince Department, Shahid Kalantary street, Mashhad, Iran, [2]Ferdowsi university of Mashhad, Animal science, Azadi street, Mashhad, Iran

The experiment was conducted to investigate the effect of processing (grinding and moist heat) of whole cottonseed (WCS) and ruminally protected lysine(Lys) and methionine(Met) on milk nitrogen(N) fractions of Holstein lactating cows during early lactation. Multiparious cows (n=12) were used in a 4×4 Latin square design. Cows were fed: 1) WCS; 2) WCS + 16g Met&20g Lys(WCS2); 3) ground cottonseed (GCS) heated in 140°C and steeped for 20 minute (GHCS1); or 4) GCS heated in 140°C and steeped for 20 minute +20g Met&30g Lys(GHCS2). MY was significantly ($P < 0.01$) affected by the diets and was greatest for HGCS2 (35.78 kg/d) and the lowest for WCS (33.07kg/d). Milk protein percent was progressively increased, averaging 3.21%, 3.30%, 3.28% and 3.48% for 1,2,3 and 4 treatments, respectively. Met and Lys supplementation of HGCS2 resulted in increased ($p < 0.05$) in casein N and in treatments of 1,2,3 and 4 were 0.37%, 0.39%, 0.39% and 0.42%, respectively. Total N and whey N showed the same pattern of response as observed for casein N. There isn't significant difference between milk NPN treatments. Physical processing of WCS and amino acid supplement can affect milk protein and N fractions.

Session 04 Poster 8

Effects of electron beam irradiation on dry matter and cell wall degradation of sugarcane bagasse

A.A. Sadeghi[1] and P. Shawrang[2], [1]Department of Animal Science, Science and Research Branch, Islamic Azad University, P.O. Box 14515.4933, Tehran, Iran, [2]Atomic Energy Organization of Iran, P.O. Box 31485-498, Karaj, Iran

In the tropical and subtropical countries, sugarcane bagasse is used in ruminant diets. Sugarcane bagasse has a high proportion of cell wall, which leads to a low degradation rate, principally, due to lignification. Electron beam has been recognized as a useful method for processing of foods and cellulose materials in paper industry. The aim of this study was to evaluate effects of electron beam at doses of 50, 100, 200 and 300 kGy on cell wall degradation of bagasse. Duplicate nylon bags of untreated or irradiated samples were suspended in the rumen of four non-lactating Holstein cows for up to 96 h. Analyses were with the GLM procedure of SAS and differences among treatments were separated using polynomial orthogonal contrasts. Electron beam irradiation increased linearly the washout fractions and decreased linearly the potentially degradable fractions of DM and NDF ($P < 0.001$). Effective degradability of DM and NDF of sugarcane bagasse increased linearly ($P < 0.001$) with increases in irradiation dose. Electron beam irradiation of bagasse at doses of 50, 100, 200 and 300 kGy increased effective degradability of DM at an outflow rate of 0.05/h by 3, 17, 28 and 33%, and of NDF by 2, 13, 25 and 29%, respectively. In conclusion, cell wall degradation of sugarcane bagasse appeared to be increased by electron beam irradiation at doses higher than 50 kGy.

Session 04 Poster 9

Monitoring the fate of gamma irradiated pea proteins in the rumen

P. Shawrang and A.A. Sadeghi, Department of Animal Science, Science and Research Branch, Islamic Azad University, P.O. Box 14515.4933, Tehran, Iran

The present study was designed to monitor the fate of γ-irradiated pea true proteins in the rumen by using SDS-PAGE methodology. Three samples (500 g each, 25% moisture content) were subjected to γ-irradiation at doses of 0, 25, 50 and 75 kGy using cobalt-60 irradiator at 25°C. Duplicate nylon bags of untreated and γ-irradiated pea were suspended into the rumen of four non-lactating Holstein cows for 0, 4, 6, 8, 12, 16, 24, 36 and 48 h. Proteins of untreated and treated bag residues were fractionated by gel electrophoresis. From gel analyses, pea proteins were composed of two major components; globulins (7S and 11S fractions) and albumins. In untreated pea, the albumin and globulin 7S subunits disappeared more rapidly than globulin 11S. Albumin subunits of γ-irradiation at doses of 0, 25, 50 and 75 kGy were degraded completely within 2, 2, 4 and 6 h and for globulin 7S within 4, 4, 8 and 12 h of incubation, respectively. The globulin 11S subunits of untreated pea were degraded slowly and represented a large proportion of the protein remaining after 12 h of incubation. Globulin 11S subunits in γ-irradiated pea at doses of 25, 50 and 75 kGy were not completely degraded after 12, 24 and 48 h of incubation. There were cross-linked products of the protein molecules that could not penetrate the running gel. In conclusion, pea proteins appeared to be effectively protected from ruminal degradation by γ-irradiation at doses higher than 25 kGy.

Session 04 Poster 10

Protein degradation kinetics of canola meal processed by combination of xylose and heat of oven or microwave source

A.A. Sadeghi and P. Shawrang, Department of Animal Science, Science and Research Branch, Islamic Azad University, P.O. Box 14515.4933, Tehran, Iran

This study was carried out to determine the effects of xylose treatment (2 g/100 g DM) in combination with oven (100°C for 2 h) or microwave heating (800 W for 2, 4 and 6 min) of canola meal (CM; 25% moisture content) on ruminal protein degradation characteristics. Duplicate nylon bags of untreated or treated CM were suspended into the rumen of four non-lactating Holstein cows for up to 48 h. Data were analyzed as a CRD design using GLM procedure of SAS. From 14% SDS-PAGE analysis, CM proteins were composed of two major components, napin (2S albumin) and cruciferin (12S globulin). In untreated CM, the napin subunits disappeared at zero incubation time. Napin subunits of xylose + oven or microwave heating for 2, 4 and 6 min were degraded completely within 2, 2, 4 and 6 h of incubation, respectively. The cruciferin subunits of untreated CM were degraded slowly and represented a large proportion of the protein remaining after 16 h of incubation. In xylose + oven or microwave heating for 2, 4 and 6 min, cruciferin were not completely degraded after 24, 24, 36 and 48 h of incubation. Xylose + oven and microwave heating for 6 min decreased ($P < 0.05$) intestinal digestibility of ruminally undegraded CP, compared with microwave heating for 2 and 4 min. In conclusion, the results indicated that xylose treatment + microwave heating for 4 min could increase digestible undegradable protein effectively.

Session 04 Poster 11

Effects of different microwave irradiation powers and periods on protein degradation kinetics and intestinal digestibility of soybean meal

P. Shawrang and A.A. Sadeghi, Department of Animal Science, Science and Research Branch, Islamic Azad University, P.O. Box 14515.4933, Tehran, Iran

This study was completed to evaluate the effects of microwave irradiation at power of 600 W for 4, 5 and 6 min, 800 W for 3, 4 and 5 min and 1000 W for 1, 2 and 3 min on protein degradation kinetics and intestinal digestibility of soybean meal (SBM). Duplicate nylon bags of untreated or irradiated SBM were suspended into the rumen of four non-lactating Holstein cows for up to 48 h and resulting data were fitted to non-linear degradation model to calculate effective degradation of CP (EPD). Intestinal CP digestibility was measured using mobile nylon bag technique. Data were analyzed as a RCB design using GLM procedure of SAS. Irradiation at 600 W over 6 min, 800 W over 5 min and 1000 W over 3 min resulted in burning of SBM. Regardless of irradiation powers and periods, washout fraction of CP, the EPD and degradation rate of CP decreased ($P < 0.05$) and the potentially degradable fraction of CP increased ($P < 0.05$). The EPD decreased as irradiation periods increased ($P < 0.05$) in different powers. Intestinal CP digestibility of 600 W for 6 min and 800 W for 5 min decreased ($P < 0.05$) compared to untreated SBM. The digestible undegradable CP values of irradiated SBM at 600 W for 5 min, 800 W for 4 min and 1000 W for 3 min were almost the same. For practical application, lower period of processing is of interest; therefore microwave irradiation of SBM at 1000 W for 3 min is the best processing condition.

Session 04 Poster 12

Effect of polyethylene glycol on the *in vitro* gas production and digestibility of fruit wastes

A.R. Safaei, H. Fazaeli, M. Zahedifar and S.A. Mirhadi, Animal Science Research Institute, Animal Science, Karaj, Iran

Twenty four samples of fruit residues collected form public places in winter season and dried at room temperature under the air condition. The dried samples were milled and sub sampled for chemical analyses. All samples were mixed and 6 sub samples were made in which, 3 samples were considered as control and the other 3 samples were treated with polyethylene glycol (PEG) where they tested for gas production and digestibility. The average dry mater of fruit residues was 23.7% and OM, CP, EE, NDF, ADF, NFC, WSC, phenolic compounds and tannins were 94.6, 5.5, 8.1, 27.3, 20.7, 53.7, 8.4, 6.9 and 5.3 percent in DM respectively. The gas yield after 24 h were 49.75 and 41.54 mL/200mg of sample for the treatment and control samples respectively that was significantly ($p < 0.05$) affected by the PEG treatment. Addition of PEG significantly ($p < 0.05$) increased the *in vitro* digestibility and ME content.

Session 04 Poster 13

Effect of barley supplementation on the utilization of *Acacia saligna* by camels

I. Awadalla, M. Mohamed and S. Abdel-Magid, National Research Center, Animal Production, Elbehoth Street, Dokki, Giza, Egypt, 12622, Dokki, Egypt

Acacia saligna were grown under desert conditions and irrigated by waste water. Fifteen growing male Maghraby camels (*Camelus dromedarious*) of 186± 5 kg body weight and about 15 months old were used in 12-week growth trial to study the effect of ground barley grain supplementation on the voluntary intake and utilization of *Acacia saligna*. Camels were divided into 3 groups (5 in each) according to weight and age, which were randomly allotted to three levels of ground barley grain supplement offered to cover 0.50 % (group A or control), 0.75% (group B) and 1.00 % (group C) of body weight. The green tree leaves and succulent stems of *Acacia saligna* were collected daily and offered *ad libitum*. Results indicated that the ground barley grain supplementation significantly improved the nutrients digestibility, nutritive values, nitrogen utilization and average daily gain. This study indicated that the utilization of *Acacia saligna* improved by camels when ground barley grain fed at a level of 1.00% of camels body weight.

Session 04 Poster 14

Investigation the use of processed spaghetti wastes with urea on controlled-release ammonia in the rumen

R. Kamali[1], A. Mirhadi[2], A. Godratnama[3], A. Toghdory[1] and M. Mohajer[1], [1]Agriculture and natural recourse research, Gorgan, 4915677555, Iran, [2]Research institute of animal science, Karaj, 31585, Iran, [3]Agriculture and natural recourse research, Mashhad, 32142, Iran

In order to investigate the effects of processed spaghetti wastes (SW) with urea on controlled-release ammonia in the rumen of fattening lambs, an experiment were conducted with 32 male lambs. Treatments include two level of processing (with and without urea) and four levels of SW (0, 7, 14 and 21%) that arranged in 2×4 factorial. Lambs were housed individually in 1.5 × 1.5 m pens. On the last day of experiment ruminal fluid samples were taken with stomach tube before feeding (0), 0.5, 1, 3 and 5 hours post feeding. The results showed that SW processing had no significant effect on ruminal ammonia-N concentration, total volatile fatty acid (TVFA) concentration and blood urea nitrogen (BUN). Concentration of ammonia-N was not affected by SW levels at 0 and 0.5 h samples, but in the 1, 3 and 5 h samples SW levels significantly decreased ammonia-N concentration in comparison with control group ($P < 0.01$). The use of SW at levels 7, 14 and 21% significantly increased TVFA concentrations in the 1 h samples, but at other times SW had no any significant effect. Concentration of BUN was decreased by using SW in the 3 h samples. In conclusion SW processing with urea had no significant effect on slow releasing urea in the rumen.

Session 04 Poster 15

Effect of ruminally protected amino acids on milk yield and composition of Holstein dairy cows fed processed cottonseed

A.R. Foroughi[1], A.A. Naserian[2], R. Valizadeh[2] and M. Danesh Mesgaran[2], [1]High educational centre of Hashemi nejad, Animal science department, Shahid kalantry street, Mashhad, Iran, [2]Ferdowsi university of mashhad, Animal science department, Azadi Street, Mashhad, Iran

The objective of this study was to evaluate the effect of processing (grinding and moist heat) of whole cottonseed (WCS) and ruminally protected lysine(Lys) and methionine(Met) on milk composition and production of Holstein lactating cows during early lactation. Multiparious cows (n=12) were used in a 4×4 Latin square design. Cows were fed: 1) WCS; 2) WCS + 16g Met&20g Lys(WCS2); 3) ground cottonseed (GCS) heated in 140°C and steeped for 20 minute (GHCS1); or 4) GCS heated in 140°C and steeped for 20 minute +20g Met&30g Lys(GHCS2). The mean DMI was significantly ($P < 0.01$) affected by diets and in treatments of 1,2,3 and 4 were 21.08, 21.19, 22.57 and 27.63 (kg/d), respectively. Physical processing of WCS did not affect ruminal pH and mean for treatments was 6.48. MY was significantly ($P < 0.01$) affected by the diets and was greatest for HGCS2 (35.78 kg/d) and the lowest for WCS (33.07kg/d). Milk fat percentage and yield were unaffected by diets. Milk protein percent was progressively increased, averaging 3.21%, 3.30%, 3.28% and 3.48% for 1,2,3 and 4 treatments, respectively. Results indicated that when cows were fed WCS and processed cottonseed associated ruminally protected lysine and methionine milk yield and composition were improved.

Session 05 Theatre 1

Modelling of manure production by pigs. Effect of feeding, storage and treatment on manure characteristics and emissions of ammonia and greenhouse gazes

C. Rigolot[1,2], S. Espagnol[2], M. Hassouna[1,3] and J.-Y. Dourmad[1], [1]INRA, UMR SENAH, Saint Gilles, 35590, France, [2]IFIP, Le Rheu, 35651, France, [3]INRA, UMR SAS, Rennes, 35000, France

A model was developed from literature data to predict ammonia and greenhouse gazes emissions (CH_4, N_2O), and the characteristics of the effluent (volume, dry and organic matter, N, K, P, Cu and Zn) produced by pigs in contrasted situations of manure collection, storage and treatment. The model was constructed in three parts. Part (1) predicts nutrient excretion according to animal performance and feeding strategy. Part (2) predicts changes in manure composition and air emissions according to storage practices (liquid or solid) and climatic conditions. Part (3) predicts the evolution of manure during biological treatment, anaerobic digestion or composting. Original equations as well as published relationships were used to build the model. Part (1) was validated using 19 experimental studies, whereas internal and expert validation was performed for part (2) and (3), and the comprehensive model. The effects of different feeding strategies and mitigation techniques during storage were tested and advantages and weak points could be identified for each alternative of manure management. Such a model can be an efficient tool to quantify and limit harmful emissions, while obtaining manure better adapted to each farming situation. This is illustrated through different examples.

Session 05 Theatre 2

Environmental impact of pig production and nutritional strategies to reduce nitrogen, ammonia, odor and mineral excretion and emission

A.W. Jongbloed, Animal Sciences Group of Wageningen UR, Animal Production Divison, Edelhertweg 15, 8219PH Lelystad, Netherlands

Raw materials for pig feeds are still often grown in regions other than where pig production takes place. This may lead to several environmental concerns regarding the quality of soil, water and air. In several cases tolerable levels may be exceeded, thus, legislation has been imposed. This paper provides an insight into nutritional approaches to reduce environmental pollution through the excretion of nitrogen (N), P, Cu and Zn by changing the diet for the animals without compromising their health and performance. In The Netherlands, the excretion of phosphorus per growing-finishing pig has more than halved over the last 30 years as a result of intensive research on P digestibility, requirements for P, and on the efficacy of microbial phytase in pig feeds. Also, N excretion, ammonia and odor emissions can be reduced substantially. Although the excretion of Cu and Zn per pig has decreased during the last decades as a result of EU legislation, it still exceeds tolerable levels in several regions. Finally, several factors that should be taken into account to judge excretion of nitrogen and minerals on a system level are discussed and conclusions drawn.

Session 05 Theatre 3

A dynamic model of ammonia emission and concentration in fattening pig buildings

J.Y. Dourmad[1], V. Moset-Hernandez[1], S. Espagnol[2], M. Hassouna[3] and C. Rigolot[1,2], [1]INRA, UMR SENAH, Saint-Gilles, 35590, France, [2]IFIP, Le Rheu, 35651, France, [3]INRA, UMR SAS, Rennes, 35042, France

The control of gas emissions from livestock buildings, especially ammonia, is important to limit the environmental impact, which depends primarily on the total emission, and to improve the welfare and health of the animals and the stockmen, which is affected by the concentration in the air. Modelling is an essential tool for a global approach of the different processes involved in the emissions. The model developed in this work aims at integrating the information and models already available in the literature in order to predict, in a dynamic way (with a 1 min time step), the gas emissions and the concentrations inside the breeding rooms. The model was validated with data from the literature. The results of this validation indicated that the model predicted in a coherent way as well the cumulated flows as the concentrations. However, we identified some lacks in knowledge, in particular concerning the estimate of the pH of the liquid manure according to the characteristics of the feed and the evolution of manure composition with time. Likewise, it appeared that the phenomena of exchanges between the air located above and below the slats must also be better specified, because they strongly influence ammonia concentration. The simulations indicated that total emission and concentration are not well correlated and are highly dependant on the ventilation system and the temperature.

Session 05 Theatre 4

Ammonia emission in organic pregnant sows with and without access to paddock

S.G. Ivanova-Peneva[1], A.J.A. Aarnink[2] and M.W.A. Verstegen[2], [1]Agricultural institute, 3 Simeon Veliki blvd., 9700 Shumen, Bulgaria, [2]Animal Sciences Group of Wageningen UR, P.O. Box 65, 8200 AB, Lelystad, Netherlands

Ammonia emission was studied during two periods in an experimental farm for organic pigs with two groups of pregnant sows, each consisting of 14 animals. One of the groups of sows had access to a paddock, and the other had not. In the first period sows had unlimited access to the paddock, while this was from 9:00 h until 15:00 h in the second period. Both measuring periods were in summer, with 45 days in between. Ammonia emissions were measured by the ventilated chamber technique at different locations inside the building and on the paved outside yard. Ammonia emission per m^2 did not differ significantly between treatments, but differed between periods and locations (inside and outside the building). When sows had limited access to the paddock, ammonia emissions were about twice lower. Sows in the paddock group had a higher total urinary-N concentration than the control sows, probably due to the consumption of clover grass in the paddock, which is rich in nitrogen. It is tentatively concluded that limited access of sows to pasture reduces ammonia emission in organic pig farming.

Session 05 Poster 5

***In vitro* fermentation characteristics of selected feedstuffs by pig faecal inocula**

S.B. Cho, D.W. Kim, O.H. Hwang, C.W. Choi, J. Hwangbo, W.T. Chung, J.H. Kwag, I.B. Chung and B.S. Lee, National Livestock Research Institute, 564 Omokchun-dong, Gweonsun-gu, 441-706, Suwon, Korea, South

Present study aimed to quantify the fermentation characteristics of feedstuffs for pigs. Different CP containing feedstuffs [soy hull (CP 13.0%), wheat (18.0%), corn germ meal (24.2%), corn gluten feed (24.8%), rapeseed meal (39.6%) and soybean meal (52.2%)] were anaerobically incubated (pH 6.2) in 5% (wt/vol) faecal slurry comprising mineral salts medium and fresh faeces from growing pigs as sources of faecal inocula. Samples were collected from the final fermented slurry at 72 h after incubation and analysed for concentrations of NH_4, volatile fatty acids (VFA) and skatole. Amino acids profiles of the feedstuffs were analysed using amino acids analyser. Increasing dietary CP level increased ($P < 0.01$) NH_4 concentration in the final slurry. Dietary valine was closely related (R^2=0.78, $P < 0.01$) to isobutyric acid concentration from the final slurry. The similar results were also found in the relationship (R^2=0.77, $P < 0.01$) between dietary leucine and isovaleric acid concentration from the final slurry. Skatole concentration from the final slurry was strongly related (R^2=0.92, $P < 0.001$) to dietary tryptophan. Present results indicated that manipulating CP level and/or amino acids profiles in diets may be needed to reduce concentrations of VFA and skatole in faeces.

Session 06 Theatre 1

The evolving role of the horse on the ancient central Asian steppe

M.A. Levine, McDonald Institute for Archaeological Research, Downing Street, Cambridge, CB2 3ER, United Kingdom

Although in recent years considerable energy has been expended trying to understand the origins of horse domestication, albeit without much success, much less effort has been put into understanding its domestic successors – the horses from the Middle Bronze Age to the Early Iron Age (c. 2000-300 BC). We do know that during this period the chariot developed and expanded westwards into Africa and Europe, and eastwards into China. This was also the time of the development of the great overland trade routes, carrying metallurgical and ceramic technologies throughout Eurasia. Sometime during this period equestrian pastoralism developed, presumably in central, but not much more is known about its origins. The political, economic and social ramifications for these developments should not be underestimated. The role played by the horse in these developments has scarcely been considered.

Session 06 Theatre 2

Rahvan (Ambling) horse breeding and sport in Turkey

U. Yavuzer, University of Harran, Animal Science, Ziraat Fakultesi Şanliurfa, 063300 Harran, Turkey

Rahvan horse that walks without shaking the rider. Rahvan (ambling) is a special gait of the horse when the horse moves the two feet on either side of its body in unison, as if running with two feet. The head of ambling hourses is large with well-placed eyes and relatively small ears; the neck is short and thick; the shoulders tend to be straight; the back is long; the croup is often quite level with a high tail-set; the quarters are slight; the legs often tend to be splayed; the hooves are vertically long and very hard in Turkey. This horse is gentle in nature and very strong and enduring. The ambling horse has adapted itself to the whims and needs of humans as easily as to its environment. He loves people. He responds well to proper treatment and discipline ambling horse races aren't as widespread as it once was, but is still played as a spectator sport, in Turkey. Societies are also attempting to keep this traditional sport alive by organising.

Session 06 Theatre 3

Safety in the human-horse relationship

J.W. Christensen and E. Søndergaard, University of Aarhus, Faculty of Agricultural Sciences, P.O. Box 50, 8830 Tjele, Denmark

When horses and humans are working together there is always a risk to the human due to the difference in size and strength between the two species, and the fact that the horse is a flight animal. In order to minimise the risk of accidents, it is very important for us to lower the risk of a horse fleeing. This may be done by training or by the way we keep the animals in general. We have shown that young group housed horses behave more adequately during training than horses housed singly. Besides horses that have been handled for a long period show less reaction to novel objects and situations than horses that have been handled only shortly as shown by the long term handled horses having lower heart rates. A low heart rate indicates that the animal is less likely to flee. Training methods to reduce fear reactions in horses have only recently been the subject for scientific investigation although the ability of a horse to habituate to a range of otherwise frightening stimuli greatly increases safety in the horse-human relationship, and finding appropriate methods for reducing fear in horses has important practical applications. We have investigated how horses respond to novel stimuli, acting on their different senses, if they generalise when meeting several stimuli, as well as how horses learn to be confident with an otherwise frightening stimulus.

Session 06 Theatre 4

Handling horses: safety and welfare issues

E.K. Visser[1], H. Roche[2], S. Henry[2] and M. Hausberger[2], [1]Animal Sciences Group, Wageningen UR, Animal Production, PO Box 65, 8200 AB, Netherlands, [2]Universite de Rennes, CNRS, Ethologie-Evolution-Ecologie, Avenue du General Leclerc, 35042 Rennes cedex, France

The interaction between horse and human can be observed in a spectrum from the short occasional interactions to a (life)long term bond. Difficulties and problems in the interspecific communication lead to accidents and decreased welfare of both humans and horses. Surveys on horse related accidents underline the need for a better knowledge and observation of the behaviour of horses in different situations and professions. The reaction of the horse on human handling depends on the human attitude, skills and experience but also on the horses' temperament. The development of methods to evaluate the way horses react to humans is rather recent and a variety of approaches have been developed, among them are observer ratings and behavioural tests. Most horse owners develop an emotional bond with their horses. The way owners consider their horse most likely has an important influence on the way they manage and handle them. The horse-rider relationship is only very poorly understood; riding accidents and severe mismatches between horse-rider and aim lead to disappointments and welfare problems. There is a strong need for specific research and training of humans working with horses in order to improve the human-horse relationship that, as shown by the high incidence of accidents and increasing number of horses with a decreased welfare is far from optimal.

Session 06 Theatre 5

How to establish a human-foal relationship based on trust?

S. Henry, M.-A. Richard-Yris and M. Hausberger, UMR CNRS 6552, Université Rennes 1, Bât. 25, 263 av. Général Leclerc, 35042 RENNES, France

Developing a positive relationship with horses is an important step for stockbreeders and users. Research studies have considered the importance of the early age in establishing a human-horse bond and have aimed to better understand what type of approaches and timing may help develop a positive relation. While it is traditionally admitted that handling of foals is a prerequisite to improve their reactions to humans, recent research revealed that forced and intensive approaches have rarely long-term beneficial effects and may even lead to short-term disturbances in the relation. Thus, even a brief handling at first suckling seems enough to induce later reluctance for human contact. On the other hand, less intrusive approaches may influence positively the foal's responses to human presence and handling. A mere exposure to humans seems to induce some habituation, but overall the use of social partners as models is of great help in establishing with foals a positive relation that allow even on a long-term basis their active cooperation during handling. Therefore, it appears important for the establishment of a bond based on trust to let foals being an active participant in the process.

Session 06 Poster 6

Safety in equestrian exhibitions

A. Checchi[1], S. Casazza[2] and F. Martuzzi[2], [1]Univ.Bologna, Agraria, V.Fanin, 40127 Bologna, Italy, [2]Univ.Parma, Med.Veterinaria, V.Taglio, 43100 Parma, Italy

The world of equestrian exhibitions aims most of its attention on the overall glamour, without focusing organizational strains on safety among public, workers and horses and preventing accidents. The not easy definition of risks in the equestrian range rises from the extreme variability of working states and the considerable number of activities which are often simultaneously performed. The sphere of these risks concern: Safety of people: buildings and structures authorized by firemen, sound fences around rings and competition fields, safety distance from public,horses and moving carriages, visible signals of routes equipped for horses. Safety of workers: specific training for horse managing employees, equipments and individual protective devices for workers, good acquaintance of expositive dynamics. Horses' physical and psychical safety: wide stalls so horses may be stabled in the most comfortable conditions; gangways/transit paths free from obstacles and unevenness that may wound horses; removal of stress from excessive lighting, ceaseless presence of noisy public, scanty moments of relax through daytime, close presence of more stallions, separation of foal from mare. As conclusion, to reach an optimized level of safety in equestrian exhibitions, fairs and expositions it is necessary to develop a wardship in the areas mentioned above. Applying some simple but basic rules, it will be possible to grant a complete and effective safety in the framework of every respectable equestrian show.

Session 06 Poster 7

Sensory sensitivity: a horse's temperamental dimension

L. Lansade, M. Lecomte and G. Pichard, INRA, Laboratoire de comportement, PRC, 37380 Nouzilly, France

Temperament is an important factor when working with horses. In order to measure the temperament more precisely, our work aimed to identify a dimension never described in horses, namely sensory sensitivity. Our study was based on the definition of a dimension as "a behavioural characteristic stable across situations and over time". We designed several tests for each sense and then determined whether the responses observed were correlated in time and between situations. The principle of the tests was to generate two stimuli of different intensities for each sense (e.g. two different sounds) and to measure the intensity of the horse's response (N=26). Using Spearman rank correlations, we tested whether the responses to these different stimuli were inter-correlated. We repeated the same tests 5 months later to determine whether the responses were correlated over time. For each sense, results show that the greater the horses' response to one stimulus, the greater their response to the other. For example, the reaction to the odour of cinnamon was significantly correlated to the reaction to lavender (R=0.53, p=0.004). However, there was no significant correlation between the responses to stimuli relating to different senses. Finally, these responses showed stability over a 5-month period (e.g. tactile test: R=0.71, $p < 0.0001$). In conclusion, our study shows that a dimension for each sense exists (e.g. tactile sensitivity) rather than a general sensory sensitivity dimension.

Session 07 Theatre 1

Livestock and environment: new challenges for a sustainable world

P. Gerber, T. Wassenaar, M. Rosales, V. Castel and H. Steinfeld, FAO, Viale delle Terme di Caracalla, 00100 Roma, Italy

Growing populations and incomes, along with changing food preferences, are rapidly increasing demand for livestock products, while globalization is boosting trade in livestock inputs and products. While growing, the livestock production is undergoing a complex process of technical and geographical change, which is shifting the balance of environmental problems that the sector causes. The livestock sector, by far the single largest anthropogenic user of land, emerges as one of the most significant contributors to global environmental issues. Livestock are estimated to be responsible for 18 per cent of greenhouse gas emissions measured in $CO2$ equivalent and are a key player in ammonia emissions. Livestock also increasingly contribute to water use, mostly for the irrigation of feed crops. It is also probably among the largest agricultural source of water pollution. Furthermore, as a consequence of the above, the livestock sector may well be the leading player in the reduction of biodiversity. On the other hand, livestock provide protein and micro-nutrients to many of the 830 million food insecure people who keep livestock, and are a key asset to almost 1,000 million rural poor. Technology options to mitigate environmental impacts are generally available, but the development and implementation of policy mixes is still missing. Policy action will require recognizing multiple development objectives in the context of variable local circumstances.

Session 07 Theatre 2

CAP reforms, land use and diversity of livestock farming systems in Europe

A. Pflimlin and C. Perrot, Institut de l Elevage, 149 rue de Bercy, 75595 Paris, France

Since 1992, the CAP has moved from the price support policy towards the single farm payment, but still gives the opportunity for different applications between EU countries. Decoupling and biofuel crops may lead to further specialisation in highly productive regions and a severe decline in less favoured areas which are also penalised by large cuts on second pillar budget. A "health check" in 2008 will bring the opportunity for further debates about full decoupling, removing milk quota and set-aside, regional premiums, cross compliance, rural development. We discuss in this paper the potential impact of these evolutions on the diversity of livestock farming systems that still characterises European herbivore husbandry. A simplified zoning, based on a few land use criteria, allows to classify the herbivore farming regions into 8 areas with specific characteristics, assets and constraints, quantifying their importance and contribution to the regional economy and environment. On the basis of this analysis, the CAP reform and the demand for biofuel can bring large changes in land use, including abandonment or forest in extensive grassland and mountain areas, where herbivore farming is a major component of the local economy. Such regions include about half of the agricultural area and more than half of the herbivore farmers in the EU-15. As the proportion and the diversity would be even higher in EU-27, the zoning proposal should be extended to all the new members as a common frame for debates about the future CAP.

Session 07 Theatre 3

Relationships between livestock production systems and landscape changes in the Belluno province, Eastern Italian Alps

G. Cocca, L. Gallo, E. Sturaro, L. Dal Compare, M. Mrad, B. Contiero and M. Ramanzin, University of Padova, viale Università 16, 35020 Legnaro PD, Italy

Natural forest expansion following abandonment of farming is believed to have originated major changes in the landscape of Italian Alps. This study was conducted in the Belluno province (3676 km^2, 1323 m a.s.l. mean elevation, SD=610 m), to determine the magnitude of landscape (forest to open areas) changes from 1980 to 2000 and to investigate on the relationships between characteristics of livestock production systems, socio-economic indicators (data from official censuses; 69 municipalities, 1444 ruminant farms) and landscape changes. Forest boundaries from regional maps (published in 1983; 1:10000 scale) were recorded by digitizing polygon features across the project area, and compared with a land-use theme created from aerial photographs (2001). Change of land use over time was obtained, for each municipality, by overlaying the 1983 and 2001 themes. From 1980 to 2000, forest areas increased by 22% and the forest/open area ratio increased from 2.5 to 6.3. The municipalities were grouped in clusters on the base of descriptors of landscape, socio-economic indicators and livestock production systems. A variance analysis showed that landscape changes differed between clusters and were less pronounced where traditional ruminant production systems had been less abandoned; socio-economic indicators were less related to landscape changes.

Session 07 Theatre 4

The perceived role of farming in a high amenity region in Ireland

B. Coutrney[1,2], W. Dunne[1] and J.J. O'Connell[2], [1]Teagasc, Rural Economy Research Centre, Malahide Road, Dublin 17, Ireland, [2]University College Dublin, Agribusiness, Belfield, Dublin 4, Ireland

The Burren region in the west of Ireland is a high amenity, semi-natural cultural landscape dependent on extensive, low-input farming and threatened by changing economic conditions both within and outside agriculture. A survey of 80 non-farmer residents was undertaken to elicit their perceptions of farming. The main findings were: 56% used Burren farmland for recreation, 88% regarded agriculture as either important or very important to their culture and heritage, 64% considered farming to be beneficial for the environment and landscape. But, 36% viewed farming as damaging, mostly to archaeological sites 46%, waterway pollution 39%, wildlife habitats 34%, stone wall destruction 20%, and on visual landscape 11%. The perceived problems faced by farmers were: inadequate farm income (58%), lack of successors (35%), uncertain future (31%), and changes in agricultural policy (16%). Forty percent consider farmers to have been strongly influenced by agricultural policy. This policy was considered positive by 44% and negative by 25%. Almost 79% considered Burren farming to be overly-reliant on financial support from EU and National government. Suggested methods for improving the environment and visual landscape were diverse: reduce the intensity of farming 34%, increase farming intensity 4%, a higher priority to Agri-environment schemes 24%, more regulation of farming 20%, and allowing Duchas (a state agency) to manage more land 14%.

Session 08 Theatre 1

The biological opportunities and limitations related to milk quality in a sustainable dairy production system

K. Stelwagen, AgResearch, Dairy Science & Technology, Ruakura Research Centre, PB 3123 Hamilton, New Zealand

The sudden lack of subsidies since the late 1980s forced the New Zealand dairy sector to undergo major changes. In a very short time frame dairy farmers had to adopt to a low-cost and low-input system in order to survive. This transformation has resulted in a very efficient, predominantly pasture-based dairy sector, reliant on seasonal calving to optimally utilise pasture growth and on minimal or no use of concentrate feeding. Such a low-input and sustainable system provides both opportunities and challenges. This overview will address the effects of sustainable farming practices on milk production, with particular emphasis on physiological aspects of milk composition and quality. In particular, data will be presented on a comparison between a low-input system (pasture) and one based on high inputs (concentrates), effects of once-daily milking, and the ability of the mammary gland to secrete bioactive components in milk (value-add).

Session 08 Theatre 2

Vitamins and selenium in bulk tank milk of organic and conventional dairy farms

U. Emanuelson and N. Fall, Swedish Univ of Agric Sci, Dept Clinical Sciences, POB 7054, SE-75007 Uppsala, Sweden

Vitamins in milk are important for the human consumer, the calf and the cow. They are also important for the dairy industry because some of them affect the risk for off-flavour. Studies indicate that milk from organic and conventional dairy farms may differ in these aspects. The aim of this study was therefore to investigate whether there are differences in the concentration of vitamins and selenium in milk between organic and conventional farms in Sweden. Bulk tank milk was sampled in 20 organic and 20 conventional dairy farms at three occasions during the indoor season. Concentrations of α-tocopherol, β-carotene and retinol were determined by HPLC and selenium by hydrid generation inductively coupled plasma atomic emission spectrometry. Herd characteristics were collected by questionnaires and from the official milk recording scheme. Multivariable linear mixed models were used to evaluate the associations between milk composition and type of farm, while adjusting for potential confounders and the repeated observations within farm. Average levels of α-tocopherol, β-carotene, retinol and selenium were 0.84 μg/ml, 0.18 μg/ml, 0.32 μg/ml and 13.6 μg/kg, respectively. There were no differences between organic and conventional farms in β-carotene, retinol or selenium concentration. Concentration of α-tocopherol was also similar, but was significantly lower in organic farms if differences in amount of roughage were accounted for in the statistical model.

Session 08 Theatre 3

Dietary fat affects plasma and milk zink content in dairy cows

J. Sehested, L. Wiking and T. Larsen, University of Aarhus, P.O.Box 50, 8830 Tjele, Denmark

The objective was to study the effect of dietary precursors for milk fat synthesis and dietary zink (Zn) level on milk free fatty acid (FFA) level. A 4x4 Latin square exp. with 16 lactating cows, 4 periods of 12 days and 4 dietary treatments was conducted. A total mixed ration (TMR) based on maize and grass-clover silages and pelleted sugar beet pulp was used on all treatments. Dietary Zn levels were 45 mg (natural content) or 100 mg (by addition of ZnO) per kg DM. A 'high de novo' milk fat diet was formulated by adding rape seed meal and molasses in the TMR, and a 'low de novo' by adding saturated fat, rape seed cake and maize. Treatments did not affect daily DM intake (23.0±0.6 kg), or yield of ECM (31±1 kg), milk fat (1.24±0.06 kg) or milk protein (1.03±0.04 kg). The 'high de novo' diet significantly increased milk content of short chain fatty acids (SCFA, C4 to C16) and decreased content of C18 and C18:1. Treatments did not influence the concentration of FFA in milk at 0 or 28 hours after milking. The 'low de novo' diet significantly increased plasma Zn (1.06 *vs* 0.90±0.04 ppm) and milk Zn (3.4 *vs* 2.9±0.1 ppm) content, whereas dietary Zn level did not in itself influence these parameters. These results do not confirm earlier findings, that high dietary Zn or a high proportion of SCFA in milk fat is followed by a decrease in concentration of free fatty acids in milk. However, the results indicate that high dietary fat might facilitate the transfer of dietary Zn into plasma and milk.

Session 08 Theatre 4

Fatty acid composition of beef: effects of production system and strategic supplementation

D.A. Kenny[1] and A.P. Moloney[2], [1]University College Dublin, Belfield, Dublin 4., Ireland, [2]Teagasc Grange Beef Cattle Research Centre, Dunsany,, Co. Meath, Ireland

Ruminant fat can contain fatty acids with putative human health benefits, including conjugated linoleic acid (CLA) and omega-3 polyunsaturated fatty acids (v-3 PUFA). The diet of the animal has a major influence on the fatty acid composition of beef and compared to concentrates, lower input systems based on grazed and conserved grass have been shown to promote tissue accumulation of CLA and v-3 PUFA. In addition, supplementation of forage-fed cattle with a blend of oils rich in w-3 PUFA and linoleic acid increases ruminal and tissue concentrations of vaccenic acid (VA), the main substrate for D-9 desaturase-catalysedtissue synthesis of the cis9, trans11 isomer of CLA. However, high tissue concentrations of v-3 PUFA may inhibit D-9 desaturase gene transcription. This has implications for the simultaneous improvement of CLA and ω-3 PUFA in bovine muscle and requires further investigation. Current intakes of health promoting fatty acids in the human diet are apparently only a fraction of that required to exert therapeutic effects. While beef, particularly from low-input systems, can make a contribution to human requirements for v-3 PUFA and CLA, clarification of inefficiencies during ruminal passage, post ruminal absorption coupled with inhibitory interactions at tissue level is required to consistently produce beef that can be considered a "functional" food with respect to fatty acid supply.

Session 08 Theatre 5

Carcass quality and fatty acid composition of intramuscular fat of Holstein bulls reared exclusively on pasture or in feedlot

H.J.D. Rosa[1], S.T. Mendes[1], O.A. Rego[1], C.C. Silva[1] and R.J.B. Bessa[2], [1]Universidade Açores, CITAA-DCA, 9700 Angra Heroismo, Portugal, [2]EZN, Fonte Boa, 2000 Santarém, Portugal

Fifty one Holstein bulls reared *ad libitum* on either grass (Azorean traditional system) or in stalls fed grass- and maize silages supplemented with 50% concentrate (intensive system) were slaughtered (23±4.2 and 17.5±5.1 months old; 272±64 and 280±49 kg carcass weight, respectively) and samples from *m. longissimus dorsi* were collected. Intramuscular fat fatty acid composition was evaluated by GC. Carcass quality was scored by the SEUROP system. Carcasses of bulls from pasture were thinner and had the lowest conformation $p < 0.05$). Fat from these animals had three fold higher PUFA n-3 and 40% lower PUFA n-6 content resulting in a n-6/n-3 ratio 5 times lower (1.46 *vs* 7.92, $p < 0.001$), and also had 80% more CLA (0.59% *vs* 0.33%, $p < 0.001$), 125% more EPA (0.79% *vs* 0.35% $p < 0.001$), 144% more DPA (0.95% *vs* 0.39%,$p < 0.001$) and 45% less TFA (3.78% *vs* 6.47%, $p < 0.001$). Concentrations of C15:0, C17:1 and C18:3 n-3 were higher ($p < 0.01$) in the pasture group while concentrations of C16:0, C18:1 *trans* and C18:2 n-6 were lower ($p < 0.05$). Concentrations of SFA, PUFA and MUFA did not differ between groups and the P:S ratio was 0.22. Significant correlations were detected between concentrations of CLA and other fatty acids, namely C18:1 *trans* 11 (r=0.57, $p < 0.001$), C14:0 (r=0.36, $p < 0.05$) and C16:0 (r=0.39, $p < 0.05$).

Session 08 Theatre 6

Growth performance, carcass quality, muscular characteristics and meat quality traits of Charolais steers and heifers

M.P. Oury[1], J. Agabriel[2], B. Picard[2], R. Jailler[2], H. Dubroeucq[2], D. Egal[2] and D. Micol[2], [1]ENESAD, BP87999, 21079 Dijon, France, [2]INRA, URH, 63122 St-Genès-Champanelle, France

In France, the decrease of bovine females number led to a deficiency of female's meat. So, the aim of this study was to compare Charolais steers (n=11) and heifers (n=11) slaughtered at the same age of 27 months to evaluate if young steers may act as an alterative to heifer's meat. Animals were only fed with grass and hay. Their growth characteristics and slaughtering results were studied (live weights, gains, fattening marks, carcass weights and compositions). Muscular composition (metabolic activity, collagen, fat) and meat quality traits of*rectus abdominis* (RA),*triceps brachii* (TB) and *longissimus thoracis* (LT) were also analyzed. Steers had a higher live weight from birth (51-41kg) to slaughter (713-625kg). Heifers carcasses had a higher fattening score at slaughtered (3.3-2.8), a higher carcass fat content (20-17%), and a higher thigh compactness than steers. Heifers meat had also less total and soluble collagen (23-29 mg/g and 17-20 %, respectively) in RA. Their muscular metabolic activity (Lactate and Isocitrate Deshydrogenase) was higher in TB and LT muscles. No significant differences were seen for meat quality traits of RA and TB. Nevertheless, LT samples of steers were more tender (P=0.09) but less juicy (P=0.07) and less flavoured (P=0.05) than those of heifers. Thus, it seems possible to replace heifers meat by the meat of young steers, slaughtered at the same age.

Session 08 Theatre 7

Genetic variation and selection for superior meat quality traits in cattle breeds under sustainable animal production systems

T. Sweeney[1], O. Aslan[1,2], B. Bahar[1], L. Pannier[1,2], A.M. Mullen[2] and R.M. Hamill[2], [1]University College Dublin, School of Agriculture Food Science and Vetrinary Medicine, Dublin 4, Ireland, [2]Teagasc, Ashtown Food Research Centre, Dublin 15, Japan

Public opinion favours a move towards a more sustainable style of farming that is focused on high-quality end products, while preserving the environment. Characteristics of the 'ideal bovine' in a sustainable production system would combine feed efficiency, methane production, disease resistance, calving ease, polledness, and meat quality characteristics. Identification of DNA markers would ensure that such traits could be enhanced through marker-assisted selection programmes. Increasing customisation of bovine 'SNP chips' will promote rapid early screening of potential breeding stock – once the relevant single nucleotide polymorphisms (SNPs) have been validated in relevant populations of cattle. Two of the most important meat quality traits are intramuscular fat and tenderness. Synonymous and non-synonomous SNPs in a number of different genes have been associated with variation in these traits. For instance, SNPs in the calpain, calpastatin and cathepsin genes have been associated with tenderness, while SNPs in the thyroglobulin, DGAT1 and FABP4 genes have been associated with intramuscular fat. It is imperative, however, that appropriate genetic-genomic-phenotypic studies are performed to assess the utility of specific DNA markers in purebred, crossbred and heritage cattle populations.

Session 08 Theatre 8

Assessment of single nucleotide polymorphisms (SNPs) in genes on bovine chromosome 14 in association with intramuscular fat (IMF) in beef

L. Pannier[1], P. Stapleton[1], A.M. Mullen[1], R. Hamill[1] and T. Sweeney[2], [1]Achtown Food Research Centre, Department of Meat Technology, Ashtown, Dublin 15, Ireland, [2]University College Dublin, School of Agriculture, Food Science and Veterinary Medicine, Belfield, Dublin 4, Ireland

Thyroglobulin (TG), diacylglycerol O-acyltransferase1 (DGAT1) and fatty acid binding protein 4 (FABP4) genes are located on bovine chromosome 14. Previous studies have indicated that SNPs in these genes are associated with IMF levels in beef. We aimed to determine the frequency of these SNPs in Irish purebred beef breeds and to determine if SNPs in these candidate genes were associated with IMF values in commercial meat samples. Blood samples (481) from pedigree bulls and *M. longissimus dorsi* samples (249) from commercial cattle were collected. SNPs were detected by PCR-RFLP genotyping. In the DGAT1 gene, two SNPs are positioned adjacent to each other, therefore samples were cloned into a pCR®4-TOPO® vector and sequenced to determine haplotypes. All three genes were found to be in linkage equilibrium when tested using the program LinkDos. Maximum likelihood haplotype frequencies were inferred for the eight possible haplotypes using the EM Algorithm implemented in Arlequin. Individual SNPs and estimated haplotypes were tested for association with IMF levels using the GLM procedure of SAS. No individual SNP or haplotype was found to be significantly associated with IMF. The results suggest that these SNPs are not influential on this complex aspect of meat quality in the tested population.

Session 08 Poster 9

Genotype frequencies of a polymorphism in the calpain I gene vary among breeds sampled from the Irish herd

O. Aslan[1,2], T. Sweeney[2], A.M. Mullen[1] and R.M. Hamill[1], [1]Teagasc, Ashtown Food Research Centre, Ashtown, D15, Ireland, [2]University College Dublin, Faculty of Veterinary Medicine, Belfield, D4, Ireland

A non-synonymous polymorphism is present at exon 9 of the calpain I gene that results in a substitution of alanine (A) for glycine (G). A significant association has previously been identified in the Irish herd between genotypes at this locus and Warner-Bratzler shear force (a measure of tenderness) in bovine *M. longissimus dorsi* aged for 14 days. GG genotypes are associated with higher shear force (decreased tenderness) than GA and AA genotypes. Here, genotypic and allelic frequencies for this locus in bulls of six of the main beef breeds in the Irish herd are presented (29 Aberdeen Angus, 10 Blonde d'Aquitaine, 38 Charolais, 48 Simmental, 24 Hereford and 61 Limousin). Samples were in Hardy-Weinberg equilibrium but overall allelic distributions differed significantly among breeds. Differentiation at the locus was moderate with global F_{ST} of 0.06. Pairwise comparisons indicated significant differences in allele frequency between some pairs of breeds but not others. In the samples tested, the allele associated with reduced shear force (A) was at higher frequencies in Angus, Charolais and Limousin bulls (0.17-0.18) than Hereford (0.04) and was absent in Blonde d'Aquitaine and Simmental bulls. This evidence suggests that potential may exist to enhance the consistency in quality of Irish beef by testing this DNA polymorphism in marker-assisted selection strategies.

Session 08 Poster 10

Assessment of single nucleotide polymorphisms (SNPs) in the bovine leptin gene in association with intramuscular fat (IMF) in beef cattle

L. Pannier[1], T. Sweeney[2], P. Stapleton[1], R. Hamill[1] and A.M. Mullen[1], [1]Ashtown Food Research Centre, Department of Meat Technology, Ashtown, Dublin 15, Ireland, [2]University College of Dublin, Faculty of Veterinary Medicine, Belfield, Dublin 4, Ireland

The leptin gene has been mapped to BTA 4 and codes for a 16-kDa protein secreted by adipose tissue. Leptin regulates adiposity and may be one of the genetic factors influencing IMF levels in meat. Our objectives were to evaluate the association of two SNPs with IMF values in Irish crossbred cattle and to characterize the frequency at which these SNPs occur in a purebred population. DNA was isolated from 249 commercial cattle and 481 pedigree bulls of nine cattle breeds. SNP alleles in the leptin gene, one in exon 2 and one in exon 3, were detected with PCR-RFLP and allele frequencies estimated for each breed. The exact test for genetic differentiation, implemented in Genepop, was significant, indicating substructure among breeds. Pairwise tests indicated that the Aberdeen Angus and Hereford bulls were significantly differentiated compared with other breeds. An association test using the GLM procedure of SAS revealed no associations between these SNPs and IMF levels although associations were found in previous studies in different cattle populations. These results suggest that these SNPs are not promising candidates for incorporation into marker assisted breeding strategies for meat quality in Irish commercial cattle populations.

Session 08 Poster 11

Elevated rate of collagen solubilization in muscles of Holstein bulls feed high energy diet

C.C.G. Silva, E.R.E. Simões, O.A. Rego and H.J.D. Rosa, CITAA, Universidade dos Açores, Angra do Heroísmo, 9700, Portugal

The aim of this study was to evaluate the effect of energy supplementation of grazing bulls on collagen content and solubility in three different muscles: *longissimus dorsi* (LD), *semitendinosus* (ST) and *supraspinatus* (SS). Thirty three Holstein bulls aged 15±2 months and weighing 387±51kg were randomly assigned to three treatment groups: 1) fed *ad libitum* on grass exclusively (control); 2) fed *ad libitum* on grass and supplemented with 4 kg/day of maize; 3) fed *ad libitum* on grass and supplemented with 8 kg/day of maize. Treatments lasted 85 days. LD of control animals had lower total collagen (6.24 mg/g) compared with ST (11.4 mg/g) and SS (14.8 mg/g). Collagen solubility was also higher ($P < 0.001$) in LD (24%) compared with the other muscles (18% and 16% in ST and SS, respectively). A similar trend was observed in supplemented animals. Total collagen did not differ among treatments. However, there were differences ($P < 0.001$) in soluble (heat-labile) collagen. The highest value of collagen solubility was found in animals supplemented with 8 kg/day of maize (34%, 23% and 25% in LD, ST and SS, respectively), compared with animals supplemented with 4 kg/day (26% in LD and 19% in both ST and SS muscles) and animals fed on grass exclusively (24%, 18% and 17% in LD, ST and SS, respectively). It was concluded that meat tenderness may be improved by supplementing grazing bulls with 8 kg/day of maize, due to an increase in heat soluble collagen.

Session 08 Poster 12

Influence of the botanical composition of grass, hay or silage on the fatty acid composition of milk

I. Morel, U. Wyss and M. Collomb, Agroscope Liebefeld-Posieux (ALP), Rte de la Tioleyre 4, 1725 Posieux, Switzerland

Milk composition in terms of components considered beneficial to human health can be directly influenced via dairy cattle feed. This applies in particular to certain fatty acids like omega 3 and conjugated linoleic acids (CLA). This study aimed to assess the effect of the botanical composition of the herbage and the conservation method on the composition of milk. Two series of two trials each including 15 dairy cows divided into three variants were conducted with grass and hay (series 1) or grass and silage (series 2). In each trial three variants were compared: a grass mixture (GR), a grass-clover mixture (GT) and a grass-alfalfa mixture (GL). In the case of green fodder, the fatty acid profile was only slightly influenced by the botanical composition. However, ensiling processes led to lower content of fatty acids, especially polyunsaturated fatty acids. This was most pronounced in the GL mixture. The presence of alfalfa in the green fodder increased the omega 3 fatty acid content of milk (especially α-linolenic acid). In hay form, the GR mixture increased the total CLA in the milk, whereas ensiling the GT mixture proved the most favourable choice in terms of milk composition (omega 3) and zootechnical parameters such as milk production and feed intake.

Session 08 Poster 13

Effects of restricted feeding and re-feeding on carcass and fatty acid composition in fat-tail Barbarine lambs

N. Atti, H. Rouissi and M. Mahouachi, National Institute of Tunisian Agronomic Research, Animal and Forage Production, INRAT rue Hédi Karray, 2080 Ariana, Tunisia

Barbarine lambs were used to study the effects of restricted feeding followed by re-feeding on carcass and meat fatty acid (FA) composition. Animals were allocated to 3 groups; one was fed on stubble grazing (low: L), another on stubble and received soya bean meal (medium: M), and the third was fed hay and 450 g of concentrate (high: H). At the end of the experimental period, 5 lambs per group were slaughtered. The remaining lambs were assigned to a high level of concentrate and hay *ad libitum* (H). At end of this compensatory period, they were slaughtered at 37.6 kg LW. After the experimental period, muscle and fat contents were higher for unrestricted (H) than for restricted lambs, but L and M lambs had relatively less fat and more muscle. FA composition was similar for all groups. After the compensatory period, all carcasses contained the same quantity of bone and muscle, but HH lambs had significantly more fat than LH and MH lambs. Carcasses of LH and MH lambs had relatively less fat (240 *vs* 310 g/kg) and more muscle (550 *vs* 500 g/kg) than HH lambs. For Meat FA composition, HH animals contained more C14:0 and C16:0, and compensating animals more C18:2 and a higher proportion of PUFA and PUFA: SFA ratio. In conclusion, feed restriction followed by appropriate re-feeding could permit production of heavy lambs with less fat and more muscle in the carcass and more meat PUFA.

Session 08 Poster 14

Relationship between raw milk chemical and hygienic quality from smallholders' herds and cattle rearing practices

M.T. Sraïri[1], H. Benhouda[1], M. Kuper[2] and P.Y. Le Gal[2], [1]Hassan II Agronomy and Veterinary Medicine Institute, Animal Production Department, P.O. Box 6 202, Madinate Al Irfane, 10 101, RABAT, Morocco, [2]CIRAD, Territoires Environnements Acteurs (TERA), TA 60/15, 34 398, Montpellier, France

In Morocco, milk production is characterised by smallholders' herds. A challenge for the dairy channel is to identify the factors affecting milk quality. In this work, 26 farms were followed up. Rearing practices (concentrates in the diet, milking hygiene ...) were monitored. Raw milk was analysed for fat and protein contents and hygienic parameters. Results show that the use of alfalfa generated acceptable values of milk fat (37.5 g/kg) and protein (30.3 g/kg). However, hygiene practices resulted in high rates of microbial contamination: 13 x 10^5 cfu/ml for the Aerobic Plate Count and 3.3 x 10^4 cfu/ml for the coliforms. A four groups' typology of rearing practices was defined with the use of multivariate analyses: *i)* clean and intensified; *ii)* clean and extensive; *iii)* dirty and intensified; *iv)* dirty and extensive. Milk samples' quality was characterised by a typology made of five groups. A contingency table of rearing practices and milk quality typologies was analysed through X square test. It revealed that reduced use of concentrates gave the highest milk fat contents ($>$ 40 g/kg). The worst hygiene practices were responsible for important microbial loads. These results suggest that under smallholders' conditions, significant margins to improve milk quality still exist.

Session 08 Poster 15

Thrice-a-day milking alleviates heat stress milk depression in Holstein dairy cows

R. Ben Younes[1], M. Ayadi[2], T. Najar[1], M. Zouari[3], A.A.K. Salama[4], X. Such[4], M. Ben M'Rad[1] and G. Caja[4], [1]Institut National Agronomique de Tunisie, 43 Av. Charles Nicolle, 1082 Tunis, Tunisia, [2]Institut Superieur de Biologie Appliquée, Route El Jorf km 22.5, 4119 Medenine, Tunisia, [3]Office des Terres Domaniales, 30 rue Alain Savary, 1002 Tunis, Tunisia, [4]Universitat Autònoma de Barcelona, Campus universitari, 08193 Bellaterra, Spain

Forty-eight Holstein cows (18.0 L/d, 3.43% fat, 3.09% protein) were blocked into a milking frequency treatment (×2 or ×3) for 2 months (16/7-19/9) in North Tunisia. Average daily temperature (T, °C), humidity (H, %) and THI index [THI = 1.8·T–(T–14.3)·(100–H)/100+32] ranged 20.9-30.2°C, 49-78% and 67-81. Heat stress symptoms on rectal T and respiratory and heart rates were detected by deriving the quadratic response models obtained when THI $>$ 64, 75 and 78, respectively. Milk yield decreased during the experiment but milk losses in ×3 cows were half of those reported in ×2 cows (–2.3 *vs* –4.7 L/d; $P < 0.08$). Milk fat content increased (16%; $P < 0.01$) but no effect in milk protein was found. In conclusion, heat stress symptoms started when T $>$ 25°C (THI $>$ 64). Best external indicator for detecting heat stress was respiratory rate which markedly increased when THI $>$ 75. Milk yield depression due to heat stress was partially alleviated by ×3 milking which may be a useful short-term strategy.

Animal feeds' microflora as a risk factor for raw milk contamination

A. Jemeljanovs, I.H. Konosonoka, J. Bluzmanis and D. Ikauniece, Latvia University of Agriculture, Research Institute of Biotechnology and Veterinary Medicine Sigra, Instituta street No 1, LV-2150, Sigulda, Latvia

The objective of the current study was to investigate the microflora of the dairy cows' feed as a source for microbial contamination of raw milk. Feed samples and bulk tank milk samples were taken obtained from 3 dairy farms in Riga region within a year. Samples were inoculated on different complex and selective culture media. Microorganisms were further identified using gram-positive and gram-negative kits of BBL Crystal Identification System. In total, 24 feed samples and 113 bulk tank milk samples were investigated. The data were analysed statistically using SPSS 11.0 software packages. Spore forming microorganisms from the genus *Bacillus* were isolated from 98.0 % of silage and 21.1 % of raw milk samples. Milk's and milk products' contaminants from the genus *Clostridium* were isolated from 29.2 % of feed samples and 5.0 % of bulk tank milk samples. Food poisoning pathogens from the genus *Listeria* were isolated from 25.0 % of feed and 13.2 % of raw milk samples. There were three species isolated from the feed samples – *Listeria monocytogenes, Listeria murrayi and Listeria grayi.* Cake samples were contaminated with fungi from the genera *Aspergillus, Fusarium, Rhizopus, Penicillum, Paecylomyces, Mucor, Alternaria, Stachybotrys and Cladosporium.* Investigations show that the feed is the potential risk factor for the raw milk contamination with pathogenic microorganisms and mycotoxins.

Changes in the composition of ewe's milk for the production of Rabaçal (PDO) cheese

O.C. Moreira[1], J.R. Ribeiro[1], J. Santos[2] and M.ªA. Castelo Branco[3], [1]INIAP-EZN, Fonte Boa, 2005-048 Vale de Santarém, Portugal, [2]DRAB, Quinta do Loreto, 3000-177 Coimbra, Portugal, [3]INIAP-EAN, Quinta do Marquês, 2784-505 Oeiras, Portugal

The objective of this study was to evaluate the characteristics of grazing ewe's milk for the production of Rabaçal (PDO) cheese. The study was conducted at two locations in the Middle Western coast of Portugal, where degraded soils were improved with swards. In the period of pasture shortage, lucerne hay was fed. At each location ten ewes were selected, for fortnight milk sampling, from 60 to 120 days after parturition. Analyses were made on whole milk (protein, fat and minerals) and after fractioning (soluble N, casein, soluble and colloidal Ca and P). Data were analysed by a mixed model using location and lactaion period and their interaction as fixed effects and considering lactaion period as repeated measurement within animal (location). Lactation period affected milk composition. Ca, P, K and fat decreased with the advance of lactation (1.94-1.51 g/l; 1.65-1.29 g/l; 1.09-0.46 g/l; 10.1-7.9 %, respectively) and Na (0.48-2.28 g/l) and protein (6.9-7.5 %) increased. Milk fractioning showed that Ca and P were mainly present in the colloidal phase. Casein fractions were higher at the beginning of lactation. It would be important to have information about ewes' milk production levels for the studied periods and to carry on this study with the cheese obtained from the analysed milks for getting a holistic understanding of this production system.

Session 08 Poster 18

The influence of different dietary vegetable oils on milk fat fatty acid composition of grazing dairy cows

O.A. Rego[1], L. Antunes[1], H.J.D. Rosa[1], A.E.S. Borba[1], C.M. Vouzela[1], C.C.G. Silva[1] and R.J.B. Bessa[2], [1]Universidade Açores, CITAA-DCA, 9700-Angra Heroismo, Portugal, [2]EZN, Fonte Boa, 2000 Santarém, Portugal

The objective was to study the effect of supplementation with vegetable oils (VO) of diverse fatty acid composition (FA) on the cow´s milk fat FA profile including CLA. Sixteen lactating Holstein dairy cows were grouped by body weight, milk yield and days in milk and randomly assigned to 4 treatments *vs* 4 periods (28 days each) in a 4x4 Latin square design. Cows were grazing as a group at 2.5 cows/ha. The treatments consisted of 4 different supplements: 5kg/d of a concentrate based on cereal (maize) mixed with minerals and vitamins (C); 4.5kg concentrate + 0.5kg of rapeseed oil per d (RS); 4.5kg concentrate + 0.5kg sunflower oil per d (SF); and 4.5kg concentrate + 0.5kg linseed oil per d (LIN). The treatments had no effect on milk production and milk protein yield and content, but milk fat yield and content decreased in SF and RS supplemented cows ($P < 0.05$). Milk fat of oil supplemented cows had lower concentration of hipercholesterolemic acids and SFA and higher concentration of UFA. In contrast with SF and LIN, the RS supplementation had no significant effect on milk fat CLA concentration. Supplementation with vegetable oils decreased the concentration of odd and branched fatty acids in milk fat.

Session 08 Poster 19

A survey of the staphylococcal flora from raw ewe´s milk and hygienic conditions on the sheep farm in the Slovak Republic

E. Dudriková, J. Filko, I. Pilipčincová, M. Húska and J. Buleca, University of Veterinary Medicine, Komenského 73, 041 81 Košice, Slovakia, Slovak Republic

Totally 264 isolates of Staphylococcus spp. were isolated from 470 individual milk samples taken from 47 sheep farms during a survey of natural microflora in ewe's raw milk. The most prevalent staphylococcal strain was S. epidermidis (90 cases) and S. caprae (51 cases). In sheep S. aureus, which is the predominant organism in clinical mastitis, was isolated only in six individual milk samples. Other coagulase-negative staphylococci traditionally considered non-pathogenic or of low pathogenicity for the mammary gland of domestic ruminants, the most frequent isolates in subclinical intramammary infections, were isolated only in a few cases. In Slovakia, mastitis is not as common in sheep as it is in dairy cattle, but outbreaks of mastitis can occur in ewes housed for lambing, probably due to contamination of the bedding from infected udder secretion. In conclusion, we could say that every day GMP and introduction of HACCP system in ewe milk production, together with the investigation of mastitis problem in ewe herds, will reduce mastitis problems. On the other hand the microbiological investigation of samples of ewe's raw milk also for other pathogenic bacteria (e. g. *E. coli*, *Salmonella* spp., and *Listeria monocytogenes*) will lead to safety and suitable milk processing of milk products made from ewe's milk because of its benefits for human nutrition.

Session 08 Poster 20

Utilization of autochthonous Zackel sheep in sustainable farming in Serbia

M.S. Savic[1], S.J. Jovanovic[1], A.V. Vranjes[2] and M.V. Vegara[3], [1]Faculty of veterinary medicine, Bul. Oslobodjenja 18, 11000, Beograd, Serbia, [2]Faculty of Agriculture, Trg D. Obradovica 8, 21000, Novi Sad, Serbia, [3]Norwegian University of Life Sciences (UMB), P.O. Box 5003, N-1432 Aas, Norway

According to the sustainable farming program, the Faculty of Veterinary Medicine is designing a plan for conservation and utilization of endangered types of native Zackel sheep in Serbia mountain ecoregion. The genetic characterization of Zackel has been performed, using microsatellite marker analysis. Morphological and productive traits and health condition of Zackel sheep reared in conventional (130 ewes) or sustainable farming systems (122 ewes) were analyzed. The main investigation topics were health and milk properties (milk constituents, suitability for cheese processing, food safety) in sustainable farming. The average milk yield was 86 kg in sustainable and 90 kg in conventional farming system. The mean content (%) of total solids, fat, protein, and casein were 18.16; 6.90; 5.81 and 4.95, respectively, in the sustainable system and 18.12; 6.80; 5.90 and 5.10, respectively, in the conventional system. The lambing rate in both systems was 115 %. Our study showed no statistically significant differences between systems in milk quality and lambing rate, but there was a decrease in mastitis, metabolic diseases and locomotory disorders and an increase in parasitosis in the sustainable compared to the conventional farming system.

Session 08 Poster 21

Effect of grass silage supplementation on performance and milk fat fatty acid composition in grazing dairy cows

C. Vouzela[1], O. Rego[1], S. Regalo[1], H. Rosa[1], A. Borba[1] and R. Bessa[2], [1]Universidade Açores, CITA-A, TerraChã, 9700 AngraHeroísmo, Portugal, [2]EZN, FonteBoa, 2000 Santarém, Portugal

To evaluate the effect of supplementation with 6 kg concentrate at 2 levels of crude protein (CP) (9 and 16% of DM) 16 dairy cows at the same lactational stage were selected and randomly assigned to the experimental groups corresponding to 4 different feeding regimens applied 4 times during 21 d periods according to a 4x4 Latin Square design. Treatments consisted of: 20h grazing + 6 kg of cereal (maize and barley) (PCE); 20 h grazing + 6 kg protein concentrate (cereal and soy) (PCP); 7h of daily grazing + 13h nocturnal grass silage + 6 kg cereal (SCE); 7h of daily grazing + 13h nocturnal grass silage + 6 kg protein concentrate (SCP) in winter. Animal performance parameters (milk production, solids, fat, protein and live weight) did not differ among treatments. Concentrate type did not significantly affect milk fatty acid (FA) profile. Supplementation with grass silage increased ($p < 0.05$) short chain FA (C_{11} to $C_{15:0}$) concentration and decreased oleic and linoleic (LA) FA concentrations. CLA, *trans*-vacenic (TVA) and linolenic acid concentrations in milk fat did not differ among treatments due to similar LA and linolenic acid (CLA and TVA ruminal synthesis precursors) concentrations in pasture and silage allowing for partial substitution of pasture by grass silage of good quality. Silage supplementation increased ($p < 0.05$) branched FA concentration in milk fat.

Session 08 Poster 22

Fatty acid profile of milk from dairy cows reared in conventional or organic farming system in mountain area

L. Bailoni[1], S. Miotello[1], V. Bondesan[2], F. Tagliapietra[1] and R. Mantovani[1], [1]University of Padova, Department of Animal Science, Agripolis, 35020 Legnaro (PD), Italy, [2]Regional Veneto Agricoltura Agency, Agripolis, 35020 Legnaro (PD), Italy

The aim of this study was to determine the fatty acid profile, n-6/n-3 ratio and conjugated linoleic acid (CLA) content in milk samples collected in 10 organic (ORG) and 6 conventional (CON) dairy farms from the same mountain area (Veneto region, NE of Italy). The ORG and CON farms were selected so they represented similar herd sizes (<5, 5-30 and >30 nr. of dairy cows), breeds (Simmental, Brown and Friesian cows) and feeding systems. Bulk milk were sampled monthly and fatty acids analysed by gas chromatographic method as the methyl ester derivates. Percentages of saturated and unsaturated fatty acids of milk samples from ORG farms were lower ($P < 0.05$) and higher ($P < 0.05$), respectively, than those from CON farms. The MUFA/PUFA ratio of ORG was double as high as that of CON. Considering nutritional recommendations, the n-6/n-3 ratio was more favourable (i.e., lower) in milk produced from ORG than CON. Finally, the CLA content was significantly higher in milk from ORG than from CON farms. The results of the present study show that the proportion of fatty acids, with beneficial effects on human health, is higher in milk obtained from organic compared with conventional farming systems.

Session 08 Poster 23

The change in the composition of fatty acids in pork as a function of CLA-enriched feed

A. Győri Boros, R. Salamon, Z. Győri, J. Gundel, S.Z. Salamon and J. Csapó, Research Institute for Animal Breeding & Nutrition, Genetics Department, Gesztenyés ut 1, 2053 Herceghalom, Hungary

The composition of fatty acids in food products is a significant factor for human health. Feeding can significantly influence the composition of fatty acids in animal fat. We analysed the effect of feeding high CLA-content (conjugated linoleic acid) feed on the composition of fatty acids in pork. The animals were grouped according to the following: Group 1) feeding an experimental ghee-mixed feed for 76 days, Group 2) feeding the same feed, but only for 33 days, Group 3) feeding sunflower-oil-mixed feed for 76 days. Ghee contains CLA in high amount. The aim of our experiment was to analyse how the high CLA content of the feed influences the fatty acid content of pork. At the end of the fattening experiment the animals were slaughtered, and samples were taken from the loin, ham, abdomen and backfat from 10 animals of each group and analysed for fatty acid content. We found significant differences in the average fatty acid content between the various tissue samples. As a result of feeding ghee-enriched feed, the CLA content of tissues significantly increased compared with the control group.

Session 09 — Theatre 1

The economic value of uniformity in slaughter pig production

C. Cornou and A.R. Kristensen, Faculty of life Sciences, University of Copenhagen, Department of Large Animal Sciences, Groennegaardsvej 2, DK-1870 Frederiksberg C, Denmark

From the point of view of the pig farmer, uniformity in slaughter pigs is a desirable property, because it reduces the period of time needed for slaughtering the pigs of a given pen or batch. In the extreme case where all pigs are identical in weight they can be slaughtered the same week, whereas a pen or batch with non-uniform pigs needs several weeks before all pigs have reached the optimal weight for slaughter. Since new piglets cannot be inserted before the pen (or section) is emptied, non-uniformity leads to lower utilisation of the finishing unit. In other words, the value of uniformity is closely related to the marketing decision and must be estimated in a way where this dependence is taken into account. A decision support system for optimization of the marketing policy is therefore needed. In this paper, a pen level model for optimization of slaughter pig marketing is presented. The model considers both aspects of the problem, i.e. when to market individual (fast growing) pigs, and when to market the remaining pigs in the pen in order to make room for insertion of a new group of piglets. It is based on a hierarchical Markov process and emphasis is put on definition of the state space in such a way that the homogeneity of the pigs is taken into account. The MLHMP software system is used for implementation of the model. By optimization of the marketing policy under different levels of homogeneity, the value of uniformity may be estimated.

Session 09 — Theatre 2

Variation of performance of a growing pig population as affected by lysine supply and feeding strategy

L. Brossard, J.Y. Dourmad and J. Van Milgen, INRA, UMR1079 SENAH, 35590 Saint-Gilles, France

Based on the performance data of a population of 192 pigs, the InraPorc® model was used to determine individual lysine requirement curves. The consequence of using different feeding strategies (i.e., 1, 2, 3 or 10 diets, with a change of diet determined by the age of the pigs) and lysine supply (ranging from 70 to 130 % of the mean requirement of the population) on performance was tested by simulation modelling. For each diet used, the lysine content in the diet was set to meet the highest mean requirement during the period. The percentage of pigs for which the lysine requirement was met increased concomitantly with increasing lysine supply, but decreased when the number of diets increased. Daily gain increased and feed conversion ratio decreased with increasing lysine supply according to a curvilinear-plateau relationship. The best performance was reached with a supply corresponding to 110 % of the mean requirement of the population and did not depend on the number of diets. The coefficient of variation of daily gain was 10%, which was inherent to this population. At lower lysine supplies, performance decreased and variability of daily gain increased with an increasing number of diets. The use of multiphase feeding system may be a means to reduce nutrient input (and excretion). However, if the nutrient supply is insufficient, this may lead to increased variability in performance. Knowledge of nutrient requirements becomes more critical when a greater number of diets is used.

Session 09 Theatre 3

The influence of increasing energy intake during gestation on litter size, piglet growth and within litter variation from birth to weaning

L. Mc Namara[1,2], P.G. Lawlor[1], P.B. Lynch[1], M.K. O'Connell[1] and N.C. Stickland[2], [1]Teagasc, Pig Production Development Unit, Moorepark Research Centre, Fermoy, Co.Cork, Ireland, [2]The Royal Veterinary College, Royal College Street, London NW1 0TU, United Kingdom

Increasing nutrition during mid gestation for sows increases the ratio of secondary to primary muscle fibres in progeny. This improvement is most pronounced in lighter littermates. The effect of five energy allowances during sow (n=238) gestation: (1) 30 MJ digestible energy (DE)/day throughout gestation, (2) 60 MJ DE/day from day 25-50 of gestation, (3) 60 MJ DE/day from day 50-80, (4) 60 MJ DE/day from day 25-80 and (5) 45 MJ DE/day from day 80-112, on litter size, piglet growth and within litter variation from birth to weaning were examined. There was no energy allowance effect on the number of piglets born alive ($P > 0.05$), birth weight ($P > 0.05$), weaning weight ($P > 0.05$) or average daily gain from birth to weaning ($P > 0.05$). However, the total number of piglets born dead was higher on Treatment 3 than either Treatment 1 or 2 ($P < 0.05$). The total number of piglets born alive per sow tended to decrease with increasing parity number ($P=0.10$). Overall, increasing energy allowance for sows during gestation had little effect on piglet birth weight and was not effective in reducing within litter variation in piglet weight. Increasing energy allowance between days 50-80 of gestation increased the number of piglets born dead per litter.

Session 09 Theatre 4

Differential effect of an antibiotic on different weaning weights

J. Deen and M. Allerson, University of Minnesota, Veterinary Population Medicine, 1988 Fitch Ave, St Paul, MN 55108, USA

The objective of this research was to examine performance in nursery pigs with varying weaning weights and to evaluate the efficacy of tulathromycin (Draxxin Injectable Solution, Pfizer Animal Health), which is indicated for the treatment of swine respiratory disease. 140 pigs were tagged, weighed, sexed, and the age of the pigs and parity of the sows were recorded at weaning. One-half (n=70) of these pigs were treated with (0.2) cc of tulathromycin. Statistical analysis consisted of multivariate logistic regression to examine predictors for poor nursery performance. Poor nursery performance was classified as a pig exiting the nursery (6 weeks post-weaning) at less than 16 Kg or dying during the nursery phase. It was found that a significant effect ($P<.01$) of the treatment was seen in pigs when pigs were weaned at less than 4.5 Kg, OR=2.3, while the effect in larger pigs at weaning was not significant, OR=1.3, p=.12. This study illustrates the differential effects that can occur in swine populations.

Session 09 Theatre 5

Impact of some sow's characteristics on birhtweight variability

N. Quiniou[1], L. Brossard[2] and H. Quesnel[2], [1]IFIP Institut du Porc, BP35104, 35651 Le Rheu cedex, France, [2]INRA, UMR SENAH, 35590 Saint-Gilles, France

Data collected from 1380 litters born between 2000 and 2004 from Large White×Landrace sows were used to quantify within litter variability of birthweight in an experimental herd. Born alive and stillborn piglets were individually weighed within the 24 h post-farrowing. For each litter, average birthweight (mBW) and its coefficient of variation (cvBW) were calculated. Litter size and parity were correlated (Pearson's correlation, r=0.13, P < 0.001). Subsequently litter size effect was tested within parity in a variance analysis. Other main effects were parity, parity within year of sow birth, year of sow birth, and season of conception. Within-litter cvBW averaged 21% and mBW 1.53 kg. The cvBW was not significantly repeatable from a parity to the following. It was significantly influenced by parity and litter size. Lowest cvBW were obtained in first and second parities (19%) and thereafter the cvBW increased by 1.42 (±0.18) point per additional litter up to the 6th. Increase in litter size within parity was associated with increased variability of BW that averaged +0.80 (±0.05) point per additional piglet. The cvBW was negatively correlated with sows'bodyweight gain during gestation (r=-0.16), but not with backfat thickness gain, except in second parity sows (r= -0.20). In fact, this latter correlation reflected a simultaneous reduction of 2nd litter size in the thinnest weaned sows.

Session 09 Theatre 6

Influencing uniformity in pigs genetically, through canalization and plasticity

M. Sancristobal, INRA, Animal Genetics, UR444, Laboratoire de génétique cellulaire, F-31326 Castanet Tolosan, France

Homogeneity of production is one of the objectives of breeders and food industrials, for obvious economical reasons. One way to achieve this goal is to control the environment. However, there are cases where the environment can not be controlled at all (open field). In any case, small variations of the environmental conditions exist, and some individuals are more sensitive to them than others. If there were any genetic control of this sensitivity to variations of the environment, then it would be possible to select for a reduced sensitivity. I will give some known examples of existence of genes affecting environmental sensitivity, and highlight some selection experiments on phenotypic plasticity, on laboratory animals. I will give some biological and statistical models that were proposed in the literature for explaining and inferring a genetic control on phenotypic or environmental variability. This last decade, since the availability of quantitative genetics models making possible to quantify the amount of genetic variance for environmental sensitivity, several data analyses and several experiments on livestock species have been carried out, giving various evidences of a genetic control.

Session 09 Theatre 7

Breeding for uniformity by exploiting genetic differences in environmental variance with an application to carcass weight in pigs

H.A. Mulder[1], P. Bijma[1] and W.G. Hill[2], [1]Animal Breeding and Genomics Centre, Wageningen University and Research Centre, PO Box 338, 6700 AH Wageningen, Netherlands, [2]Institute of Evolutionary Biology, School of Biological Sciences, University of Edinburgh, West Mains Road, EH9 3JT Edinburgh, United Kingdom

Genetic variation in environmental variance may be utilized to improve uniformity in livestock populations by selection. The objective of this study was to investigate effects of genetic parameters and breeding goal on selection responses in mean and variance when applying index selection. Both means and environmental variances were treated as heritable traits. Economic values for mean and variance were derived when profit is based on thresholds with an intermediate optimum. Carcass weight in pigs was used as an illustration, where the highest price is paid for pigs between 80 and 98 kg carcass weight. The ability to change the variance in the desired direction depended on the genetic correlation between the breeding values for mean and environmental variance and on the economic values in the breeding goal. After one generation of selection, the proportion of pigs in the optimum range increased from 80% to 82 – 86%, depending on the heritability of environmental variance. Consequently, profit increased by € 0.19 to € 0.83 per pig. It is concluded that genetic variation in environmental variance can be exploited to select for increased uniformity in pigs, resulting in higher profit.

Session 09 Theatre 8

Influencing uniformity in pigs through selection of parent stock boars: an industry viewpoint

E.B.P.G. van Haandel[1], P.K. Charagu[2] and A.E. Huisman[1], [1]Hypor B.V., P.O. Box 30, 5830AA Boxmeer, Netherlands, [2]Hypor Inc., 402 McDonaldstreet, S4N6E1 Regina, Canada

Breeding objectives for finishing pigs are influenced by different feeding regimes and production environments found around the world. Payments, however, will ultimately depend on the packing plants' grid score where carcass uniformity is a critical factor. Choice of terminal sire has been shown to affect the performance of finishing pigs. Tailoring terminal sires by the use of either differentiated breeds/lines or divergent breeding goals within lines have both been shown to have influence on growth and carcass performance of finishing pigs. Genetic variation depends on the character of the traits and is expressed as heritability. Generally a large part of the variation is due to environmental influences. A consistent selection objective or a narrowly selected parent-group within a terminal sire line can reduce remaining environmental variation in offspring till probably 80 percent of its original variation. Synthetic lines will have an advantage over crossbred lines, since progeny from the latter might elicit more variation due to segregation. However, selection in subsequent generations on multiple traits can eliminate limiting factors on more than one trait and consequently have impact on total variation. Ultimately, variation is partly influenced by the variation in maternal parent stock. Other factors such as parental imprinting can also significantly affect variation in offspring and have a substantial impact on product uniformity.

Session 09 — Theatre 9

Is genetic resistance to Salmonella uniform in pigs?

I.H. Velander[1], B.J. Nielsen[1], J. Boes[1], B. Nielsen[2] and B. Guldbrandtsen[3], [1]Danish Pig Production, Axeltorv 3, Copenhagen, Denmark, [2]DMA, Axeltorv 3, Copenhagen, Denmark, [3]University of Aarhus,Faculty of Agricultural Science, Blichers Allé, Tjele, Denmark

Previous experimental salmonella infection studies in pigs in Denmark have shown considerable differences in antibody response. Some pigs remained seronegative after infection. To investigate the level and uniformity of salmonella resistance and to test if antibody response was affected by genetics in crossbred pigs an experiment was conducted. In total 600 salmonella seronegative offspring out of Duroc sires and L×Y sows were orally infected with *S. Typhimurium* via the feed at 15-20 kg lives weight. On day 15 post inoculation, all pigs were blood sampled and sera were tested for Salmonella antibodies. Pigs with low antibody response were re-tested twice. Almost 7% of the 600 pigs showed no or very low antibody response, indicating a possible genetic component to resistance. An animal model with repeated measurements of antibody response was used to calculate genetic parameters and heritability significantly greater than 0 was detected. To investigate whether salmonella negative status in pigs is related to a single recessive allele, a chi-square test was conducted. The hypothesis of a single recessive gene causing resistance to salmonella was rejected. The results indicate that salmonella resistance in pigs has a genetic component. The existence of genetic variation allows the use of selection to improve level and uniformity in resistance to salmonella in pigs.

Session 09 — Poster 10

The influence of gestation feed allowance for sows and piglet birth weight on pig growth performance to slaughter

P.G. Lawlor[1], P.B. Lynch[1], M.K. O'Connell[1], L. Mcnamara[1,2], P. Reid[3] and N.C. Stickland[2], [1]Teagasc, Pig Production Development Unit, Moorepark Research Centre, Fermoy, Co.Cork, Ireland, [2]The Royal Veterinary College, Royal College Street, London NW1 0TU, United Kingdom, [3]Teagasc, Ashtown Food Research Centre, Ashtown, Dublin 15, Ireland

Sows (n=80) blocked at mating on parity and weight were assigned to the following gestation feed allowances: (1) 30 MJ DE/day throughout gestation, (2) 60 MJ DE/day from day 25 to 50, (3) 60 MJ DE/day from day 50 to 80, (4) 60 MJ DE/day from day 25 to 80 and (5) 45 MJ DE/day from day 80 to 112. Sows on treatments 2 to 4 were fed 30 MJ DE/day at all other times during gestation. At weaning, three pigs (heavy, medium and light birth weight) of the same gender (entire male or female) were selected from each litter and penned individually to slaughter at c.157 days of age. Gestation feeding regime had little effect on pig weight ($P > 0.05$), daily gain ($P > 0.05$) or feed conversion efficiency ($P > 0.05$). Carcass back-fat thickness was 10.6, 9.9, 9.2, 11.0 and 10.6 (s.e.d. 0.39mm; $P < 0.05$) for Treatments 1 through 5 respectively. Birth weight was 1.81, 1.51 and 1.15 (s.e.d. 0.025kg; $P < 0.001$) and carcass weight was 72.2, 72.8 and 66.2 (s.e.d. 1.61kg; $P < 0.01$) for heavy, medium and light birth-weight pigs, respectively. In conclusion, gestation feeding allowance did not affect the performance of progeny but did reduce carcass fatness. Carcass weight was less for light birth-weight pigs than medium and heavy birth-weight pigs.

Session 10 Theatre 1

Breed and sex differences in genetic parameters for behaviour test results in German Shepherd Dogs and Labrador Retrievers in Sweden

E.H. van der Waaij[1], E. Wilsson[2] and E. Strandberg[3], [1]Wageningen University, Animal Breeding and Genomics Centre, PO Box 338, 6700 AH Wageningen, Netherlands, [2]The Swedish Armed Forces, Dog instructor Center, Box 194, SE 195 24 Marsta, Sweden, [3]Swedish Univ. of Agricultural Sciences, Dept. of Animal Breeding and Genetics, PO Box 7023, S-75007 Uppsala, Sweden

Behaviour tests on 2756 German Shepherd Dogs (GS) and 1812 Labrador Retrievers (LR) included observations on courage, defence drive, prey drive, nerve stability, temperament, cooperation, affability, and gun shyness. Apart from affability in the LR, all traits had non-zero heritabilities, ranging from 0.07 for defence drive in GS to 0.56 for gun shyness in LR. There was no common env. effect for any of the traits. The maternal genetic effect was only significantly different from zero for defence drive in GS. Genetic correlations ranged from close to unity between courage and hardness, to almost zero between gun shyness and defence drive. The genetic correlations between hardness and nerve stability in LR (0.87) and GS (0.57) were significantly different. Genetic correlations between males and females were significant non-unity (>2 SE) for defence drive and cooperation in the GS, and for hardness in the LR. Combining the results of both breeds in a meta-analysis resulted in increased significance (3 SE) for cooperation, whereas the other correlations became non-significantly different from unity.

Session 10 Theatre 2

Joint genetic analysis of male and female fertility after AI in sheep

I. David, L. Bodin, C. Leymarie, E. Manfredi and C. Robert-Granié, INRA, UR631 SAGA, chemin de borde rouge, F-31320 Castanet Tolosan, France

The outcome of an insemination may be viewed as a combination of 2 main traits; one relative to the female (i.e. female fertility), the second relative to the male (i.e. male fertility). Nevertheless, studies generally focussed on the estimation of the genetic fertility of only one sex. Using 147,018 AI records of the Manech tête rousse breed located in the south-west of France, we propose a joint genetic evaluation of male and female fertility. The result of an insemination was defined as a success (Y=1) when lambing occurred 144 to 158 days after insemination (gestation period), elsewhere it was a failure (Y=0). This binary variable was analysed under the assumptions of the threshold model. One way to associate male and female fertility is to consider that they act additively on the underlying scale. Using Bayesian methods, heritabilities obtained with such approach were 0.056 (credibility interval: [0.049-0.063]) and 0.007 (credibility interval: [0.003-0.010]) for female and male fertility, respectively. The genetic correlation between the two traits was 0.14 and not different from 0. Others assumptions may be considered for the association between fertility traits. For instance the probability of AI success may be viewed as the product of the male by female phenotypic fertility defined as two unobserved binary variables. Limits and advantages of such models will be discussed.

Session 10 Theatre 3

Genetic structure of Italian Alpine goat populations based on microsatellites markers

F. Panzitta[1], A. Stella[1], A. Montironi[1,2] and G. Gandini[2], [1]Parco Tecnologico Padano, Statistical Genetics and Bioinformatics, via Einstein-Polo Universitario, 26900 Lodi, Italy, [2]University of Veterinary Medicine, VSA, via Celoria 12, 20133 Milano, Italy

A great variety of local goat populations are present in the Italian Alps. From these populations, some breeds have been recently defined purely on the basis of their geographical distributions and morphological aspects. The genetic structure of these goat breeds (Lariana, Ciavenasca, Trentina, Verzaschese, Vallesana, Passiria and Roccaverano) was analysed by genotyping 26 microsatellite markers. The expected heterozygosity ranged from 0.62 to 0.71. Genetic differentiation was moderate with an overall value F_{ST} = 0.03. The Neighbour-Joining tree was built using the Nei Standard and the Reynold's genetic distances. Clusters corresponded to the geographical distribution of the breeds, grouping together, respectively, Lariana and Ciavenasca, Trentina and Passiria, Verzaschese and Vallesana breeds. Ignoring the originally defined structure of the breeds, a Bayesian-based clustering analysis reorganized the seven breeds into five different groups, and only 63% of individuals were correctly assigned to the breed of origin. The genetic analysis suggests to reconsider the breed structure defined on the basis of geographical and morphological elements, however culture differentiation among breeds should be considered.

Session 10 Theatre 4

Breed specific differences in the expression of 3β-hydroxysteroid dehydrogenase (HSD3B) and sulfotransferase 2B1 (SULT2B1) in boars with high and low backfat androstenone levels

M. Moe[1,2], E. Grindflek[1,3] and O. Doran[4], [1]NORSVIN, 2304, Hamar, Norway, [2]Department of Animal and Aquacultural Sciences, Norwegian University of Life Sciences, 1432, Ås, Norway, [3]CIGENE, Norwegian University of Life Sciences, 1432, Ås, Norway, [4]School of Clinical Veterinary Science, University of Bristol, Langford, Bristol BS40 5DU, United Kingdom

Boar taint is primarily caused by the excessive production or deficient degradation of androstenone and skatole. We have studied the relationship between backfat androstenone levels and expression of hepatic and testicular 3β-hydroxysteroid dehydrogenase (HSD3B), cytochrome P45017 (CYP17) and sulfotransferase 2B1 (SULT2B1), enzymes which are involved in androstenone synthesis and/or catabolism. The enzyme expression was analysed by Western blotting using microsomes isolated from liver and testis of pure bred Landrace and Duroc boars with various androstenone levels in backfat. A negative relationship between the hepatic but not testicular HSD3B and androstenone level was found for Landrace but not Duroc pigs. In Duroc but not Landrace pigs the negative relationship was observed between adipose tissue androstenone and the level of both hepatic and testicular SULT2B1 proteins. There was no significant relationship between backfat androstenone and expression of either hepatic or testicular CYP17A1 in the breeds studied. The results suggest that the mechanisms regulating androstenone deposition in pig adipose tissue are breed-specific.

Session 10 — Theatre 5

Comparison of statistical models to analyze the genetic effect on within-litter variance in pig

D. Wittenburg, V. Guiard, F. Teuscher and N. Reinsch, Forschungsinstitut für die Biologie landwirtschaftlicher Nutztiere (FBN), Wilhelm-Stahl-Allee 2, 18196 Dummerstorf, Germany

Genetics not only influences the weight of piglets at birth, it also affects the variability of birth weights within litter. Previous studies on this topic assigned the sample standard deviation of piglets' birth weights within litter as observation to the sow. However, the sex effect on piglet's phenotype has been neglected so far. This work deals with the genetic effect on within-litter variance when different statistical models are used and considers the sex effect and appropriate weights per trait. Traits were formed from the pooled sample variance of male and female birth weights within litter. The logarithmized within-litter variance and the sample standard deviation were suitable variables for a linear model approach. A generalized linear model with log-link function was applied for the untransformed sample variance. Models were compared by analyzing data from 9439 litters from a Large White and Landrace line of BHZP's breeding programme. The estimates of heritability ranged from 7-11%. Although the generalized linear mixed model is preferred from a mathematical view, the rank correlations between breeding values of the linear mixed models and the generalized linear mixed model were relatively high, i.e. 94% to 98%. Finally, this study showed, that a significant sex effect on the piglet's phenotype but not on its variability can be adequately handled by the pooled within-litter variance.

Session 10 — Theatre 6

Environmental and genetic effects on longevity of sows

A.R. Sharifi[1], H. Henne[2], H. Simianer[1], M. Tietze[1] and V. Ducrocq[3], [1]Institut of Animal Breeding and Genetics, Albrecht-Thaer-Weg 3, 37075 Göttingen, Germany, [2]BHZP, an der Wassermühle 8, 21368 Ellringen, Germany, [3]UR337,INRA, Domaine de Vilvert, 78352 Jouy en Josas, France

The longevity of breeding sows is a trait with a significant impact on profitability of pig production because of decreasing replacement costs of sows. The purpose of this study was to analyse the influence of several explanatory factors on the functional length of productive life. In addition to this, the intention was to estimate the genetic parameters of the trait longevity. The length of productive life was defined as lasting from first farrowing to culling. The data of 95000 sows from nucleus and multiplier farms were used. A piecewise Weibull baseline model was used to analyse the data with parity*stage of farrowing interval. Included in this model were the time-dependent covariates number of live born piglets and year*season as well as the effect of the time independent variables foot and leg score, daily gain, back fat thickness, age at first farrowing, line and stage (nucleus or multiplier). All explanatory factors included in the model except daily gain and line had a significant effect on longevity. The hazard decreases with the increase in the number of live born piglets and it increases with age at first farrowing as well as with the foot and leg score. Nucleus sows have a higher culling risk than the multiplier sows due to the more stringent selection in the nucleus stage. The heritability of longevity was estimated to be 0.12.

Genetics of growth in pigs under different heat loads

I. Misztal[1], B. Zumbach[1], S. Tsuruta[1], J.P. Sanchez[1], M.J. Azain[1], W. Herring[2], J. Holl[2] and T. Long[2], [1]University of Georgia, Athens, GA 30605, USA, [2]Smithfield Premium Genetics, PO Box 668, Rose Hill, NC24458, USA

Data included carcass weights of 21,653 crossbred pigs (Duroc x [Landrace x Large White]) in North Carolina. Monthly heat loads were calculated as degrees of average temperature-humidity index (THI) in °C over 23. Assumed heat load (H) was a sum of heat loads 4 months prior to harvest. Variance components were estimated with 3 models: univariate - not accounting for heat stress, 2-trait, and linear random regression using linear splines (RRMS). The 2-trait model treated observations in July-October ("hot") and December-June ("cold") as separate traits. RRMS added a random regression on heat load for the sire effect.. The estimate of heritability in the univariate model was 0.17. In the 2-trait model the estimates were 0.15±0.01 for "cold" and 0.34 for "hot"; the genetic correlation was 0.08. The same estimates with RRMS were 0.15, 0.59 and -0.45. A separate line may be justified for optimum performance under high heat loads.

Genetic parameters for direct and associative effect on survival time in three strains of laying hens

E.D. Ellen and P. Bijma, Wageningen University, Animal Breeding and Genomics Centre, Marijkeweg 40, 6709 PG, Netherlands

In the future, problems like mortality due to cannibalism may increase due to prohibition of beak-trimming and traditional battery cages, resulting in a decrease in survival rates. Survival in laying hens not only depends on the individuals own phenotype, but also on phenotypes of group members. Traditional animal models are therefore inadequate, because associative contributions of other genotypes in the group are not taken into account. This paper presents estimated genetic parameters both for the direct effect only and for the combination of direct and associative effects on survival days in three purebred strains (W1, WB, and WF) of laying hens from Hendrix Genetics. For the analysis, 16,984 non-beak trimmed laying hens were used. Heritability of survival days for the three strains using the classical animal model ranged from 1.7% (WF) through 9.7% (WB). The proportion of heritable variation of survival using the model which contains both the direct and associative effect ranged from 5.2% (WF) through 19.2% (W1). Those results show that the amount of heritable variation in survival days is substantially larger than suggested by traditional animal models. Consequently, prospects for reduction of mortality using genetic selection are quite good. Genetic selection targeting both direct and associative effects is expected to substantially reduce one of the major welfare problems in egg production.

Session 10 Theatre 9

Heterogeneity of genetic parameters for calving difficulty in Holstein heifers in Ireland

J.M. Hickey[1,2,3], M.G. Keane[1], D.A. Kenny[2], A.R. Cromie[4], P.R. Amer[5] and R.F. Veerkamp[3], [1]Teagasc, Grange Research Centre, Dunsany, Co. Meath, Ireland, [2]University College Dublin, School of Agriculture, Food and Veterinary Medicine, Dublin 4, Ireland, [3]Animal Sciences Group, ABGC, 8200 AB, Lelystad, Netherlands, [4]ICBF, Bandon, Co. Cork, Ireland, [5]Abacus Biotech Ltd, PO Box 5585, Dunedin, New Zealand

The objective of this study was to estimate genetic parameters for calving difficulty in first parity Holsteins and to determine if these differed with age of heifer at calving. Calving difficulty records for 18,806 Holstein heifers which calved between January 2002 and May 2006 were analyzed with univariate, multitrait and random regression (RR) linear sire-maternal grandsire models. Model comparison, using likelihood ratios, Akaike and Bayesian information criterion, suggested that heterogeneity of (co)variance existed for the direct genetic, but not the maternal genetic or residual components. RR model estimates of direct genetic variance and heritability were heterogeneous along the dam age trajectory, decreasing initially with dam age before subsequently increasing. Heritability estimates ranged from 0.11 to 0.37 and were higher for records with younger and older dams at parturition. Genetic correlations between the direct components of calving difficulty were from unity to 0.5. Re-ranking and re-scaling of bulls for the direct component of calving difficulty occurred along the dam age trajectory.

Session 10 Theatre 10

Do type traits have an economic value in dairy cows?

B. Lind[1], S. Schierenbeck[1], F. Reinhardt[2] and H. Simianer[1], [1]Institute of Animal Breeding and Genetics, Albrecht-Thaer-Weg 3, 37075 Göttingen, Germany, [2]United Data Systems for Animal Production (VIT), Heideweg 1, 27283 Verden, Germany

Economic values in breeding programs are necessary to weigh the traits according to their commercial relevance in a total merit index. Type traits have a potential economic impact in three areas: i) they contribute to functional traits of cows, e.g. longevity or udder health; ii) they have an effect on the price of sold heifers; iii) they allow a more cost-efficient labour in the dairy unit, e.g. through better milkability or locomotion. In assessing the economic values of different type traits double counting must be avoided, i.e. it is not appropriate to assign a value to a trait like udder depth which contributes to longevity, if longevity as such already has an economic weight. Concerning ii), we estimated effects of type traits on auction prices in a comprehensive multiple regression analysis of auction data. Due to the low rate of realisation of heifer sales (on average 0.065 heifers are sold per cow and year) type traits have a combined economic value of 0.48 € per genetic standard deviation. Thus the economically justified relative index weight for type traits, reflecting the economic benefit for the dairy farm, is just 1%. It is demonstrated to what extent genetic progress in other important breeding goals like productivity, longevity and fertility will be reduced if, for marketing reasons, type traits are assigned a higher relative index weight than economically warranted.

Session 10 Theatre 11

Milk-fat composition of dairy cows can be improved by use of genetic variation

A. Schennink[1], W.M. Stoop[1], M.H.P.W. Visker[1], J.M.L. Heck[2], H. Bovenhuis[1], J.J. van der Poel[1], H.J.F. van Valenberg[2] and J.A.M. van Arendonk[1], [1]Wageningen University, Animal Breeding and Genomics Centre, PO Box 338, 6700 AH Wageningen, Netherlands, [2]Wageningen University, Dairy Science and Technology, PO Box 8129, 6700 EV Wageningen, Netherlands

Dietary fat may play a role in the etiology of many chronic diseases. Milk and milk-derived foods contribute substantially to dietary fat, but have a fat composition that is not optimal for human health. We measured the fat composition of milk samples in 1,918 Dutch Holstein Friesian cows in their first lactation. Substantial genetic variation in milk-fat composition was found: heritabilities were high for short- and medium-chain fatty acids (C4:0-C12:0, C14:0 and C16:0) and moderate for long-chain fatty acids (saturated and unsaturated C18). Our analysis revealed that the *DGAT1 K232A* polymorphism, which is known to affect milk-fat percentage, has a clear influence on milk-fat composition. The *A* allele of *DGAT1* is associated with less saturated fat; a smaller fraction of C16:0; and larger fractions of C14:0, unsaturated C18, and CLA. These results show that selective breeding can make a significant contribution to change the fat composition of cow's milk.

Session 10 Theatre 12

Genotype by housing interaction for conformation and workability traits in Danish Holstein cattle

J. Lassen[1] and T. Mark[2], [1]DJF, AU, 8830, Denmark, [2]LIFE, KU, 1870, Denmark

The aim of this study was to investigate the extent of G*E interaction for conformation and workability traits registered in free stalls and tie stalls in Danish Holstein and to investigate consequences for selection. Information about housing systems for Danish dairy cattle herds from 1997 to 2002 were merged with records for 22 conformation and 2 workability traits from the same period. In total, 51363 cows with records were extracted for first parity cows and 30187 cows were available for analyses after data editing. The pedigree file included 169041 animals. Variance components and breeding values were inferred using a bivariate animal model, assuming performance in each housing system to be different, but correlated traits. The model included fixed effects of herd, year*season, inspector and days in milk and age at calving as well as a random animal and residual effect. Limited evidence of G*E were found in this study. A genetic correlation significantly different from unity was only found for body width (0.87), rear leg rear view (0.92), milking speed (0.94) and temperament (0.95). However, heritabilities were in general lower in free stalls than in tie stalls (17 % lower on average). This illustrates that it is more difficult to classify cows in free stalls and that the records obtained in free stalls are less valuable for genetic evaluation. Similar work will be performed and presented for fertility and health data and consequences for selection will be quantified.

Session 10 — Theatre 13

Divergent selection for length of productive life in rabbit

C. Larzul[1], H. Garreau[2] and V. Ducrocq[1], [1]INRA, UR337 SGQA, Domaine de Vilvert, 78352 Jouy-en-Josas, France, [2]INRA, SAGA, Chemin de Borde Rouge, 31356 Castanet-Tolosan, France

In several species, concerns about functional traits increase in relation to robustness. Studies have been carried out mainly in cattle but also in pig, poultry and rabbit. Longevity, or length of productive life, is analysed using survival analysis techniques. Genetic evaluations are performed and estimated breeding values can be included in selection objectives. However, one-way direct selection on functional longevity has been rarely carried out in farm animals.
The aim of the study was to perform a divergent selection for functional longevity, based on genetic merit estimated through survival analysis techniques. The experiment was conducted in rabbit to estimate the efficiency of such a selection and to analyse the consequences on other reproduction traits. Given the herd management, length of productive life was measured as the number of artificial inseminations for rabbit does. A total of 48 males were progeny tested based on the longevity of ten daughters. Based on their estimated genetic merit, 5 "high longevity" and 5 "low longevity" males were selected and produced a new generation. Each one of the new 48 males produced ten daughters to estimate the response to selection on longevity and other reproductive traits. Though preliminary, differences observed for survival rates between the two lines were encouraging.

Session 10 — Poster 14

Genotype by environment interaction for yield traits in Holstein cattle in Slovenia using reaction norms

B. Logar[1], S. Malovrh[2] and M. Kovač[2], [1]Agricultural Institute of Slovenia, Animal Science Department, Hacquetova 17, 1001 Ljubljana, Slovenia, [2]University of Ljubljana, Biotechnical Faculty, Zootechnical Department, Groblje 3, 1230 Domžale, Slovenia

The objective of the study was to evaluate genotype by environment interaction for yield traits in Holstein cattle in Slovenia using reaction norms. A linear random regression model with heterogeneous environmental variances and regression of phenotypic observations of daughters within sire on herd environment was used. A total of 85515first to third lactation records of 49947 cows offspring of 179 sires were included in the data. The phenotypic measures were 305 days milk, fat and protein yield. The environmental value was the herd-year average of each trait. Estimated correlation between level and slope of reaction norms were 0.65, 0.72 and 0.67 for milk, protein and fat yield, respectively. Residual variances increased approximately linearly with increasing value of the herd environment. The heritabilities as a function of the production environment ranged from 0.20 to 0.25. The lowest heritability was found in the environments one standard deviation below the average of environmental variables analyzed. Distinctive crossing of sires' reaction norms for protein and fat production occurred. That is in agreement with the estimated rank correlations between the predicted offspring performances in average and deviating environments (higher than 0.89). For milk yield, crossing of reaction norms was less distinctive.

Session 10 Poster 15

Assessment of dairy cow profitability

S. König[1,2], J. Fatehi[1], H. Simianer[2] and L.R. Schaeffer[1], [1]Department of Animal and Poultry Science, University of Guelph, N1G2W1 Guelph, Canada, [2]Institute of Animal Breeding and Genetics, University of Göttingen, 37075 Göttingen, Germany

A simulation program was developed to generate the day by day operations in a dairy cattle herd for a population of 100,000 cows. A standard AI stud, i.e. 120 young bulls per year and 35% matings of cows to young bulls, was created as a source of sires for breeding. Annual genetic changes were incorporated and 20 new young bulls from 20 elite cows were continuously entered on a monthly basis. Proven bulls were selected based on a total merit index of 100 daughters. For the day by day operations in a total of 1450 herds, different sizes and types of operations within herds, e.g. breeding strategies, age structures and culling decisions, were considered according to probabilities of occurrence. The individual variability of cows was simulated in 212 different traits and changes during life in production, growth, feed intake and some type traits were modelled through random regression coefficients. The influence of management practices and environmental effects on all these traits and diseases and their interactions in combination with prices and costs over the cow's whole lifespan was used to determine net returns per cow and day. Net returns per cow and day were regressed on their true breeding values to determine relative economic values for each trait. Sensitivity analyses incorporating different parameters were applied to find the most relevant traits for dairy cattle producers.

Session 10 Poster 16

Genetic parameters for body weight of the Goettingen minipig estimated with multiple trait and random regression models

F. Köhn[1], A.R. Sharifi[1], S. Malovrh[2] and H. Simianer[1], [1]Institute of Animal Breeding and Genetics, Albrecht-Thaer-Weg 3, 37075 Göttingen, Germany, [2]Animal Science Department, Biotechnical Faculty, Groblje 3, 1230 Domzale, Slovenia

The Goettingen minipig is a laboratory animal especially developed for medical research. For an easy and comfortable handling during experiments and for low costs a low body weight is essential. To breed for an even smaller minipig, genetic parameters for body weight were estimated using a multiple trait model (MTM) and a random regression model (RRM). Regressions for the random effects in the RRM were modelled using Legendre polynomials. The data set focused on the time period from 30 to 400 d of age. Eight age classes were built which resulted in a MTM with 8 traits and a RRM with 8 classes to consider heterogeneous residual variances. The heritabilities were moderate with slightly higher values estimated with the RRM. The major eigenfunction estimated with the RRM showed positive values throughout the whole trajectory, i.e. a selection for low body weight has positive effects on this trait throughout the whole range of time. It is concluded that the RRM is too susceptible for the routine application of estimation of breeding values in regular time intervals because of a more complex model selection that has to be carried out for every new data set. However, RRM are a useful tool for the examination of changes in body weight due to selection via the estimated eigenfunctions and their associated eigenvalues.

Session 10 Poster 17

Short-term selection for body weight and growth rate in Japanese quail

G.S. Farahat and E.A. El Full, Faculty of Agriculture, Fayoum Univ., Poultry Production Dept., Fayoum, Egypt., 63514, Egypt

Three quail lines were used in this work to study the effects of mass selection for increased body weight or growth rate: HBW_{42} line was selected for high body weight at 42 day of age, HGR_{1-42} line was selected for high growth rate during the period from one-day to 42 day of age and randombred control line (RBC) was maintained as non-selected pedigreed population over three generations of selection. There was an asymmetry in the direct response to selection in the male and female BW and GR favoring females. After three generations of mass selection, the HBW_{42} line exceeded the RBC by 8.40g (4.81%) and 13.66g (7.53%) for males and females, respectively. Females had significantly higher response on generation number either as a deviation or as a percent deviation from the controls. The ratios of effective to the realized selection differentials for the selected lines and sex groups were more than unity, except for females in the 1st generation of selection for HGR_{1-42}. Selection yielded, with few exceptions, consistently lower heritability estimates in the selected lines, regardless of the estimation method. It can be concluded that the direct response for HBW_{42} line was greater than those of the line selected for HGR_{1-42} compared to their controls.

Session 10 Poster 18

Use of IGF2 gene tests for carcass quality and sow productivity in Canadian pig populations

P.K. Mathur[1], Y. Liu[1], L. Maignel[1], B.P. Sullivan[1] and S. Chen[2], [1]Canadian Centre for Swine Improvement, Central Experimental Farm, Building#54, Ottawa, ON K1A 0C6, Canada, [2]University of Guelph, 95 Stone Road West, Guelph, ON N1H 8J7, Canada

The efficiency of pork production depends upon the ability of sows to produce large number of piglets with good pork quality at a lower cost. The carcass payment system favouring leanness has led to an increase in the leanness of the sows. As a result some sows have very low backfat and poorer body condition, which adversely affects their fertility and productivity. IGF2 (Insulin-like growth factor 2) gene has been shown to play an important role in mammalian growth, influencing foetal cell division and differentiation, and postnatal muscle growth. The A allele is associated with lower backfat, higher lean yield and larger loin eye area which is favourable for market hogs. New results in Canadian pig populations have suggested that sows with paternal G allele had significantly higher total number of pigs at birth useful for sow productivity. As a special case, IGF2 gene can then be used to produce leaner pigs from fatter dams. Sires homozygous for A allele and dams with G allele can be used to produce larger numbers of leaner hogs because the IGF2 allele from the dam has no effect in progeny due to its imprinting mechanism. Therefore, sows can have better body condition and produce more as well as better quality hogs. Results on different sow productivity and carcass quality traits will be presented.

Session 10 Poster 19

Inbreeding of Polish Holstein-Friesian cattle and its impact on production traits

T. Jankowski and T. Strabel, Agricultural University of Poznan, Dept. of Animal Breeding and Genetics, Wolynska 33, 60-637 Poznan, Poland

Pedigree of over 5.5 million cows of Polish Holstein-Friesian breed born in years 1960-2006 was traced back using national pedigree database and information from Canadian, US and Interbull pedigree files. Inbreeding coefficients were computed using the method of VanRaden, which accounts for missing pedigree information. The method to measure completeness of five-generation pedigrees was applied as proposed by Cassell. Pedigrees completeness was relatively low – below 53% in average. For birth years 1986 through 2006, cows inbreeding has steadily increased from 0.11% to 1.87%. For the same period and cows with the minimum of 70% pedigrees completeness inbreeding increased from 0.54% to 3.01% with a high annual rate of 0.14 % per year. The effects of inbreeding on milk, fat and protein yield was examined using first lactation daily yields recorded between 1995 and 2006. The observations belonged to over 721 thousands cows with pedigrees >70% complete. The test day model included inbreeding percentage as a covariate. The estimated effect was negative and equal 16.6 kg milk, 0.55 kg fat and 0.37 kg protein in lactation per 1% increase in inbreeding.

Session 10 Poster 20

Reaction norms for litter size in pigs

P.W. Knap[1] and G. Su[2], [1]PIC International Group, Ratsteich 31, 24407 Schleswig, Germany, [2]University of Aarhus, Faculty of Agricultural Sciences, Blichers Allé, 8830 Tjele, Denmark

GxE can be quantified by reaction norm (RN) analysis, regressing phenotypic expression of a genotype on environment. The slope (b_1) quantifies environmental sensitivity, e.g. for progeny groups. Coherent estimation of environmental levels and RN parameters and their variance components is critical. We used an MCMC algorithm (Su *et al.*, 2006 JAS 84:1651) in DMU to simultaneously estimate these parameters for litter size of sows of pure lines (A, B) and their cross (AB) in 3 steps: (i) parity 1 litters of B sows (33641 recs, 33641 sows, 792 sires, 66 farms); (ii) all parity litters of B sows (113961, 52120, 1091, 93); (iii) all parity litters of A, B and AB sows (297518, 121104, 2040, 144). We used HYS as a covariate of RN, with heterogeneous residual variance. Heritabilities of intercept (b_0) were 0.1 as expected for this trait; h^2 of b_1 were (i) 0.15, (ii) 0.08, (iii) 0.02, with strongly decreasing s.e. in that order. Genetic correlations between b_0 and b_1 were (i) –0.09, (ii) +0.26, (iii) +0.69, ditto. Std of sire b_1 were (i) 0.47, (iii) 0.43, (iii) 0.30. Increasing the informativeness of the system (data volume, sire number, HYS spread) leads to better (lower s.e.) estimation of GxE, which in this case seems less serious (lower std of sire b_1 and h^2 of b_1) and shows less re-ranking of genotypes (higher r_G between b_0 and b_1). Alternatively, parity 1 and/or pure lines may be more environmentally sensitive than higher parities and crossbreds; this requires further study.

Session 10 — Poster 21

Estimation of genetic parameters for economic characters in three Chinese lines of silkworm *Bombyx mori* L.

S.Z. Mirhosseini[1,2], M. Ghanipoor[2], A.R. Seidavi[3], A.A. Shadparver[1] and A.R. Bizhannia[2], [1]Guilan University, Rasht, 4185743999, Iran, [2]Iran Silkworm Research Center, Rasht, 4185743999, Iran, [3]Islamic Azad University, Rasht Branch, Rasht, 4185743999, Iran

An experiment was conducted for estimating of heritability(H), genetic correlation(GC), environment correlation(EC) and phenotype correlations(PC) of 3 traits including cocoon weight(CW), cocoon shell weight(CSW), and cocoon shell percentage(CSP) in 3 commercial lines including 110, Koming1 and Y on the basis of 8 successive generations' data. Variance and covariance using BLUP method was estimated by means un-derivation algorithm on the basis of full-sib data. GC, EC and PC of CW-CSW for 110 line were 0.936, 0.448 and 0.727, for Koming1 were 0.827, 0.575 and 0.681 and for Y line were 0.904, 0.440 and 0.752. GC, EC and PC of CW-CSP for 110 line were 0.301, -0.267 and -0.108, for Koming1 were -0.137, -0.199 and -0.170 and for Y line were 0.031, -0.364 and -0.217 respectively. GC, EC and PC of CSW-CSP for 110 line were 0.879, 0.387 and 0.608, for Koming1 were 0.844, 0.391 and 0.647 and for Y line were 0.754, 0.363 and 0.553 respectively. The highest and lowest H for CW belonged to Koming1 (0.486) and 110 (0.349) line. Meanwhile the highest and lowest H for CSP belonged to Koming1 (0.226) and 110 (0.163) lines respectively. Results were showed CW and CSW having high H and CSP have lower H. These results shows additive gene effects affect on CW and CSW traits. But CSP controlled by non-additive gene effects.

Session 10 — Poster 22

Study on genetic variation of reproductive characteristics in Iranian native silkworm gene pools

M. Mavvajpour[1], S.Z. Mirhosseini[1,2], M. Ghanipoor[1] and A.R. Seidavi[3], [1]Iran Silkworm Research Center, Rasht, 4185743999, Iran, [2]Guilan University, Rasht, 4185743999, Iran, [3]Islamic Azad University, Rasht Branch, Rasht, 4185743999, Iran

Reproductive characteristics play an important role in survival of silkworm gene pools and economic efficiency of cocoon and egg producing centers. Five Iranian native silkworm breeds including Baghdad (B), Khorasan Orange (KO), Gilan Orange (GO), Khorasan Pink (KP) and Khorasan Lemon (KL) (12 family in each breed) were selected and reared during 3 generations. Genetic variance components for these traits were estimated by REML method in animal model based on full sib information and heritability coefficients of the traits were derived. Average for fecundity in B, GO, KL, KO and KP was estimated to be 500.8, 500.9, 414.2, 511.9 and 403.8, for fertility was 91.07, 86.73, 81.55, 92.01 and 85.05 and for hatchability was 87.10, 87.86, 76.22, 86.08 and 84.56, respectively. Heritability for fecundity in B, GO, KL, KO and KP was estimated to be 0.086, 0.0514, 0.1213, 0.2549 and 0.0123, for fertility was 0.1057, 0.0647, 0.0208, 0.0598 and 0.4638 and for hatchability was 0.0567, 0.1973, 0.5412, 0.5066 and 0.825, respectively. High values of heritability for some traits indicate that the direct selection cause an appropriate responses of these characteristics. Heritability which is defined as genetic to total variance ratio, is an indicator of genetic variation.

Session 10 — Poster 23

Genetic parameters of the major fatty acid (FA) contents in cow milk

H. Soyeurt[1,2], A. Gillon[1], S. Vanderick[1], P. Mayeres[3], C. Bertozzi[3], I. Misztal[4] and N. Gengler[1,5], [1]Gembloux Agricultural University, Animal Science Unit, Passage des Déportés 2, 5030 Gembloux, Belgium, [2]FRIA, Rue Egmont 5, 1000 Brussels, Belgium, [3]Walloon Breeding Association, Rue des Champs Elysées 4, 5590 Ciney, Belgium, [4]University of Georgia, Animal & Dairy Science Unit, 425 River Road, 30605 Athens, USA, [5]FNRS, Rue Egmont 5, 1000 Brussels, Belgium

Genetic parameters were estimated for fractions of saturated, monounsaturated (MONO) FA, C12:0, C14:0, C16:0, C18:0, C18:1 and C18:2 *9-cis,12-cis* in milk. 7,700 samples collected from April 2005 to May 2006 were scanned by MilkoScan FT6000. The FA contents were predicted from spectral data. The final data set included the FA contents as well as milk yields, %fat and %protein for 40,007 records on 2,047 cows in 7 breeds. Analyzes were by a multi-trait model with effects herd*test date*lactation number, lactation stage*lactation number, age*lactation number, regression on breed fractions, permanent environments within and across lactations and animal effect. The heritabilities of FA ranged from 5 to 38%. Heritability for SAT and MONO in milk were 36.28% and 14.73%, respectively. The genetic correlations between C12:0 and C14:0, C16:0, C18:0, C18:1, C18:2 in milk were 0.94, 0.89, 0.93, 0.53 and 0.61, respectively. An index including FA with a similar mammary origin could be used to modify the FA composition.

Session 10 — Poster 24

An approach to derive economic weights in breeding objectives using partial profile choice experiments

H.M. Nielsen[1] and P.R. Amer[2], [1]Norwegian University of Life Sciences, Department of Animal and Aquacultural Sciences, P.O. Box 5003, NO-1432 Aas, Norway, [2]Abacus Biotech Limited, P.O. Box 5585, Dunedin, New Zealand

The aim was to show how choice experiments can be used to derive economic weights in breeding objectives. We simulated a partial profile choice experiment with four different attributes (traits) each at three levels. In a partial profile design only a subset of traits is used for each comparison, making participation in the experimental process less onerous. Three different choice designs were compared. In the first design, respondents were presented with all four traits simultaneously. In the second and third designs, respondents were only presented with three or two out of the four traits per choice set. All three designs included 4 traits each at 3 levels and respondents should choose between two alternative genotypes. The effectiveness of different designs was evaluated based on comparisons between true and simulated preferences. Regression coefficients from the conditional logit model were used to estimate the relative economic weights of traits. Designs involving 20 choices based on a subset of two traits at each choice, and over 30 respondents provided relatively accurate estimates of respondent preferences. The method can be used to define economic weights for use in animal breeding selection indexes where traditional approaches such as profit equations and bioeconomic models are not practical.

Session 10 — Poster 25

Goat casein haplotypes of two distant European breeds compared

R. Finocchiaro[1], B.J. Hayes[2], J.B.C.H.M. van Kaam[3], M. Siwek[1], B. Portolano[1] and T. Ådnøy[4], [1]University of Palermo, Department S.En.Fi.Mi.Zo., Viale della Scienze, 90128 Palermo, Italy, [2]Department of Primary Industries, Animal Genetics and Genomics, 475 Mickleham Rd, Attwood VIC 3049, Australia, [3]Istituto Zooprofilattico Sperimentale della Sicilia "A. Mirri", Via Gino Marinuzzi 3, 90129 Palermo, Italy, [4]Norwegian University of Life Sciences, Department of Animal and Aquaculture Sciences, P.O. Box 5003, N-1432, Ås, Norway

The diversity among haplotypes derived based on 22 SNPs and one deletion within four casein genes in two geographically distant goat populations, the Sicilian Girgentana breed and the Norwegian goat breed, has been characterized in this work. Genotypes on 40 male and female individuals belonging to the Girgentana breed and 436 Norwegian bucks were available. A proportion of casein loci haplotypes were found to be identical in both Norwegian and Girgentana goats, despite the large geographical distance, and phenotypic divergence between the breeds. Given the fact that the casein genes occur within a 250kb fragment of the genome, this suggests that these haplotypes originated from a common ancestral population, rather than appearing subsequently to breed formation. Minimum Spanning Trees were constructed among the SNP haplotypes of the casein genes. The separation between south European and north/central European goat populations was further supported by the presence of a null allele at CSN2 in Girgentana and its absence in the Norwegian goats.

Session 10 — Poster 26

Large scale simulation with a complex breeding goal; future sheep breeding schemes for Norway

L.S. Eikje[1], L.R. Schaeffer[2], T. Ådnøy[1] and G. Klemetsdal[1], [1]Department of Animal and Aquacultural Sciences, Norwegian University of Life Sciences, P.O. Box. 5003, 1432 Ås, Norway, [2]Centre for Genetic Improvement of Livestock, Department of Animal and Poultry Science, University of Guelph, Guelph, ON, N1G 2W1, Canada

National sheep breeding schemes were evaluated by stochastic simulation, for all 9 traits that make up aggregate genotype, considering also rate of inbreeding. The schemes were: (i) Test rams and elite rams used by natural mating in ram circles, with 120 000 ewes, as a reference, and for same number of ewes: (ii) Test rams in natural mating in ram circles and elite rams in AI within and across circles, and: (iii) Test rams and elite rams in AI. Within schemes (ii) and (iii) experimentation was done for: 1) number of ewes mated / inseminated per test ram, 2) number of ewes inseminated per elite ram and 3) percent ewes mated to elite rams versus test rams. Largest genetic response was found in scheme (ii). Most advantageous results were with on average 50 ewes mated naturally to each test ram, on average 1100 ewes inseminated to each elite ram and using an elite mating percentage of 30, resulting in a response about 40 per cent larger than in the reference scheme (scheme i). Correspondingly, rate of inbreeding was 0.42 per cent per generation. For scheme (iii), largest response was obtained for an average of 100/1100 ewes inseminated to each test ram/elite ram, for an elite mating percentage of 10, giving rate of inbreeding of 0.51 per cent per generation.

Session 10 Poster 27

Genetic parameters for components of boar taint and female fertility

K.A. Engelsma, R. Bergsma and E.F. Knol, IPG, Institute for Pig Genetics, P.O. Box 43, 6640 AA Beuningen, Netherlands

Of an anticipated 2000 entire males we analysed the first 693 animals for incidence of androstenone (AND) (1.67 ±1.49 µg/g), skatole (SK) (0.082 ±0.09 µg/g) and indole (IND) (0.065 ±0.07 µg/g). Boars were culls from a synthetic sire line, slaughtered at 124 ±12 kg. Traits were log transformed and analysed using ASReml, with pen, age at start test, days tested and days end till slaughter as fixed effects, and slaughter day and animal effect as random. Resulting heritability estimates: AND (0.75 ±0.14), SK (0.53 ±0.13) and IND (0.29 ±0.10). Genetic correlation between SK and IND was high (0.75 ±0.10), for AND with SK and IND 0.22 ±0.19 and 0.30 ±0.20, respectively. Parameters for female fertility traits were estimated using 5320 litters of 1800 sows. Litter size (h^2: 0.15 ±0.03) and age at first insemination (h^2: 0.20 ±0.06) had low genetic correlations ($-0.15< r_g <0.05$) with AND, SK and IND. Genetic correlation between interval weaning to 2nd insemination and AND was negative (-0.58 ±0.34), but positive with SK and IND (0.62 ±0.38 and 0.62 ±0.37, respectively). Next step is the estimation of genetic correlations with finishing traits. Tissue of animals and their dams was stored for QTL studies. Given current results a substantial change in levels of boar taint seems to be feasible. To shift the distribution towards a zero boar taint perception for all consumers might be a bridge too far.

Session 10 Poster 28

Three versus four traits random regression test day model genetic evaluation for the Holstein breed in Italy

F. Canavesi, S. Biffani and E.L. Nicolazzi, ANAFI, R&D, Via Bergamo 292, 26100 CREMONA, Italy

Genetic evaluation for production traits in the Holstein breed in Italy is based on a Random Regression Test Day Model (RRTDM) since November 2004. More specifically the model is a multiple lactation, multiple traits (milk, fat, protein and somatic cell count) RRTDM, similar to the model used in Canada for the official genetic evaluation. Fixed regression curve effects include time, region, age at calving, parity and season of calving. An improved model uses test days pre-adjusted for number of days of pregnancy and include the effect of year of production instead of time in the fixed effects definition. A new set of genetic parameters was estimated in December 2006 without somatic cell count. As a result genetic correlations across lactations and within trait increased by 0.06-0.10. Data from February 2007 and November 2002 evaluation were used to assess the different ability of the two systems (four traits versus three traits) to predict future breeding values. Correlations among proofs on the same data set ranged from 0.99 to 0.98. Rank correlation was also very high. Rank correlations of the top 250 bulls comparing November 2002 and February 2007 proofs in the three traits model were 0.01-0.02 higher than with the four trait model. Research is still ongoing in order to evaluate the impact of the three trait evaluation system when new lactation test day records are added from run to run.

Session 10 Poster 29

Environmental effects on lactation curves included in a test-day model genetic evaluation

H. Leclerc, D. Duclos and V. Ducrocq, INRA, UR337 Station de Génétique Quantitative et Appliquée, 78 350 Jouy en Josas, France

Implementation of the French genetic evaluation for production traits in dairy cattle based on a random regression test day model is in progress. Four fixed effects were included: month and age at calving, length of dry period were modelled each through 6-knot regression splines on days in milk and gestation through 4-knot regression splines on days carried calf. Data were test-day records on 6 production traits (milk, fat and protein yields, fat and protein percentages and somatic cell score) from the first 3 lactations of French Montbéliarde collected from 1988 to 2005. Curves by region-year-parity class of fixed effects were fitted, but interactions between year and region-parity class effects on the lactation shape were not significant. Thus, the fixed year effect was modelled as a constant over the whole lactation. To reduce computational requirements and time, data were pre-adjusted (2-step procedure) for effects with no year interaction (22 splines coefficients). At a 4-year interval, precorrection factors for lactation curves estimated were consistent, since correlations for genetic and permanent environment effects obtained with the 2-step procedure were at least 0.9999 and 0.9995 respectively. The 2-step procedure does not have any impact on estimation of genetic and permanent environment effects: correlations with the 1-step procedure were at least 0.998 and 0.997 respectively. The use of a 2-step procedure reduces computation time by a factor 4.

Session 10 Poster 30

Genetic determinism of metabolic resource allocation in pigs: a study in a Duroc population

N. Ibáñez-Escriche[1], J. Soler[2], L. Varona[1], J. Reixach[3], J. Tibau[2] and R. Quintanilla[1], [1]IRTA-Lleida, Genètica i Millora Animal, Av. Rovira Roure, 191, 25198, Spain, [2]IRTA, Control Porcí, Monells, 17121, Spain, [3]Selección Batallé S.A., Riudarenes, 17421, Spain

One of the main objectives in pig production is to improve feed efficiency. We have studied the genetic determinism of feed efficiency, by analysing the metabolic resource allocation for body weight maintenance and growth. The individual feed intake and body weight gain have been controlled in a population of 370 castrated males belonging to a commercial *Duroc* line. A procedure based on a hierarchical Bayesian analysis has been used for the analyses. The first stage of the model consisted of a multiple regression model of feed intake on metabolic body weight and on body weight gain. The second stage of the model included a multivariate animal model for both partial regression coefficients. Posterior means for the proportion of variance due to additive genetic variance were 0.43 and 0.37 (with posterior standard deviations of 0.18 and 0.16) for regression coefficients of feed intake on, respectively, body weight and weight gain. These results allow us to conclude that the metabolic resource allocation in pigs has an important genetic determinism and could be improved by selection.

Fertility in beef cows

F. Phocas, INRA, UR337 Station de génétique quantitative et appliquée, 78352 Jouy-en-Josas, France

Genetic parameters were estimated for three traits for assessing fertility of artificially inseminated (AI) heifers and primiparous cows: calving success to first A.I. (CS1), overall calving success (CS), either from the initial AI or a backup bull mating, and the number of days from initial AI to calving (DC1), dealing with censored data. Records were registered from 1998 to 2005, in 292 herds dedicated to the on-farm French Charolais progeny test. Data consisted of 85,397 cow-year records, including 30,460 inseminated heifers that were bred by 1,844 sires. The average heifer CS1 and CS rates were 56.4% and 85.3%, respectively; for primiparous cows, the corresponding means were 51.8% and 84.9%. Strong unfavourable environmental effects were estimated: heifers inseminated before 19 months had 7% less chance to calve to first AI; primiparous cows had 30% less chance to calve to first AI when their first calving was a caesarean; they had also 14% less chance to calve to first AI after a heat synchronization treatment. Heritability estimates were very low, ranging from 0.014 for CS1 to 0.020 for DC1 under linear models and from 0.018 to 0.028 under discrete models. No significant genetic correlations were estimated between heifer and corresponding primiparous cow traits. It made it difficult to propose an efficient way of selection for high female reproductive performance.

Genetic parameters for measures of residual feed intake and growth traits in Duroc pigs

M.A. Hoque and K. Suzuki, Tohoku University, Faculty of Agriculture, Sendai, 981-8555, Japan

Genetic parameters for residual feed intake (RFI), daily feed intake (FI), daily gain (DG), backfat thickness (BF), and loin eye area (EMA) on 1642 Duroc pigs in seven generations were estimated with REML. Four measures of RFI were estimated from models that included initial test age and weight with DG (RFI1); with DG and BF (RFI2); with DG and EMA (RFI3); and with DG, BF, and EMA (RFI4). Heritabilities for RFI measures were moderate (ranged from 0.22 to 0.38). The corresponding estimates for FI, DG, and EMA were also moderate (ranged from 0.45 to 0.49), while for BF was high (0.72). Genetic correlations of BF with RFI measures were higher when BF was not included in the estimation of RFI (ranged from 0.77 with RFI1 and 0.76 with RFI3 *vs* 0.11 with RFI2 and 0.07 with RFI4). Genetic correlations of EMA with measures of RFI were all negative (ranged from -0.30 to -0.60). Selection for desired gain has resulted small but favourable genetic changes in RFI measures. Phenotypic correlations between RFI measures were zero, and genetic correlations between them were low (ranged from 0.17 to 0.23). FI was strongly correlated with all the RFI measures, both genetically (ranged from 0.56 to 0.77) and phenotypically (ranged from 0.56 to 0.66). The results suggested that selection against RFI may cause reduction in FI that does not go to productive purpose. BF would also decrease, and EMA would increase. The amount of change in BF or EMA would be varied depending on whether RFI was adjusted for BF.

Session 10 Poster 33

Estimation of breeding values for test day somatic cell score and clinical mastitis using threshold model

K. Vuori, E. Negussie, I. Strandén and E.A. Mäntysaari, MTT Agrifood Research Finland, Biotechnology and Food Research, Biometrical Genetics, H Building, 31600 Jokioinen, Finland

Breeding values were estimated using test-day (TD) somatic cell score (SCS) and lactation average (LA) clinical mastitis (CM) data by a bivariate linear-threshold sire model (LT-TD). CM was a categorical trait and the model was solved by an empirical Bayes approach. Model estimates were compared to two bivariate linear-linear sire models, where SCS were either TD or LA records (named LL-TD and LL-LA models, respectively). Data consisted of about 1.5 million first lactation TD records of Finnish Ayrshire cows (N=323336) with first calving from 1995 to 2006, and pedigree had 6790 sires. Solving times were <1min, 5min and 30min for LL-LA, LL-TD and LT-TD models, respectively. Comparisons of CM estimates were made for sires born 1996-1998 and with at least 20 daughters. There were 32, 31 and 37 common sires in top50-list between models LL-LA and LL-TD, LL-LA and LT-TD, and LL-TD and LT-TD, respectively. Prediction ability as correlations of evaluations from two data sets, created by splitting original data into half within daughter groups, were 0.55, 0.60 and 0.62 for LL-LA, LL-TD and LT-TD models, respectively. According to the results, TD models were better than LA model, and the threshold model showed little advantage to linear model.

Session 10 Poster 34

Genetic variation in measures of feed efficiency and their relationships with carcass traits in Duroc pigs

M.A. Hoque[1], K. Suzuki[1], H. Kadowaki[2] and T. Shibata[2], [1]Tohoku University, Faculty of Agriculture, Sendai, 981-8555, Japan, [2]Miyagi Prefecture Animal Industry Station, Miyagi, 989-6445, Japan

Genetic parameters for feed efficiency and carcass traits on 1642 pigs in 7 generations of Duroc population were estimated. Feed efficiency traits included feed conversion ratio (FCR) and residual feed intake (RFI). Carcass traits were loin eye muscle area (EMA), backfat thickness (BF), intra-muscular fat, and meat tenderness. Three measures of RFI, i.e., phenotypic (RFI_p), genetic (RFI_g), and nutritional (RFI_n) RFI, were calculated by the difference between actual and predicted FI. The RFI_p and RFI_g were estimated, respectively, by the residual of phenotypic and genetic regressions of FI on MWT and daily gain. The RFI_n was estimated by the difference between actual and expected FI that predicted by the nutritional requirement. The mean values for RFI_p and RFI_g were close to zero, and for RFI_n was negative (-0.11 kg/day). All the measures of feed efficiency were moderately heritable (h^2 = 0.34, 0.41, 0.42, and 0.28 for RFI_n, RFI_p, and RFI_g, and FCR, respectively). The heritability estimates for all carcass traits were moderate (ranged from 0.38 to 0.46), except for BF, which was high (0.72). Measures of RFI were favourably correlated with EMA (ranged from -0.56 to -0.61). BF was more strongly correlated with RFI ($r_g \geq 0.76$) than with FCR ($r_g = 0.58$). This study provides evidence that selection against either RFI_p or RFI_g would give similar correlated responses in carcass traits.

Session 10 Poster 35

The "Julius Kühn Museum and Collection": Animal breeding history and scientific resource

J. Wussow, R. Schafberg and H.H. Swalve, Institute of Agricultural and Nutritional Sciences, Animal Breeding, Adam-Kuckhoff-Str. 35, 06108 Halle(Saale), Germany

The first agricultural university college in Germany was established under the guidance of Julius Kühn (1825-1910) at the University of Halle-Wittenberg in 1863. As an aid for teaching and research, in 1864, the "livestock garden" which presented livestock of various species and breeds *in vivo* was implemented. The "Kühn Collection" drew its specimens from the animals of the livestock garden which existed for more than 100 years. The collection nowadays comprises osteological specimens of various livestock species and breeds, e.g. 670 pig skulls, 700 cattle skulls and 2000 sheep skeletons. The uniqueness of the collection not only arises from its extent but also from the fact that specimens stem from exactly defined (e.g. age) animals kept under specific and documented environmental conditions and with known pedigrees. Besides this animal specific data, hair samples, dermoplastic preparations and photos are kept. The collection provides a sustainable and solid basis for scientific research on aspects of evolutionary biology, animal breeding history, heredity, genetic diversity, livestock housing and nutrition as well as for studies on the reproduction of inter-species crosses. Highlights of the collection are specimens from starvation – fattening experiments with pig full-sibs to assess phenotype by environmental interactions, undertaken in 1913, and samples from the still existing experimental flock of Karakul sheep.

Session 10 Poster 36

Effect of two-way selection on egg composition by means of the TOBEC method, on the hatchability of hen's eggs, and on the body weight and body composition of hatched chicks

G. Milisits, E. Kovács, L. Locsmándi, A. Szabó, G. Andrássy-Baka, O. Pőcze, J. Ujvári, R. Romvári and Z. Sütő, University of Kaposvár Faculty of Animal Science, Guba Sándor u. 40., 7400 Kaposvár, Hungary

Pushing the advantage of the so-called TOBEC (Total Body Electrical Conductivity) method, allowing the determination of egg composition *in vivo*, correlations between egg composition, hatchability, live weight and body composition of hatched chicks were studied. Altogether 1.500 hens' eggs were measured by TOBEC, and the extreme and average 15-15% were chosen for incubation based on the measured values (E-value). During the incubation it was observed, that eggs with high E-value/egg weight ratio fall out in a significantly higher ratio at first candling and at hatching, than eggs with low E-value/egg weight ratio. The E-value/egg weight ratio as independent variable in a linear regression model resulted in a 41.3% accuracy in the prediction of hatching weight of the chicks. Examining the separate effect of egg weight and egg composition on the hatching weight it was shown, that the increase of E-value/egg weight ratio at the same egg weight, and the increase of egg weight at the same E-value/egg weight ratio resulted in an increase in the hatching weight. All traits of main body composition (dry matter, crude protein, and crude fat) of the chicks hatched from eggs with low E-value/egg weight ratio showed higher values than that of the chicks hatched from eggs with high E-value/egg weight ratio.

Session 10 Poster 37

First approaches for a combined use of microsatellites and pedigree data to estimate relationships

E. Bömcke[1] and N. Gengler[2], [1]Gembloux Agricultural University (FUSAGx), FRIA, Passage des Déportés 2, 5030 Gembloux, Belgium, [2]Gembloux Agricultural University (FUSAGx), FNRS, Passage des Déportés 2, 5030 Gembloux, Belgium

The main objective of conservation is the preservation of genetic diversity. Among several tools recommended to measure genetic variability relationship coefficients are often used. They can be estimated from pedigree but also from direct knowledge of the genotype. The Skyros pony is an example of a small breed concerned by the management of genetic variability. However, the analysis of the pedigree showed its very poor quality. A part of the population was therefore genotyped for 16 microsatellites and two estimators of relationships were tested: the Lynch & Ritland (L&R) (only based on genotypes and described as having the best performances for all population compositions of the mammal species) and a new estimator (developed for this study and combining pedigree and microsatellite information). In order to compare the results, a Principal Components Analysis (PCA) was performed. The new estimator explained a higher percentage of information within the 3 principal factors of the PCA (41.99%) than the L&R estimator (20.61%). The graphic PCA representation showed a better separation between the reference group and the rest of the population in the case of the new estimator. The new estimator showed interesting preliminary results and results support interest of combining information in case of incomplete pedigrees and/or use of a limited number of markers.

Session 10 Poster 38

Comparing the effectiveness of different models in accounting for the cyclic effect of months

O. Sasaki[1], Y. Nagamine[1], K. Ishii[1] and C.Y. Lin[2], [1]National Institute of Livestock and Grassland Science, Tsukuba, 305-0901, Japan, [2]Agriculture and Agri-Food Canada, (Dept. of Animal and Poultry Science, University of Guelph), Guelph, N1G 2W1, Canada

Yearly effect on milk yield of Japanese Holstein cows was linear whereas monthly effect was cyclic. These results were used to simulate the data with or without interaction between year and season. Four different models were compared: (1) x=F+Y+P+e, (2) x=FY+P+e, (3) x=F+YP+e, and (4) x=FP+Y+e where F= farm effect, Y=year effect, and P= period effect. The length of period is every 15 days (P_{15}), 30 days (P_{30}) or 90 days (P_{90}). R^2 adjusted for degree of freedom (R^{*2}) for all models were larger with P_{15} than with P_{30} or P_{90}. Model 1 with P_{15} effect had the largest R^{*2} in the absence of interaction whereas Model 3 with P_{15} effect had the largest R^{*2} in the presence of interaction. R^2 from regressing the predicted on the true residuals (R_r^2) was the largest for Model 1 with P_{30} effect in the absence of interaction but was the largest for Model 3 with P_{90} effect in the presence of interaction. The results suggest that the choice of a model for statistical analysis depends upon the length of the periods, the data structure, and the nature of interaction. Model 1 best fitted the simulated data in the absence of interaction whereas Model 3 best fitted the simulated data in the presence of interaction. The fitting of P_{15}, P_{30}, or P_{90} depends upon the cyclic nature of the period within year.

Session 10 Poster 39

Assessment of heterogeneity of residual variances in an autoregressive test-day multiple lactations model

J. Vasconcelos[1], A. Martins[2], A. Ferreira[2] and J. Carvalheira[1], [1]CIBIO-U.Porto, Vairão, 4485-661 Vairão, Portugal, [2]ABLN, Vila do Conde, 4480 Vila do Conde, Portugal

The autoregressive test-day animal model (ARTD) is a class of TD models where the within animal covariance follows 1st order autoregressive processes. Animal correlations are fitted within and across lactations (L) using short (STE) and long (LTE) term environmental effects. TD yields are treated as independent random variables with heterogeneous (HE) variance for each L and describe STE effects. LTE effect accounts for the animal correlation across L. Results from a previous study suggested that the usual assumption of a homogeneous (HO) residual variance (RV) component may not hold. To assess HE of RV across L, 5 random samples of the Portuguese Holstein milk database were extracted and analyzed by ARTD models differing on the RV structure (HO *vs* HE). Likelihood ratio tests indicated a better fit (P<.01) for models with HE RV. Heritabilities were similar, averaging.32,.21 and.19 for 1st, 2nd and 3rd L, respectively. Correlations between the animal's EBV, PEV, accuracy of prediction (ACC) and rank were high (>.99) suggesting that HE residuals will not change the genetic rank of sires and cows. Differences among EBV were on average not significant (P>.05). On the other hand, differences among PEV and ACC (P<.001) favoured the HE RV model, implying greater potential for genetic progress. These advantages (better fit and higher ACC) arc going to be explored in future genetic evaluations in Portugal.

Session 10 Poster 40

Association between somatic cell score and fertility traits of Polish Holstein-Friesian cows

A. Zarnecki[1], M. Morek-Kopec[2] and W. Jagusiak[2], [1]National Institute of Animal Production, Balice k. Krakowa, 32- 083, Poland, [2]Agricultural University, Al. Mickiewicza 24/28, 30-059 Krakow, Poland

Somatic cell score shows a high genetic correlation with clinical mastitis and other bacterial infections which negatively influence production and reproduction of cows. This study estimated the association between somatic cell score (SCS) and two fertility traits: nonreturn to estrus by 56 days (NRR), and number of days from calving to first insemination (CFI). Data were records of 298016 Polish Holstein Friesian cows and contained identification, parity, test day SCS, and insemination and calving dates. The model used to evaluate the effect of preceding test day SCS on NRR included random herd effect and fixed effects of parity, year of calving, lactation stage at first service, month of first service, interval between preceding test day SCS and day of first service, and regression of NRR on SCS. A similar model was applied to analyse the relation between test day SCS and CFI. All fixed effects except number of days from preceding test day SCS to first insemination significantly affected NRR. Highly significant but small regression coefficients of NRR on SCS were found for intervals from 1 to 49 days between preceding SCS and first insemination. Higher SCS within these intervals was associated with lower NRR. Similarly, fixed effects significantly influenced CFI. Significant regression and quadratic regression of SCS on CFI showed that higher SCS was related to longer CFI.

Session 10 — Poster 41

On asymptotic identifiability in animal models with competition effects

R.J.C. Cantet and E.P. Cappa, Fa.Agronomía, Universidad de Buenos Aires; CONICET, Av San Martín 4453, C1417DSQ Buenos Aires, Argentina

There is an increased interest in estimating the (co)variance components of additive animal models with direct and competition effects (AMC). However, all attempts to estimate the dispersion parameters in differents animal species faced problems of convergence and highly variable estimates. The trouble is related with the lack of asymptotic identifiability in certain AMCs. This property is observed when building the 4×4 information matrix (I) for the AMC REML likelihood, and its smallest eigenvalue goes to zero. The singularity of I is due to confounding between the fixed contemporary group (pen) effects and the additive competition effects. The incidence matrix of additive competition effects (Z_c) can be written as a function of the "intensity of competition" (IC) among animals in a contemporay group. The IC values can be interpreted as weighting factors expressing how intense a pair of animals compete in relation to the remainder competitors. The set of IC values in any row of Z_c should add up to 1 (standardization) in order for the phenotypic variance of any given observation not be affected by the number of competitors. Moreover, data sets to estimate the (co)variance components in the AMC should be obtained with some sort of design in which rank[$X Z_c$] is equal to rank[X] for the (co)variance components to be asymptotically identifiable.

Session 10 — Poster 42

Doppel gene polymorphisms and their relation to scrapie susceptibility in Portuguese sheep breeds

P. Mesquita, M. Batista, M.R. Marques, I.C. Santos, J. Pimenta, I. Carolino, F. Santos Silva, C.O. Sousa, A.E.M. Horta, J.A. Prates and R.M. Pereira, Estação Zootécnica Nacional, Fonte Boa, 2005-048 Vale de Santarém, Portugal

Doppel gene (*Prnd*) is located downstream from cellular prion protein (PrPC) gene (*Prnp*) and probably results from its ancestral duplication. PrPC is involved in transmissible spongiform encephalopathies aetiology, including scrapie disease in sheep. Scrapie susceptibility has been related to some *Prnp* gene polymorphisms. Doppel protein (Dpl) is significantly homologous to PrPC but it is minimally expressed in central nervous system and does not seem crucial for prion disease. Instead, Dpl is thought to have an essential role in male fertility. However, in *Prnp* (°/°) mice, ataxia occurs due to Dpl up-regulation. This ataxic phenotype is corrected after introducing a*Prnp* wild-type transgene, suggesting an interaction between Dpl and PrPC. The aim of this work was to characterise *Prnd* gene coding region polymorphisms in 8 Portuguese sheep breeds and to establish associations with *Prnp* genotypes, previously determined by SNP analysis and grouped according to 5 grades of scrapie susceptibility. PCR-SSCP analysis revealed a synonymous polymorphism (G to A substitution) in codon 26 of *Prnd* gene. We found a statistically higher frequency of heterozygotes associated with the ARQ/AHQ *Prnp* genotype (p=0.003) (Genepop v. 3.4). In conclusion, there is preliminary evidence for a correlation between *Prnd* and *Prnp* genotypes, demanding further investigation.

Session 10 Poster 43

Effects of management intensity and genetic value of the sire on milk yield and form of lactation curve of first-lactation Simmental cows

A. Gerber, D. Krogmeier, R. Emmerling and K.-U. Götz, Bavarian State Research Center for Agricuture, Institute for Animal Breeding, Prof.-Dürrwaechter-Platz 1, 85586 Grub, Germany

Under continued selection for milk yield practitioners frequently raise the question whether the use of bulls with high breeding values for milk yield on extensive dairy farms is appropriate. The present investigation analysed the effect of breeding value for milk yield on first lactation yield and on the shape of lactation curve on farms with different intensity of management. Data comprised 278 bulls of birth years 1993 to 1994 and their daughters (327.509 cows) in Bavaria. Daughters were divided into 3 groups with respect to relative breeding value for milk yield of their sire and to intensity of management. Milk, fat and protein yield and lactation curves were compared between groups and intensities. As could be expected, the advantage in milk, fat and protein yield of daughters of high yielding bulls is significantly higher under intensive conditions. Under intensive management, lactation curves of daughters of high yielding sires showed a steep slope up to lactation peak. Peak milk yield was higher and day of peak milk yield was later in lactation. Under extensive conditions lactation curves were much flatter, indicating better persistency. The hypothesis that daughters from high yielding bulls under extensive conditions will start high into lactation followed by a drastic collapse in milk yield due to metabolic stress was not confirmed.

Session 10 Poster 44

Characterization of DGAT1, GHR and PRLR mutations in the Reggiana cattle breed and associations with milk production traits

L. Fontanesi[1], E. Scotti[1], A. Bagnato[2] and V. Russo[1], [1]University of Bologna, DIPROVAL, Sezione di Allevamenti Zootecnici, Via F.lli Rosselli 107, 42100 Reggio Emilia, Italy, [2]University of Milano, Dep. VSA, Via Celoria 10, 20133 Milano, Italy

The identification of quantitative trait loci (QTL) and the subsequent characterization of quantitative trait genes (QTG) have been the objective of several studies in dairy cattle usually using cosmopolitan and widely spread breeds. The effects of these QTG have been only in few cases investigated in local breeds. Reggiana is a local dairy cattle breed that account less than 2000 cows reared in the North of Italy, mainly in the province of Reggio Emilia. Its milk is used for the production of Parmigiano Reggiano cheese. We analysed in 129 Reggiana sires mutations in genes that other studies in other breeds have shown to affect milk production and composition traits. PCR-RFLP protocols were set up to genotype the *DGAT1 K232A*, *GHR F279Y*, *PRLR S18N* and *PRLR P186L* mutations. Fragment analysis was used to genotype the *DGAT1 VNTR* mutation. Alleles *A* (0.83), *3* (0.51), *F* (0.92), *S* (0.60) and *L* (0.74) were the most frequent at the *DGAT1 K232A*, *DGAT1 VNTR*, *GHR F279Y*, *PRLR S18N* and *PRLR P186L* mutations, respectively. Association study between the genotypes at these loci and sire estimated breeding values for several milk production traits showed significant effects ($P < 0.05$) of the *DGAT1 K232A* mutation on protein yield, protein percentage and fat percentage.

Amiata donkey local breed: genetic and morphological characterization

R. Ciampolini[1], F. Cecchi[1], E. Ciani[2], E. Mazzanti[1], M. Tancredi[1] and S. Presciuttini[3], [1]University of Pisa, Animal Production, Viale delle Piagge 2 - Pisa, 56124, Italy, [2]University of Bari, General and Environmental Physiology, Via Amendola 165/a - Bari, 70126, Italy, [3]University of Pisa, Centre of Genetical Statistics, SS Abetone e Brennero 2 - Pisa, 56127, Italy

The genetic and morphological characterization of the local Amiata donkey breed has been considered. The genetic structure was investigated using information from pedigrees (608 subjects) and using STR markers (18 microsatellite loci were analysed in 50 unrelated individuals), while the morphological structure was investigated through the data reported in the Anagraphic Register (10 scores, evaluated by certificated experts of the breed and measurements of withers height, chest and cannon circumference). Genealogical analysis reveal that the inbreeding level seems to be rather acceptable, contrary to the expectation for a population that has suffered a severe bottleneck in recent years. However, the lack of genealogical records for a substantial number of individuals may have caused a possible underestimation of the inbreeding coefficient. In fact, a substantial increase in the average inbreeding coefficient is observed in those subjects with the maximum number of traced generations. On the contrary molecular data evidenced an elevated genetic homogeneity. From the point of view of the morphological analysis the presence of two well separate morphological groups and the clean uniformity is evident inside the breedings.

The influence of milk protein polymorphisms on morphometric characteristics of milk fat globules in Italian Friesian dairy cow

M. Martini[1], F. Cecchi[1], C. Scolozzi[1], F. Salari[1], F. Chiatti[2], S. Chessa[2] and A. Caroli[3], [1]Università di Pisa, Dip. Prod. Anim., Viale delle Piagge, 2, 56124, Italy, [2]Università di Milano, Dip. VSA, Via Celoria, 10, 20133, Italy, [3]Università di Brescia, Dip. Sci. Biomed. Biotec., Viale Europa, 11, 25123, Italy

The influence of b-casein (b-CN), k-casein (k-CN) and b-lactoglobulin (b-LG) genotype on morphometric characteristics of milk fat globules was evaluated on 89 milk samples from Italian Friesian pluriparous cows reared in the same herd in a medium-temperate climatic zone in the province of Pisa. Individual milk samples were collected and analyzed in a triplicate. Milk protein genotypes were determined by isoelectric focusing analysis of individual milk samples. The morphometrical analysis of fat globules (number of milk fat globules/ml and diameter) was performed. The frequency distribution of total measured milk fat globules was evaluated according to their size. Fat globule diameters were divided into nine classes of 1.5 micron width: from class 1 (1.5-3) to class 9 (>13.5). The analysis of variance was carried out to evaluate the effect of b-CN, k-CN, and b-LG genotype on all the evaluated parameters with simultaneous adjustment for sampling season, parity, and days from calving as a covariate. Significant effects were observed mainly for b-LG genotype, with AA genotype associated with a higher proportion of fat globules greater than 6 micron than AB and BB.

Session 10 Poster 47

Influence of G77A polymorphism in CATD gene on skeletal muscle lysosomal proteolysis and some basic composition of meat from bulls of different breed

S.J. Rosochacki[1,2], E. Juszczuk-Kubiak[2], L. Barton[3], T. Sakowski[2] and M. Matejczyk[1], [1]Bialystok Technical Univ., 45E Wiejska str, 15-351, Poland, [2]Inst. Genetics & Animal Breeding, PAS, Jastrzębiec, 05-552 Wólka Kosowska, Poland, [3]Inst. Animal Sciences, 815 Pratelstvi, 104-00 Praha, Czech Republic

Polymorphism in procathepsin D gene was evaluated with proteolytic enzymes involved in protein catabolism, fibril size and basic meat traits. MLLT samples were cut out from 34 Limousine (LIM) and 41 black-white (BW) 12-15 months old bulls. CATD gene fragment was amplified and analyzed by RFLP/ApaI. Cathepsin D (CatD), pepstatin sensitive (PSCatD) and leupeptin-insensitive (LIA) acid autolytic (AAA) activities were measured. The muscle structure was done on frozen stained cuttings. Some meat quality traits were evaluated. The data were analyzed by analysis of variance. Genotype frequencies in G77A mutation in CATD gene were: in BW - GG 0.634; in LIM - GG 0.618, no AA genotype. Morphology showed more fibrils of higher diameter and homogeneity in LIM than BW bulls. CatD, PSCatD and LIA were higher in GA genotype BW bulls muscles by 6.5%, 7.1%, 11.3% ($p < 0.05$). No such differences were found out in LIM, but AAA was higher by 8.9% in GG muscle. BW meat of GA animals had higher fat content and SF48, but GG genotype meat had better WHC48 ($p < 0.05$). In LIM bulls, there was no influence of G77A mutation on meat traits. These results indicate on genotype dependent parameters concerning the protein turnover in BW as compared to LIM bulls.

Session 10 Poster 48

Bayesian inference of inbreeding effects on litter size and gestation length in Duroc pigs

J. Farkas[1], I. Curik[2], L. Csato[1], Z. Csörnyei[3], R. Baumung[4] and I. Nagy[1], [1]University of Kaposvar, 40 Guba S. Street, 7400 Kaposvar, Hungary, [2]University of Zagreb, 25 Svetosimunska Street, 10000 Zagreb, Croatia (Hrvatska), [3]National Institute for Agricultural Quality control, 24 Keleti K. Street, 1024 Budapest, Hungary, [4]Univ. of. Nat. Resources and Applied Life Sciences, 33 Gregor Mendel Str., 1180 Vienna, Austria

The effects of litter and dam inbreeding on the number of piglets born alive (NBA) and on the gestation length (GL) were analyzed using the data of 11141 farrowings of 3196 Duroc sows that were collected between 1982 and 2003. Total number of animals in the pedigree file was 4819; the average number of known ancestors was 21.6. Average complete generation equivalents of litters and dams were 2.18 and 1.94, respectively. To estimate the effects of inbreeding bivariate repeatability animal models were used, applying a Bayesian approach. Litter and dam inbreeding effects (per 1% inbreeding) on NBA resulted in 95% high posterior density intervals from –0.315 to 0.137 and from –0.173 to 0.069. The effects of litter and dam inbreeding on GL were negligible (high posterior density intervals at a 95% probability were: -0.128 to 0.175 and -0.144 to 0.033, respectively).

Session 10 Poster 49

Estimation of genetic parameters for udder health with use of a test-day model

K. Grandinson[1], E. Strandberg[1] and K. Johansson[2], [1]Swedish University of Agricultural Sciences, Dept. of Animal Breeding and Genetics, P.O. Box 7023, 75007 Uppsala, Sweden, [2]Swedish Dairy Association, SLU, P.O. Box 7023, 75007 Uppsala, Sweden

A random regression test-day model was used to estimate genetic parameters for two measures of udder health: the logarithm of the concentration of cells in the milk (somatic cell score, SCS) and the logarithm of total amount of cells in the milk (CELL). Monthly test-day records were available from 52 261 first parity cows calving between 1999 and 2000. The data included on average 9.6 records per cow. A sire model was used, including fixed effects of calving age and herd-test-day, a random permanent cow effect and a random genetic effect of sire. The best model used a third-order Legendre polynomial for the sire effect and a fourth-order Legendre polynomial for the cow effect. The heritabilities ranged from 0.07 to 0.10 over lactation, and were similar for the two traits - slightly higher for CELL than for SCS. For both traits, the heritability increased towards the end of lactation. The genetic correlation between different days in lactation varied from 0.37 to 1 for SCS and from 0.53 to 1 for CELL. Preliminary results indicate that the total amount of cells in the milk may have a stronger genetic correlation with mastitis, compared with somatic cell score. We will continue to investigate this relationship further.

Session 10 Poster 50

Heritability of supernumerary teats in Turkish Saanen goats

M. Brka[1], N. Reinsch[2], C. Tölü[3] and T. Savas[3], [1]Faculty of Agriculture and Food Sciences, University of Sarajevo, Institute for Animal Breeding, Zmaja od Bosne 8, 71000 Sarajevo, Bosnia and Herzegowina, [2]Research Institute for the Biology of Farm Animals (FBN), Genetics & Biometry, Wilhelm-Stahl-Allee 2, 18196 Dummerstorf, Germany, [3]Çanakkale Onsekiz Mart University, Department of Animal Science, Campus, 17020, Çanakkale, Turkey

Supernumerary teats (hyperthelia) are an undesired condition because of interferences with the milking process and possible negative side-effects on udder health. In dairy goats there is, however, not t much research on the incidence and heritability of SNT. The incidence of supernumerary teats was investigated on 398 Turkish Saanen Goats, all of them from a single experimental herd. Animals were progeny of 25 bucks and born in a nine-year period from 1997 to 2006. Inspection for the presence of supernumerary teats was done at an early age of one to two weeks. The total number of affected animals was 68 (17%), 28 of them were male and 39 were female. The average inbreeding coefficient off the affected animals was 11%. Presence or absence of supernumerary teats was trated as a binary trait. A mixed linear model was used in order to estimate its heritability in the experimental population. REML-estimates were obtained by employing the VCE-package. The estimate for the heritability was 0.34 with a standard error of 0.07. This is in a similar range as has been reported earlier for dairy cattle.

Session 10 Poster 51

Genetic diversity of six Italian local chicken breeds

E. Zanetti, C. Dalvit, M. De Marchi and M. Cassandro, University of Padova, Department of Animal Science, Viale Università 16, 35020, Legnaro, Italy

Molecular genetic methods are more and more involved in preservation of endangered breeds of domestic animals. Genetic relationships of various chicken breeds, native of Veneto Italian region, were studied on the basis of microsatellite DNA polymorphisms. A total of 329 DNA samples, extracted from blood obtained from 6 local chicken breeds (45 Robusta Lionata, 43 Robusta Maculata, 45 Ermellinata di Rovigo, 45 Pèpoi, 90 Polverara, 49 Padovana) and a commercial broiler line (12 Golden Comet), were analysed using 19 microsatellite markers included in the list of recommended primers for chicken analysis suggested by the FAO. The PCR primers, used to detect short repeat length polymorphisms, were synthesized and typed by PCR amplification and electrophoresis using a DNA sequencer. Based on the frequencies of alleles, observed and expected heterozygosity and deviations from Hardy-Weinberg equilibrium were calculated. In addition, Nei's standard genetic distances between breeds were estimated. The results suggested that microsatellite DNA markers are a useful tool to investigate genetic variability and to support its maintenance, moreover they permit to study the genetic relationships among the local chicken breeds examined, and to implement genetic traceability systems.

Session 10 Poster 52

Influence of sire genotype for the beef production sub index on progeny performance

A.M. Clarke[1,2], M.J. Drennan[1], D.A. Kenny[2] and D.P. Berry[3], [1]Teagasc, Grange Beef Research Centre, Dunsany, Co. Meath, Ireland, [2]School of Agriculture, Food Science & Veterinary Medicine, University College Dublin, Belfield, Dublin 4, Ireland, [3]Teagasc, Moorepark, Dairy Production Research Centre, Fermoy, Co. Cork, Ireland

Irish beef sires are ranked using genetic indexes on component traits for profitability. The aim of this study was to quantify potential differences in progeny of high and low genetic merit sires ranked for overall beef production. In 2005, progeny of 22 sires of high (n=11) and low (n=11) merit were purchased at weaning from commercial beef herds. Their bull progeny were finished on an ad-lib barley diet and slaughtered at 16 months of age. Phenotypic progeny performance for feed intake, live-weight, residual feed intake and carcase traits were regressed, using mixed models, on sire genetic merit for the performance index and sire predicted transmitting ability for carcass weights (PTA_{cwt}); sire was included as a random effect. An increase in index value (€) or PTA_{cwt} had no significant effect on energy intake, residual feed intake, average daily gain, carcass conformation and fat score, carcass meat and fat proportion or killout. However, a 1 kg increase in PTA_{cwt} resulted in an increase ($P < 0.05$) of 2.38 kg in slaughter weight and 1.31kg in carcass weight. Although not significant increases in the index also had positive effects on slaughter and carcass weight.

Session 10 Poster 53

QTL detection for muscle cholesterol content and fatty acid composition in a Duroc population

R. Quintanilla[1], I. Díaz[2], D. Gallardo[3], J. Reixach[4], J.L. Noguera[1], O. Ramírez[3], L. Varona[1], R. Pena[1] and M. Amills[3], [1]IRTA, Genètica i Millora Animal, Rovira Roure 191, 25198 Lleida, Spain, [2]IRTA, Tecnologia dels Aliments, Finca Camps i Armet, 17121 Monells, Spain, [3]UAB, Ciència Animal i dels Aliments, Edifici V, 08193 Bellaterra, Spain, [4]Selección Batallé SA, Segadors s/n, 17421 Riudarenes, Spain

The amount and composition of intramuscular fat (IMF) have a strong influence on the sensory, nutritional and technological properties of meat. We have carried out a QTL study in pigs dealing with fatty acid composition and cholesterol content of muscles *Longissimus Dorsi (LD)* and *Gluteus Medius (GM)*. A population of 370 castrated males belonging to five half-sib families was generated from a commercial *Duroc* line. Percentage of IMF, fatty acid composition in the C12-C22 interval and cholesterol content were determined in samples of *LD* and *GM* of each individual. All individuals plus the five parental boars were genotyped for 110 microsatellites covering the whole genome. QTL analyses revealed several genomic regions with significant effects on both *LD* and *GM* fat: on SSC 1, 3 and 6 for IMF percentage, on SSC 11 for cholesterol and palmitic content, on SSC 6 and 7 for vaccenic, and on SSC 12 and 14 for estearic acid. Moreover, significant QTL related with oleic and linoleic acids content were detected on SSC 3 and 15 for *GM*, and on SSC 2 and 1 for *LD*. Highly significant QTL involved in *GM* IMF were also shown on SSC 7, 8 and 18, and related with *LD* cholesterol content on SSC 6 and 18.

Session 10 Poster 54

Candidate genes affecting morphology and life performance traits in cattle

F. Napolitano[1], G. Catillo[1], G. De Matteis[1], L. Pariset[2] and A. Valentini[2], [1]CRA-Istituto Sperimentale per la Zootecnia, via Salaria 31, 00016 Monterotondo, Italy, [2]Università della Tuscia, via S. C. De Lellis, 01100 Viterbo, Italy

We selected 9 candidate genes, involved in the regulation of lipid metabolism, that were assumed to affect growth, conformation and life performance parameters: Signal Transducer and Activator of Transcription 5b(STAT5b), Activin A Receptor (ACVR2B), Peroxisome Proliferator-Activated Receptor-Alpha (PPARA) and Insulin-like Growth Factor (IGF1), for their involvement in the regulation of growth, skeletal and muscle development; Diacylglycerol O-acyltransferase2 (DGAT2), Diacylglycerol O-acyltransferase 1 (DGAT1) and Stearoyl-CoA Desaturase (SCD), for their role in mediating and regulating fatty acid synthesis; Fatty Acid-Binding Protein 4 (FABP4) and Plasminogen Activator Inhibitor 1 (PAI1), for their role in regulating adipogenesis, feeding intake and obesity. In each gene, the presence of at least one Single Nucleotide Polymorphism (SNP) had been identified. Two groups of sires were selected in the database of the International Genetic Evaluation among those who had either the highest (13) or lowest (14) genetic merit for milk fat percentage. We tested the allelic effect at each SNP on the bull genetic merit of type and functional traits. We found that the SNP in SCD and PAI1 influenced udder functionality; the SNP in FABP4 influenced rump depth and milkability; the SNP in DGAT1, PPARA and ACVR2B influenced longevity; the SNP in DGAT2 influenced calving ease.

Association between LGB polymorphism and bovine milk fatty acids composition: preliminary results

G. Conte[1], M. Mele[1], S. Chessa[2], A. Serra[1], B. Castiglioni[3], A. Caroli[4], G. Pagnacco[2] and P. Secchiari[1], [1]Univ. of Pisa, DAGA, via del Borghetto,80, 56100 Pisa, Italy, [2]Univ. of Milan, VSA, via Celoria,10, 20133 Milano, Italy, [3]CNR, IBBA, via Bassini,15, 20133 Milano, Italy, [4]Univ. of Brescia, DSBB, v.le Europa,11, 25123 Brescia, Italy

LGB polymorphism was mostly studied, to check particular effects on ruminant milk production traits. Results obtained were controversial and did not resolve the effective biological role of BLG. The similarity of BLG with lipocalins suggests a possible role of BLG as apolar molecules carrier, particularly fatty acids (FA) and retinol. Two forms of BLG protein are the more diffuse in bovine milk: A and B. *LGB*A* differs from *LGB*B* in the amino acid sequence at position 64 ($Asp_A \rightarrow Gly_B$) and 118 ($Val_A \rightarrow Ala_B$). The aim of this paper was to check a possible association between BLG polymorphim and bovine milk FA composition. Milk BLG phenotyping was obtained by IEF method. We classified 468 Holstein Italian Frisian into three phenotypes: AA, AB and BB. Their frequencies were 25.85%, 46.15%, 28% respectively. The phenotypes were compared with individual milk FA composition. Milk from cows with BLG AA phenotype showed a higher satured (SFA) and medium chain FA (MCFA) content. These differences were due mainly to palmitic acid (PA) content, the most representative milk fatty acid of these FA classes. The significant association between phenotype and milk FA suggests a possible role of BLG as carrier of FA in mammarian cells, especially for PA.

Breeding for health traits in Norsvin Landrace and Norsvin Duroc

I. Andersen-Ranberg and H. Tajet, Norsvin, Department of Animal and Aquacultural Sciences, Norwegian University of Life Sciences, P.O. Box 5003, 1432 Ås, Norway

In January 2007, Norsvin started a four year project, “Breeding for better health in pigs”. One of the objectives of this project is to estimate genetic parameters for some selected health traits in pig production. In the first study, four health traits were analyzed; scrotal hernia, umbilical hernia, cryptorchism, and arthritis in piglets. Approximately 614 000 records from Norsvin Landrace nucleus herds were analyzed with single trait linear animal model. The animals were born between 1993 and 2006. Absence and presence of each of the four traits were scored as 0 and 1, respectively. In each of the four models, effects of herd-year, parity, month of birth, number of live born piglets in each litter and animal were included. Heritabilities for scrotal hernia, umbilical hernia, cryptorchism, and arthritis in piglets were 0.26, 0.09, 0.09 and 0.30, respectively. The results are probably influenced by the accuracy of the recordings. Higher accuracy could increase the heritability estimates. In March 2007, scrotal hernia, umbilical hernia, cryptorchism, and arthritis in piglets were introduced as part of the breeding goal for Norsvin Landrace and Norsvin Duroc.

Association between a leptin gene polymorphism with milk production traits in dairy cattle

L. Orrù, F. Napolitano, G. De Matteis, G. Catillo, M.C. Savarese, M.C. Scatà and B. Moioli, Istituto Sperimentale per la Zootecnia (CRA), Via Salaria 31, 00016 Monterotondo (Roma), Italy

Leptin is an hormone that plays an important role in energy metabolism, growth and reproduction. Evidence of association between C1180T single nucleotide polymorphism (SNP), a transition responsible of the Arg25Cys substitution, with productive traits has been reported in literature. In the case of milk and protein yield however, contradictory results were obtained by Buchanan *et al.* (2003), who showed higher milk yield for TT *vs* CC genotypes, and Madeja *et al.* (2004), who did not found any association with either milk or protein yield. In the present study, 75 cows of the Jersey, Piedmontese and Valdostana breeds, as well as 27 Holstein bulls, were genotyped at this SNP. In cows, a significant positive effect of allele C was made evident on daily milk yield (+1.42 kg; $P < 0.01$) but not on protein and fat percentage. A positive, although not significant, effect of allele C was also evident on the bull genetic merit for milk yield (+243 kg). The contradictory results between the present results and Buchanan's findings could be explained because our trials were performed on different breeds, therefore some interaction between C1180T polymorphism and breed might be inferred.

Study of genetical habit of whole bood potassium and gene frequency of L&H and it's relationship with other blood electrolytes in Zel sheep in Iran

H. Moradi Shahrbabak[1], M. Moradi Shahrbabak[1] and G.H. Rahimi[2], [1]University of Tehran, Faculty of agriculture, Animal Science, p.o.box:4111, 31587-11167 Karaj, Iran, [2]University of Mazandran, Faculty of agriculture, Animal Science, 31587-11167 Sari, Iran

The whole blood potassium concentration has shown the bimodal distribution in sheep, which has been classified into LK and Hk types. HK allele is recessive to LK with a single gene inheritance. This polymorphism showed different behavior in different environment, Which could be due to adaptation process. This Research was conducted on the Zel breed sheep, Which has been located in Mazandran(Sari) with a moderate climate.Blood potassium concentration ranged from 183/15 to 480/1ppm in Zel sheep and the Mean value of blood potassium concentration of Zel LK Animals was 277/37.The ferquency of LK gene was found to be 1 in Zel sheep.The relationship between potassium and sodium concentration in whole blood of sheep was significant. And negative estimated correlation around –0.35 which was significant.The mean of whole blood sodium concentration was 2806/1 ppm for LK sheep.

Session 10 Poster 59

Effect of the TG gene polymorphism on the intramuscular fat deposition in Hungarian beef cattle

I. Anton[1], K. Kovács[1], L. Fésüs[1], L. Lehel[1], Z. Hajda[1], P. Polgár[2], F. Szabó[2] and A. Zsolnai[1], [1]Research Institute for Animal Breeding and Nutrition, Genetics, Gesztenyés út 1., 2053 Herceghalom, Hungary, [2]University of Veszprém, Georgikon Faculty, Deák Ferenc u. 16., 8360 Keszthely, Hungary

Intramuscular fat content, also known as marbling of meat, represents a valuable beef quality trait. The effect of a 5'-polymorphism of TG (thyroglobulin) gene has been proved to affect intamuscular fat content in cattle. The objective of this study was to evaluate the effect of the TG locus on beef quality traits in some cattle breeds in Hungary. 60 blood samples have been collected from Red Angus, Hungarian Fleckvieh, Charolais and Limousin breeds and genotypes were determined by PCR-RFLP assay. Animals had been slaughtered, beef quality traits data were determined and statistical analyses have been carried out to find association between genotypes and intramuscular fat deposition. This project was supported by the Hungarian Scientific Research Fund (Project T048947).

Session 10 Poster 60

Genetic analysis of superovulatory response in Swedish Red and Swedish Holstein heifers

S. Eriksson[1], M. Häggström[1] and H. Stålhammar[2], [1]Swedish University of Agricultural Sciences, Dept. of Animal Breeding and Genetics, P.O. Box 7023, SE-750 07 Uppsala, Sweden, [2]Svensk Avel, Örnsro, 532 94 Skara, Sweden

The response to superovulatory treatment in terms of total number of embryos and number of transferable embryos per flush was studied for heifers in the Swedish nucleus herd Viken. The data was recorded during 2004 to 2006 and contained 761 treatments on 423 animals (212 Swedish Red, and 211 Swedish Holstein), sired by 90 bulls. Information on other fertility traits was available for 418 of these animals. The average was 6.11 (SD 5.63) embryos and 3.55 (SD 4.03) transferable embryos per flush, non-responders included. Both number of embryos and quality of embryos increased in the herd during the studied period. The effect of breed was significant, with higher total number of embryos in Swedish Red than in Swedish Holstein. The data was analysed using a linear animal model including fixed effects of breed, age, treatment number and year-season. Estimated repeatabilities were 0.40 and 0.44 for number of embryos and number of transferable embryos, respectively. Corresponding heritabilities were estimated at 0.11 and 0.12. Genetic and environmental correlations between total number of embryos and number of transferable embryos were strong, whereas weak correlations were estimated between number of transferable embryos and number of discarded embryos. No clear phenotypic relationships between superovulatory response and other fertility traits were detected in this data.

Session 10 Poster 61

Principal components approach for estimating heritability of mid-infrared spectrum in bovine milk

H. Soyeurt[1,2], S. Tsuruta[3], I. Misztal[3] and N. Gengler[2,4], [1]FRIA, Rue Egmont 5, 1000 Brussels, Belgium, [2]Gembloux Agricultural University, Animal Science Unit, Passage des Déportés 2, 5030 Gembloux, Belgium, [3]University of Georgia, Animal & Dairy Science Unit, 425 River Road, 30605 Athens, USA, [4]FNRS, Rue Egmont 5, 1000 Brussels, Belgium

Mid-Infrared (MIR) spectrometry can be used for prediction of milk composition. The data included 9,663 records on 1,937 cows in 1 to 12 parity recorded from April 2005 to May 2006. Each sample was scanned by MilkoScan FT6000 into 1,060 points. Principal components approach (PCA) indicated that 48 principal components (PC) described 99.02% of information. The next analysis considered PC constructed for 2,850 first lactation records of 738 cows in 7 breeds from 26 herds. PC were analyzed by multi-trait REML using the canonical transformation..Effects included were herd*test date, lactation stage, permanent environmental and animal. Heritabilities varied from 0.005% to 57.20% for different pin numbers. Spectral regions with heritability greater than 5% were located between 1 to 181; 194 to 558 and 709 to 1,060 pin numbers. PCA involving points in those regions demonstrated that only 9 PC explained 99.23% of information. MIR spectrum contains specific regions with substantial genetic information potentially useful for selecting improved milk quality directly on spectral data.

Session 10 Poster 62

Variation of lactoferrin (LTF) content predicted by mid-infrared spectrometry (MIR)

H. Soyeurt[1,2], F. Colinet[1], V. Arnould[1], P. Dardenne[3], I. Misztal[4] and N. Gengler[1,5], [1]Gembloux Agricultural University, Animal Science Unit, Passage des Déportés 2, 5030 Gembloux, Belgium, [2]FRIA, Rue Egmont 5, 1000 Brussels, Belgium, [3]Agricultural Walloon Research Centre, Quality Unit, Chaussée de Namur 24, 5030 Gembloux, Belgium, [4]University of Georgia, Animal & Dairy Science Unit, 425 River Road, 30605 Athens, USA, [5]FNRS, Rue Egmont 5, 1000 Brussels, Belgium

Methods to measure LTF do not permit the use of this data in genetic. The aim of this study was to create an equation to predict LTF (pLTF) in milk by MIR. Calibration with Partial Least Squares on 69 samples showed a cross-validation coefficient of determination equal to 75.01%. So, pLTF for 7,690 samples collected from April 2005 to April 2006 were predicted. Milk yields, %fat, %protein and somatic cells count (SCS) were added. 39,441 records on 1,910 cows were analyzed by a multi-trait mixed model including as effects: herd*test date*lactation number, lactation stage*lactation number, age*lactation number, regression on breed fractions, permanent environments within and across lactations and animal effect. Variances and standard errors were estimated by AI-REML. Number and stage of lactation influence pLTF. A difference across 7 breeds was observed. Heritability for pLTF was 19.70%. So, the genetic variability for LTF exists. The phenotypic correlation observed between SCS and pLTF was equal to 0.26. As SCS increases in presence of mastitis, this observation seems to indicate that pLTF could be an indicator of mastitis.

Session 10 Poster 63

Pedverif: pedigree verification with multiple candidate parents

J.B.C.H.M. Van Kaam[1], R. Finocchiaro[2], B. Portolano[2] and S. Caracappa[1], [1]Istituto Zooprofilattico Sperimentale della Sicilia "A. Mirri", Via Gino Marinuzzi 3, 90129 Palermo, Italy, [2]University of Palermo, Department S.En.Fi.Mi.Zo., Viale della Scienze, 90128 Palermo, Italy

A new computer program named Pedverif is able to rapidly verify normal pedigrees as well as pedigrees with multiple candidate parents and report problems. Genotype information can be used if available. The following actions are undertaken: - If needed set the sex of specified animals. - Check if sex codes need to be converted to another format. - Check if animals do not have identification equal to the code for missing identifications. - Check for and add absent candidate sires and candidate dams within a record. - Check if animals do not have a candidate parent occurring multiple times. Remove duplicate occurrences. - Sort candidate sires and candidate dams per record. - Check if each animal has just one pedigree record. Identical duplicate records are removed. Non-identical duplicate records are reported. - Check if all parents have a pedigree record. Add parents with a missing record as base parents to the pedigree. - Check and add sex of all sires and dams. - Verification of a pedigree for loops: animals who are mistakenly their own ancestor. - Verification for non-Mendelian inheritance of alleles on multiple loci. - Add a missing parent if an animal has no possible combination of the given candidate parents. - Produce an output file with all possible combinations of given (and if necessary missing) candidate parents for each animal.

Session 10 Poster 64

Genetic distances estimated from two partly overlapping microsatellite marker data sets

H. Täubert[1], D. Bradley[2] and H. Simianer[1], [1]Institute of Animal Breeding and Genetics, University of Göttingen, Albrecht-Thaer-Weg 3, 37075 Göttingen, Germany, [2]Smurfit Institute of Genetics, Trinity College Dublin, Dublin 2, Ireland

This project is the first application of our new algorithm to combine two partly overlapping microsatellite marker data sets. It is a linear regression based procedure to estimate single marker contributions to genetic distances. Two data sets with 37 and 40 breeds, typed in 19 and 20 microsatellite markers were used. The overlap between the two sets were 8 common breeds and 5 common markers. Three genetic distances were estimated: Cavalli-Sforza's chord distance, Reynolds' distance and Nei's standard distance. In the first step only breed combinations with complete marker information are used to assess the contributions of each marker to the distance. Based on these contributions linear regression factors for each marker are calculated. For the combination of breeds with missing marker information these regressions are use to estimate the missing contributions iteratively in upgrading regression coefficients with new estimated contributions. A FORTRAN-program has been developed to perform the estimations and has already been tested on simulated data. We can show in comparing phylogenetic trees based on the distance matrices of the single and combined data sets the successful estimation of missing marker contributions. The derived genetic distances correspond with the expected phylogeny of the breeds in both data sets.

Session 10 Poster 65

Bovine leptin promoters identified with different levels of expression

B.M. Kearney[1,2], M. Daly[1], F. Buckley[3], T.V. Mc Carthy[2], R.P. Ross[1] and L. Giblin[1], [1]Moorepark Food Research Centre, Teagasc, Fermoy, Co.Cork, Ireland, [2]Biochemistry Dept., UCC, Cork, Co.Cork, Ireland, [3]Moorepark Dairy Production Research Centre, Teagasc, Fermoy, Co.Cork, Ireland

Dairy cows with poor body condition after parturition are less likely to cycle and are at an increased risk of disease. Body condition score (BCS) is the appraisal of a cow's fat reserves. BSC has a high heritability and efforts are underway to unravel the genetic component of this complex phenotype. One such candidate is the adipocyte signal, leptin which signals the level of the body's fat reserves to the hypothalamus. Single nucleotide polymorphisms (SNPs) in this gene have been associated with intramuscular fat, postpartum luteal activity, body weight and feed intake. In a preliminary study we sequenced 2.4kb of the leptin promoter in 15 Holstein Friesian bulls from the Irish national herd. These sires ranked highly or lowly for BCS based on evaluations' of their daughter progeny. From the data generated, we have identified two major haplotypes which represent two distinct leptin promoters. Reverse transcriptase RT-PCR was performed on mammary tissue, liver and perirenal fat of an individual animal carrying both leptin promoter types and heterozygous for the C73T SNP in exon 2. Using allele specific primers for the C73T SNP, mRNA from each promoter could be distinguished and quantified. The promoter linked to the C allele repeatedly associated with higher leptin expression.

Session 10 Poster 66

estimates of genetic relationship among traits of growth, meat productivity and the number of teats during post weaning period in purebred Berkshire

M. Tomiyama[1], T. Oikawa[1], T. Kanetani[2] and H. Mori[2], [1]Okayama University, Agriculture, 1-1-1 Tsushimanaka, 7008530, Japan, [2]Okayama Prefectural Center for Animal Husbandry and Research, 2272 Misaki-chou, 7093494, Japan

Pork produced from three-way cross is common for daily meat market because cross bred animals have high performance due to hybrid vigor. However, meat from Berkshire is popular in Japan because of consumer's strong demand for high quality pork. The aim of this study is to initiate genetic improvement of a population of Berkshire by estimating genetic parameters for meat production traits during post weaning period. Records on 4,650 purebred Berkshire (2,393 males, 2,257 females) pigs from Okayama prefecture were used in this study. Traits studied included body weight, DG, meat productivity and number of teats. Genetic parameters were estimated by REML. Estimates of heritability for direct genetic effect for body weight, back fat thickness and loin eye area (LEA2) measured at 60 days of age, and DG from weaning to 60 days, were 0.22, 0.49, 0.22 and 0.25, respectively. For the same traits, estimates of permanent environmental effect (c^2) were 0.12, 0.18, 0.21 and 0.13, respectively. Thus, it is indicated that c^2 at A60 has large effect on those traits. Estimates of genetic correlation between loin eye area in days at finish (LEA2) and LEA1, and between LEA1 and back fat thickness in days at finish (BFT2) were 0.80 and -0.84, respectively. Thus, it is suggested that LEA1 should be included to a selection criteria.

Session 10 Poster 67

Gene expression analysis of pig muscle associated to cholesterol and fat parameters

A. Cánovas[1], R. Quintanilla[1], L. Varona[1], I. Díaz[2], J. Casellas[1] and R.N. Pena[1], [1]IRTA-Lleida, Genètica i millora animal, Rovira Roure, 191, 25198 Lleida, Spain, [2]IRTA-Monells, Tecnologia dels aliments, Finca Camps i Armet, 17121 Monells, Spain

In order to detect and identify genes involved in lipid metabolism in pigs we have used a microarray approach over muscle samples from an experimental Duroc population of 370 castrated males distributed in five half-sib families. A total of 70 *Gluteus medius* muscle samples were processed, from animals with the most extreme levels (HIGH and LOW lines; 35 animals per line) for cholesterol and fat parameters such as plasma lipoprotein and triglycerides levels, percentage of intramuscular fat and fatty acid composition in muscle. Each sample was individually hybridized using *GeneChip Porcine Genome®* arrays*(Affymetrix)*. After normalizing data with the RMA algorithm, comparison between lines was performed with a standard t-test and a Bayesian analysis by means of a mixed model with heterogeneous residual variances. The t-test results showed a total of 1007 genes significantly expressed (p-value<10^{-7}) between the HIGH and LOW lines, and 140 of these had a ratio of differential expression superior to 1.5. The mixed model analysis resulted in a total of 500 genes differently expressed at a significant level (p-value<10^{-9}), 158 of which showed a ratio between classes greater than 1.5. There was a high degree of coincidence between both analyses. The biological functions were related to a variety of processes such as: transcription factor, lipid metabolism and RNA processing.

Session 10 Poster 68

Estimation of genetic parameters for yield and marbling traits using a random regression model on slaughter age in Japanese Black cattle

T. Osawa[1], K. Kuchida[1], S. Hidaka[1] and T. Kato[2], [1]Obihiro University of A&VM, Obihiro, 080-8555, Japan, [2]Tokachi Federation of Agricultural Cooperative, Obihiro, 080-0013, Japan

The slaughter age of Japanese Black cattle in Japan ranges from about 24 to 35 mo of age, and genetic effects on the meat yields or quality might vary depending on age. The purpose of this study was to estimate genetic parameters for yields and marbling traits using a random regression model on slaughter age. 5,722 Japanese Black cattle were obtained from a carcass market, and digital images of rib eye were taken between the 6-7th rib by a photographing device. The carcass weight (CW) and rib eye area (REA) as yield traits, and marbling score (MS) by a grader, fat area ratio (FAR) and overall coarseness of marbling (OCM) by image analysis as marbling traits were analyzed. Genetic parameters were estimated with the GIBBS3F90 program. The model included sex, slaughter age and day at carcass market as fixed effects. Fattening farm was a random effect along with additive genetic effects with random regression on slaughter age using second-order Legendre polynomials and residual effects with heterogeneous variances. Heritability estimates were 0.23~0.74, 0.42~0.54, 0.32~0.59, 0.55~0.67 and 0.27~0.48 for CW, REA, MS, FAR and OCM, respectively. Genetic correlations among age were 0.87~1.00 for CW. The correlations for other traits increased with age. The minimum correlations were 0.39, 0.57, 0.61 and 0.28, for REA, MS, FAR and OCM, respectively.

Session 10 Poster 69

MC4R and FUT1 genes in sows of Large White breed in the Czech Republic

I. Vrtkova[1], J. Dvorak[1], V. Matousek[2] and N. Kernerova[2], [1]Mendel University Brno, Zemedelska, 613 00 Brno, Czech Republic, [2]University of South Bohemia, Studentska, 370 00 Ceske Budejovice, Czech Republic

A porcine melanocortin 4 receptor (*MC4R*) variant (Asp298Asn) on pig chromosome 1 is considered to be economically important. There were analysed the association of this *MC4R*polymorphism with back fat and proportion of lean meat. The oedema disease and post-weaning diarrhoea in 4-12 weeks old pigs is associated with the colonization of the small intestine by toxigenic *Escherichia coli* strain with fimbriae F18. The *FUT1* gene has been determined as a marker gene for *E. coli* F18 receptor locus. The M307 polymorphism of the *FUT1* gene influences susceptibility to adhesion of *E. coli* F18 to intestinal mucosa and an outbreak of illness. We analysed the association of *FUT1* gene with death loss of piglet. There were analysed 73 sows of Czech Large White (CLW) breed with benefit litter size. Samples of blood of pigs were analysed by PCR–RFLP methods. Frequences of genotypes of marker *MC4R* were: AA 0,44; AB 0,52; BB 0,04. Average height of back fat (mm) was: AA 9,08; AB 8,01. Average proportion of lean meat (%) was: AA 60,94; AB 61,88; BB 61,27. Frequences of genotypes of marker *FUT1* were: AA 0,04; AG 0,26; GG 0,70. Death loss of piglet on the first litter was: AG 3,16 and GG 2,29 piglets. Death loss of piglet on the second and the subsequent litters was: AG 2,28 and GG 2,51 piglets. The study was supported by grants of Ministry of Agriculture of the Czech Republic 1G58073 and QG60045 and MSM 6007665806.

Session 10 Poster 70

Genetic relationships among year classes for marbling traits in Japanese Black cattle

T. Osawa[1], K. Kuchida[1], S. Hidaka[1] and T. Kato[2], [1]Obihiro University of A&VM, Obihiro, 080-855, Japan, [2]Tokachi Federation of Agricultural Cooperative, Obihiro, 080-0013, Japan

Marbling scores (MS) in Japan are visually classified into 12 levels by grader. Thus, the scale of the classification might change by year. The purpose of this study was to estimate genetic parameters for marbling traits using a multiple-trait (MT) model and a random regression (RR) model on year. 5,986 Japanese Black cattle were obtained from a carcass market, and digital images of rib eye were taken between the 6-7th rib. The MS, fat area ratio (FAR) and overall coarseness of marbling (OCM) by image analysis were analyzed. The year effect was sorted into 5 classes (Apr/00–Mar/01, Apr/01–Mar/02, Apr/02–Mar/03, Apr/03–Sep/04 and Oct/04–Feb/06). Genetic parameters were estimated with the AIREMLF90 program for the MT model and the GIBBS3F90 program for the RR model. The MT model included sex, slaughter age and day at carcass market as fixed effects, fattening farm and additive genetic effects as random effects, and residual effects. The RR model included the same fixed effects as the MT model, random effects in the MT model with random regressions on year using second-order Legendre polynomials, and residual effects assuming heterogeneous variances. Variance component estimates were similar, and genetic correlations among year classes were high in both models. The correlations from the RR model were 0.95~0.97, 0.93~0.97 and 0.80~0.99 for MS, FAR and OCM, respectively. MS could be treated as same traits through years.

Session 10 Poster 71

The genetic polymorphism at the β-lactoglobulin gene in proximal region in the Czech goat populations

Z. Sztankoova, V. Matlova and G. Mala, Institute of Animal Science, Molecular genetics, Pratelstvi 815, 104 00 Praha Uhrineves, Czech Republic

The genetic polymorphism at the β-lactoglobulin gene in proximal region in the Czech goat populations. Beta lactoglobulin (β – Lg) is a major whey protein in the milk of ruminants, a globular protein belonging to the lipocalin family. The purpose of this work was to evaluate the genetic polymorphism of beta lactoglobulin gene in the proximal promoter region in the Czech goat breed using PCR-RFLP method. We detected the genetic polymorphism, nucleotide substitution, at position -60 (C → T) and -341 (T → C) in the proximal promoter region at β – Lg gene. The amplified fragment with size of 710 bp and 555 bp was digested by two restriction enzymes SmaI and FspbI, respectively. A total of 114 goats belonging to White Shorthair (50, WSH) and Brown Shorthair (64, BSH) goat were analyzed. Molecular analysis showed predominant variant C (0.71, WSH) and T (0.52, BSH) at position -60, and in both goat breeds, was predominant variant T (0.90, WSH and 0.87, BSH) at position -341. Experiment is in progress to evaluate the effect of these variants in the expression of influence β – Lg gene and therefore in the milk composition. This work has been supported by the Ministry of Agriculture of the Czech Republic (Institutional Programme) MZE 0002172701401 and National agency for Agriculture research (IG57051).

Session 10 Poster 72

Genetic polymorphism at CSN1S1, CSN2 and CSN3 loci in the Holstein and Czech Fleckvieh breeds kept in the Czech Republic

Z. Sztankoova, J. Soldat, T. Kott and E. Kottova, Institute of Animal Science, Molecular genetics, Pratelstvi 815, 104 00 Praha Uhrineves, Czech Republic

Genetic polymorphism at *CSN1S1*, *CSN2* and *CSN3* loci in the Holstein and Czech Fleckvieh breeds kept in the Czech Republic. The dairy industry continually strives to improve the quality of its products. The genes that encode the major milk proteins are thought of as candidate genes for the observed in protein composition. The bovine casein locus contains four milk protein genes which are closely linked, in the order *CSN1S1*, *CSN2*, *CSN1S2* and *CSN3*. Several polymorphisms have been found for each casein gene. The objective of this work was to determine genetic polymorphism at the candidate gene *CSN1S1*, *CSN2* and *CSN3* in the Holstein and Czech Fleckvieh breeds population kept in the Czech Republic. A total of 686 animals belonging to 184 Holstein (H) and 502 Czech Fleckvieh (CF) were determined by PCR-RFLP methods and using Light Cycler analysis. The results showed allele*B* was predominant in both breeds (0.965 and 0.852) at locus *CSN1S1,* allele *A2* was predominant in both breeds (0.818 and 0.644) at the *CSN2*, as well as at the locus *CSN3,* allele *A* was predominant in both breeds (0.788 and 0.640), respectively. This work has been supported by the MZE 0002172701401.

Session 10 | Poster 73

Investigation of the genetic differentiation between Greek sheep breeds for conservation purposes

C. Ligda[1], J. Al Tarayrah[2] and A. Georgoudis[2], [1]National Agricultural Research Foundation, P.O. 60 458, 57 001 Thessaloniki, Greece, [2]Aristotle University of Thessaloniki, Dept. of Animal Production, 54 006 Thessaloniki, Greece

The aim of this work was to explore the genetic diversity between the Greek sheep breeds in order to have an overview of the current status of their diversity and make suggestions for conservation strategies. The work was carried out in the frame of the Econogene project, funded by the European Union within the Quality of Life V framework programme. The genetic diversity of 10 sheep breeds was investigated on the basis of allele frequencies at 28 microsatellite markers. The Nei's coefficient of differentiation Gst, over all breeds and loci was 0.031, indicating that crossbreeding and common breeding practices, reduced the value of the coefficient differentiation. The phylogenetic tree constructed using the Reynolod's distance showed three clusters of breeds. The first group consisted of Anogeiano and Sfakia breeds, from the Crete island, the Skopelos breed formed a separate group, while the remaining breeds were grouped together. The Structure analysis, shown 7 ancestral populations, clustering in separated groups the breeds of Crete and the other island breeds, Kymi, Kefallenias, Lesvos, Skopelos and also the Peliou breed, which is an isolated population not mixed with other breeds of the region. The results presented are in accordance with the geographical location of the breeds, the history of the origin of the breeds and the breeding practices.

Session 10 | Poster 74

Genetic parameters for chronic respiratory diseases and immune traits in Landrace pigs

W. Onodera[1], T. Kachi[1], H. Kadowaki[2], C. Kojima[2], E. Suzuki[2] and K. Suzuki[1], [1]Tohoku University, Faculty of Agriculture, Sendai, 981-8555, Japan, [2]Miyagi Prefecture Animal Industry Experiment Station, Miyagi, 989-6445, Japan

This study was conducted to estimate genetic parameters for the change to a morbid state by atrophic rhinitis (AR) and mycoplasma pneumonia (MPS), and immune traits on 807 Landrace pigs. Immune traits viz. phagocytic capacity (PC), complement alternative pathway activity (CAPA), total leukocyte (WBC), and ratio of granular leukocytes to lymph cells (RGL) were measured from blood collected at 7-w of age and 105-kg body weight. Sheep erythrocyte specific IgG (sIgG) and cortisol level (CL) were also examined at 105-kg body weight. The morbid state of AR and MPS were visually evaluated after slaughtering the pigs at 105-kg body weight. Heritability for AR was moderate (0.36) and for MPS was low (0.14). Immune traits at 7-w of age were influenced by common environmental effect. Heritabilities for immune traits at 7-w of age and at 105-kg body weight were low to moderate (0.04 to 0.37). Genetic correlation of AR with MPS was moderate (0.32). Genetic correlations of AR and MPS with immune traits at 7-w of age ranged from -0.43 to 0.33. Genetic correlation between AR and CAPA was negative and high (-0.52), and between MPS and CL was positive and high (0.60) at 105-kg body weight. Present results indicate that immune traits are largely influenced by common environment at 7-w of age, and CAPA and CL are important traits that were genetically correlated with diseases.

Joint effects of CSN3 and LGB loci on breeding values of milk production parameters in Czech Fleckvieh

J. Matejickova[1], A. Matejicek[2], E. Nemcova[1], O.M. Jandurova[3], M. Stipkova[1] and J. Bouska[1], [1]Research Institute of Animal Production, Pratelstvi 815, 104 01 Prague – Uhrineves, Czech Republic, [2]University of South Bohemia, Studentska 13, 370 05 Ceske Budejovice, Czech Republic, [3]Research Institute of Crop Production, RSV Karlstejn 98, 267 18 Karlstejn, Czech Republic

The aim of this study was to estimate the joint effects of *CSN3* and *LGB* loci on breeding values of milk production parameters. *CSN3* (kappa-casein) and *LGB* (beta-lactoglobulin) genotypes of 120 Czech Fleckvieh sires were detected using the PCR-RFLP method. Breeding values of sires were obtained from the Official Database of Progeny Testing. Ten genotype combinations were detected. Genotypes *ABAB* (25.0 %), *ABAA* (13.3 %) and *ABBB* (13.3 %) were the most frequent. Significant effects of genotype combinations on breeding values for fat and protein content were found. The highest breeding values for milk (+621 kg) and protein (+15.8 kg) yields were associated with genotype combination *ABAA*, while the highest breeding values for content parameters (+0.15 % for protein content and +0.55 % for fat content) were associated with genotype combination *BBAB*. In comparison with the results of our previous studies focusing on separate testing of *CSN3* and *LGB* loci, the results of the current study refer to the advantage of comparing the effects of all genotype combinations. This finding brings a clearer view on loci effects and helps to simplify decisions useful for breeding. This study was supported by project MZE 0002701402.

Investigation of MEL1 gene in different sheep breeds: new polymorphisms and haplotype reconstruction

M. Árnyasi[1], G. Novotni Dankó[1], S. Lien[2], L. Czeglédi[1], G.Y. Huszenicza[3], A. Kovács[1] and A. Jávor[1], [1]University of Debrecen, Animal Sciences, Böszörményi str. 138, 4032 Debrecen, Hungary, [2]Norwegian University of Life Sciences, Centre of Integrative Genetics, P.O.Box 5003, 1432 Aas, Norway, [3]Szent István University, Obstetrics and Reproduction, István str. 2, 1078 Budapest, Hungary

In Hungary most of the sheep breeds have cyclic ovarian activity from August till January. It means that the milk and meat production is not continuous during the year even though the market would require it. Therefore use of a polymorphic genetic marker in the selection program for continuous milk and meat production would be particularly advantageous in sheep farms. Melatonin is known to be a key factor influencing seasonal reproduction in sheep. Our aim was to search new polymorphisms, which can be used in the selection. DNA samples were collected from nine different breeds with different capability for aseasonality. The sequencing of 294 individuals identified a total of 16 single nucleotide polymorphisms (SNPs) of which only 10 have been previously reported. Six of the 16 SNPs cause amino acid changes in the protein. Nineteen haplotypes were reconstructed from the sequence data. The frequencies of the haplotypes were calculated in case of each breed. Further study will be to genotype the ewes for these SNPs and make association between the haplotypes and the out of season ovarian function.

Session 10

Poster 77

Strain of Holstein-Friesian cow affects peripartum immune function

G. Olmos[1,2], L. Boyle[1], B. Horan[1], D.P. Berry[1], J.F. Mee[1] and A. Hanlon[2], [1]Moorepark Dairy Production Research Centre, Teagasc, Fermoy, Co. Cork, Ireland, [2]School of Agriculture, Food Science & Veterinary Medicine UCD, Belfield, Dublin 4, Ireland

Haematological profile and acute phase proteins (APP) are non-specific immune indicators of animal health. The objective of this study was to evaluate the effect of three strains [selected on an economic breeding index (EBI)] of spring-calving Holstein-Friesian cows (n=126) on peripartum APP and haematological profiles. Strains were fed either a high concentrate (2.8 LU/ha and 1,200 kg concentrate/cow/year) or low concentrate (2.6 LU/ha and 500 kg concentrate/cow/year) grass-based diet. Blood samples were taken at -24, +3 and +35 days relative to calving. Box-cox transformations were used to normalize data when necessary. Data analysis was done using mixed models for repeated measures. Fixed factors included were: strain, feed system and parity. Neither strain nor feed system had an effect on APP ($P > 0.05$). However, the effect of strain on white blood cell counts (WBC) and total neutrophils differed depending on time relative to calving ($P < 0.05$); in contrast to the cows in the lower EBI strains, WBC and total neutrophil counts increased from pre-calving to calving in the highest EBI strain cows. Thus it was concluded that high EBI strain cows had improved peripartum immune function.

Session 10

Poster 78

Evidence of domestication of Korean wild boars

I.-C. Cho[1], S.-H. Han[1], S.-S. Lee[1], M.-S. Ko[1] and J.-T. Jeon[2], [1]National Institute of Subtropical Agriculture, Livestock division, 1696, O-deung dong, Jeju, 690-150, Korea, South, [2]Gyeongsang National University, Division of Applied Life Science, 900, Gajwa dong, Jinju, 660-701, Korea, South

The mtDNA D-loop region of Korean wild boar and Jeju (Korea) native pig was sequenced in order to determine evidence for pig domestication in the Northeast Asia area. Sequences of Asian wild boar and Chinese pig breeds reported in previous studies were also used. A total of 68 Asian wild boars and 60 Asian domestic pigs were analyzed in this study. Genetic variations were detected at 82 positions. Seventy two (72) haplotypes were identified within wild boar and domestic pig and two haplotypes were shared. The greatest distance between domestic and wild boar was found between haplotype D8 and W20, while the shortest was found between haplotypes D16 and W35, and haplotypes D32 and W36 (0, 100% identical). This study proposed a clear evidence for pig domestication to have occurred from wild boar subspecies in Northeast Asia. Particularly, two Jeju native pigs showed 100% sequence identical with six Chinese domestic pigs and one Chinese wild boar. Also, one of the four groups of Korean wild boar was similar with Asian domestic pigs taking into consideration sequence polymorphism and genetic distance. Consequently, this indicates that only minimum number of many wild boar sub-species inhabiting the Asia Continent contributed to pig domestication.

Session 10 Poster 79

A novel mutation in the promoter region of A-FABP gene and association with carcass traits in pigs

S.H. Han, I.C. Cho, S.S. Lee and M.S. Ko, National Institue of Subtropical Agriculture, Rural Development Administration, Livestock Division, San 175-6 Odeung-dong, Jeju, 690-150, Korea, South

The association was tested between the genotypes of porcine adipocyte fatty acid binding protein (*A-FABP*) gene and the carcass traits in commercial pigs. Interestingly, only two kinds of genotypes (-406*T/T* and *T/C*) for *A-FABP* gene were detected using *Hin*f-RFLP for the substitution mutation T-406C on the promoter region of *A-FABP* gene in the pigs examined but the homozygote -406*C/C* did not. The reason of the lack of the genotype -406 *C/C* is due to the breeding system because only Duroc which usually used as a terminal sire has highly polymorphic in this gene among three breeds involved in pig production. The pigs containing the genotype -406*C/*- were significantly associated with increase the intramuscular fat content and carcass weight ($p < 0.05$), but did not with the other carcass traits(fat composition, color score, texture score, moisture, and separation score between muscles) tested. This study suggested that the genotype -406C/- of porcine *A-FABP* gene may not only be an useful molecular marker for the intramuscular fat but also contribute for improvement of meat quality by production of well-marbled pigs in the molecular breeding using the animals containing this genotype, especially Duroc as a terminal sire for commercial pigs.

Session 10 Poster 80

An optimized marker set for paternity control in sheep

J. Buitkamp and J. Semmer, Bavarian State Research Center for Agriculture, Institute of Animal Breeding, Prof.-Duerrwaechter-Platz 1, 85586 Poing, Germany

In the past paternity testing in sheep mainly relied on blood typing in Germany. Since the information contents of the systems available were not satisfying, an ovine microsatellite set for paternity control was developed. The criteria for selecting markers were cost efficiency, robustness of typing, distribution over different chromosomes and degree of polymorphism in the main German breeds. Microsatellites were chosen mainly from the marker panel typed on the International Mapping Flock (IMF, AgResearch, NZ) and tested on reference samples of unrelated sheep from the main German breeds. Finally, eleven microsatellites were selected according to their exclusion probabilities, absence of 0-alleles and suitability for multiplex PCR as well as the feasibility of automated allele calling on a fluorescent sequencer. The whole set of markers is amplified in two multiplex reactions that can be pooled and analyzed by one capillary injection on a sequencer. Data were generated from several hundred paternity cases on an ABI3100 genetic analyzer. Exclusion probabilities were calculated from parental data. The total exclusionary powers were well above 0.995 (first and second parent as well) for all main breeds (Merinolandschaf, Suffolk, Bavarian Schwarzkopf) investigated.

Session 10 Poster 81

Relationships between morphological characteristics of teat and mastitis in three dairy cattle breeds in the tropics

M. Riera Nieves[1], J.C. Alvarez[2] and R. Rizzi[3], [1]Universidad del Zulia, Av. Goajira, Maracaibo, Venezuela, [2]ASOCRICA, Av. F. Miranda, Carora, Venezuela, [3]University of Milan, Via Celoria 10, Milan, Italy

Morphological characteristics of teats and the presence of subclinical mastitis were studied in 46 Holstein, 33 Jersey and 439 Carora cows in eleven farms of Venezuela during a period of ten months. Teat shape was classified as funnel, cylindrical and bottle and teat-end was classified as round, pointed, plane prolapsed disk, inverted and mixed. The length of the teat was measured from the base of the teat to the teat apex. The diameter of teat was measured at middle high on the teat. The presence of subclinical mastitis was determined by California Mastitis Test. A total of 1983 records were processed by an animal model including fixed factors (breed, farm, calving month, age, stage of lactation, udder quarter) and random factors (additive genetic effects and non-additive plus permanent environmental effect). Heritability estimates of presence of mastitis, teat shape, teat-end shape, teat length, teat diameter were.33, 0.25, 0.03, 0.07, 0.08, respectively. The genetic correlations between the presence of mastitis with teat-end shape, teat length an teat diameter were 0.30, 0.18 and-0.23, respectively. Results suggest that cows with teat-end prolapsed or inverted or with long and thin teats are prone to mastitis. A negative correlation (-0.39) was found between teat length and teat diameter.

Session 10 Poster 82

Influences on the response to selection for a QTL trait in a multiple-trait scenario

K.F. Stock and O. Distl, University of Veterinary Medicine Hannover, Institute for Animal Breeding and Genetics, Bünteweg 17p, 30559 Hannover, Germany

Simulated data were used to compare responses to different modes of single- (STS) and multiple-trait selection (MTS) with regard to a binary QTL trait in dependence on heritability level and proportion of genetic variance explained by the QTL. Fixed effects, residual and additive genetic variances for one continuous trait (T1) and liabilities for four categorical traits (T2 to T5) were simulated. QTL effects were simulated for T2 with QTL variance explaining 50% or 10% of the total genetic variance. Heritabilities of 0.50 (T1), 0.20 or 0.40 (T2) and 0.20 (T3, T4, T5) and additive genetic correlations of +0.40 (T1-T2) and ±0.20 (T2-T3, T2-T4, T4-T5) were simulated. Prevalences of dichotomized traits were 0.25 (T2, T3, T5) and 0.10 (T4).Phenotypes or phenotypes and genotypes of 700 individuals and their parents were considered for genetic analyses in mixed linear-threshold animal models using Gibbs sampling. Higher heritability increased selection response on T2, but the impact of relatively larger QTL variance was similar at both levels of heritability. Using RBV only, about 50% of sires were selected and T2 prevalence was reduced by 46-55% (STS) and 26-31% (MTS). Relative decrease of T2 prevalence was largest if genotype and polygenic RBV were jointly considered for sire selection (STS 58-76%, MTS 32-48%), but only 12-15% of sires had above-average RBV and homozygous genotype for the favorable QTL allele.

Session 10 | Poster 83

Power of QTL mapping using both phenotype and genotype information in selective genotyping

S. Ansari Mahyari and P. Berg, University of Aarhus, Dept. Genetics and Biotechnology, Research Center Foulum, 8830 Tjele, Denmark

Selective genotyping is traditionally only based on the extreme phenotypes (EP). EP was compared with criteria based on both phenotypic and genotypic information to find informative animals for genotyping. Alternative strategies were to minimize the error of the estimated QTL effect (MinERR) and to maximize the likelihood ratio test criteria (MaxLRT). In a stochastic simulation using 30 half-sib families and 120 daughters per family in a daughter design, a QTL was positioned within a previously defined genomic region with 10, 20 and 30% genotyping levels. Power to detect QTL was significantly higher for both MinERR and MaxLRT compared to EP and random genotyping methods (either across or within family), at all genotyping levels. Power (α=0.01) with 20% genotyping for MinERR (MaxLRT) was 80 (75)% of that obtained with complete genotyping compared to 70 (38)% for EP within and across families. With 30% genotyping the powers were changed to 78 (83)% and 78 (58)%, respectively. Power of these proposed strategies showed significantly increased compared to random genotyping and genotyping the extremes across families, and marginally increased power compared to genotyping extremes within family. The lowest sensitivity in detecting the correct location was with random genotyping and the maximum was obtained with the MaxLRT strategy. However, MinERR approach showed more percentage of simulations with a significant QTL (α = 0.01) after 20% genotyping.

Session 10 | Poster 84

Ant colony optimization as a method for strategic genotype sampling

M.L. Spangler[1], K.R. Robbins[1], J.K. Bertrand[1], M.D. Macneil[2] and R. Rekaya[1], [1]University of Georgia, Animal and Dairy Science, Department of Animal and Dairy Science, Athens, GA 30602, USA, [2]USDA-ARS, Fort Keogh Livestock and Range Research Laboratory, Miles City, MT 59301, USA

A simulation study was carried out to develop an alternative method of selecting animals to be genotyped and to compare the proposed method to other known procedures. The simulated pedigrees included 5,000 animals and were assigned genotypes for a bi-allelic gene based on assumed allelic frequencies of 0.7/0.3, and 0.5/0.5. Real beef cattle pedigrees of varying sizes were used to test selected methods using simulated genotypes. The proposed method of ant colony optimization (ACO) was evaluated based on the number of alleles correctly assigned (AK_P), the probability of assigning true alleles (AK_G), and the probability of correctly assigning genotypes (APTG). The proposed technique was compared to other known methods of genotype sampling, such as selection based off of the diagonal element of the inverse of the relationship matrix (A^{-1}). Comparisons of these two methods showed that ACO yielded an increase in AK_P ranging from 4.98 to 5.16% and an increase in APTG from 1.6 to 1.8% using simulated pedigrees. Gains in real beef cattle pedigrees were slightly lower. These results suggest that ACO can provide an optimal genotyping strategy with different pedigree sizes and structures.

Session 10 Poster 85

Considerations on experimental design of a laying hen performance field test

H. Glawatz and N. Reinsch, Research Institute for the Biology of Farm Animals, Wilhelm-Stahl-Allee 2, 18196 Dummerstorf, Germany

International laying hen performance test results are published in company reports, capacities for independent line-comparisons are limited. Possible genotype-environment-interactions must be taken into account when interpreting station data. Therefore field tests are discussed for evaluating the suitability of layers for egg production, particularly in non-cage housing systems such as organic farms. We consider some options for experimental field test designs with regard to practical limitations and statistical power. A block design with farm as a block effect seems to be adequate. Taking the usual size of organic farms and other organisational aspects into account, a block size of two different origins seems to be feasible. Using this block size the comparison of three lines on 32 farms detects a line difference of one standard deviation (experimental power: 80 %, type I error: 5%). With 40 farms a difference of 0.8885 s is reached. A test of two origins would need 18 farms assuring one s. With four origins we needed 48 farms for one s difference. As there seem to be no data for the size of one s under organic farming conditions, station data from floor-housing experiments may be comparable. For mortality, egg number per hen and egg mass one s may be taken as 10 %, 20 eggs, and 0.73g. In conclusion field tests require a substantial organisational effort in order to reach a satisfactory level of statistical power and should be supplemented by station tests.

Session 10 Poster 86

Diversity of single nucleotide polymorphisms (SNPs) in the promoter region of the bovine neuropeptide Y (NPY) gene

B. Bahar and T. Sweeney, University College Dublin, School of Agriculture, Food Science & Veterinary Medicine, Belfield, Dublin 4, Ireland

Neuropeptide Y is a peptide hormone involved in the neuronal regulation of appetite and feeding behaviour in animals. Mutations in the NPY gene could potentially influence the feeding characteristics of cattle. Our objective was to determine the diversity of SNPs in the promoter region of bovine NPY gene. Blood samples from 11 different breeds of *Bos taurus* (n=34) comprised of Limousin (6), Simmental (6), Charlois (5) Aberdeen Angus (4), Hereford (3), Parthenais (3), Salers (3), Aubrac (1), Belgian Blue (1), Blonde d'Aquitaine (1) and Shorthorn (1) were collected and used for extraction of genomic DNA. Two partially overlapping genomic regions spanning a total of 1.5 kb length upstream of the translation initiation site of the gene was amplified by PCR. The nucleotide sequence of the PCR product was determined using both forward and reverse primers. Comparative sequence analysis identified 20 SNPs in the 1.5kb promotor region of the NPY gene which comprised of one each of A>T and G>A, two each of 1 base deletion and 1 base addition and fourteen heterozygotes. Phylogenic analysis showed that among the breeds, Simmental and Aberdeen Angus were most conserved, whereas, Limousin and Charlois were highly divergent. It is concluded that the diversity of SNPs existed in the promoter region of NPY gene of cattle breeds and the functional role of these SNPs in feeding behaviour of cattle warrants further study.

Session 10 Poster 87

Effects of allele variants at the E-Locus on production, fertility and conformation traits in Holstein dairy cattle

S. König[1], D. Andresen[2], W. Wemheuer[2] and B. Brenig[2], [1]University of Guelph, Department of Animal and Poultry Science, Gordon Street, N1G2W1 Guelph, Canada, [2]University of Göttingen, Institute of Veterinary Medicine, Burckhardtweg 2, 37077 Göttingen, Germany

At the Extension (E) locus, the E^D (dominant black), the E^+ (combination of black and red), and the e (recessive red) alleles are segregating. *Bos Primigenious*, the archetype of dairy cattle, is supposed to be of genotype E^D/ E^+, i.e. changing the red coat colour at birth into black later in life. This could be due to natural selection for environmental adaption and suggests the hypothesis that bulls carrying the E^+ allele are charcterized by improved fertility, persistency, longevity, or functional type traits as well. In total, 299 sons of 4 heterozygous E^D/E^+ bull sires were genotyped. 53.2% of sons were of genotype E^D/E^D and 46.8% of genotype E^D/E^+ indicating that all bull dams were homozygous E^D/E^D. National EBVs of bulls were de-regressed and then analyzed with the GLM procedure including the fixed effect of the bull sire, the regression coefficient for the number of E^+ alleles representing half of the gene substitution effect, and the random residual effect consisting of polygenic and environmental effects. The E^+ variant was associated ($p < 0.01$) with improved maternal non-return rate and maternal calving ease, and highly significant for rump angle ($p < 0.001$), i.e. favouring animals with low pin bones. For all other traits, no significant impact of the different allele variants was found.

Session 10 Poster 88

Genetic characterization of two sheep breeds from Southern Italy by STR markers

R. Ciampolini[1], F. Cecchi[1], E. Ciani[2], E. Mazzanti[1], M. Tancredi[1], E. Castellana[2], S. Presciuttini[3] and D. Cianci[2], [1]University of Pisa, Animal Production Department, Viale delle Piagge 2, 56124 Pisa, Italy, [2]University of Bari, General and Environmental Physiology Department, Via Amendola 165/a, 70126 Bari, Italy, [3]University of Pisa, Centre of Statistical Genetics, S.S.12 Abetone e Brennero 2, 56127 Pisa, Italy

Altamurana and Leccese are triple-purpose sheep breeds from Southern Italy that suffered in recent years a marked population size reduction. Aim of this work was to evaluate the level of within and among-breed genetic variability, as a contribution in conservation decisions. In the whole, 182 animals were typed for 19 STR markers. Data analysis was carried out by the software Arlequin. The mean number of alleles per marker was 11.3 (range 4 to 17). Several "private" alleles were observed, mainly in Leccese. Significant ($P < 0.01$) deviation from Hardy-Weinberg proportions was present for some markers, with excess of homozygous genotypes. More than 50% of all pair-wise marker comparisons showed gametic unbalance in both breeds. Allele-sharing similarities were 0.369 and 0.317 within Altamurana and Leccese, respectively, whereas it was 0.298 between breeds, indicating a significant level of genetic differentiation; the mean value of F_{ST} was 0.044. The breed assignment test allocated all individuals to their true breed of origin, except for one Altamurana and one Leccese individual. The present results may contribute to design a more appropriate management of these genetic resources.

Session 10 Poster 89

New genetic parameters for simultaneous breeding value estimation of production and litter size traits of Large White pigs in Slovakia

D. Peškovičová[1], P. Demo[1], E. Groeneveld[2] and L. Hetényi[1], [1]Slovak Agricultural Research Centre, RIAP, Hlohovská 2, 94992 Nitra, Slovakia (Slovak Republic), [2]Institute for Animal Breeding, FAL, Mariensee, 31535 Neustadt, Germany

New covariance matrices for 7-trait animal model used in the national genetic evaluation of pigs in Slovakia were estimated for Large White breed using REML method. Simultaneous breeding value estimation and selection of pigs based on four field test traits (lifetime average daily gain, ultrasound backfat thickness, number of piglets born alive in the 1st litter and number of piglets born alive in 2nd and later litters) and three station test traits (average daily gain, valuable cuts content, backfat thickness) was introduced in 2000. Covariance matrices for complex model were estimated using the data from performance testing since 1995 to 2005 (376000 records). Comparing to currently used covariance parameters, slightly lower genetic correlations among field and station test traits were estimated. Genetic correlations among backfat thickness in both testing environments (field and carcass station) and litter size were low but comparable in magnitude and sign (-0.19 and -0.16 for the 1st litter, -0.07 and -0.12 for the 2nd and later litters). Positive genetic correlations among litter size and valuable cuts content at the station (0.14 and 0.09) were estimated. Genetic correlations between litter size and average daily gain in both testing environments were close to zero.

Session 10 Poster 90

Genetic characterization of the Czech Spotted cattle breed using panel of 10 microsatellite markers

I. Manga, L. Putnova, J. Riha, I. Vrtkova and J. Dvorak, Mendel University Brno, Zemedelska, 613 00 Brno, Czech Republic

The aim of our work was an evaluation of molecular-genetic characterization of the Czech Spotted cattle breed. We performed the analysis of 10 microsatellite loci: BM1824, BM2113, ETH3, ETH10, ETH225, INRA023, SPS115, TGLA122, TGLA126, TGLA227. The genotyping of microsatelitte markers was performed by fluorescent fragment analysis on ABI PRISM 310 Genetic Analyzer (Applied Biosystems) and detected by software GeneScan 3.7 NT. In 240 unrelated individuals, the allele frequencies, observed and expected heterozygosity, test for deviations from Hardy-Weinberg equilibrium and Polymorphism information content (PIC) were calculated. We found out gene diversity (H_E) = 0.745. The observed heterozygosity (H_0) across all loci was 0.749, the average PIC was 0.709. The parameters of genetic diversity were evaluated by PowerMarker v3.28 analysis tools. After that, we used PCA and hierarchical clustering method to evaluate the genetic homogenity. The joining distance of clustering, by which the cluster contained 95% of all data combinations obtained by microsatellite loci analyse, was 0.319, opposite to even 0.788 by 100% data combinations. This refers to high genetic homogenity of the Czech Spotted cattle.
Supported by the Czech Science Foundation no.523/03/H076 and Ministry of Agriculture of the Czech Republic 1G58073.

Session 10 Poster 91

Estimation of genetic parameters of *Heterakis gallinarum* resistance in laying hens

E. Moors[1], A. Kanan[1], H. Brandt[2], S. Weigend[3], G. Erhardt[2] and M. Gauly[1], [1]Institute of Animal Breeding and Genetics, University of Goettingen, Albrecht-Thaer-Weg 3, 37075 Goettingen, Germany, [2]Institute of Animal Breeding and Genetics, University of Giessen, Ludwigstrasse 21b, 35390 Giessen, Germany, [3]Institute for Animal Breeding Mariensee, Federal Agricultural Research Centre, Hoeltystrasse 10, 31535 Neustadt, Germany

Infections with gastrointestinal parasites constitute a major health problem and economic importance in poultry production. The most common gastrointestinal nematodes are A. galli and H. gallinarum. In the present study two groups of White Leghorn and New Hampshire (male and female) were compared regarding the worm burden. Half of the animals were orally infected with 100 embryonated H. gallinarum eggs at an age of eight weeks, while the other group was kept as uninfected control. Eight weeks after infection animals were slaughtered to count the worms in the intestine. Body weights were recorded at the age of 8 and 16 weeks. Daily weight gains between time of infection and slaughtering were significant higher in the uninfected animals than in the infected ($p=0.038$). Worm burden (log) was significant higher in male animals compared to female animals ($p < 0.001$). White Leghorn had a significant higher worm burden than the New Hampshire ($p < 0.13$) for the New Hampshire, respectively.± 0.09) for the White Leghorn and 0.31 (±0.001). Estimated heritabilities were 0.41 (It can be concluded that the selection for H. gallinarum resistance should be possible in White Leghorn and New Hampshire.

Session 10 Poster 92

Investigation into the posibility of a GXE in the South African Jersey population

D.J. V Niekerk[1], F.W.C. Neser[1] and J. Vd Westhuizen[2], [1]University of the Free State, PO Box 339, 9300 Bloemfontein, South Africa, [2]ARC Animal Improvement Institute, P Bag X2, 0001 Irene, South Africa

The possibility of a genotype by evironment interaction for milk production in the South African Jersey population were investigated by grouping 54 864 lactation records completed between 1980 and 2002 into different production or environmental levels. The dataset consists of 301 herds and 884 sires. Three different sets of criteria were used to group the animals. Firstly a cluster analysis was applied using different management, climatic and genetic factors. This analysis resulted into four different clusters. Secondly the herds were devided in four categories according to feeding systems and geographic location. All the herds using a Total Mixed Ration (TMR) were placed in one group while the pasture herds form a second group. The herds that are situated in the warmer northern areas of South Africa were placed into a third group and the herds in the Overberg area that do not have access to irrigation water formed a fourth group. Lastely the herds were devided into four groups according to production levels. A bivariate animal model was used to determine the genetic correlations between each group in the different scenarios. The correlations varied between 78% and 99%. The lowest correlation was between the warmer Northern areas and the Overberg area. The highest correlation was between Cluster 3 and Cluster 4 as well as between the low production group and the medium-low production group.

Session 10 Poster 93

Construction of a numerator relationship matrix and its inverse including genetically identical animals for mixed model analysis

K. Yasuda and T. Oikawa, Okayama university, Agriculture, 1-1-1 Tsushimanaka Okayama-city, 7008530, Japan

In the field of animal breeding, estimation of genetic parameters and prediction of the breeding values for animals are routinely conducted using animal model. Therefore, incorporating the inverse of a numerator relationship matrix (NRM) into a mixed model is requisite. In breeding population of beef cattle in Japan, the number of genetically identical animals (GIA) is growing because performance test for meat quality traits is systematically performed incorporating GIAs. However it was not possible to include GIA in NRM because it is not possible at present to directly calculate inverse of NRM including GIA. Therefore, our scheme is to develop a procedure incorporating GIA into NRM. In an assumed K matrix, diagonal elements are set to 1, the non-diagonal elements between GIAs are set to 1-x and the other element are set to 0, where x is a arbitrary value near 0. The inverse of the K matrix is calculated by a simple formula. Thus applying Famula's method, the inverse of the A matrix is calculated by the products of the lower triangular matrix which identifies the parents of each individual, its transposed matrix, the inverse of the K matrix and the inverse of the D matrix. The application of this procedure is discussed by using a numerical example.

Session 10 Poster 94

Genetic variability three autochthonous cattle breeds in Croatia

J. Ramljak[1], A. Ivankovic[1], I. Medugorac[2], M. Konjacic[1] and N. Kelava[1], [1]University of Zagreb, Faculty of Agriculture, Department of Animal Production, Svetosimunska 25, 10000 Zagreb, Croatia (Hrvatska), [2]Faculty of Veterinary Medicine, The Ludwig-Maximilians-University Munich, Institute for Animal Breeding, Veterinaerstr. 13, 80539 Munich, Germany

Autochthonous cattle breeds Croatian Busha (CB), Slavonian Syrmian Podolian Cattle (SSP) and Istrian Cattle (IC) are part of the national and global genetic resource whose number drastically decreased in the nineteen years the twentieth century. All of three breeds are critically endangered and are included in organized protective programmes. The aim of this study is determination genetic variability within and among autochthonous Croatian cattle breeds. Blood samples were collected from 157 animals and genotyped with 20 microsatellite markers of the Utrecht List. The number of alleles was in range between 6 (BM1824, ILSTS006, INRA023) to 17 (INRA063). The highest number of polymorphic alleles is determined in CB (8,8) and the smallest number in SSP (4,5). The highest value of observed and expected hetreozygosity was determined in CB (H_O=0,622, H_E=0,733) and the smallest in SSP (H_O=0,531, H_E=0,538). The parameters of population subdivision shows that 8,4% of the total genetic variability were due to differences between breeds and the rest due to individual differences. The smallest genetic distances estimation was between CB and IC and the biggest between SSP and IC.

Session 10 — Poster 95

Population bottlenecks in the Croatian donkey breeds

A. Ivanković[1], J. Ramljak[1], N. Kelava[1], M. Konjačić[1] and P. Mijić[2], [1]Faculty of Agriculture, Department of Animal Science, Svetošimunska 25, 10000 Zagreb, Croatia (Hrvatska), [2]Faculty of Agriculture, Department of Zootechnics, Trg Sv. Trojstva 3, 31000 Osijek, Croatia (Hrvatska)

For thousands of years donkeys have inhabited the Mediterranean part of Croatia. At the end of the 20th century donkey breeds in Croatia had started being viewed as endangered. During the last decades the number of animals in the Istrian and North-Adriatic donkey population decreased drastically. The population size in the Littoral-Dinaric donkey is relatively stable. From a conservation perspective, detection of dramatic changes in population bottlenecks is another important aspect of population monitoring programme. Presence of possible genetic bottlenecks in the population was tested on the basis of biochemical (protein polymorphism) and DNA markers (microsatellites). Infinite allele model, Stepwise mutation model and Two phase model were used. Indicated that expected number of loci with heterozygosity excess is the biggest in the population of Littoral-Dinaric donkey and the smallest in the population of Istrian donkey. The Sign test and Wilcoxon sign-rank test under bottleneck hypothesis detected significant departure from mutation-drift-equilibrium in the population for part of the loci studied implying genetic bottleneck in the population. The results may form the basis for the future breeding strategy and management of the donkey breeds in Croatia.

Session 10 — Poster 96

Sperm-mediated gene transfer in poultry. 1. The relationship with cock sperm viability

A.Y. Gad, E.A. El-Gendy and A. Mostageer, Faculty of Agriculture, Cairo University, Animal Production, 7 El-Gamaa st. Giza, 12613, Giza, Egypt

The cock spermatozoa were used in this study as vectors for gene transfer. The objectives of the study were to assess the efficacy of the sperm to uptake exogenous DNA in relation to sperm viability. Two trials were carried out. Trial 1, was achieved to assess the effects of semen dilution (4 µl diluent/1 µl semen), heat incubation (exposure of semen to 37°C for 30 minutes) and the addition of lipofectin on sperm characteristics. No significant effects of heat incubation or semen dilution were mostly observed in the percentages of live, dead and abnormal sperm. The addition of lipofectin at 5% concentration to the diluted semen then heat incubated, showed the lowest detraction in semen characteristics. Trial 2, was designed to assess the effects of lipofectin (5%) addition on the fusion of DNA into the sperm and sperm viability. The exogenous DNA used was the plasmid pUC18. The plasmid DNA was successfully internalized into the sperm treated and un-treated with lipofectin. However, the DNA fragment recognized in the sperm incubated with lipofectin was apparently of higher yield. It was concluded that lipofectin stabilizes and facilitates the fusion of DNA into the sperm.

Session 10 Poster 97

Genetic diversity in the genetic resource of Old Kladruber Horse using microsatellite DNA markers

L. Putnova[1], I. Vrtkova[1], P. Horin[2], J. Riha[1] and J. Dvorak[1], [1]Mendel University Brno, Zemedelska, 613 00 Brno, Czech Republic, [2]University of Veterinary and Pharmaceutical Sciences Brno, Palackeho, 612 42 Brno, Czech Republic

The Old Kladruber horse is the most important genetic resource in the Czech Republic. Seventeen microsatellites were genotyped in this breed using a panel of microsatellites recommendedfor routine parentage testing by the ISAG Equine Genetics Standing Committee. The analysis of all loci in the population (n=153) revealed 105 alleles. The mean number of alleles per locus was 6.18, ranging from 4 (HTG7) to 10 (ASB17). Allele frequencies for the most frequently allele were as follows: AHT4 (H=0.37), HMS7 (O=0.31), HTG4 (M=0.27), VHL20 (Q=0.38), AHT5 (N=0.47), ASB2 (M=0.35), ASB23 (J=0.30), HMS6 (L=0.47), HTG6 (O=0.59), ASB17 (R=0.44), CA425 (N=0.58), HMS1 (M=0.60), LEX3 (L=0.61), HMS2 (K=0.46), HMS3 (P=0.45), HTG10 (O=0.48), HTG7 (O=0.87). The highest heterozygosity (above 70%) was observed for HTG4 (78.47), ASB2 (77.95), VHL20 (77.77), ASB23 (77.05), HMS7 (76.93) and AHT4 (71.68). The lowest heterozygosity (below 50%) was found for HTG7 (22.47). The value of inbreeding coefficient was zero. In the Hardy-Weinberg equilibriumconformity test only HMS3, LEX3 and HTG7 were not in equilibrium ($P < 0.01$). The average probabilities of paternity exclusion/one parental genotype unavailable/parentage exclusion (CEP1-3) estimated for this panel were 99.81%/99.59%/99.99%.
Supported by Ministry of Agriculture of the Czech Republic 1G58073 and GA CR 523/06/1402.

Session 10 Poster 98

Preimplantational genetic diagnosis in bovine embryos

D.I. Ilie[1], A.C. Stanca[2], V.B. Carstea[2], G. Ghise[3], E. Gocza[2] and I. Vintila[3], [1]Victor Babes University of Medicine and Pharmacy Timisoara, Imunology, Eftimie Murgu 2, 300014, Romania, [2]Agricultural Biotechnology Centre Gödöllo, Department of Animal Biology, Szent – Györgyi Albert u.4., 2100, Hungary, [3]Banat's University of Agricultural Sciences and Veterinary Medicine, Faculty of Animal Science and Biotechnology, Timisoara, Calea Aradului 119, 300645, Romania

Preimplantation genetic diagnosis (PGD) is a diagnostic procedure that enables the diseases detection into embryos conceived through *in vitro* fertilization (IVF). This technique was first undertaken for X-linked diseases in 1989 (Handyside *et al*; Lancet 1989;1:347-349) and for autosomal recessive conditions in 1990 (Verlinsky Y.; Hum Reprod 1 990;5:826-829). In the present study we used Polymerase Chain Reaction (PCR) that enables us to analyze the CD18 gene (Bovine leukocyte adhesion deficiency - BLAD) and to detect the sex of bovine embryos conceived through *in vitro* fertilization. The technique based on PCR method allowed as identifying the gene of economic importance and embryos gender. Embryos were produced by *in vitro* maturation of oocytes, *in vitro* fertilization and *in vitro* culture. For the PCR we used a simple DNA extraction method from embryos and gene amplification by PCR (Polymerase Chain Reaction). BLAD genotyping were done using RFLP technique (Restriction Fragment Length Polymorphism) with HaeIII enzyme.

Session 10 Poster 99

Genetic parameters for milkability from the first three lactations in Fleckvieh

J. Dodenhoff and R. Emmerling, Bavarian State Research Center for Agriculture, Institute of Animal Breeding, Prof.-Duerrwaechter-Platz 1, 85586 Poing-Grub, Germany

Test-day records from the routine dairy recording from Bavarian Fleckvieh cows were analysed. Average flow rate (AFR) was derived from the milk flow curve based on threshold flow rates. Two observations per test-day were available (morning milking, evening milking). Several data sets with observations from approximately 35,000 cows each were sampled from the total data set. For each of the first three lactations six time periods with up to 30 days were defined with the number of days depending on the stage of lactation. Morning milkings and evening milkings were considered as different traits, i.e., each cow had only one observation per time period. Estimates of (co)variances for log-transformed AFR from these 36 time periods were obtained by REML using an average information method. For each data set several multiple trait analyses with four traits were run for various combinations of these 36 traits. An algorithm for iterative summing of expanded part matrices was applied in order to combine the estimates. Estimates of heritability were around.28 in lactation 1 and slightly higher in lactations 2 and 3. Within lactations the estimates decreased towards the end of lactation. Estimates of the genetic correlations between lactations 2 and 3 suggest that they could be considered as one trait. All estimates of the genetic correlation between morning milking and evening milking from the same test-day were above.90.

Session 10 Poster 100

Exploring the relationship between the polymorphism of the melatonin receptor gene and the variability of sheep seasonality

E. Ugarte[1], I. Beltrán De Heredia[1] and L. Bodin[2], [1]NEIKER-Tecnalia, Health and Production Production, Granja Modelo Arkaute. Apdo 46., 01080. Vitoria-Gasteiz, Spain, [2]INRA, SAGA, UR631, F-31326 Castanet-Tolosan, France

In sheep, seasonal reproductive activity is related to melatonin. Specific melatonin receptors MT1 and MT2 have been identified and the association between reproductive seasonality and polymorphism of the gene for the MT1 receptor was widely studied. In Latxa breed, as in other breeds, a RFLP site has been found in Exon II of the MT1 receptor gene. This silent mutation in position 612, also presents a complete linkage desiquilibrium with a non silent mutation at position 706. Several experiences have been made to check in Latxa the relationship encountered in other breeds between this coding mutation and seasonality. Five half-sisters families were created proceeding of heterocigotous males which for the RFLP site and the associated non silent mutation. For several measurements related to natural seasonality, no significant differences were found according of the inherited allele. However, after sequencing a large part of the Exon II of this gene, we found in Latxa a similar pattern as in other breeds. For the allele associated with higher seasonality in other breeds there is a small no polymorphic DNA fragment around the non silent mutation which can define an haplotype, while for the other allele the same fragment is highly polymorph. These results, point out that this haplotype may be important in the regulation of out of season reproduction.

Session 10 — Poster 101

Estimates of variance components for test-day models with Legendre polynomials and linear splines

J. Bohmanova[1], F. Miglior[2,3], J. Jamrozik[1] and I. Misztal[4], [1]University of Guelph, CGIL, Department of Animal and Poultry Science, Guelph, ON, N1G2W1, Canada, [2]Agriculture and Agri-Food Canada - Dairy and Swine Research and Development Centre, Sherbrooke, QC, J1M1Z3, Canada, [3]Canadian Dairy Network, Canadian Dairy Network, N1GT42, Canada, [4]University of Georgia, Department of Animal and Dairy Science, Athens, GA, 30601, USA

Genetic parameters were estimated for test-day (TD) milk, fat and protein yield and somatic cell score for the first three lactations of 6,094 Holstein cows with 96,756 TD yields using six random regression models. Only TD with DIM $\leq$ 365 and all traits present on a test-day were included. Legendre polynomials of order four and linear splines with four to seven knots were fitted for the fixed regression and for additive genetic and permanent environmental effects. The same type of function was applied for both fixed and random regression. Residual variance was modeled by a step function with either four or twelve intervals. A single chain of Gibbs sampler with 100,000 samples was generated, with 10,000 samples as burn-in, in order to obtain posterior distribution of parameters. Both models with Legendre polynomials and linear splines estimated larger additive genetic variances at the beginning and at the end of lactation. However, smaller variances at the extremes of lactation were estimated by models with linear splines. Models were compared using Akaike's information criterion, Bayesian information criterion, Bayes factor and Deviance information criterion. All the criteria favored the spline model with seven knots, which was the most complex model.

Session 10 — Poster 102

Estimation of variance and (co)variance components of egg and chick weight in the Oudtshoorn ostrich population

M.D. Fair[1], J.B. Van Wyk[1] and S.W.P. Cloete[2], [1]University of the Free State, PO Box 339, 9301 Bloemfontein, South Africa, [2]Elsenburg Agricultural Development Institute, Private Bag X1, 7607 Elsenburg, South Africa

Data of pedigree and performance for 77 241 individual egg records were obtained from a large pair breeding flock maintained at KKADC, near Oudtshoorn, South Africa. The data was collected from 1991 to 2005 and comprised the progeny of 414 dams and 441 sires, a total of 343 breeding pairs. Data involving records of egg weight (EGWT) and chick weight (CHWT), were analysed using REML. All traits were treated as trait of the individual. An animal model with fixed effects, year (1991-2005), month (Jun-Jan), age of service sire (AS) and age of dam (AD)(2-11 years), sequence of the egg laid within season (seq) (1-120) and random effects of direct genetic effect (a), dam (m) genetic maternal effect of the dam, common environment (ce) defined as a unique hen within a year, permanent environment (pe) defined unique hen over years, paddock (p) paddock that the breeding pair were kept in and seq was used for univariate analyses of the traits. Heritability estimates (±s.e.) were 0.09 ± 0.03 and 0.15 ± 0.04 for EGWT and CHWT. Corresponding estimates for maternal effects were 0.18 ± 0.06 and 0.28 ± 0.08. Effects of ce, pe, p and seq were significant but relatively low for all traits. The results indicate that genetic improvement in production traits in ostriches is possible.

Session 10 Poster 103

Sustainability of Hungarian Grey cattle's production during centuries

I. Bodó[1], I. Gera[1], B. Béri[2] and A. Radácsi[2], [1]Hungarian Grey Cattle Breeders Association, Lőportár Street 16., 1134 Budapest, Hungary, [2]University of Debrecen, Institute of Animal Science, Böszörményi Street 138., 4032 Debrecen, Hungary

Many centuries ago the Hungarian Grey cattle was a beef producing breed which was exported walking to Western European countries from Hungary. Its beef quality was appreciated. Caused by the wars, some eastern diseases and custom policy the good business stopped and then the high working quality of the breed was proven in both agriculture and transport. At the second half of the twentieth century the development of mechanization made the animal power unnecessary and the breed declined dramatically. Half a century ago the population size of cows was not more than 160, however after a bottle neck it reached 6000 for 2006. In production of beef quantity the Hungarian Grey is not competitive with the modern beef breeds because of the slight dressing percentage, smaller daily gain and by extensive breeding conditions. Carcass traits and meat quality (protein, fat content, different fatty acids and UFA, PUFA) of intensively and semi-intensively fattened Hungarian Grey and Holstein-Friesian bulls were evaluated. Hungarian Grey cows were crossed by Charolais, Blonde d'Aquitaine, Chianina, Piemontese bulls. Chemical composition of the meat was investigated. The difference between the meat quality of weaned calves was not significant.

Session 10 Poster 104

Estimates of heritability of and correlations for milk and growth traits in Egyptian Zaraibi goats

I. Shaat[1], M. Shaaban[2], A. Abdel-Hakim[1] and A. Hamed[1], [1]Animal Production Research Institute, Sheep and Goats, 4, Nadi El-Said Street, Dokki, Cairo, Egypt, [2]Faculty of Agriculture, Al-Azhar University, 2Animal Production Department, Nasr-City, Cairo, Egypt

A total of 2363 lactation records obtained from 975 does progeny of 99 sires and 571 dams were collected from 1995-2003. Milk traits were, total milk yield (TMY), milk yield at 90 days (MY90) and lactation period (LP). Growth traits data were collected from 6755 Zaraibi kids progeny of 110 sires and 1331 dams during 1990-2003. Measured traits were weight at birth (WB), 90 days (W90-D), 180 days (W180-D), 365 days (W365-D) of age. Mixed model methodology based on a multi-trait animal model was used to estimate genetic parameters. Heritability estimates (h^2) for milk traits were 0.35, 0.27 and 0.15 while the repeatability estimates were 0.43, 0.33 and 0.22 for TMY, MY90 and LP, respectively. Genetic correlation (r_G) was high for TMY with MY90 and LP being 0.89 and 0.80, respectively, and low between MY90 and LP (0.46). For growth traits, h^2 were 0.21, 0.16, 0.12 and 0.12 for WB, W90-D, W180-D and W365-D, respectively. r_G for WB with W90-D, W180-D and W365-D were 0.42, 0.47 and 0.62 respectively and for W90-D with W180-D and W365-D were 0.77 and 0.88 while it was 0.77 for W180-D with W365-D. The relatively high estimate of h^2 for total milk yield indicates that selection for this trait will be effective especially with the high and positive r_G between this trait with the other milk production traits.

Investigation of candidate genes affecting boar taint

E. Grindflek[1,2], M. Moe[1,2], I. Berget[3], C. Bendixen[4] and S. Lien[1,3], [1]Norwegian University of Life Sciences (UMB), Dept. Animal and Aquacultural Sciences, POBox 5003, 1432 Ås, Norway, [2]Norsvin, POBox 504, 2304 Hamar, Norway, [3]UMB, CIGENE, POBox 5003, 1432 Ås, Norway, [4]Danish Institute of Agricultural Sciences (DIAS), POBox 50, 8830 Tjele, Denmark

Boar taint in entire mail pigs is primarily caused by the two compounds androstenone and skatole. Previously we have conducted a cDNA microarray experiment to examine the transcription profiles in boars with extreme high and low levels of androstenone. Some of the most promising results from this study, together with a number of other candidate genes taken from literature, were chosen for further investigations. Expression patterns of 20 genes were investigated in 200 Duroc and Landrace boars with extreme values of androstenone using real-competitive (rc) PCR on the MassARRAY platform (Sequenom, USA). Additionally a PCR resequencing effort generated a number of SNPs which facilitated allelc expression or association studies for the majority of these genes. The quantitative gene expression and allele expression was analysed with the HPRT as housekeeping gene.Results showed that CYP11A1, CYP17A1, CYB5, DHRS4, FTL, STAR and SULT2A1 were all highly upregulated ($p < 0.001$) in boars with high androstenone levels for both breeds, whereas CYP19A2 was upregulated in Duroc only ($p < 0.05$). On the contrary TEGT was upregulated in Landrace only ($p < 0.05$). Overall results show large differences between gene expression level in Duroc and Landrace boars with extreme levels of androstenone.

Genetic characterization of four Romanian horse breeds using 12 microsatellites

S.E. Georgescu, M.A. Manea and M. Costache, University of Bucharest, Molecular Biology Center, 91-95 Spl.Independentei, Bucharest 5, 050095, Romania

One of difficulties in implementating a selective breeding program in horse stocks is loss of genetic variability and increases in inbreeding as a result of the unintended mating of related individuals. The effects of inbreeding in horses will result in a decrease in genetic variability, which will limit the potential for genetic gain from artificial selection.The PCR technology provides a sensitive method for parentage verification and individual identification. It can also be used to screen for markers linked to performance traits or genetic disorders. Genetic characterizations of four Romanian horse breeds (Thoroughbred, Arabian, Romanian Sport Horse and Hucul) were made using 12 microsatelite markers (AHT4, AHT5, ASB2, HMS2, HMS3, HMS6, HMS7, HTG4, HTG6, HTG7, HTG10, VHL20). Amplification of the STR loci was realized by multiplex PCR using StockMarks for Horse Paternity PCR Typing kit (AppliedBiosystems). High levels of polymorphism were observed over populations. A number of loci had shown different alleles frequencies among the horse populations. No significant differences were observed in average between the four populations suggesting that intensive breeding practices will not generate a decrease of genetic variability. In horse breeding this technology has the potential to be of great use in monitoring levels of genetic variation within stocks as well as for parentage and relatedness purposes.

Session 11 — Theatre 1

Crossbreeding in beef production: some principles and tools

G. Simm, T. Roughsedge and E.A. Navajas, SAC, Sustainable Livestock Systems Group, West Mains Road, Edinburgh EH9 3JG, United Kingdom

Crossbreeding is used mainly for: (i) improving the efficiency of a production system by crossing breeds with high genetic merit in different traits (*complementarity*), (ii) producing individuals of intermediate performance between that of more extreme parent breeds, (iii) grading up to a new breed, (iv) creating a composite breed, (v) introducing a gene for a favourable characteristic (*introgression*), or (vi) exploiting *heterosis*. Heterosis is the advantage in performance above the mid parent mean. It is most useful when it leads to the average performance of crossbred animals exceeding that of the best parent breed. Individual heterosis directly influences the performance of the crossbred animal itself. Maternal and paternal heterosis arise when dams or sires are crossbred, and the effects are often measured in terms of improved reproductive efficiency or improved performance of offspring. Different systems of crossing lead to different proportions of individual, maternal and paternal heterosis being maintained. However, the most appropriate system of crossing depends not only on this, but also on the additive merit of the breeds available – which changes - and the absolute level of performance of crossbreds. Hence, the optimum choice of breeding method and breeds is a dynamic one, even for a constant breeding objective. We discuss tools available to help decision making, and how changing objectives – such as addressing environmental impact – might influence choices.

Session 11 — Theatre 2

Genetic evaluation of growth of dual-purpose bulls

J. Pribyl[1], H. Krejcova[1], J. Pribylova[1], I. Misztal[2], N. Mielenz[3], J. Kucera[4] and M. Ondrakova[4], [1]Inst.Anim.Sci., Uhrineves, 104 00, Czech Republic, [2]Univ. Georgia, Athens, GA 30602, USA, [3]Martin-Luther Univ., Halle-Wittenberg, 06001-06132, Germany, [4]Czech Fleckvieh Breeders Assoc., Prague, 170 41, Czech Republic

Growth is important in genetic evaluation of cattle. Before entering artificial insemination centres, young Czech Fleckvieh (Simmental type) bulls are performance tested for growth. They enter the test station at an early age and are selected out at about 13 months of age. They are weighed at monthly intervals. Nutrition is regulated to achieve a mean daily gain of 1.2 kg. The evaluation database contains 8,158 young bulls and pedigree data on 16,154 animals. Because of compensation between different growth phases, there are low negative correlations between daily gains (-0.11 to -0.03 between 100-day segments) for different periods. This implies that evaluation of growth cannot be just an analysis of average daily gain for the total period or body weight. RR-TDM was used to evaluate the trajectory of daily gains. The genetic component had high positive correlations, but the animal's permanent environment showed high negative correlations between different growth phases. For the 100-day segments from 100 to 400 days of age, correlations between breeding values ranged from 0.71 to 0.77. Average reliability for the sum of the breeding values from 100 to 400 days of age was 0.51.

Session 11 Theatre 3

Problems of conservation and crossing of Hungarian Grey cattle

I. Bodó[1], I. Gera[1], A. Radácsi[2] and B. Béri[2], [1]Hungarian Grey Cattle Breeders Association, Lőportár street 16., 1134 Budapest, Hungary, [2]University of Debrecen, Böszörményi Street 138., 4032 Debrecen, Hungary

The Hungarian Grey is a traditional cattle breed which has been saved from extinction. Half a century ago the population of females was not more than 160 but now it is more than 6000. As the preservation policy of pure bred animals was on-going the sale of their products became more and more important. Many experimental crossings were carried out, showing that beef production from Hungarian Grey cattle can be increased by cross breeding. Data from the first production cycle using Charolais, Blonde d'Aquitaine, Simmental, Piemontese, Chianina and Belgian Blue are summarized. The Charolais was the most popular crossing partner and the F1 cows out of Charolais bulls were again crossed with Charolais bulls. Another 3 breed cross ((Hungarian Grey x Simmental) x Charolais) was compared with pure bred Simmental and Simmental x Hereford cows. The calf crop per cow and per year for the Hungarian Grey crossed herd gave the best results when reproduction was also taken into consideration. It is not easy to harmonize the pure bred preservation policy with cross breeding. Only commercial crossing is allowed and the terminal progeny must all be slaughtered. In the herd book of the Hungarian Grey Cattle Breeders' Association, there is a separate section (D) for registration of crossbred animals, but in order to maintain the purity of the breed there is no transition to the other sections.

Session 11 Theatre 4

Crossbreeding in temperate grazing conditions: breed and heterotic effects for beef tenderness in steers

E.A. Navajas[1], I. Aguilar[2], J. Franco[3], O. Feed[3], S. Avendaño[4] and D. Gimeno[5], [1]SAC, West Main Rd, Edinburgh, EH9 3JG, United Kingdom, [2]INIA, Ruta 48 km10, Canelones, Uruguay, [3]EEMAC, Ruta 3 km365, Paysandu, Uruguay, [4]Aviagen, Newbridge, Midlothian, EH28 8SZ, United Kingdom, [5]SUL, Rbla Brum 3764, Montevideo, Uruguay

Crossbreeding (CB) is used to combine desirable traits from two or more breeds for specific environments and markets, and to exploit heterosis. Defining the most profitable CB systems requires CB parameters for the economically relevant traits, including meat eating quality (MEQ). Individual breed effects (a^I) and heterosis (h^I) were estimated for tenderness in Hereford (H), Angus (A), Salers (S) and Nellore (N). It was assessed by shear force measured in m. longissimus samples from 68 purebred H and 213 F1, 35 F2 and 67 backcrosses between H, A, S and N produced in a CB experiment designed to estimate CB parameters, which was carried out in Uruguay. Least square means were estimated for all genotypes using a mixed model including sire as random effect (73), and year (4) and genotype (13) as fixed effects. The a^I and h^I were estimated by the appropriated contrast between means. The a^I effects were +0.79 kg, +0.93 kg and +4.14 kg for A, S and N, compared to H, whilst h^I values were very small and not significant ($P > 0.05$). Beef from A and H crosses was tender and S crosses had intermediate values. Decreasing percentages of N improved tenderness to values close to H. These results confirm the relevance of including MEQ when evaluating CB systems.

Session 11 Theatre 5

Liveweight breeding values for dairy breeds and crosses in New Zealand

J.E. Pryce, S. Sim, B.L. Harris and W.A. Montgomerie, LIC, Strategy and Growth, Private Bag 3016, 3240 Hamilton, New Zealand

Genetic evaluation of liveweight in lactating dairy cattle has been practised in New Zealand since 1993. In dairy farm profit, two relevant aspects of revenue from liveweight include beef production and feeding costs. In the New Zealand economic index for dairy cattle, the costs associated with feeding outweigh the benefits of increased carcass value of the cow herself, so liveweight has a negative weighting in the index. Nevertheless, the genetic trend for liveweight shows a small positive increase as a correlated response to selection for milk production traits. The New Zealand dairy cow population is 4 million, around 35% of which are crosses between Holstein-Friesians and Jerseys. Heterosis estimates between Holsteins, Friesians and Jerseys range between 5.0 and 10.4 kg, depending on the cross, breed difference estimates are up to 90 kg. We have recently begun performance testing future sires of New Zealand dairy cattle using liveweight measurements from arrival at our facilities to around 400 days with the intention of estimating 200 d and 400 d breeding values. In practice, calf-rearers and beef producers prefer Holstein-Friesians over Jerseys, as they are larger and faster growing and because the influence of angular Holsteins is less than in other countries.

Session 11 Theatre 6

Quality beef production from pure and crossbred dairy calves

D.C. Patterson, L.E.R. Dawson and T. Yan, Agri-Food and Biosciences Institute, Agriculture Branch, Large Park, Hillsborough, BT26 6DR, United Kingdom

This paper compares the efficiency of production and product quality of dairy beef with suckler beef, and also compares pure dairy breed beef with beef x dairy breed beef. In Europe, beef from the dairy herd constitutes a major proportion of the total beef output and the paper assesses dairy beef in terms of both output and quality. It also compares beef from dairy calves with beef from suckler herd calves in terms of: 1.assessments of meat quality and sensory characteristics, 2. biological efficiency of the total system in terms of nutrients required per kg carcass, 3. potential environmental impact per kg carcass on water and air quality. The beef characteristics of calves of a range of both pure dairy and beef breeds x dairy breeds are compared in terms of growth rate, feed conversion efficiency, killing out proportion, carcass traits and meat quality. The comparisons presented are taken from adequately replicated experiments carried out in Ireland, where the predominant production base is heavily reliant on grass, either grazed in summer or conserved as silage for winter feed, and supplemented with moderate inputs of concentrates, with slaughter age typically ranging from 18 to 30 months of age for steers.

Session 11 Theatre 7

Comparison of Friesian, Aberdeen Angus x Friesian and Belgian Blue x Friesian steers finished at pasture or indoors

M.G. Keane, Teagasc, Grange Beef Research Centre, Dunsany, Co. Meath, Ireland

About 50% of Irish dairy cows are bred to beef sires. The objective of this study was to compare Friesian (FR), Aberdeen Angus x Friesian (AA) and Belgian Blue x Friesian (BB) spring-born steers slaughtered off pasture at the end of the second grazing season or finished indoors in the second winter. Those slaughtered off pasture received 3.65 kg/day supplementary concentrates for the final 105 days. The indoor finished group was housed when the pasture finished group was slaughtered and offered grass silage + 5 kg/day concentrates for 141 days. Mean slaughter weights and carcass weights per day of age for FR, AA and BB were 852, 802 and 834 (s.e. 13.1) g, and 427, 412 and 452 (s.e. 7.3) g, respectively. Corresponding kill-out, carcass conformation score and carcass fat score values were 501, 514 and 542 (s.e. 2.4) g/kg, 1.90, 2.15 and 2.89 (s.e. 0.073), and 3.09 3.27 and 2.59 (s.e. 0.122). Responses to concentrates at pasture were 101 g live weight and 83 g carcass weight per kg dry matter. All three breed types produced acceptable carcasses following indoor finishing but only AA were acceptably finished off pasture. For acceptable finish off pasture minimum carcass weights of 250, 280 and 340 kg are required for AA, FR and BB, respectively.

Session 11 Theatre 8

Extensive steer fattening including a summering period and different finishing strategies

A. Chassot[1], P.-A. Dufey[1] and J. Troxler[2], [1]Agroscope Liebefeld-Posieux (ALP), Tioleyre 4, 1725 Posieux, Switzerland, [2]Agroscope Changins-Wädenswil (ACW), Changins, 1260 Nyon, Switzerland

Crossbred steers (Limousin x Red Hostein) were used to investigate the effect of a reduced stocking rate of unfertilised mountain pastures on current and subsequent finishing performance. Different feeding strategies were tested during the finishing period. Stocking rate had a marked effect on animal performance in summer. By decreasing it, it was possible to extend the grazing period and increase growth rate. As a consequence, pasture productivity in terms of live weight gain per ha decreased less than proportionally. At the end of the summer period, carcass quality generally did not meet market requirements with respect to fat score and carcass conformation (meatiness), but the different finishing treatments corrected this after one to two months. During the finishing period, the daily growth rate of the steers was more than double that at pasture, independent of live weight and finishing treatment. The phenomenon of compensatory growth is the most likely explanation for this exceptional performance. In conclusion, the extensification of mountain pastures and their utilization for fattening steers or heifers seems to be a promising alternative to abandoning these areas as pastures for dairy cattle. The high potential for compensatory growth results in a very efficient finishing and has therefore to be considered.

Session 11 Theatre 9

A comparison of high and standard dairy genetic merit, and Charolais x dairy male animals for beef production

M. Mcgee, M.G. Keane and A.P. Moloney, Teagasc, Grange Beef Research Centre, Dunsany, Co. Meath, Ireland

The increased proportion of Holstein genes in the dairy herd has consequences for beef production. A total of 72 spring-born male calves, comprised of 24 Holsteins (HO), 24 Friesians (FR) and 24 Charolais × Holstein-Friesians (CH) were reared from calfhood to slaughter in a 3 breeds (HO, FR and CH) × 2 production systems (intensive 19-month bull beef and extensive 25-month steer beef) × 2 slaughter weights (560 and 650 kg) factorial experiment. Liveweight gain, carcass gain, kill-out proportion, carcass conformation and carcass fat score were 830, 811 and 859 (s.e. 14.9) g/day, 540, 533, 585 (s.e. 7.7) g/day, 526, 538 and 561 (s.e. 3.0) g/kg, 1.51, 2.18 and 2.96 (s.e. 0.085), and 3.40, 4.25 and 4.06 (s.e. 0.104) for HO, FR and CH, respectively. Carcass measurements were greater for HO than FR and for FR than CH. Carcass measurements were also greater for animals on the intensive system (bulls) than the extensive system (steers) in absolute terms, but the opposite was so when they were expressed relative to liveweight.. It is concluded that HO grew as fast as FR but had a lower kill-out, while CH grew faster than both dairy strains and had a higher kill-out. There were large differences between the breed types in body and carcass measurements, and hence in carcass shape and compactness but differences in tissue distribution across the pistola were small.

Session 11 Theatre 10

Effect of crossbreeding on market value of calves from dairy cows

R. Dal Zotto, M. De Marchi, P. Carnier, M. Cassandro, L. Gallo and G. Bittante, University of Padova, Animal Science, Viale Universita 16, 35020 Legnaro, Italy

Market value (MV) of purebred Brown Swiss (BS), Holstein-Friesian (HF), Simmental (SI), and Alpine Grey (AG) calves and crossbreds from Limousin (LI) and Belgian Blue (BB) sires mated to BS, HF, SI, and AG cows were studied to evaluate crossbreeding effects. A total of 58,877 records of calves marketed in 143 weekly auctions from 2003 to 2005 were analyzed using a linear model that included the effects of genetic group, sex and age of calf, year-month of auction, and all two-way interactions. The model had an R^2 value of 0.78. Genetic group and sex had the greatest effects on MV. For all genetic groups, male calves had greater MV than females (€356 v. 290) with the exception of HF and BS (€175 v. 179 and €167 v. 171, respectively). Male and female HF and BS calves are all used for veal production, whereas with the other breed types, females are mostly used for veal production while males are fattened for beef. BBxSI calves had the highest MV (€489) whereas BS and HF purebreds had the lowest MV (€170 and 177, respectively). When used as a sire breed, BB increased MV of the progeny more than LI. The effects of BB and LI on MV were greater when mating was to SI and AG cows rather than to BS and HF cows. For purebreds, SI had the highest MV (€303). In conclusion, BB sired the highest MV crossbred calves.

Session 11 Theatre 11

Forage and concentrate based finishing strategies for cull dairy cows

W. Minchin[1,2], M.A. O'Donovan[2], D.A. Kenny[1], L. Shalloo[2] and F. Buckley[2], [1]University College Dublin, School of Agriculture, Food science & Veterinary Medicine, UCD, Belfield, Dublin 4, Ireland, [2]Teagasc, Dairy Production Research Centre, Teagasc, Moorepark, Fermoy, Co Cork, Ireland

An experiment was conducted to evaluate four fattening treatments for cull dairy cows based on grass silage and concentrate. Sixty-eight multiparous Holstein-Fresian cull spring-calving dairy cows were randomised and assigned to a four treatment (n=17) finishing experiment. The four treatments were: *ad-libitum* grass silage (GS), GS + 3 kg/d concentrate (GS+3C), GS + 6 kg/d concentrate (GS+6C) and GS + 9 kg/d concentrate (GS+9C). Liveweight, body condition score, back fat, skeletal size, carcass classification and group dry matter intake (DMI) were measured. Individual DMI was measured once (week 5), to allow calculation of residual feed intake (RFI). Finishing targets were set to ensure that cows reached the carcass criteria required by the abattoir to optimise carcass value. These were: carcass weight > 272 kgs, fat score 3 or 4L and carcass conformation P+ or O. In all treatments evaluated the total feed budget was 1.4 t DM/cow. As the level of concentrate increased there was a linear decrease in days to slaughter. There was a linear response to concentrates up to 6 kg/d. The RFI (kg) DMI values for the GS, GS+3, GS+6 and GS+9 treatments were -2.7, -2.3, -2.0 and -1.6 which were significantly different between treatments ($P < 0.05$).

Session 11 Poster 12

Effect of diet containig grape skin on productive performances of Podolica and crossbred baby-beef

G. Marsico, S. Dimatteo, A. Rasulo, S. Tarricone and R. Celi, University of Bari, Department of Animal Production, Via Amendola 165/A, 70126 Bari, Italy

Meat production is affected by many factors such as genotype and feeding. Some Italian autochthonous bovine breeds, like the Podolica, have a strong rustic nature. Because of this they can use some agro-industrial by-products as alternative low-cost high quality feedstuffs. Thus, the effect of including 20% grape skin in the ration of Podolica and cross-bred (Marchigiana x Podolica and Chianina x Podolica) baby beef was evaluated. Podolica, Marchigiana x Podolica and Chianina x Podolica calves (n=8) were used. Each genotype was divided into two groups. One group was offered a commercial feed, while the second group was offered a feed containing 20% grape skin. The animals were weighed every month to measure growth rate, and feed intake was recorded to measure Feed Conversion Index (FCI). Use of grape skin in the ration did not significantly affect final live weight or slaughter measurements, but it negatively affected FCI due to increased feed intake. The half-carcass of grape skin-fed animals had a higher shoulder proportion and a lower proportion of belly. It is concluded that further research is warranted on the role of grape skin for rustic bovine feeding.

Session 11 Poster 13

Effect of breed on growth rate and body measurements in young steers

B. Campion[1,2], M.G. Keane[1], D.A. Kenny[2] and D.P. Berry[3], [1]Teagasc, Grange Beef Research Centre, Dunsany, Co. Meath, Ireland, [2]School of Agriculture, Food Science & veterinary Medicine, University College Dublin, Belfield, Dublin 4, Ireland, [3]Teagasc, Moorepark Dairy Production Research Centre, Fermoy, Co. Cork, Ireland

Growth rate influences profitability in beef production and linear body measurements are useful indicators of growth rate and live weight (LW). The objective of this study was to compare growth rate and body measurements of Aberdeen Angus × Holstein-Friesian (AA; n=59), Belgian Blue × Holstein-Friesian (BB; n=59), Holstein (HO; n=28) and Friesian (FR; n=28) male cattle in their first year of life. At housing, body measurements, namely height at withers, chest girth, length of back, chest depth and pelvic width were recorded. BB were heavier ($P < 0.05$) at housing than AA but otherwise there were no differences ($P > 0.05$) between the breeds. From arrival to housing (mean = 186 days), FR grew faster ($P < 0.05$) than both AA and BB, while HO was not different. FR and HO were taller than the AA and BB and also had greater chest depth ($P < 0.05$). However, when measurements were expressed per 100 kg LW, AA had greater ($P < 0.05$) values for all body measurements than BB with FR and HO intermediate and not different ($P > 0.05$) from the beef crosses. There were no differences between FR and HO for any of the traits measured. It is concluded that BB were most, and AA were least compact and both growth rate and linear body measurements were influenced by breed.

Session 12 Theatre 1

Impact of nutrition and feeding practices on equine health and performance

P. Harris, WALTHAM Centre for Pet Nutrition, Equine Studies Group, Freeby Lane, Waltham on the wolds, Leics, LE14 4RT, United Kingdom

For the horse, what and when they are able to eat is now predominantly determined by man and we therefore have to take responsibility for the effects that our choice of managemental practices have on their health and welfare. Nutrition can also have profound effects on their performance and, while good nutrition cannot improve the intrinsic ability of the horse (and rider), inappropriate or inbalanced nutrition may impose limitations. Especially in developed countries we therefore should no longer consider the feeding of horses purely as a means to provide the basic nutrients that they require. Instead we should be looking at how the type and amount of nutrients provided can help optimise performance, reduce the risk of disease, minimise any adverse welfare effects and, where necessary, provide the best possible support for the ill or convalescing horse. This paper will highlight the impact that diet, and the way we feed can have both positively and negatively on equine performance and health. It will highlight recent work on the role of antioxidants in respiratory health, in particular the role of vitamin C in recurrent airway obstruction, as well as energy sources in the growing and exercising horse. The potential importance of determining the glycaemic response to a diet will be considered with reference to laminitis, obesity, DOD and insulin resistance

Session 12 Theatre 2

Impact of nutrition on skeletal development in the growing horse

I. Vervuert and M. Coenen, Institute of Animal Nutrition, Nutrition Diseases and Dietetics, Gustav-Kühns-Str. 8, D-4159 Leipzig, Germany

In order to produce healthy sport horses it is necessary to begin feeding according to nutrient requirements for the growing horse early in life to ensure an optimal skeletal development. Milk intake meets foal´s nutritional needs for about the first two months of life, but in the course of lactation period, energy, protein and mineral supply by milk does not cover foal´s nutrient requirements. Energy and protein intake are the primary factors influencing the growth rates of young horses. Furthermore, the quality of energy intake (e.g. carbohydrates with a high glycaemic and insulinaemic load) raises greater awareness in the last few years as insulin metabolism might affect cartilage differentiation. The most obvious function of minerals in the body is to provide structural support of the skeleton, but maintaining acid-base, water balance, membrane permeability as well as regulatory functions on cell replication, differentiation and other cell processes should not be neglected. Functions of minerals are interrelated against each other (e.g. Ca and P); which makes it necessary to have a well balanced composition of each mineral in the diet. An undersupply, but also an excess of minerals affect skeletal development adversely (e.g. lack of Ca and P can cause bone deformities). The role of Vitamin A and Vitamin D in the growing horse is not completely clear, however an excessive intake of those vitamins is supposed to effect skeletal development.

Session 12 Theatre 3

Effects of feeding and management practices on developmental orthopaedic disease (DOD) in French foals

J. Lepeule[1], N. Bareille[1], C. Robert[2], J.M. Denoix[2] and H. Seegers[1], [1]Animal Health Management Group, BP 40706, 44307 Nantes, France, Metropolitan, [2]Biomécanique et Pathologie Locomotrice du Cheval, 7 avenue du Général de Gaulle, 94700 Maisons-Alfort, France, Metropolitan

The aim of this study was to assess the effects of feeding and management practices on DOD prevalence at weaning. 392 foals from 3 breeds (French Trotter, Selle Français, Thoroughbred) were included in a cohort study. To determine the statuses of foals regarding DOD, they were x-rayed at weaning on the front- and hind-limb digits, carpi, hocks and stifle joints. Logistic regression models were run and included adjustment terms for individual characteristics (breed, month of birth, etc...). Studied feeding practices (for the mares during the late pregnancy and early lactation and for the foals) were: mean daily amount of concentrates, duration of their distribution, and calcium/phosphorus and copper/zinc ratios. Studied management practices were: age at grazing, batch size, surface and slope of pastures, regularity and duration of exercise (access to pasture), and frequency and regularity of handling. Prevalence of DOD was 46.3%. The significant ($p < 0.10$) risk factors of DOD identified were: a daily amount of concentrates distributed to the mare comprised between 4 and 7 kg compared to less than 4kg (Odds Ratio (OR) =2.0); a large surface of pasture offered during the early period (until 2 months of age) (OR=4.8); irregularity of exercise during the early period (OR=2.0).

Session 12 — Theatre 4

Influence of management on growth and development of the Lusitano horse

M.J. Fradinho[1], L. Mateus[1], R. Agrícola[2], M.J. Correia[3], M.J.C. Vila-Viçosa[4], M.F. Silva[1], G. Ferreira-dias[1] and R.M. Caldeira[1], [1]Faculdade de Medicina Veterinária, TULisbon, CIISA, Av. Universidade Técnica, 1300-477, Portugal, [2]Coudelaria Nacional, Fonte-Boa, Santarém, Portugal, [3]Coudelaria de Alter, C. Arneiro, Alter-do-Chão, Portugal, [4]U Évora - Dep. Med. Vet., Herd. Mitra, Évora, Portugal

The aim of this study was to characterize body condition changes in Lusitano mares under extensive management conditions, and its influence on foals growth and development, from birth to weaning. Body condition score (BCS) and body weight (BW) were monthly assessed on mares from the 9th month of gestation to weaning. Foals were monthly weighed and withers height (WH), girth (G) and cannon circumference (CC) were measured from birth to weaning. Foals blood was collected for determination of bone alkaline phosphatase, osteocalcin, IGF-I and leptin, and superficial cortical bone properties were assessed by QUS. Statistical analysis of biometrical parameters was performed with SAS by PROC GLM. Blood indicators and QUS data were analyzed by use of ANOVA for repeated measures. No significant changes were found in BCS throughout the study, although differences between farms were detected ($P < 0.05$). Quadratic models were best fitted to describe the variation of BW, G, WH and CC of foals. The effect of group was significant for BW, G, and CC ($P < 0.0001$). Blood indicators results suggest similar age-related changes observed in other light breeds. QUS measurements provide preliminary data on cortical bone assessment at this age.

Session 12 — Theatre 5

Blood bone markers related to birth date in a Hanoverian foal population

I. Vervuert[1,2], S. Winkelsett[1], L. Christmann[3], E. Bruns[4], B. Hertsch[5] and M. Coenen[1,2], [1]Institute of Animal Nutrition, Bischofsholer Damm 15, D-30173 Hannover, Germany, [2]Institute of Animal Nutrition, Nutrition Diseases and Dietetics, Gustav-Kühn-Str. 8, D-04159 Leipzig, Germany, [3]Hanoverian Breeders Association, Lindhooper Str. 65, D-27283 Verden, Germany, [4]Institute for Animal Breeding and Genetics, Albrecht-Thaer Weg 3, D-37075 Göttingen, Germany, [5]Clinic for Horses, Surgery and Radiology, Oertzenweg 19, D-14163 Berlin, Germany

There is growing interest in the healthy skeletal development in the early phase of foals life. 284 foals were selected for monitoring bone markers in blood. Foals were subdivided according to birth date and housing management into early born and late born foals. Plasma osteocalcin and PICP were analysed as markers of bone formation, and ICTP was determined as a marker of bone resorption. Animals were x-rayed for information on osteochondrosis between the fifth and tenth months after birth. Osteocalcin, PICP and ICTP in plasma decreased with age ($P < 0.05$), but these changes were more distinct in the late born foals than in the early born ones. Neither sex nor the predisposition to osteochondrosis affected the course of bone markers in either group. As expected, there was an age-related decrease in bone markers during the first 200 days of life. Surprisingly, there was a correlation between the fall in bone markers and date of birth, indicating reduced rates of skeletal remodeling in late born foals.

Session 12 Theatre 6

How can we predict an increased risk of Laminitis and what are the main Nutritional countermeasures to Laminitis

P. Harris, WALTHAM Centre for pet nutrition, Equine Studies Group, Freeby Lane, Waltham on the wolds, Leics, LE14 4RT, United Kingdom

Laminitis occurs all around the world in horses and ponies and has major welfare implications. Recognising and treating the condition in its early stages so that the pain and suffering is kept to a minimum is obviously important. However, it would be preferable to be able to recommend certain interventions/ countermeasures that avoid or prevent the condition from occurring in the first place. Determining which animals have an increased risk of suffering from this condition enables these countermeasures to be targeted at the most vulnerable group. New diagnostic techniques with the potential to identify at risk animals will be discussed. The main potential nutritional and managemental countermeasures that may help to reduce the incidence or severity of this important condition, in particular those that improve insulin sensitivity will be highlighted.

Session 12 Theatre 7

Se enriched fertilizers to naturally increase the Se content of cereals, hay and straw used in a horse diet: effects on the antioxydant status

L. Istasse[1], S. Paeffgen[2], O. Dotreppe[1] and J.L. Hornick[1], [1]Faculty of Veterinary Medecine, University of Liege, Animal Production, Nutrition Unit, B43, Boulevard de Colonster, 20, 4000 Liège, Belgium, [2]Kemira GrowHow, Avenue Einstein, 11, 1300 Wavre, Belgium

Se is a trace element of importance implicated in processes such as antioxidant mechanisms, immune response, reproduction, Se could be supplemented either on an organic or on an inorganic form. Selenate contained in fertilizers is naturally transformed in organic forms (selenomethionine fe) by the plants. Three g Se/ha were applied with the fertilizer in a pasture for hay production. Similarly 4g Se/ha were spread with the third nitrogen application on winter barley and on spelt fields. Those feedstuffs were incorporated in a horses ration containing 57% concentrate and 43% roughage. The Se concentrations were 63 and 297 µg/kg DM in the control and the Se enriched diets. Six adult horses trained 4 days a week were used in a cross – over design. In the control group, plasma Se content was 112.3 µg/l while it was 146.0µg/l in the Se group ($P < 0.001$). The Se plasma concentration increased with time in the Se group. The oxydative burst, a measure of blood neutrophils activity was also affected by the dietary Se inclusion. It was concluded that the use of Se enriched fertilizers is a good technique to improve the Se content of roughages and of cereals for horses diet with as results improvement of the antioxydant status.

Session 12 Theatre 8

Genetics and environment in equine health

V. Gerber[1], A. Ramseyer[1,2], M. Mele[1,2] and D. Burger[2], [1]Equine Clinic University of Berne, Vetsuisse-Faculty, Laenggassstrasse 124, 3012 Bern, Switzerland, [2]Swiss National Stud, Les Longs-Prés, 1580 Avenches, Switzerland

The original debate of nature *versus* nurture has evolved into understanding that it is neither genetics nor the environment that are exclusively responsible for an individual's phenotype. In developmental orthopaedic disease as well as in polysaccharide storage myopathy, alimentation, prolonged inactivity/intense activity as well as genetic factors contribute to disease manifestation. Equine Sarcoid and Insect Bite Hypersensitivity (IBH) are skin diseases for which the manifestation appears to depend on a genetic predisposition and exposure to an allergen (culicoides) or pathogen (*Bovine papilloma* virus), respectively. The manifestation of IBH in Icelandic horses suggests that the presence of culicoides can not only have a triggering, but also a protective effect: exposure early in life, or possibly passive transfer of maternal immunity, seems to suppress the allergic response. Furthermore, there is increasing evidence that genetic and management factors also influence the manifestation of recurrent airway obstruction as well as the oral stereotypies, crib-biting and wind-sucking. In the future, interdisciplinary approaches will be necessary to more specifically and quantitatively define the nature, strength and interactions (synergistic, multiplicative, additive or antagonistic, protective) of the genetic and environmental factors.

Session 12 Theatre 9

Genetics and management in equine recurrent airway obstruction

V. Gerber[1], J. Klukowska-Roetzler[1], G. Dolf[2], D. Burger[3] and A. Ramseyer[1], [1]Equine Clinic, Vetsuisse-Faculty, University of Berne, 3012 Bern, Switzerland, [2]Institute of Genetics, Vetsuisse-Faculty, University of Berne, 3012 Bern, Switzerland, [3]Swiss National Stud, Les Longs-Prés, 1580 Avenches, Switzerland

Equine recurrent airway obstruction (RAO) is characterized by clinically evident increased breathing effort due to cholinergic bronchospasm, coughing and airway hyperreactivity as well as neutrophil and mucus accumulation in the airways. Owners of descendants of two RAO affected founders (sire 1 and sire 2) as well as age-matched controls (all horses 6 years old or older) were contacted by phone to gather information by a standardized questionnaire: frequency and severity of respiratory symptoms; management, feeding, deworming as well as other diseases and the use of the horse. Sire (4.1-5.5 fold increased risk), hay-feeding and age (in decreasing order of strength), but not time-spent-outside, were associated with more severe clinical signs. We further found some preliminary evidence for an effect of high deworming frequency on severity of clinical signs and an association of high HOARSI with chronic recurrent urticaria. Moreover, a strong genetic linkage and increased gene expression of the interleukin 4 receptor α chain (*IL4RA*) gene was detected in the sire 1 family, but not in the other. We propose that there is a inverse relationship between predisposition / susceptibility to allergic disease and parasites and that complex gene-environment interactions influence RAO manifestation through the IL4Rα pathway.

Session 12 Theatre 10

Lower critical temperature of competition horses

K. Morgan, L. Aspång and S. Holmgren, Swedish Equestrian Educational Centre, Ridsportens Hus, 734 94 Strömsholm, Sweden

The common theory, contradictory to horse practice, is that the active competition horse is considerable more resistant to ambient climate than the horse on maintenance, since the active competition horse get more feed and therefore has a higher energy intake. The aim of the study was to investigate if there was any difference in lower critical temperature, LCT, between the active competition horse and the horse on maintenance. The practical aim was to improve management of competition horses in relation to climate. The hypothesis was that there is a difference in LCT, which is due to the difference in feed intake. In the study we used a computer model for estimation of LCT in three different types of individuals used for equitation; pony, warmblood and thoroughbred. We studied horses in winter coat and clipped horses. The intensity of feeding were maintenance and competition. A relative part of heat production of the metabolisable feed intake energy was estimated. Input data were all based on previous research. The result showed a variation in LCT; for the pony from 1,4°C to10,8 °C, the thoroughbred from –2,1°C to 7,9 °C and the warmblood from –3,4°C to 7,4°C. The competition horse had a higher energy intake in total, that gave a lower LCT in absolute value. The span between LCT was more narrow with this new model compared to previous model with a constant part of heat production. In conclusion, the climate resistance of the competition horse has been overestimated in the earlier model.

Session 12 Theatre 11

Management factors and behaviour in horses

I. Bachmann[1] and E. Søndergaard[2], [1]Haras national suisse, CP 191, 1580 Avenches, Switzerland, [2]University of Aarhus, Faculty of Agricultural Sciences, P.O. Box 50, 8830 Tjele, Denmark

Housing and management conditions influence most of the functional systems of the normal behaviour of horses, e.g. feeding, locomotion and social behaviour. In nature, locomotion and feeding behaviour are closely connected as feeding means grazing and walking at the same time. When housing horses, the two behaviours are separated and none of them occur in the natural way. Horses are often fed with limited amounts of roughage, and they have only limited access to exercise and some only to forced exercise. Most horses are housed singly, thus their opportunities for social behaviour are very limited. Horses are capable of adapting to many circumstances, but sometimes limits are reached and behaviour problems appear, often as stereotypies like crib biting, box walking or weaving. Stereotypies are a sign that welfare problems are or were present. It is thus expected that also the performance of the horses may be affected but our knowledge in that area is very scarce. However, we know today that e.g. the social environment of young horses can influence the human-animal relationship and as a consequence the training ability and learning aptitude. For the well being and probably also for the performance of horses, more attention should be given to the housing of horses in a behaviour adapted way. Housing and management conditions should allow tactile contact to other horses, daily free movement, as well as the provision of high amounts of roughage.

Session 12 Poster 12

Signs of hereditary diseases in three-year-old franches-montagnes horses

M. Mele[1,2], V. Gerber[1], R. Straub[1], C. Gaillard[3], L. Jallon[4] and D. Burger[2], [1]Equine Clinic, Vetsuisse-Faculty, University of Berne, 3012 Bern, Switzerland, [2]Swiss National Stud, Les Longs-Prés, 1580 Avenches, Switzerland, [3]Institute of Genetics, Vetsuisse-Faculty, University of Bern, 3012 Bern, Switzerland, [4]Swiss franches-montagnes horse breeding association, Les Longs-Prés, 1580 Avenches, Switzerland

The objective of this study was to investigate clinical signs of diseases with known or suspected hereditary components like equine sarcoid, insect bite hypersensitivity (IBH), osteochondrosis, podotrochleosis, prognatism and wind-sucking in the franches-montagnes (FM) horse. We performed a clinical examination on 702 three-year-old FM horses, which were shown at the Swiss-Field-Tests in 2004. A questionnaire on health, environment and feeding habits of the animals was completed. In 11.9% of the horses, sarcoids were detected. The prevalence was higher in chestnuts (16.6%) than in bays (10.1%). The prevalence of sarcoids in offspring from sires with known sarcoids was not significantly higher than in descendants from stallions without a known history of sarcoids. Clinical signs of IBH were only found in six horses (0.9%). In 12.0% of the animals, hoof abnormalities were recorded. The angle between hoof base and hoof wall was 56.7 ± 0.1°, the average hoof width was 13.7 ± 0.3cm in the front feet. We found no significant difference between left and right feet. With the exception of a high prevalence of sarcoid, our results indicate that the FM horse is overall a healthy breed.

Session 12 Poster 13

Signs of hereditary diseases in three-year-old Swiss Warmblood horses

S. Studer[1], V. Gerber[2], R. Straub[2], W. Brehm[2], C. Gaillard[3], A. Lueth[4] and D. Burger[1], [1]Swiss National Stud, Les Longs-Prés, 1580 Avenches, Switzerland, [2]Equine Clinic, Vetsuisse-Faculty, University of Berne, 3012 Bern, Switzerland, [3]Institute of Genetics, Vetsuisse-Faculty, University of Berne, 3012 Bern, Switzerland, [4]Swiss Sporthorse Breeding Association, Les Longs-Prés, 1580 Avenches, Switzerland

The objective of this study was to investigate clinical signs of hereditary diseases like equine sarcoid, osteochondrosis (OC) and idiopathic laryngeal hemiplegia (ILH) in relation to environment, management and conformation of the horses. For this purpose, we analyzed veterinary examinations of 403 stallions at their approvals since 1994 and examined 493 three-year-old Swiss Warmblood horses, which were shown at the Swiss-Field-Tests in 2005. A questionnaire on health and management of the animals was completed. In 11.5% of horses sarcoids were found. The prevalence of sarcoids in offspring of sires with known sarcoids was not significantly higher than in descendants from stallions without a known history of sarcoids. We found distended joints as a possible symptom of OC in 11.4% of the horses. We did not find a relationship between enlarged joints in the offspring and the presence of OC in the sires. While we found a high number of sarcoid affected horses compared to other studies, the prevalence of enlarged joints was low and very few horses displayed abnormal respiratory noise at work. Furthermore, we found no correlation between conformation and the horse's general health.

Session 12 Poster 14

Relationship between activity, growth and endurance of trotters

L. Voswinkel[1], K.-H. Tölle[2], D. Hinrichs[1], K. Blobel[3] and J. Krieter[1], [1]Institute of Animal Breeding and Husbandry, Christian-Albrechts-University, Olshausenstr.40, 24098 Kiel, Germany, [2]Chamber of Agriculture Schleswig-Holstein, Futterkamp, 24327 Blekendorf, Germany, [3]HIPPO-Blobel, Klaus-Groth-Str.52, 22926 Ahrensburg, Germany

The aim of the study was to examine the influence of the activity of juvenescent horses on growth parameters and on endurance. Data of 30 trotters from one stud were available. The weanlings were kept together in two groups –seperated by sex- in loose housing systems with direct access to pasture. Activity-measurements were taken using pedometers. For intervals of ten weeks, development in height at withers, cannon bone diameter, weight and diameter of flexortendons were recorded. X-rays of the epiphysis of the distal radius were also taken. During training heart rate, speed and distance were recorded by a combined heart rate/Global Positioning System. Five minutes after load, blood samples were taken to measure the lactate value. Statistical analysis was performed with mixed linear models. Positive phenotypic correlations were found between activity and diameter of superficial digital flexortendons (r=0.27) as well as diameter of deep digital flexortendons (r=0.24). Hardly any relationship could be determined between the activity during rearing and the heart rate. A negative correlation was estimated between activity and lactate value (r=-0.40). These results show the importance of activity during rearing as a factor which enables horses to become most resistant to injury during adulthood.

Session 13 Theatre 1

Farmers'decisions: it is not only profit and production that counts

S.J. Oosting[1] and B.B. Bock[2], [1]Wageningen University, Animal Production Systems Group, P.O. Box 338, 6700 AH Wageningen, Netherlands, [2]Wageningen University, Rural Sociology Group, Hollandseweg 1, 6706 KN Wageningen, Netherlands

Economic theory presumes rational reasoning as the dominant driver of farmers' decision making. To maximize profits farmers are supposed to decide on basis of rational calculation of availability and quality of resources. But farmers are no economists and it is not only money that counts. We present examples from the Netherlands that illustrate subjective and personal motives of farmers' decisions. Among dairy farmers, for instance, we find those who are inspired by passion for cattle and others driven by best grassland management and a third group taking their technological equipment as basis for decisions. Under present day conditions, however, only a relatively small group is able to meet ends by specialized dairy farming alone. Farmers who do not follow the development pathway of up scaling and intensification create alternative farm types motivated by a great variety of often very personal values: organic farmers start from a specific attitude to land and animals, while green care farmers do so from a drive to help people. Of course they also want to make money but economic reasoning follows value-based decisions. And the decisions they take vary as well, demonstrating that there is not one optimal decision: green care farmers maintain a slightly adapted, but strong livestock component, while those with recreational activities gradually reduce farming to a minimal level.

Session 13 Theatre 2

Representing farmers' objectives in integrated models: trying to hit a moving target for agricultural development

P.K. Thornton, ILRI, PO Box 30709, Nairobi 00100, Kenya

Livestock keepers everywhere are facing considerable pressures of change. Human populations in many developing countries are still growing rapidly. Increases are likely in the consumption of animal products as a result of population growth, higher incomes, increased urbanisation, and changing dietary preferences. The impacts of climate change on agriculture will add significantly to the challenges of development. There may be opportunities for some smallholders to benefit from increases in demand for livestock products, but for many, the risks associated with climate change and increased climate variability and land pressure will be difficult to cope with. Livestock research has a substantial contribution to make in helping millions of people adapt to change. However, to be effective, research outputs need to be targeted appropriately, a key part of which is understanding the objectives of agricultural households. ILRI and partners have undertaken integrated modelling activities in the livestock systems of East Africa over several years. Some results of this work are presented, in which household objectives vary widely. Different types of household models are needed to investigate different questions. Integrated modelling is a tool of great utility for assessing impacts and adaptation strategies, but there is still much to do in modelling the decision-making of livestock keepers, in the search for livelihood options that can help to alleviate poverty and increase food security.

Session 13 Theatre 3

A proposal of a grid to analyse farmers' informational activity

M.-A. Magne, S. Ingrand and M. Cerf, UMR Métafort, TSE, INRA Theix, 63122 Saint Genes Champanelle, France

The farmers' decisional process is generated by the fit or the misfit between farmer's goals and his actual working situation. Understanding the informational activity (i.e. what occurs before the final decision) is a major stake to better understand the decisional process and then to be able to help farmers to take the "good" one. We assume that modelling the farmers' informational activity to be able to represent the diversity of how they take their decision, is an appropriate methodology. In this perspective, we propose to model the informational activity by using a pyramidal diagram linking four elements: the farmer, the fields concerned by the farming activity, the events concerned by the decisions and the informational resources. Four hypotheses are tested to explain the different relationships within this analysis framework and then the whole system of information. We present i) the different elements relevant for understanding the structure and the functioning of the system; ii) the different criteria explaining how the system is activated by different kind of events. These two types of results will be mobilised with our partners working in advising structures to think about new ways and methodologies to reach farmers' requirements.

Session 13 Theatre 4

Management tools coupled with a database to support dairy farmers in decision making

S. Karsten[1], E. Stamer[2], E. Kramer[1], W. Junge[1] and E. Kalm[1], [1]Institute of Animal Breeding and Husbandry, Hermann-Rodewald-Str. 6, 24118 Kiel, Germany, [2]TiDa Tier und Daten GmbH, Bosseer Str. 4c, 24259 Brux, Germany

Technologies like electronic identification and sensor technique have been developed to record individual cow performances. Day-to-day decision making is mainly based on that information. For improvement of production efficiency and profitability, unexpected and undesirable change in individual performance as well as in the production process should be quickly identified. But the amount of data recorded is huge and not easy to evaluate by humans. Hence, tools are needed to integrate and process the data contemporary. For utilisation by farmers, the tools must be easy to use and the results should be plain. For integration of the data of the on-farm herd management software, process techniques and off-farm data sources, the database "KuhDaM" has been developed. "KuhDaM" processes nearly all information collected on dairy farms. Data are transferred into the database at least weekly to ensure its appropriateness as management tool. To help farmers with day-to-day decision making and management decisions, Internet-based analysis tools have been coupled with the database to monitor the production process and the functioning of the equipment. Cusum control charts represent management data in a graph and are therefore easy to understand and to interpret. Examples for applications are the support of estrus detection by using activity measurements or the monitoring of feed intake.

Session 13 Theatre 5

Multipurpose fodder trees in the Ethiopian highlands: farmers preference and relationship of indigenous knowledge of feed value with laboratory indicators

A. Mekoya[1], S.J. Oosting[1], S. Fernandez-Rivera[2] and A.J. van der Zijpp[1], [1]Wageningen University, Animal Production Systems Group, Marijkeweg 40, 6700 AH Wageningen, Netherlands, [2]International Livestock Research Institute, Addis Ababa, P.O.Box 5689, Ethiopia

In Ethiopia, the introduction of exotic multipurpose fodder trees (MPFT) started in 1970s for livestock feed and soil conservation. However, the adoption of exotic MPFT was limited. The objectives of this study were to assess farmers' preference criteria, compare their preference between exotic and local MPFT and evaluate the relationship of farmers' knowledge of feed value with laboratory indicators. Focus group discussions and preference ranking and scoring were conducted in two production systems in the Ethiopian highlands. Farmers preferred local MPFT to exotics for biomass production, multi-functionality, life span, and compatibility to the cropping system. In terms of feed value, ease of propagation and growth potential, farmers ranked local MPFT lower than or comparable to exotics. There was strong correlation between farmers feed value score and laboratory results. Farmers were able to discriminate effectively MPFT species that had high and low protein and fibre content. We concluded that incorporating locally available MPFT, farmers' indigenous knowledge and preference criteria at the research and development inception process is vital to offer a wider dimension of opportunities for acceptability of a technology by farmers.

Session 13 Theatre 6

Differences of technology adoption and objectives of mountain cattle farmers depending on continuity prospects

A.M. Olaizola[1], A. García-Martínez[2] and A. Bernués[2], [1]Universidad de Zaragoza, Agricultura y Economía Agraria, Miguel Servet 177., 50013-Zaragoza, Spain, [2]CITA-Gobierno de Aragón, Tecnología en Producción Animal, Apdo. 727, 50080-Zaragoza, Spain

Continuity of farms is a central issue when assessing the sustainability of agro-ecosystems in mountain and other less favoured areas in the short-medium term. Continuity chances of a sample of 71 farms in 3 valleys of the Spanish Pyrenees was determined for the next 15 years in terms of farmer age and presence of descendants willing to continue in agriculture. Data was collected trough direct questionnaires to farmers. Differences in technology adoption during a 5-year period previous to the interview and in farmer's intentions of adoption in the 5-year period after the interview were analysed. Main differences in the previous 5-years referred to: improvement of facilities and buildings, use of larger grazing areas, increment of fenced grazing areas and integration into product quality schemes, significantly more frequent in farms with good chances of continuity. For the next 5-year period, further fencing of grazing areas and diversification towards tourism activities were the changes most frequently envisaged by farmers with continuity. Farmer's objectives were also analysed; the most significant differences were: re-investment in agriculture, reduction of debts, increasing physical size of the farm and maintaining the farm in good farming conditions (more important for farmers with continuity).

Session 13 Theatre 7

Low stocking rate as a source of flexibility in beef-farming systems: the analysis of a 15-year trajectory for seven farms in the Limousin region

L. Astigarraga and S. Ingrand, INRA, SAD, UMR1273- Equipe TSE, 63122 St Genes Champanelle, France

The aim of this work was to study the technological practices adopted by beef producers in the Limousin region (France) and their production system capacity to adapt to market prices and climate fluctuations. A 15-year trajectory was surveyed on seven farms in the early nineties, selected because of their extensive conditions of production (<1.0 UGB/ha, >70 ha). Quantitative (economic results and performance) and qualitative analysis (based on interviews) were utilized to characterize the evolution of herd management and land use. The principal changes were observed on the final weight and age of livestock categories, obtaining heavier (grass-fed calves) and older (heifers for breed and bull-calves) animals. These changes were concordant with an increase in pasture areas and consequent reduction of areas for crops (cereals), the introduction of corn for silage and an earlier calving season in the fall. The adoption of these practices is thought to constitute structural-adaptation strategies to avoid the risk of drought. The consequences of these adaptive changes in livestock farming are further discussed.

Session 13 — Theatre 8

Choice of a suckler cattle farming system: a decision support tool confronted with the farmers' behaviour

P. Veysset, M. Lherm, D. Bébin and K. Bensaid, INRA, UR506, Laboratoire d'Economie de l'Elevage, 63122 St Genes Champanelle, France

The complexity of the farms' socio-economic environment, as well as the great diversity of existing production systems, makes strategic decision-making very difficult. INRA's Livestock Farming Economics Unit developed a linear programming model, called Opt'INRA, to optimize diverse activities in suckler cattle farms. An analysis of several farm components (structural, human, livestock and plant production) was carried out and the outputs of the model were compared with reality by confronting optimal solutions with specific assets in 22 farms. Concerning livestock and plant production, the deviations from reality referred either to structural constraints (land use, soil and climate conditions) or to farmer's behaviour: anticipation of future markets (stability of Italian market), know-how (fattening), social environment (producers group) and risk aversion. Labour management was considered in farmers' decisions according to the historical stage of the farm (farmer establishment, middle and end of career), the structure of the farm and other factors of production (buildings, machinery). Decision support tools should be designed and developed together with end-users for them to be usable and useful. Such tools encourage the dialogue between the adviser and the farmer, highlighting the most sensitive activities and the farmer's perception of his/ her profession.

Session 13 — Theatre 9

Do labour productivity and work pace expectations affect reproduction management and performance in pig farms?

G. Martel[1,2], J.-Y. Dourmad[1] and B. Dedieu[2], [1]INRA, UMR 1079, SENAH, Domaine de la Prise, 35590 Saint Gilles, France, [2]INRA, UMR 1273, Métafort, INRA, 63122 Saint Genès Champanelle, France

Increases in labour productivity are an essential lever, as well as technical effectiveness, for competitiveness of pig farming. However, the search for controlled daily work or available days for vacation also becomes important. The objective of this study is to explore how these expectations about work might bring specific combinations of practices or affect performance. The study was carried out by direct investigation of stockbreeders from Brittany (France). Factorial analysis was used to identify relationships between practices, labour productivity, sows productivity and work pace. Results showed independence between sow productivity and labour productivity. Three independent types of work pace expectations were identified: the limitation of daily work density and the avoidance of activities during the weekend, either insemination or farrowing supervision. Finally, results indicated a relationship between weaning, oestrus detection and insemination techniques, and work pace. A relationship is also seen between farrowing and cross-fostering techniques, and work and sow productivity. Results suggest that work pace expectations influence the choice of reproduction practices especially at a family farm. Although labour productivity is affected by some specific techniques at farrowing, the results also indicate that it was mainly related to the size of farrowing batches.

Session 13 Theatre 10

Dynamics of farming styles in pig farmers' decision making in France

M.A.M. Commandeur and F. Casabianca, INRA LRDE, SAD, Quartier Grossetti, 20250 Corte, France

A comparison study was done in Brittany (region of strong pig production) and in Midi Pyrenees (region of declining pig production). Field surveys were conducted in both regions among pig farmers, who are implicated in both farrowing and finishing. Five dimensions were identified that best describe the contrasts in farmers' logic: *i* Animals and Technology, *ii* Labour and Investments, *iii* Ambitions, *iv* Socio-Professional Integration, and *v* Images of Methods and Products. Comparison of the datasets of the two regions reveals that the farmers are surrounded by different *spaces of information*, which appeal differently to their specific modes of rationalization. Issues like plural activity and multi functionality were hardly relevant in Brittany, but very relevant in Midi Pyrenees. In Brittany two styles of *entrepreneur* were identified, alongside *craftsman*, *inheritor*, and *stockman*; whereas only the latter three styles were identified in Midi Pyrenees. Perspectives for pig farming in Brittany depend largely on the farmers' capacity to use the opportunities inside the production basin for manipulating economic margins. In contrast in Midi Pyrenees, the perspectives depend on the cooperative capacity to restore a link of pig production to regional quality features and to implement subsequent projects. The farming styles approach to study farmers' decision making reveals dominant influences of the local spaces of information and the dynamics of the specific production basins.

Session 14 Theatre 1

Artificial insemination and embryotransfer in sheep and goats: state of the art

W. Holtz and M. Gauly, Institute for Animal Husbandry and Genetics, Goettingen University, Albrecht-Thaerweg 3, 37075 Goettingen, Germany

Semen collection, requiring a teaser and artificial vagina, is an established technique in sheep and goats. Under certain conditions electroejaculation is preferred. In sheep, artificial insemination (AI) usually implies the use of fresh semen. Few authors report satisfactory results with cryopreserved semen deposited in the vagina or cervix. The limiting factor is the inability to penetrate the cervix. In some countries laparoscopic intrauterine insemination is common. In goats, generally penetration of the cervix is possible. With fresh semen conception rates resembling the results of natural mating are achievable. Some groups report good conception rates with cryopreserved semen, others do not and resort to laparoscopy. A recently established new technique permits intrauterine insemination of goats by transcervical route with satisfactory results. Embryotransfer comprises collection and transfer of morulae or blastocysts following superovulation. Until recently, in sheep and goats collection and transfer involved surgery. A few specialists collect embryos laparoscopically. Not long ago a nonsurgical way of collecting embryos has been devised for goats. Transfer of embryos requires at least a mini-invasive intervention, sometimes supported by laparoscopy. Cryopreservation and a range of manipulations on embryos from microsurgery to cloning and generation of transgenic stock offer tremendous potential for genetic improvement and pharmaceutical purposes.

Comparison of artificial insemination methods in sheep using semen from *Ovis g. musimon*

G.M. Vacca, V. Carcangiu, M. Pazzola, M.L. Dettori, S. Luridiana and P.P. Bini, Dipartimento di Biologia Animale, via Vienna 2, 07100 Sassari, Italy

Reproductive biotechnologies find application both in conservation of endangered species and to improve production from domestic animals. The cross between Mouflon and domestic sheep could be exploited to produce suckling lamb, a foodstuff with high organoleptic qualities and that could meet modern consumer demand. Mouflon semen was collected from two adult males using an artificial vagina during the sexual season. 80 ewes were synchronized by insertion of intravaginal sponges (40 mg FGA), and by an intramuscular injection of 400 IU of PMSG at sponge removal. 40 ewes (Group A) were inseminated into the cervix using refrigerated semen (+4°C, 0.25 ml and 800 million spermatozoa *per dose*) 56 hours after the sponge removal. 40 ewes (Group B) were inseminated into the uterus using the laparoscopic technique 54 hours after sponge removal using cryopreserved and later thawed semen (0.25 ml and 400 million spermatozoa). Chi-squared test showed values significantly higher ($P < 0.01$) in fertility rate in Group A (47.5% *vs* 32.5%). In both groups prolificity rate was similar (1.1%). In conclusion, intracervical method with the use of fresh semen, gives better reproductive performances, and it is easy to apply for obtaining cross lamb on a large scale. On the other hand, result regarding laparoscopic-intrauterine technique could be considered good and the method is very useful because it allows to use semen in a rational way and in a longer lapse of time.

The development of artificial insemination of sheep and goats in Iceland

O.R. Dýrmundsson[1], T. Ólafsson[2] and J.V. Jónmundsson[1], [1]The Farmers Association of Iceland, Bændahöllin v/Hagatorg, IS-107 Reykjavík, Iceland, [2]Southram Artificial Insemination Centre, Austurvegur 1, IS-800 Selfoss, Iceland

Artificial insemination in sheep has been practiced in Iceland since 1939 and is now probably more widely applied there than in any other European country, except France. While fresh semen in 0.1 ml doses has been used to a large extent, and only by vaginal insemination, without a duckbill speculum, there has been a promising development since 1979 in using frozen semen. Most of the insemination work is carried out by the sheep farmers themselves receiving processed semen in straws from two AI centres in the country, by both road- and air transport. Furthermore, Icelandic frozen ram semen has been exported to the USA since 1998. While average conception rates of 70% for fresh semen and 50% for frozen semen are generally achieved there is considerable variation in fertility results, mainly ranging from 60-80% for individual rams. Rams of high breeding merit, normally progeny tested, are selected into the AI centres according to strict sanitary standards. The use of AI is of utmost importance in the breeding work on sheep farms in all parts of the country and most of the flocks are included in the individual recording scheme supervised by the Farmers Association of Iceland. In addition to making better use of valuable genetic material, AI benefits the national scrapie eradication scheme. Insemination in goats is still in its infancy in Iceland, beginning in 1998 and only applied on a small scale as yet.

Session 14 Theatre 4

Results of artificial insemination of ewes in various Hungarian sheep flocks

S. Kukovics, T. Németh and A. Molnár, Research Institute for Animal Breeding and Nutrition, Gesztenyés ut 1, 2053 Herceghalom, Hungary

The artificial insemination (AI) was quite common in Hungary during the 1960's when more than 63% of the total ewe population (approximately 1.4 million heads) was inseminated. Fresh semen collected on the farms and transported semen from regional AI centres were almost equally used. After the beginning of 1990's, the ratio of inseminated ewes were reduced and nowadays its level only 2-3%. In our present study the use of AI was evaluated and the practice of AI in three different sheep farms was studied in the years 2003-2006: a corporation farm with 5,000 heads and two private farms (300-300 heads). Selection of ewes on heat was carried out in the morning and two (in corporation) or three (other two farms) AI was administered 12 hours apart. Oestrus synchronisation also used in some parts of the populations (Chronogest sponge for 14 days followed by 500 IU PMSG), and AI administered 48-60 hours after the removal. Salamon type diluting liquid (ratio 1:5-1:10 depending the quality) was mainly applied to the locally collected fresh semen. Dosage of AI was 0.1 ml carrying approximately 50 million live sperms. Effectiveness of AI was a little bit different farm to farm: the best pregnancy rate was 90% (on smaller farms, cervical AI) and the lowest one was 40% (on corporation farm, vaginal AI). The returned ewes were mated by rams. Similarities and differences among the farms were summarised in tables and figures.

Session 14 Theatre 5

Effect of different hormonal treatments on reproductive activity of Sarda ewes

V. Carcangiu, G.M. Vacca, M.C. Mura, M. Pazzola, M.L. Dettori, A.M. Rocchigiani and P.P. Bini, Dipartimento di Biologia Animale, Via Vienna 2, 07100 Sassari, Italy

The aim of the study was to evaluate the effect of treatments with melatonin in association with progestinic drugs, or with melatonin alone, or progestinic drugs alone, on reproductive activity of Sarda ewes. The test was made on 200 pluriparous ewes that were in the same lactation state. The animals were subdivided into four groups that were homogeneous about the age and productivity levels. Group A was treated with melatonin, group B was treated with progestinic drugs (FGA); group C was treated with melatonin and progestinic drugs together; group D was the control group. On 26th of March the animals of A and C groups were treated with a slow release implant containing 18 mg melatonin. On 16th of April the groups were separated and the animals of B and C groups received vaginal sponges containing 40 mg FGA. On 1st of May the sponge were removed and were administered 400 UI of PMSG. In each group on the same day were introduced three rams. From 1st of May to 30th of June, every week from each animal a blood sample was drawn for progesterone dosage. Groups B (60%) and C (65%) showed the greatest number of pregnant ewes on 30th of May respect to the animals of A (50%) and D (40%) groups ($P < 0.01$). At the end of the test groups A (85%) and C (90%) showed the greatest number of pregnancies respect to B (70%) and D (68%) groups ($P < 0.05$). Results show that melatonin alone or in association had a positive influence on reproductive activity.

Session 14 Theatre 6

Synchronization of estrus in indigenous Kilis goat

Ü Yavuzer, University of Harran, Faculty of Agriculture, Dept. of Animal Science, Şanlıurfa, 63040, Turkey

This study has been conducted in order to compare fertility of indigenous Kilis goat on which have been performed estrus syncronization during their breeding season by using different hormonal treatments. A total of 75 female Kilis goats (2 to 6 years) were diveded randomly into three equal groups. Syncronized estrus was induced in indigenous Kilis Goats, breeding season, using a two dose $PG_{2\alpha}$ treatment 13 days apart in Group 1. The females in Group 2 were treated with 30 mg florogeston asetat (FGA) for 18 days and injected with 100 IU eCG at sponge withdrawal in animals. 25 goats which were not treated with any hormones formed the group, Group 3. Goats was checked two days after the last treatments with the aid of aproned bucks. Goats were hand-mated within two days after the last treatments. The female: male ratio was 5:1 during mating. Pregnancy diagnosis were performed 40 days after mating using ultrasound. Estrus response was found 75%, 80% for Group 1 and Group 2 respectively. There was no significant difference in estrus responce between Group 1 and Group 2. Prolificacy were 1.45, 1.55,1.50 for Group 1,Group 2 and Group 3 respectively. Prolificacy was not significantly affected by different treatments.

Session 14 Theatre 7

Male effect in Churra Galega Bragançana and Suffolk ewes under long-day artificial photoperiod

J. Azevedo[1], T. Correia[2], R. Valentim[2], J. Almeida[1], J. Simões[1], L. Galvão[2], H. Velasco[2], R. Maurício[2], P. Fontes[1], A. Mendonça[2] and M. Cardoso[2], [1]UTAD-CECAV, Apartado 1013, 5301-911 Vila Real, Portugal, [2]IPB-ESA, Apartado, 5301-855 Bragança, Portugal

This study aimed to evaluate the male effect in Churra Galega Bragançana (CGB) and Suffolk (S) ewes under artificial long-day photoperiod (16L:8D). On March 21st, 34 CGB and 27 S ewes, 2-5 years old, were allocated in light control facilities. Two months later, ovarian activity was registered by progesterone concentrations in blood plasma twice weekly. Ovarian activity was controlled by vaginal sponges (FGA). Vasectomised aproned rams were used to induce ram effect and estrus detection. Ewes were observed for estrus twice daily. Ewes presenting ovarian activity were identified by endoscopy. Chi-square tests were performed to compare proportions and Student's t-tests to compare means of the ovulation rate. After 2 months in 16L:8D photoperiod, 81.5% of Suffolk and 64.7% of CGB ewes were in anestrous ($P \leq 0.01$). Male effect was influenced by breed (S *vs* CGB) – Ovulation: 18.2% *vs* 63.6% ($P \leq 0.001$) and Estrus: 50.0% *vs* 75.0% ($P \leq 0.001$) –, except for ovulation rate (1.0±0.0). Data suggests that male effect is more effective in CGB than in Suffolk ewes.

Session 14 — Theatre 8

Hormonal stimulation and oocytes retrieval in FSH stimulated goats

G.M. Terzano, E.M. Senatore, M.C. Scatà and A. Borghese, Istituto Sperimentale per la Zootecnia, Animal production, Via Salaria 31, 00016 Monterotondo Scalo-Roma,Italy, Italy

The number and quality of laparoscopically recovered oocytes were compared in 42 multiparous goats subjected to 2 different superovulatory treatments. In group A (n=23) estrous was synchronized by the insertion of 45 mg FGA intravaginal sponges (Chronogest, Intervet, Italy) for 10d and 250IU of pFSH injected (im) 12h apart in a decreasing dosage over 4d, starting 72h before sponges removal; the sponges were removed 12 hours before the last pFSH injection and the aspirations were performed 12h later. Group B (n=19) were treated similarly to the group A with pFSH injection starting on day 4 and implants were not removed before the aspiration. A higher (P < 0.01) number of detected and aspirated follicles was found in group B (20.2-18.2) than in group A(16.4-14.9). A higher (P < 0.01) percentage of small follicles was found in group A(24.0) than in group B(8.8); a higher (P < 0.05) percentage of medium follicles was found in group B(79.5) than in group A(62.1). There was no difference in the number of large follicles (13.4 *vs* 11.8, respectively) and recovered oocytes (11.2 *vs* 12.5). When oocyte quality was concerned a higher rate of grade A(P < 0.05) and B(P < 0.01) oocytes was observed in group B(31.3-52.6) than in group A(7.6-29.1). The oocytes maturational competence was also significantly higher in group B (85.1%) than in group A (71.2%). The results suggest that treatment B improves follicular growth and oocytes quality in goats.

Session 14 — Theatre 9

Ovarian function and conception rate in protein-overfed lactating Awassi ewes at the beginning of the breeding season

A. Márton[1,2], V. Faigl[1], M. Keresztes[1], M. Kulcsár[1], L. Pál[2], F. Husvéth[2] and G.Y. Huszenicza[1], [1]Szent István University Faculty of Veterinary Sci., István u. 2., 1078 Budapest, Hungary, [2]University Pannonia Faculty of Agricultural Sci., Deák Ferenc u.16., 8360 Keszthely, Hungary

Interrelations of the ovarian response to a standard chronogest+eCG treatment with plasma levels of hormones (insulin, IGF-I, thyroids) and metabolites (non-esterified fatty acids, bOH-butyrate, urea-N, PUN) related to the energy and protein metabolism were studied in protein-overfed lactating Awassi ewes (n=105) in mid August. The ewes were inseminated with fresh-diluted semen after the gestagen removal, and mated thereafter; 26 of them conceived at the fixed-time AI (calculated from lambing dates). The ovarian function was monitored by milk progesterone (P4) profiles. The ovarian function before-treatment was still acyclic and already cyclic in 33 and 72 ewes, and 29 and 43 of the cyclic animals were in the follicular and luteal phases, respectively. After the gestagen removal almost all (n=104) ewes ovulated, although at AI elevated P4 levels related to presence of partially luteinized follicles, and short-lived CL-s were observed in 10 and 5 animals (none of them conceived). The cyclic ewes had higher insulin and IGF-I levels than the acyclic animals, and those not conceived had higher PUN than the pregnants. The other metabolic parameters did not differ. Neither the conception rate, nor the ovarian response was influenced by the pre-treatment ovarian function.

Dietary effects on the semen quality and freezing ability of Dorper rams

L.M.J. Schwalbach, N. Bester, J.P.C. Greyling, H.J. van der Merwe and K.C. Lehloenya, University of the Free State, Department of Animal, Wildlife and Grassland Sciences, P.O. Box 339, Bloemfontein, 9300, South Africa

A study to evaluate the effects of high dietary energy levels on the quality and freezing ability of ram semen was conducted. Twenty-four 11 to12 month old Dorper rams were randomly allocated to two groups of 12 rams each and fed two energy levels: Low Energy (LE; 6.52 MJ ME/kg DM) and High Energy (HE; 9.39 MJ ME/kg DM) diets for 127 days. At the end of the trial, semen was collected from all rams with the aid of an artificial vagina and frozen using a one-step dilution (1+4) technique using Salomon's medium containing 5% Glycerol. The fresh and frozen semen samples were evaluated using standard laboratory techniques and the results compared statistically. Results indicate that although higher energy levels accelerate body (ADG 229 *vs* 112g/ram/day) and testicular development (SC 35.5 *vs* 29.5cm) and resulted in scrotal fat deposition (280 *vs* 106g). However, these induced changes had no significant effect on the semen quality and quantity. Similar values were recorded (semen volume 1.2 *vs* 0.8 ml, motility 77 *vs* 88% and 91 *vs* 96% normal sperm cells, for the HE and LE groups, respectively). Similarly no significant differences were noted on the freezing ability evaluated post thawing (52 *vs* 60% live; 57 *vs* 60% motile and 91 *vs* 96% normal sperm cells). It was concluded that conditioning rams for a period of 4 months with a diet containing up to 9.39 MJ ME/kg DM does seem to be a safe practice.

Sexual performance of rams sequentially exposed to short-tailed and fat-tailed ewes

R.T. Kridli[1], A.Y. Abdullah[1], M. Momani Shaker[2] and K.Z. Mahmoud[1], [1]Jordan University of Science and Technology, Irbid, 22110, Jordan, [2]Czech University of Agriculture, Suchdol, Prague, Czech Republic

The objective of the study was to evaluate whether prior exposure of rams to short-tailed (ST) females would enhance their mating ability when later exposed to fat-tailed (FT) females. Twenty two yearling, Awassi (A; n=7), F_1 Charollais x Awassi (CA; n=7) and F_1 Romanov x Awassi (RA; n=8) rams were individually subjected to sexual performance tests on six 20-minute occasions; the first three of which were with Romanov x Awassi ewes (ST) and the last three were with Awassi ewes (FT). Mounting frequency was influenced by ram group x ewe type interaction ($P < 0.01$) being greater when rams were exposed to FT ewes. The ability of Awassi rams to raise the tail of FT ewes was greater ($P < 0.001$) than CA and RA rams (4.1±0.4, 0.7±0.4 and 0.96±0.4 times/20 min in A, CA and RA rams, respectively). Mating frequency was influenced by ram group ($P < 0.01$), ewe type ($P < 0.05$) and their interaction ($P < 0.01$). Mating frequency was greater when rams were exposed to ST ewes (0.9±0.1 and 0.5±0.1 matings/20 min with ST and FT ewes, respectively), while Awassi rams were more capable of mating with FT ewe than the other ram groups. The results of the present study indicate that exposing crossbred rams to short-tailed ewes does not appear to improve their mating ability when later exposed to fat-tailed ewes. Sexual performance traits may be affected by other factors rather than just heterosexual experience.

Session 14 Theatre 12

Fertility index for Austrian sheep and goats

B. Fuerst-Waltl and R. Baumung, University of Natural Resources and Applied Life Sciences Vienna, Department of Sustainable Agricultural Systems, Gregor Mendel-Str.33, 1180 Vienna, Austria

In sheep and goats, being usually kept in low input systems, functional traits may have a higher impact on profit than performance traits. One of the most important functional trait complexes, reproductive performance, can be assessed by the total number of offspring. An improvement of this trait may thus be achieved by optimizing single traits like age at first lambing, lambing interval, litter size, stillbirth rate and postnatal lamb losses as well as longevity. A new fertility index was developed including those traits except longevity for all Austrian sheep and goat breeds. Target trait is (number of lambs born + survived until 48h)/2 within breed and age of ewe. Not only the animal's own performance but also the performance of its dam and its paternal granddam are taken into account. Index weights are derived by assuming a heritability of 0.10 and a repeatability of 0.30 and that dam and paternal granddam are unrelated. The fertility index of an animal thus changes every time the animal itself, its dam or its paternal granddam lambs. It may also be calculated if only ancestor performances are available, i.e. for young male and female animals at auctions. The new fertility index is a simple breeding value estimation without taking environmental effects into account and is intended for supporting management decisions. Future developments should however focus on a breeding value estimation based on an animal model.

Session 14 Poster 13

Cryopreservation of goat semen without permeating cryoprotectants

S. Becker Silva and W. Holtz, Institute for Animal Husbandry and Genetics, Goettingen University, Albrecht-Thaerweg 3, 37075 Goettingen, Germany

Extenders for cryopreservation of semen generally contain permeating (cytotoxic) cryoprotectants; in sheep and goats usually glycerol (Gly). Here goat semen was frozen without Gly by exposing the cells to nonpermeating disaccharides. Ejaculates (n= 19) from 4 males were diluted at 30°C with 4 parts of Tris-egg yolk (TY) extender. After cooling to 4°C within 2h, equal aliquots were distributed to 4 groups. In Group 1 (Control, TYGly), 6.8% Gly was added; in Group 2, 300mM sucrose (Suc) and 3.4% Gly; in Group 3, 300mM Suc and in Group 4, 300mM trehalose (Tre). After 5 min of equilibration, semen in 0.25mL straws was frozen at -120°C and stored in liquid nitrogen. Straws were thawed at 38°C for 30s and their content was diluted with TY to near isoosmolality either in 1 or 5 steps at 20°C. Motility (Mot %) was judged after 6h of incubation at 38°C and post-thaw membrane integrity (MI) after eosin-nigrosin staining. MI was expressed as % of the TYGly group. Thawing regimen had no effect on Mot% or MI. The average Mot% after 6h of incubation for Groups 1, 2, 3 and 4 were 26, 44, 44 and 55%, respectively. Corresponding IM were 100, 220, 257 and 301%. Extenders containing disaccharides were superior to the traditionally used TYGly-extender ($P < 0.05$, Tukey test). There was little difference between sucrose extenders with or without a small amount of Gly. Tre proved to be more suitable than Suc. This *in vitro* study needs to be followed up by insemination trials.

Session 14 — Poster 14

Transcervical insemination in goats and sheep attempted

W. Holtz, B. Sohnrey and M. Gauly, Institute for Animal Husbandry and Genetics, Goettingen University, Albrecht-Thaerweg 3, 37075 Goettingen, Germany

A technique for inseminating goats with cryopreserved semen by transcervical route has recently been described (J Anim Sci 83, 1543 – 8, 2005). It comprises grasping of the external cervical os with sharp-pointed forceps, passing of a catheter with a stylet inserted through the cervix and, after withdrawal of the stylet, advancing it as deep as possible before passing a PE tube containing semen through it to accomplish deep uterine insemination. The catheter is then guided to the other side with a finger in the vaginal fornix, so semen is deposited in both horns. This technique is routinely used in our breeding flock whenever frozen semen is to be introduced. Since publication of the initial results (71% pregnancies, 1.76 kids/doe) another 133 does have been inseminated, yielding 83 (62.4%) pregnancies with 1.96 kids born. With the acquired skill and similar equipment it was attempted to AI sheep: multiparous ewes of the German Blackheaded Mutton breed brought into estrus by treatment with progestagen sponges and eCG. Numerous efforts to penetrate the cervical canal were futile. Neither attempts to ease the cervical lock by applying estradiol-17beta or prostaglandin E dissolved in medical gel to its external orifice at various times before attempted penetration nor i.m. or i.v. application of up to 800 i.u. of oxytocin were of any avail. In conclusion, attempts to apply the newly established AI method for goats to sheep, were not met with success.

Session 14 — Poster 15

The effect of using honeybee royal jelly for ram semen dilution and freezing

Y. Jafari Ahangari, University of Agricultural Sciene & Natural Resources, Animal Science, Basij square, 49178-Gorgan, Iran

The effect of using honeybee royal jelly as a protein source in tris extender on motility and survival traits of spermatozoa was investigated. Semen samples from three Dallagh rams were pooled and subjected to six treatments as follows; tris extender containing 0, 1, 2, 3, 4 and 5 percentages of royal jelly, in a randomized complete block design with sub-samples in an experimental unit. Diluted semen samples (1:4) were cooled to 5°C before freezing in liquid nitrogen. Data were analyzed by using Minitab programme and Duncan's multiple range tests was used for comparisons of means. Results showed that differences between control treatment and those treatments with royal jelly on survival traits of spermatozoa were significant ($p < 0.01$). Percentages of live and motile spermatozoa for six treatments of royal jelly at pre-freezing stage were 75.67 and 72.66, 76.33 and 73.23, 76.89 and 73.89, 77.67 and 74.56, 78.44 and 75.22, 78.56 and 75.33; at post-freezing stage were 34.00 and 31.33, 34.77 and 32, 35.56 and 32.77, 36.56 and 33.56, 37.33 and 34.44, 37.55 and 34.56 respectively. Survival and motility rates of spermatozoa were not improved with further increases of royal jelly in diluent. The effect of tris extender with four and five percentages of royal jelly on survival rates of spermatozoa were not significant ($p > 0.01$). Therefore an addition of four percent of royal jelly in tris extender is recommended for dilution and freezing of ram semen.

Session 14 Poster 16

Sexual performance of yearling Awassi and crossbred rams

R.T. Kridli[1], M. Momani Shaker[2], A.Y. Abdullah[1] and M.M. Muwalla[1], [1]Jordan University of Science and Technology, Irbid, 22110, Jordan, [2]Czech University of Agriculture, Suchdol, Prague, Czech Republic

This study was conducted to evaluate the sexual performance of 10-month-old, ram lambs of different breed groups. Eight ram lambs of each Awassi (A), F_1 Charollais x Awassi (CA) and F_1 Romanov x Awassi (RA) breed types were subjected to sexual performance tests by being individually exposed to two estrous Awassi ewe lambs for four, 20-min periods. Bouts of leg kicking and anogenital sniffing were similar among breed groups. Mounting frequency was greater ($P < 0.05$) in RA than in A and CA ram lambs (20.9±3, 14.6±3 and 13.2±3 mounts in RA, A and CA rams, respectively). Tail-raising was greater ($P < 0.05$) and mating rate tended to be greater ($P < 0.10$) in A than in RA and CA ram lambs (0.4±0.1, 01.±0.1 and 0.1±0.1 matings in A, CA and RA rams, respectively). The number of mounts per tail-raising (efficiency) was influenced by breed group and test day ($P < 0.05$). Awassi ram lambs maintained the best efficiency throughout the experiment. Efficiency in RA and CA ram lambs improved with each test day. Results of the present study indicate that RA ram lambs have greater mounting frequency than A and CA, while Awassi are more capable of mating with fat-tailed females than the CA and RA ram lambs.

Session 14 Poster 17

Diagnosis of early pregnancy in Awassi sheep

Ü Yavuzer[1], F. Aral[2] and R. Demirkol[3], [1]Harran University, Faculty of Agriculture, Dept. of Animal Science, Şanlıurfa, 63040, Turkey, [2]Harran Universitesi, Faculty of Veterinary Sciences, Dept. of Reproduction and Artificial Insemination, Şanlıurfa, 63040, Turkey, [3]Harran University, Faculty of Veterinary Sciences, Dept. of Reproduction and Artificial Insemination, Şanlıurfa, 63040, Turkey

This study was carried out to determine the best predicting techniques for diagnosis of early pregnancy in Awassi sheep. Adult Awassi sheep (n=70) were syncronized during the breeding season with PGF_{2a}. The ewes were treated PGF_{2a} at 10 days interval. Cervical artificial insemination with fresh semen was applied. Jugular blood samples were collected 17 days after artificial insemination and serum was analyzed by RIA(radioimmunassay) to indicate pregnancy. Pregnacy diagnosis was made by transrectal ultrasonograpy (Scanner 100 Vet, 6/8 Mhz transducer) on d 40 of gestation. Pregnancy diagnosis was verified on birth. Of the techniques available for diagnosis of pregnancy and the determination of pregnant Awassi ewes, transrectal ultrasonograpy detecs pregnancy with an accuracy of 96%. The accuracy of early pregnancy diagnosis on day 17 after artificial insemination using RIA kits was 100%. The accuracy of progesterone assay for determining fetal numbers is relatively higher than transrectal ultrasonagrapy. RIA tecnique is the best to predict pregnancy for Awassi sheep.

Session 14 Poster 18

Environmental and genetic factors affecting male and female fertility in dairy sheep

I. David, C. Leymarie, E. Manfredi, C. Robert-Granié and L. Bodin, INRA, UR631 SAGA, Chemin de borde rouge, 31320 Castanet Tolosan, France

Artificial Inseminations (n=677 095) recorded during 5 years in "*Lacaune*", "*Manech tête rousse*", "*Manech tête noire*" and "*Basco béarnaise*" sheep were analysed to determine environmental and genetic factors affecting insemination result. Each on-farm recorded insemination were matched to the corresponding ejaculate produced at the AI centre and to the corresponding outcome which is a binary response observed at lambing of either success (1) or failure (0). Separate analyses within breed were performed using a linear model which estimates jointly male and female fertility. After selection of significant fixed effects ($P < 0.05$) the remaining environmental female effects were age, synchronisation (0/1) on the previous year, total number of synchronisations during the female reproductive life, time interval between previous lambing and insemination, already dry or still lactating (0/1) when inseminated, milk quantity produced during the previous year. Environmental male effects were motility and concentration of the semen. Non-sex specific effects were the inseminator, the interaction herd*year nested within inseminator considered as random effects and the interaction year*season considered as a fixed effect. Heritability estimates varied from 0.001 to 0.005 for male fertility and from 0.040 to 0.078 for female fertility. Repeatability estimates varied from 0.007 to 0.015 for male fertility and from 0.104 to 0.136 for female fertility.

Session 14 Poster 19

The effect of different times of eCG hormone injection on fertility traits of Kordian ewes

Y. Jafari Ahanagri and S. Hassani, Gorgan University of Agricultural Science & Natural Resources, Animal Science, Basij Square, Gorgan, 49178, Iran

60 Kordian ewes of 3 to 5 years old were selected and divided into four groups. CIDR was inserted into vagina of each ewe and then after 13, 12 and 11 days, all ewes in each group received an injection of 400 IU eCG hormones intramuscularly, as treatments 1, 2 and 3 respectively. Ewes in group four (control) also received an injection of 2 ml of saline after 13 days. All CIDRs were withdrawn after 13 days and rams entered the flock for mating. Results showed that all ewes were in heat after 48 hours of CIDR withdrawal. Different times of eCG injection had no significant effect on estrus synchronization of ewes ($p > 0.05$). Non-return rates of ewes were 73, 67, 60 and 73%, pregnancy rates 73, 67, 60 and 73%, fecundian rates 118, 130, 133 and 118%, lambing rates 87, 87, 80 and 87% and length of gestation period 149, 148, 149 and 148 days for treatments 1, 2, 3 and control respectively. Twinning rates of ewes for treatments 1, 2, 3 and control were 18, 30, 33 and 18% respectively. Differences of twinning rates between treatments two and three versus one and control were significant ($P < 0.05$). In conclusion, an injection of eCG hormone, at 24 and 48 hours before CIDRs withdrawal, was not beneficial for Kordian ewes fertility traits, except for twinning rate.

Session 14 Poster 20

Effect of dopamine agonist (bromocriptine) on reproductive responses of ewes synchronized to estrus using CIDR-G

M.Q. Husein, H.A. Ghozlan and J.S. Issa, Jordan University of Science and Technology, Faculty of Agriculture, Department of Animal Production, P.O.Box 3030, 22110 Irbid, Jordan

The objective was to evaluate the effects of bromocriptine (BR) on reproductive performance of ewes induced to estrus using CIDR-G. In May, 37 ewes received intravaginal CIDR-G for 12 d and were randomly assigned to 4 groups and treated twice daily with 2.5 mg BR tablets from d –2 to d 0 (group A), –2 to 3 (group B), 0 to 3 (group C; d 0= day of CIDR-G removal) and those in group D served as controls. Blood samples were collected for progesterone (P_4) and prolactin (PRL) analysis. Fertile rams were joined at 0 d and estrus was monitored at 6-h intervals for 5 d. Estrus expression and intervals to estrus were similar ($P > 0.05$) among groups. Following CIDR-G insertion, P_4 levels rose and reached 4.9±0.4 ng/ml 2 d post insertion and then declined gradually to 2.3±0.2 ng/ml on d 0. PRL levels during this period were similar and averaged 7.7±0.4 ng/ml and were not influenced by P_4 emanating from CIDR-G. During the periods of BR treatment, PRL decreased gradually among groups and increased after d 0 in a fluctuated manner ($P > 0.05$). Pregnancy rates based upon d 19 P_4 were similar among groups and were 67, 62.5, 60 and 70% respectively. Results report for the first time PRL levels of <10 ng/ml in Awassi ewes during May. PRL was not influenced by CIDR-G during the 12-d period of insertion. BR treatment administrated orally did not either affect P_4 levels or influence reproductive performance of ewes following d 0.

Session 15 Theatre 1

Aspects of milk quality and milk products in relation to sustainable production in sheep and goats

R. Rubino and V. Fedele, CRA, ISZ-PZ, Via Appia - Bella Scalo, 85054 Muro Lucano (PZ), Italy

To define the limits and parameters of quality evaluation is not quite simple. In the industrial countries the quality was defined by the legislation. Consumers demand for another dimension of quality: health protection and sensory characteristics. Each plant affected differently milk and dairy products' quality. In comparison to grazed herbage, the preserved one (hay) impoverished the products in terpenes content and modified their sensory characteristics. The cheese from grazing system was well distinguished and more accepted. The highest contribution to the diets of some plants enriched milk in aromatic compounds. From winter to summer no large variation was observed in milk monoterpenes compounds, while sesquiterpenes increased eightfold. Summer milk was characterised mostly by resinous and citrus odours; mint, green-herbaceous and fruity were also perceived by panellists. Milk and cheese from grazing system were characterised by higher vitamin, CLA and omega3 content. Tocopherol and retinol content in cheese from grazing goats reached values 25-30% higher than cheese from goats fed hay, and also the highest degree of antioxidant protection. According to the seasonal evolution of CLA's precursor in the grazed herbage, this qualitative parameter was higher in winter and spring milk. The grazing systems have a favourable impact on global quality of milk and dairy products and this has to be incentive for their promotion and valorisation.

Session 15 Theatre 2

Differentiating in milk: an example of producing healthy milk on a sustainable basis

A.K. Schaap, Campina, Member Services, Postbus 2085, 5300 CB Zaltbommel, Netherlands

The co-operative dairy company Campina announced the introduction of an innovative concept for milk based on a differentiation on farm level. Aim is to strengthen the market position of the members of Campina by anticipating on modern consumers needs and taking a pro-active approach towards corporate social responsibility issues. Special arrangements are made with more than 500 farmers of Campina and through separate logistics and processing the largest differentiated milk stream for branded fresh products is realized per April 2007. Cornerstones of the concept are "nearby", "healthy", "sustainable" and "natural". Co-makership between farmers, feed industry and dairy company is an essential part. One of the goals is to produce 'even healthier milk' by raising the content of unsaturated fatty acids by 20% and doubling the content of Omega 3 on a year-round basis through a feeding programme in combination with outdoor grazing. Realizing all goals is a management challenge for all partners. A monthly standard for the content of unsaturated fatty acids is set and milk is tested on farm bulk tank level every delivery. Farmers receive a cost based fee according to the monthly average content in the milk delivered. For the Omega 3 content a fixed standard is practised and milk is tested every three months. Possible side effects on animal health, animal welfare, fertility and environmental impact are monitored intensively. With endorsement of several NGO's the use of sustainably produced soy is introduced.

Session 15 Theatre 3

Implementing successful milk quality improvement programs on farms

P.L. Ruegg, University of Wisconsin, Dept. of Dairy Science, 1675 Observatory Dr., Madison, 53706, USA

Production of high quality milk is the primary objective of most dairy farmers. In Wisconsin, dairy farmers can enroll in a program ("Milk Money") to create a targeted milk quality plan. During "Milk Money", farmers meet for 4 months with a self selected team of advisors to focus on issues that affect milk quality. At each meeting, a list of actions to complete before the next meeting is developed, responsibilities are assigned and methods to assess progress are agreed upon. At subsequent meetings, the action list is reviewed and people are held accountable for their tasks. This process has been amazingly successful and the average team reduces their SCC by about 80,000 cells/ml. Data demonstrate that adoption of recommended practices is enhanced when a clearly defined milk quality plan is collaboratively developed and clearly communicated to people that are held accountable for implementation. The use of standardized milking practices and frequent training has been shown to be critical for improving milking performance and reducing the rate of clinical mastitis. At the completion of the Milk Money program, 63% of herds reported that they had achieved their milk quality goals and 83% indicated that they believed that participation in the Milk Money program had a long-term positive benefit on milk quality. Herds that did not achieve their goals, reported that lack of time, other farm problems, lack of focus, seasonal influences and poor choice of goals were possible reasons for failure to meet goals.

Session 15 Theatre 4

Aspects of meat quality and meat products in relation to sustainable production in sheep and cattle

K.R. Matthews, Meat and Livestock Commission, Winterhill House, Milton Keynes, MK6 1AX, United Kingdom

Sustainable animal production depends on providing products that have a market and therefore on product quality. This paper reviews recent research on beef and lamb focussing on aspects that affect sensory and nutritional quality. Production system, and in particular diet, has an important influence on fatty acid composition and flavour. There is also increased understanding of the precursors in meat that are important for flavour development during cooking. These can be modified by diet but further research is needed in this area. Animal handling preslaughter and electrical inputs during post slaughter operations have important effects on the pattern of pH fall post slaughter. Understanding of the combined effects of modern processing operations is increasing and improved monitoring of pH and temperature is required to optimise quality. In response to consumer need, the meat industry in the UK has been challenged to reduce the contribution of meat to sodium in the diet. Sodium (particularly sodium chloride) has important functions in many meat products which will be briefly described. A joint industry action plan has been developed to reduce sodium levels. In conclusion, a number of factors through the supply chain for beef and lamb have an influence on the consumer perception of the resulting products. Recent research and industry development has identified means of optimising quality helping to maintain a sustainable production sector.

Session 15 Theatre 5

Contribution of animal feeds to the quality of animal products

L.A. den Hartog and R. Sijtsma, Nutreco, Agriculture R&D and Quality Affairs, P.O. Box 220, 5830 AE Boxmeer, Netherlands

The quality of animal products is highly influenced by the quality of the animal feeds. This is the case for both "negative" and "positive" quality attributes, such as safety and product quality, respectively. Quality assurance programs based on GMP and more recently HACCP were introduced. Because of the risks are related with the use of feed ingredients, risk assessment and management of feed ingredients are receiving much attention. Another trend is to insure product liability. In this way, downstream partners have the guarantee that in case of feed safety incidents, damage is paid. Despite all these efforts, feed safety is presumed to give only limited competitive advantage. For that reason, the European animal feed industry is focusing on other quality attributes which can create added value. With animal feed and feeding programs, the composition and sensory quality of animal products can be steered in a way that it matches with specific consumer demands. In particular the present health consciousness of consumers is offering opportunities for marketing animal products with for instance modified fatty acid composition or fat level. Information needs to be exchanged in order to secure traceability, quality assurance and control, and production of animal products fulfilling the consumer demands. Joint efforts by the feed industry and other chain partners also offer new opportunities. New concepts can be developed based on a combination of several quality attributes, brought in by the chain partners.

Session 15 — Theatre 6

PDO and sustainable development: targeting the average production as a way to question productivity?

A. Lambert-Derkimba[1], J.M. Astruc[2], D. Regaldo[3], F. Casabianca[1] and E. Verrier[4], [1]INRA, LRDE, Quartier Grossetti, 20250 CORTE, France, [2]I.E., Génétique, BP 52627, 31326 Castanet-Tolosan, France, [3]I.E., Génétique, 149, rue de Bercy, 75012 Paris, France, [4]INRA, UMR Génétique et Diversité Animales, 16, rue Claude Bernard, 75231 Paris, France

Protected Designation of Origin is an efficient tool for sustainable development preserving know-how, landscapes, and genetic resources. In France, local breeds are often included in the rules of PDO cheeses. In some cases, an upper limit of the herd average milk yield is fixed. Motivations and consequences of such rules were studied in the cases of 2 French mountainous areas with a large PDO cheeses production: Northern Alps (3 dairy cattle breeds) and Western Pyrenees (3 dairy sheep breeds). Evolution of the involved breeds and strategy of their breeders are analysed, putting emphasis on the balance between the milk yield and protein/fat contents. Two analyses were performed: 1- Evolution over time of the recorded performances for the different traits considered. 2- Increase of the number of traits submitted to genetic evaluation and the successive definitions of the aggregate genotype. We discuss consequences of a milk yield limit on the breeding goals as the disputes among producers about such a rule. The role of animal breeding in situations where productivity is questioned by farmers themselves is linked with sustainability of low-input systems, landscape management and preservation of agricultural activities in mountain areas.

Session 15 — Theatre 7

A comparison between Norwegian and Dutch dairy production systems with regard to their socio-cultural sustainability

B. Boogaard[1], B. Bock[1], E. Krogh[2] and S. Oosting[1], [1]Wageningen University, P.O.Box 338, 6700 AH Wageningen, Netherlands, [2]University of Life Sciences, P.O.Box 5003, 1432 Ås, Norway

On average, the number of dairy cows at a dairy farm is 65 in The Netherlands and 17 in Norway. Although land use is different between both countries, the objective of dairy production systems is not only food production, and demand for sustainable agriculture is a general expression of collective concern. The present study had the objective to assess and compare relevant socio-cultural themes in dairy production systems in The Netherlands and Norway, essential for design of sustainable future dairy production. The research method was similar for three regions in The Netherlands and one region in Norway: Two citizen panels visited two dairy farms per region. On farm, each respondent filled out a questionnaire about their individual perception (hearing, smelling, seeing and feeling) and recorded valuable aspects per farm with a digital camera. After qualitative analysis, Norwegian and Dutch socio-cultural themes were compared. This paper describes several differences and similarities. For example, the importance of 'quietness of the countryside' in contrast to 'daily life stress', as illustrated by the quotation *"The peaceful and relaxing quietness gives freedom to mental reflection and relieve from everyday stress"*. Other examples of socio-cultural themes are: farm activities, farm income, animals, food production, landscape, culture, and public services.

Session 15 Theatre 8

A comparison between housing systems of dairy cows with regard to milk quality, animal welfare and animal health

M. Klopčič, M. Čepon, J. Osterc and D. Kompan, University of Ljubljana, Biotechnical Faculty, Zootechnical Department, Groblje 3, 1230 Domžale, Slovenia

The quality of animal products is highly depend on farming system which include housing system, quality of feed, grazing or not grazing, level of production etc. More as 5000 dairy farmers (which are included in milk recording) respondent filled out a questionnaire (inventory) about housing system, milking system, way of producing and gathering of forage, level of using concentrate, using of manure, health and fertility problems.

Session 15 Theatre 9

Electronic identification and molecular markers for beef traceability from farm to retailer

J.J. Ghirardi, G. Caja, M. Hernández-Jover, N. Jiménez and A. Sánchez, Universitat Autònoma de Barcelona, Campus universitari, 08193, Spain

Use of low frequency (134.2 kHz) electronic boluses and DNA microsatellites (n = 12) was evaluated for tracing beef. Calves (n = 3,657) were identified with official (OE, Allflex, n = 7,314) and biopsying (E1, Biopsytec, n = 2,562; E2, TypiFix, n = 1,095) ear tags, and boluses (B1, 75 g, n = 3,057; B2, 73 g, n = 600, Rumitag). Calves were intensively fed and slaughtered before 1 yr of age. Blank read-write high frequency (13.56 MHz) inlays were attached to the calf shank before hide removal and bolus codes automatically read and transferred to inlays in the slaughtering line. Carcass samples (n = 900) were taken using biopsy tubes (n = 357, Biopsytec) and plastic sticks (n = 543, Identigen). Sticks were also used to sample 30 meat cuts randomly taken in 9 butcheries. On-farm traceability for B1 (99.8%) and B2 (100%) were greater than for all ear tags (OE, 96.4%; E1, 98.4%; E2, 99.1%). On-line readings failed in 37% of cases at the start of the experiment, requiring adaptations to abattoir conditions. In the rest of calves (n = 2,058) data transfer to carcasses was 98.6% successful. Tracing back of carcasses to calves in 176 pairs of samples showed 5 not matching (2.8%). Retailer matching was 100%. In conclusion, the electronic and DNA tracing system used showed 97.2% of traceability.

Session 15 — Theatre 10

Electronic identification and molecular markers for lamb traceability

G. Caja[1], J.J. Ghirardi[1], M. Hernández-Jover[1], J.C. Pozo[2], D. Albardonedo[3] and A. Sánchez[1], [1]Universitat Autònoma de Barcelona, Campus universitari, 08193 Bellaterra, Spain, [2]Ovino del Suroeste, SCL, 06700 Villanueva de la Serena, Spain, [3]Asociación Española Criadores Ovino Precoz, Castelló 45, 28001 Madrid, Spain

Use of low frequency (134.2 kHz) electronic mini-boluses and DNA microsatellites (n = 12) was evaluated for traceability. Lambs (n = 1,908) were identified with ear tags (VE, visual, n = 1,908; BE, biopsying, n = 980) and mini-boluses (B1, 9 g, n = 1,091; B2, 20 g, n = 817). Lambs were slaughtered (3 mo of age) in 2 abattoirs. Bolus codes were read before evisceration and transferred to high-frequency (13.56 MHz) read-write inlays recorded which were fixed to carcass shank. Carcass samples (n = 868) were also taken. On-farm traceability was lower for VE (96.8%) than for BE (99.7%), B1 (98.4%) and B2 (100%). Data transfer to carcasses was 98.9% successful. Abattoir and total traceability differed between B1 (97.7 and 96.1%) and B2 (99.9 and 99.9%), respectively. Tracing back of carcasses to lambs in 50 pairs of samples showed 1 not matching (2.0%). In conclusion, the dual electronic and DNA tracing system used showed high efficiency (98.0%) under practical conditions, although user-friendly reading equipment is needed in practice.

Session 15 — Poster 11

The effect of intramammary infusion of tilmycosin as a dry cow therapy on the rate of sub-clinical mastitis

M. Mohammadsadegh and S. Lotfollahzadeh, Islamic Azad Uniiversity Garmsar Branch, Veterinary Medicine, Engelab Street, Garmsar, Semnan Province, Iran

To evaluate tilmycosin efficacy in dairy cows to reduce the rate of intra mammary infections in dry period and new infections in post partum period, sixty quarters of 30 Holstein cows in the dairy farms of fashapoye in Tehran suburb infected by gram positive bacteria were divided equally in two, test and control groups. Bacterial examination and somatic cell count were examined at the first day of drying, and then, the animals of test group were infused by tilmycosin (5 ml, intra mammary 30%, razak, Tehran, Iran) and control group by cloxacillin (10 ml,500 mg, cloxalmo, DC oint; zistkimia. CO.). Bacterial examination at 7 days and somatic cell count at 10 days after parturition were examined repeatedly. The rates of intra mammary infections and new infection rates were compared by chi square and student T test in two groups. Results showed that tilmycosin has less effect on reduction of intra mammary infection due to chorine bacterium bovis and has no effect on streptococcus agallactiae but has the same effect as cloxacillin against coagolase negative staphylococcus and has more effect against staphylococcus aurous ($p < 0.01$). Tilmycosin were more effective than cloxacillin against new infections. In conclusion, tilmycosin is a suitable DC treatment.

Session 15 — Poster 12

Milk production and body dimensions of Balkan goats

V. Bogdanovic[1], I. Djordjevic[2] and I. Djurdjevic[3], [1]Faculty of Agriculture, University of Belgrade, Animal Science, Nemanjina 6, 11080 Zemun-Belgrade, Serbia, [2]Assocciation for Animal Welfare "Herd-man", Nis, 18000, Serbia, [3]Extension Service, Aleksinac, 18220, Serbia

Balkan goat represents the autochthonous breed of goat in Serbia. This breed is spread out not only in Serbia but also in other Balkan's and Southeast European countries. Like any other local, autochthonous breed of domestic animals, Balkan goat is characterised by relatively low production level. Nevertheless, this breed of goat in Serbia is more and more interesting from conservation and traditional livestock point of view. In order to obtain data on milk production and body dimensions, 120 Balkan goats and total of 445 lactation records were analysed. Average±SD length of lactation was 256.3±22.4 days. Average±SD milk yield and milk-fat content were 378.46±93.77 kg and 3.71±0.25%, respectively. It should be pointed out that variability in milk production traits represents good biological basis for improvement. According to body dimensions, Balkan goat is breed of medium size. Average±SD height at withers, body length, width of chest, depth of chest and body weight are 70.4±3.92cm, 78.8±6.24cm, 18.3±2.12cm, 33.1±2.63cm and 49.4±7.09kg, respectively.

Session 15 — Poster 13

Fatty acid profile of milk and intramuscular fat depending on PUFA n-3 sources in the diets

M.B. Zymon and J.A. Strzetelski, National Research Institute of Animal Production, Department of Animal Nutrition and Feed Science, ul. Sarego 2, 31-047 Kraków, Poland

The aim of this study was to determine the effect of different sources of PUFA n-3 in cows and calves' diet on fatty acid profile of milk and meat fat. The 1. experiment was performed on 12 dairy cows, allotted to 3 groups. Cows were fed for 8 weeks maize-lucerne silage and concentrates supplemented with flax seeds, fish oil or without additives (control). In 2. experiment 24 calves were divided into 3 groups and fed from 7 to 90 day of age concentrates without additives (control) or supplemented like in 1. experiment. Treatments had no effect on feed intake both in cows and calves, milk production and nutrient content of milk and meat. Feeding flax seeds improved calves daily weight gains and feed utilization ($p \leq 0.05$). Compared with control group fish oil increased proportion of CLA in milk fat more than flax seeds, while there were no differences between groups in CLA content of intramuscular fat. Flax seeds improved the health quality of milk and meat by decreasing SFA content, increasing the level of $C_{18:3}$ n-3 and reducing the PUFA n-6/n-3 ratio. There were also higher contents of MUFA and EPA in milk from cows feeding flax seeds. Fish oil reduced the level of $C_{18:0}$ both in milk and meat fat and significantly increased the content of EPA and DHA in meat fat, while in milk fat there was only a little rise in DHA content. Furthermore, fish oil most of all decreased the PUFA n-6/n-3 ratio in the intramuscular fat.

Session 15 Poster 14

Comparison of two different methods to determine meat quality

Y. Bozkurt, S. Ozkaya and B. Kilic, Suleyman Demirel University, Faculty of Agriculture, Department of Animal Science, Dogu Kampus- Isparta, 32260, Turkey

The objective of this study was to determine muscle color of beef carcasses using digital image analysis. Fourteen beef carcasses were selected from slaughterhouses. Data collected on these carcasses included colorimeter measurements and digital images and measurements of muscle colour (L*, a*, b* values) and muscle pH from longissimus muscle at 24 hours after slaughtering. The discrepancies between colorimeter and digital image analysis values of L*, a*, b* were large (25.6±3.37, 3.01±3.38 and 2.25±3.56, respectively). There were significant differences between L* values ($P < 0.05$) but there were non-significant differences between a* and b* values ($P > 0.05$). The correlation coefficient was found significant ($P < 0.05$) between pH and a* values (r=0.83). The results showed that prediction ability of digital image analysis was low for prediction of muscle colour. However, it was concluded that red value (a*) can be predicted by digital image analysis and there is a need for further studies in order to develop better techniques to use for prediction.

Session 15 Poster 15

Estimates of heritability of beef quality traits in Piemontese cattle

A. Boukha, A. Albera, M. De Marchi, L. Gallo, G. Bittante and P. Carnier, University of Padova, Animal Science, Viale Universita 16, 35020 Legnaro, Italy

The aim of this study was to estimate heritability of beef quality traits in Piemontese cattle. A total of 804 young bulls, progeny of 109 AI sires, were sampled from 124 fattening farms (FF) and slaughtered in different days at the same commercial abattoir. At slaughter, bulls were 523±73 d old and average carcasses weight (CW) was 417±45 kg. Carcasses were scored for fleshiness (EUS) and fatness. An individual beef sample was collected from Longissimus Thoracis 24 h after slaughter and held refrigerated at 4 °C for 8 d. Measured traits were pH at ageing (pH_{8d}), beef colour (L^*, a^*, b^*, Hue and Chroma), shear force (SF), drip (DL) and cooking loss (CL). A mixed model, including the fixed effects of FF, slaughter age and CW class and the random effect of the bull, was analysed using Bayesian methodology and the Gibbs sampler. The estimated heritability for EUS, SF, DL, Hue, L^* and a^* was of intermediate magnitude (from 0.22 to 0.49) whereas heritabilities for all other traits were low (from 0.04 to 0.16).

Effect of pregnancy and fetal development on carcass weight and classification of Azorean bovine

O. Abaurrea, A. Borba and F. Moreira Da Silva, University of the Azores, DCA - Animal Reproduction, Angra do Heroísmo, 9700, Portugal

During pregnancy, several factors and metabolites such as insulin, catecholamine, foetal and placental proteins, steroidal hormones and growing factors play an important role in mobilisation and deposition of lipids and proteins in carcasses with the consequent higher final weight. For those reasons, in Terceira Island, a high number of dairy cows are slaughtered during pregnancy due to the higher weight registered in these situations. In the present study, carcasses of all cows slaughtered in the Terceira Island from April to July were weighted and classified according to EUROP – 1,2,3,4 and 5. After evisceration, reproductive organs were separated and if cows were pregnant the crown-rump length were measured in all foetuses. From the 699 cows slaughtered, 297(42.5%) were pregnant while 402 (57.5%) were non pregnant. For pregnant animals, 86 (29%) were slaughtered to three months pregnancy, 123 (41%) from three to six months pregnancy and 88 (30%) were more than six months pregnant. Results showed that, on average, the weight of carcasses of pregnant cows was 244,241 kg, (± 48,72 kg) while the weight of non pregnant cows was 228,407 kg, (± 56,36 kg). When cows were separated according pregnancy time, it has been observed that, independently of age, carcasses of pregnant cows were heavier than those of non pregnant cows ($R2 = 0.999$) ($p < 0.001$). For carcass classification, no statistical differences were observed between pregnant/non pregnant cows.

Feeding level and method effects on carcass traits and muscle chemical composition of steers

M.G. Keane[1] and L. Lescure[2], [1]Teagasc, Beef Production, Grange Beef Research Centre, Dunsany, Co. Meath, Ireland, [2]Ecole Nationale Superieure Agronomique, Toulouse, 31326 Castenet, France

The effects of level and method (separate feeds or total mixed ration (TMR)) of offering concentrates with grass silage on muscle chemical composition of steers were evaluated. There were 6 feeding treatments for 132 days: (i) silage *ad libitum* (SO), (ii) SO plus low concentrates offered separately (LS), (iii) SO plus low concentrates offered as a TMR (LM), (iv) SO plus medium concentrates offered separately (MS), (v) SO plus medium concentrates offered as a TMR (MM), (vi) concentrates *ad libitum* (AL). Low and medium target concentrate levels were 3 and 6 kg dry matter per head daily, respectively. A sample of *m. longissimus* was chemically analysed. Mean carcass weights and carcass fat scores for SO, LS, LM, MS, MM and AL were 305, 350, 348, 367, 360 and 374 (s.e. 7.4) kg, and 2.0, 3.4, 3.3, 3.5, 3.6 and 3.5 (s.e. 0.18), respectively. Corresponding values for muscle moisture and lipid were 749, 739, 737, 729, 732 and 733 (s.e. 3.2) g/kg, and 21, 28, 32, 36, 34 and 34 (s.e. 3.1) g/kg, respectively. Muscle moisture concentration decreased and lipid concentration increased with increasing concentrate level. It is concluded that carcass weight and measures of carcass and muscle fatness increased curvilinarly with increasing concentrate level but there was no effect of feeding method.

Session 15 Poster 18

Quality of meat from purebred French Alpine Kids and Boer Crossbreeds

H. Brzostowski, J. Sowińska, Z. Tański and Z. Antoszkiewicz, Department of Sheep and Goat Breeding, Oczapowskiego 5, 10-719 Olsztyn, Poland

Some quality indices of meat from purebred French Alpine kids and Boer crossbreeds aged 50 days were evaluated in the study. Samples of *m. quadriceps femoris* were taken to determine the chemical composition and physicochemical properties of meat as well as a water-to-protein ratio, energy value, levels of cholesterol and amino acids in protein, and fatty acid concentration in intramuscular fat. It was found that meat from crossbred kids, compared to meat from purebred kids, contained more intramuscular fat, cholesterol and vitamin A, had a higher calorific value, a brighter color, a lower water-holding capacity, a higher level of physiological maturity (measured as the value of a water-to-protein ratio), and got higher scores for tenderness and juiciness. The protein of meat from crossbred kids had a more desirable ESSA/NEAA ratio, while intramuscular fat contained less OFAs and had more desirable UFA/SFA and DFA/OFA ratios. Due to a high protein content (19.44 and 19.74%), low levels of fat (1.67 and 1.96%) and cholesterol (48.76 and 56.63 mg/100g), a low energy value (96.36 and 101.47 kcal/100g), a high concentration of essential amino acids, a desirable fatty acid profile and high scores for sensory properties, meat from purebred French Alpine kids and (especially) Boer crossbreds may be recommended as a valuable component of a low-fat diet.

Session 15 Poster 19

Microbial analysis of meat from Pomeranian lambs, stored under modified atmosphere conditions

Z. Tański, H. Brzostowski, J. Sowińska and Z. Antoszkiewicz, Department of Sheep and Goat Breeding, Department of Sheep and Goat Breeding, Oczapowskiego 5, 10-719 Olsztyn, Poland

The growth of mesophilic bacteria was studied in samples of meat from Pomeranian lambs aged 50 and 100 days, stored under modified atmosphere (80% N_2 and 20% CO_2) conditions. Samples of*m. quadriceps femoris* were taken for microbiological analysis. Bacterial counts were determined in fresh meat samples prior to packaging (48 h postmortem) and in modified atmosphere packaged (PA/PE) meat samples stored for 10, 20 and 30 at 1^{O}C. It was found that modified atmosphere (MA) reduced the rate of mesophilic bacteria growth over storage. The rate of bacterial growth was slower in meat from lambs aged 100 days, compared to meat from 50-day-old lambs, despite the fact that contamination rates in fresh meat prior to packaging were higher in the former case. The bacterial count (CFU/g) in unpacked meat from lambs aged 50 days was $2.98 x 10^4$, and during MA storage gradually increased, to $1.67 x 10^5$ on day 10, to $3.97 x 10^5$ on day 20 and to $1.47 x 10^6$ CFU/g on day 30. The bacterial count in meat from lambs aged 100 days was $3.05 x 10^4$, $1.62 x 10^5$, $3.95 x 10^5$ and $1.23 x 10^6$ CFU/g, respectively. The count of mesophilic bacteria determined in lamb meat, including samples stored under MA for 30 days, did not affect its suitability for human consumption.

Session 15 Poster 20

Productive characteristics in organic rabbit

C. Russo, M. D'agata, C. Mozzoni, G. Preziuso and G. Paci, University of Pisa, Department of Animal Production, Viale delle Piagge. 2, 56124 Pisa, Italy

The aim of this trial was to test the effect of different housing systems, organic and intensive, on productive characteristics of a rabbit local population of Tuscany characterized by slow growing. 84 rabbits of a local population were housed in colony cages under organic system (Group A), according to an official organism of certification which states the exclusive use of local strains; 72 rabbits of the same population (Group B) and 72 hybrids (Group C) were housed in colony cages under conventional system. The rabbits were fed *ad libitum* with an organic diet. At the weight of 2400g but at different ages (local population: 102 days; hybrids: 90 days) 30 animals of each group were slaughtered. The local population showed the best productive performances: Groups A and B had the lower mortality (Group A 12% and Group B 10% *vs* Group C 28%), the higher slaughter live weight (Group B 2539g *vs* Group A 2433g *vs* Group C 2306g) and the higher weight gain (Group A 27g and Group B 28g *vs* Group C 25g). Group A and B carcasses had the higher dressing percentage (Group A 61% and Group B 61% *vs* Group C 58%) due to the lower full gastrointestinal tract (Group A 20% and Group B 19% *vs* Group C 22%)and the higher commercial carcass percentage (Group A 59% and Group B 59% *vs* Group C 57%). These results promote meat production obtained by slow growing rabbit population fed with organic diet, characterized by low energy content and without any pharmacological supplementation.

Session 15 Poster 21

Effect of lactation period prolongation on marketing weight of growing rabbit puny

W.H. Kishk, Suez Canal University, Animal Production, Suez Canal University, Faculty of Agriculture, Ismailia, Egypt, 41522, Egypt

A total of 150 of growing rabbit puny after birth of White New Zealand rabbit were divided into two groups. The first group was weaned at 35 days of age (control group), while the second group was weaned at 60 days of age. Both groups were subjected to growth rate measurements, mortality rate, blood metabolites levels, and marketing weight at 10 weeks of age. Data showed that there were significant differences ($P < 0.05$) between the two groups in most of studied criteria. Weaning at 60 days of age enabled growing rabbit pups to reach a higher body weight at marketing age (10 weeks). Marketing body weight averaged 1.845 ± 0.48 (Kg) and 1.650 ± 0.35 Kg for delayed weaning and control group, respectively. It could be concluded that weaning at 60 days of age as for growing rabbit is better to get a higher marketing weight for rabbit meat production especially for extensive and semi-intensive rabbit breeding programs.

Session 15 Poster 22

Effect of lactation period on body physical characters in growing rabbit puny

W.H. Kishk, Suez Canal University, Animal Production, Suez Canal University, Faculty of Agriculture, Ismailia, Egypt, 41522, Egypt

One hundred new born of White New Zealand rabbit were divided into two groups. The first group was weaned at 35, the second group at 60 days of age. During lactation period and until marketing age both groups were subjected to measurements of body physical characters. These characters were lengths of ear loop, head, body, tail, fore legs, hind legs in addition thoracic, abdomen circumference and marketing body weight at 10 weeks of age. The results showed that there were significant differences ($P < 0.05$) in marketing body weight, thoracic, abdomen circumference and body length between two studied groups. While there were no significant differences in lengths of ear loop, head, body, tail, fore legs, hind legs between two investigated groups. It could be concluded that it is better to prolong lactation period in growing rabbits to get better physical body characters and marketing weight.

Session 15 Poster 23

Adaptation of new born camel calves to desert climate

A. Magdub, T. Abdusalm and A. Ruk, univ. of Elfateh, Animal production, tripoli, 218, Libyan Arab Jamahiriya

This Study was conducted on 8 newel new born calf-camels (4 males and 4 females) in the desert of Libya to study the effect of age, sex and season of the year on the physiological responses during the period from birth to age of 1 year. The measurements included: Body wt, rectal temp., respiration rate, and heart rate in addition to blood constituents of glucose, urea, creatinine, protein, NA, K Thyroglobulin's(TBG), and thyroid hormones. Temperature humidity index(THI) was calculated using weather data of temperature and humidity to study the effect of season. The results showed an increase in body wt for both sexes (0.47 kg/day) during the 1st month of age (spring season), followed by retarded growth (0.225 kg/day) during 6-10 months of age (summer season), the males had higher growth rate than the females. The overall growth rate averaged 0.335 kg / day. Body temperature, heart rate, respiration rate, urea, creatinine and glucose levels were higher at first month and started to decline thereafter. Proteins, Na, K, TBG levels were lower during 1st month of age, then started to elevate. Thyroid hormones (T4, T3 free and bound) were slightly higher at birth, then declined to consistent levels as calves got older. The results of this study indicated that the levels of glucose, thyroid hormones and proteins and their changes provided some features of how these animals get adapted from the early age to the adverse desert conditions.

Session 16 Theatre 1

Neutrophil transcriptome analysis during transportation stress

K.R. Buckham Sporer[1,2,3], B. Earley[1], M. Crowe[2] and J.L. Burton[3], [1]Teagasc, Grange Beef Research Centre, Co. Meath, Ireland, [2]UCD, School of Agriculture, Food Science and Veterinary Medicine, Dublin, Ireland, [3]Michigan State University, Dept. Animal Science, East Lansing, MI, USA

Severe respiratory disease (BRD) often occurs following transportation of feedlot calves. Neutrophils play a critical role in immune defense against BRD pathogens but also contribute to lung damage if not tightly regulated. Bovine neutrophils are sensitive to glucocorticoid, the main steroid hormone of stress that has pronounced genomic effects in target cells. We hypothesized that truck transportation causes a stress response significant enough to impact expression of inflammation-regulating genes in circulating neutrophils. Transportation for 9 h, even at optimal stocking density and with a rest stop, caused a classical stress response [cortisolemia and neutrophilia] that was accompanied by expression changes for dozens of neutrophil genes responsible for such inflammatory behaviors as transendothelial migration, degranulation, bactericidal capacity, tissue softening, fibrosis, and apoptosis. Expression levels for most genes showed only weak correlations with plasma cortisol but, for a key cluster of affected genes, expression was highly correlated with blood neutrophil count, haptoglobin, and IL-8, each which correlated positively with plasma cortisol. Transportation stress modified the neutrophil transcriptome into one suggestive of heightened inflammatory potential of the cells, possibly linked to the pathobiology of BRD.

Session 16 Theatre 2

Cardiac responses to stress during transport and housing of farm animals

E. Von Borell, Martin-Luther-University Halle, Institute of Agricultural & Nutritional Sciences, Adam-Kuckhoff-Str. 35, 06108 Halle, Germany

Heart rate (HR) has been used as an indicator for stress and welfare in applied farm animal research. However, increases in HR often just reflect a rise in physical activity which is characterised by decreasing parasympathetic (vagal) and increasing sympathetic tone. An increase in physical activity and thus sympathetic tone is thus a well adaptive response to challenges such as transport and grouping of farm animals. At rest, an increase of HR may result from reduced vagal activity as well as from increased sympathetic activity or from a combination of concurrent changes of activity within both branches of the autonomic nervous system (ANS). HR therefore provides little information for the assessment of sympathovagal balance. In contrast, heart rate variability (HRV=variation of inter-beat intervals) analysis allows a more accurate determination of the functional regulatory characteristics of the ANS. Healthy cardiac function is characterized by irregular time intervals between consecutive heart beats and decreased HRV has been implicated in situations of stress and reduced welfare. HRV is a particularly good indicator for the non-invasive assessment of ANS activity in response to psychophysiological stress as recently documented for farm animals by a group of the EU concerted action (COST 846). This paper is intended to provide insight on the use and limitations of HR and HRV as indicators for acute and chronic stress in farm animals.

Effect of loading activities on cattle welfare when transporting from farm to abattoir

G. Gebresenbet, Swedish University of Agricultural Sciences, Department of Biometry and Engineering, Box 7032, 750 07 Uppsala, Sweden, Sweden

During handling and transport from farm to abattoir animals are subjected to series of events that induce stress on animals compromising their welfare. Earlier studies have confirmed that loading and un-loading are among the most stress inducing factors on animals during transport (Gebresenbet, *et al.*, 2004; Gebresenbet and Ericsson, 1998; Trunkfield and Brom, 1990; Warriss,1990). The main objective of the current project is to improve loading methods and facilities to minimize the prevailing adverse effects on animal welfare by identify which of the loading activities are most that induces stress on animal, and developing procedures and facilities. To achieve the stated objectives, continuous measurement of heart rate were made on 153 cows and heifers from 51 farms. In addition, video filming and close observations have been made. Preparation for transport, farms with and without loading facilities, and methods of loading were also studied and related to response of animals.

The welfare of weanling heifers transported from Ireland to Spain

B. Earley, D.J. Prendiville and E.G. O'Riordan, Teagasc, Grange Beef Research Centre, Dunsany, Co. Meath, Ireland

Forty continental × weanling heifers (245, s.d 32.2kg) were transported (T) from Ireland to France on a roll-on roll-off ferry at a stocking density of 0.93m^2/animal and then by road for 9-h to a French lairage. Twenty T heifers were unloaded (ULT) and rested for 12 h in the lairage and the remainder rested (RT) on the transporter. All heifers had access to hay and water. After the rest period, the heifers were re-loaded. The subsequent journey by road from France to Spain was 9 h travel, 7 h rest (on the transporter) and a further 7 h travel. All T heifers were blood sampled prior to transport (day (d) 0), on arrival in the French lairage (d 4), on arrival at the farm in Spain (d 6) and on d 8, 10, 12 and 36. Twenty continental × weanling heifers (247, s.d. 36.0 kg) remained in Ireland as controls (C) and were blood sampled at the same times as T heifers. Heifers were weighed on d 0, 4, 6, 12 and 36 of the study. Heifers transported to France lost 6.2% of their live weight while C heifers lost 2.1%. Both ULT and RT heifers had lower ($P < 0.05$) live weight than C heifers on d 6. During the sea crossing (24 h) from Ireland to France, heifers spent 39% of time lying. Neutrophil % was higher ($P < 0.05$) at d 6 in RT heifers remaining on the transporter (in France) than ULT heifers. In conclusion, there was no welfare advantage in resting animals on a transporter during the rest period (in France). Transport had no adverse effect on the welfare of weanling heifers transported from Ireland to Spain.

Session 16 — Theatre 5

Improving production and welfare of livestock through good human – animal interactions

M.A.W. Ruis[1], H.A.M. Spoolder[1], S. Waiblinger[2], X. Boivin[3] and G.J. Coleman[4], [1]Animal Sciences Group, Wageningen UR, Animal Production, P.O. Box 65, 8200 AB Lelystad, Netherlands, [2]University of Veterinary Medicine, Veterinärplatz 1, A-1210 Wien, Austria, [3]INRA, Clermont-Ferrand-Theix, 63122 St-Genès Champanelle, France, [4]Monash University, Wellington Road, Clayton 3800 Vic, Australia

Training packages for livestock handlers to improve human –animal interactions are currently set up in Europe (within Welfare Quality®). Abundant research in Australia has shown that positive handling by humans is beneficial for production and welfare of livestock. Therefore, in Australia, cognitive-behavioural intervention programmes were developed for pig and cattle farmers ('Prohand'), presented in multimedia format. Researchers in France and Austria are currently developing a cognitive intervention program for dairy and beef cattle. Dutch researchers are doing the same for pigs and laying hens. In 2007, scripts will be established – by pilot training sessions with farmers - in the native and English languages. Scripts, together with photo/video material, will be assembled by Australian researchers into multimedia programmes. A stockperson attitude questionnaire is part of the training package, and it is aimed to improve the handler's knowledge on animal fear and species-specific behaviour, and to give tools for appropriate handling procedures. In 2008, field tests will be conducted with groups of stockpeople. Using their feedback, final training programmes will be ready at the beginning of 2009.

Session 16 — Theatre 6

The effect of floor type on the welfare and performance of finishing beef steers

D.J. Prendiville[1], B. Earley[1], B. Mc Donnell[1], C. Molloy[1] and M.A. Crowe[2], [1]Teagasc, Beef Research, Grange,Dunsany, Co.Meath, Ireland, [2]University College Dublin, Food Science and Vetinary Medicine, Belfield, Dublin4, Ireland

The objective was to investigate the effects of floor type on welfare and performance of finishing beef steers over 150 days. Continental crossbred (n=124) and Holstein-Friesian (n=20) beef steers (total = 144; mean body weight (BW) = 503 ± (s.e. 4.29kg) were blocked by breed and BW and were randomly assigned to one of four treatments: 1) Concrete slats alone, 2) Mat 1 (EasyFix Rubber Products), 3) Mat 2 (Irish Custom Extruders Ltd) and 4) Wood-chips. There were four pens with nine steers each at a mean space allowance of 2.73m^2/head. Dirt scores and live weights were measured. A pathological examination of the four hooves of each animal was recorded. Behavioural observations were recorded over 5 time periods. Animals had *ad libitum* access to a total mixed ration of grass silage and rolled barley on 50:50 dry matter basis. The number of lesions on the hooves of animals on Mat 1 and Mat 2 and wood chip treatments were greater ($P < 0.05$) than the animals on concrete slats. Live weight gain and carcass characteristics did not differ ($P > 0.05$). Lying and standing times did not differ ($P > 0.05$) for animals on concrete slats or wood-chips. The proportion of animals lying was greater on Mat 1. In conclusion, maintaining finishing beef steers either on a concrete slatted floor, or the placing of mats or wood-chips over the concrete slats was not detrimental to animal performance or welfare

Effect of transport on rabbit's welfare: serum corticosterone determination

E. Cavallone[1], F. Luzi[2], C. Lazzaroni[3], M. Bianchi[4] and M. Verga[2], [1]Univ. of Milano, Dept. Vet. Clin. Sci., via Celoria 10, 20133 Milano, Italy, [2]Univ. of Milano, Ist. Zootecnica Vet. Med. Fac., via Celoria 10, 20133 Milano, Italy, [3]Univ. of Torino, Dept. Animal Science, via L. da Vinci 44, 10095 Grugliasco, Italy, [4]Univ. of Bologna, Dept. Food Science, piazza Goidanich 60, 47023 Cesena, Italy

Mammals react to adverse situations with the activation of hypothalamus–hypophysis–adrenal gland axis, which promotes the synthesis of corticosteroids. The aim of this study was to study the influence of transport (1 *vs* 3h) and lairage time (0 *vs* 5h) on serum corticosterone concentration in rabbits (n=70). Blood samples were collected before (basal level) and after transport and lairage. The serum was analysed using a commercial kit for mouse and rat based on RIA competition method validated for rabbit according to NCCLS. Corticosterone values after transport and lairage (*vs* basal level) were: 39.1±24.4 *vs* 35.0±23.7 µl/ml in short transport without lairage; 47.8±18.2 *vs* 40.5±26.7 µl/ml in short transport with lairage; 27.0±12.8 *vs* 44.3±27.8 µl/ml in long transport without lairage and 46.0±24.5 *vs* 38.4±29.2 µl/ml in long transport with lairage. These results confirmed the effect of environmental conditions as stressors and the opportunity to find a non-invasive method to measure stress in animals. The results showed a high individual variability due more to animal manipulations and environment than transport and lairage.

Effect of transport on rabbit's welfare: serum lysozyme determination

F. Servida[1], F. Luzi[2], C. Lazzaroni[3], M. Petracci[4] and M. Verga[2], [1]Univ. of Milano, Dept. DIPAV, via Celoria 10, 20133 Milano, Italy, [2]Univ. of Milano, Ist. Zootecnica Med. Vet. Fac., via Celoria 10, 20133 Milano, Italy, [3]Univ. of Torino, Dept. Animal Science, via L. da Vinci 44, 10095 Grugliasco, Italy, [4]Univ. of Bologna, Dept. Food Science, piazza Goidanich 60, 47023 Cesena, Italy

The aim of this study was to determine the influence of transport and lairage on serum lysozyme concentration to study animals' stress. Lysozyme is considered as a component of the earlier protective mechanisms. In this work the effects exerted on welfare by different times of journey (1 *vs* 3h) and lairage (0 *vs* 5h) were tested on 70 rabbits. Two blood samples have been taken before and after the animal transport and lairage. Lysozyme concentration was determined using a micromethod assay set up at DIPAV starting from the lysoplate Ossermann's method. In short transport without lairage (14.7±8.8 *vs* 11.8±6.6 µg/ml) and short transport with lairage (8.9±3.0 *vs* 6.0±3.6 µg/ml) groups, a decrease of lysozyme activity was observed after transport, whereas in long transport without lairage (7.4±2.3 *vs* 9.4±3.9 µg/ml) and long transport with lairage (13.7±11.3 *vs* 16.8±13.0 µg/ml) groups, an increased activity after transport was noticed. Probably transport leads to changes in neuroendocrine and immune system-derived substances, which influence the innate immune factors, but an explanation of these mechanisms needs some additional studies.

Genetic relationship between the behaviour and the constitution of sows

B. Hellbrügge[1], K.-H. Tölle[2], J. Bennewitz[1], U. Presuhn[3] and J. Krieter[1], [1]Institute of Animal Breeding and Husbandry, Christian-Albrechts-University of Kiel, Olshausenstr. 40, 24098 Kiel, Germany, [2]Chamber of Agriculture, LVZ Futterkamp, 24327 Blekendorf, Germany, [3]farm concepts, Heidmühlener Straße, 23812 Wahlstedt, Germany

Records were available from 13,971 piglets and 1,538 purebred Landrace litters. Before penning into the farrowing compartment, sows were newly mixed for washing with 5 sows per group and were scored for aggressive behaviour on a linear scale. The maternal ability was analysed with two separation tests and a screaming piglet test. 18 % of the sows showed aggressive behaviour towards other sows. During lactation, the sows reacted more active in the separation tests than in the screaming piglet test. The heritabilities were estimated with threshold models. For aggressive behaviour the heritability was $h^2 = 0.32$ and for the tests during lactation, values ranged from $h^2 = 0.06$ to $h^2 = 0.13$. The constitution was described with the health status (MMA treatment), backfat thickness at the beginning and end of lactation as well as the exterior using a linear scoring system. The heritabilities were $h^2 = 0.08$ for the health status and $h^2 = 0.17$ to 0.26 for backfat thickness. The genetic correlations indicated, that healthy sows had fewer piglet losses and react more active in the behaviour tests. Additionally the backfat thickness had a significant influence on the responsiveness of sows whereas the exterior had no significant impact.

Comparison of maternal abilities of Meishan and Large White breeds in a loose-housing system

L. Canario[1], Y. Billon[2] and J.P. Bidanel[1], [1]INRA, UR337 SGQA, F-78350 Jouy-en-Josas, France, [2]INRA, UE967 GEPA, F-17700 Surgères, France

The maternal abilities and farrowing behaviour of 16 Large White (LW) and 16 Meishan (MS) gilts were compared in a loose housing system. Females were inseminated with semen from the other breed in order to produce the same litter genetic type, i.e. LW x MS. Farrowing events occurred over four successive batches. Sow behaviour was analysed over the first three hours after the onset of farrowing, by use of video recordings. LW gilts produced larger litters (15.2 *vs* 12.9 piglets born in total; $P < 0.05$) as well as heavier piglets and more heterogeneous litters (mean and within-litter standard deviation of birth weight of, respectively, 1.33 *vs* 1.14 kg; $P < 0.05$ and 0.28 *vs* 0.19 kg; $P < 0.01$). The number of stillbirths, farrowing duration and birth to weaning survival did not differ between breeds (0.6 stillborn piglet/litter in both breeds; 3.1 *vs* 3.5 h, P=0.56; 90 *vs* 84 %, P=0.15, in MS and LW sows, respectively). LW sows had a similar colostrum production, but produced more milk and lose more weight during lactation than MS sows. Breeds differed in their behaviour at farrowing and during lactation. MS gilts spent more time in a standing position and performed more nesting behaviour at farrowing than LW gilts, which spent more time sitting ($P < 0.05$). Piglet behaviour at birth (vitality, time to first colostrum intake, nose contacts with their dam) and in early lactation (nursing duration and frequency) was also investigated.

Session 16 Poster 11

The transport stress response in lambs of different genetic groups

J. Sowińska, H. Brzostowski and Z. Tański, Department of Sheep and Goat Breeding, Department of Sheep and Goat Breeding, Oczapowskiego 5, 10-719 Olsztyn, Poland

The present study was carried out on 100-days-old ram lambs of Pomeranian Sheep breed (P) and cross-breds F_1 of Pomeranian ewes with Berrichon du cher (PBCh and Charolais (PCh) rams (12 in each group). Blood samples were collected twice: before (term I) and after transport to the slaughterhouse (term II). The level of cortisol was determined in the blood serum by means of adioimmunoassay. Transport caused a significant increase in cortisol level in each genetic group. The lambs of P group were more sensitive for the transport (cortisol level increased 3.86-times: from 19.79 to 76.52 nmol/l) as compared to the groups of cross-breds PBCh (cortisol level increased 2.49-times: 27.07 – 67.58 nmol/l) and PCh (2.19-times: 24.49 – 53.76 nmol/l, respectively.

Session 17 Theatre 1

How number of starts inform about the selection bias when using performers for breeding evaluation: the example of French trotters

A. Rose, C. Blouin and B. Langlois, INRA, GA-SGQA, Domaine de Vilvert, 78 350 Jouy en Josas, France

For trotters, qualification allows only 40% of horses to participate in races. The question of a selection bias, when using competition data for breeding value estimation, is therefore raised. To analyse this matter we looked at the number of starts of 2 to 5 year-old French trotters born between 1996 and 2000. Four variables were studied: -1- a none or all variable (starter/non starter) and three truncations of the number of starts;-2a- all data including zero; -2b- only starters excluding zero; -2c- limited to earning horses. Sire, Sire-mother and animal model were applied using REML. Corrections were made for age, sex, year, breeding area, and category of the breeder according to the number of horses produced. The range of the estimations of heritability are for 1: 0.42-0.57; for 2a: 0.18-0.34; for 2b: 0.02-0.06; for 2c: 0.04-0.18. A rupture clearly appears when taking into account the zero starts or not, indicating two phenomena: being prepared to participate in races and when prepared, to take part in a different number of races. If sources of over estimations can be discarded, the first phenomenon appears highly heritable. The second phenomenon, however, seems highly environmental or trainer dependant. The distribution of the number of starts for horses earning money compared to those not earning anything allows estimating the number of horses prepared to race but that never started. It leads to correct the apparent selection rate from 25% to 85%.

Session 17 — Theatre 2

Genetic Evaluation of Spanish Purebred Trotters' performance

M.D. Gómez[1], I. Cervantes[1], A. Molina[1], P. Moll[2] and M. Valera[3], [1]University Cordoba, C.U. Rabanales, 14071 Cordoba, Spain, [2]Astrot, Hip. Son Pardo, 07004 Palma De Mallorca, Spain, [3]University Seville, Ctra Utrera km1, 41013 Sevilla, Spain

A repeatability multivariate BLUP Animal Model, using Groeneveld's VCE (version 5) software program, was developed to estimate breeding values of Spanish Purebred Trotters. Logarithm of annual earnings (measured in euro), annual best racing time (measured in seconds) and square root of proportion of races placed 1-4 (measured as the percentage related to the total number of stars) of 5,067 animals were taken. The data file had 293,366 racing performances from 1990 to 2006. The pedigree file had 10,789 animals and was generated using four generations of the participant horses. Year of race (from 1990 to 2005), sex (male, gelding and female) and total number of starts in a year (covariate for earnings) were included as fixed effects. The random variables were individual additive and permanent environmental effect. The heritability values ranged between 0.14 for the percentage of races placed 1-4 and 0.45 for annual earnings. They were a little higher than those indicated in other trotter breeds, because of the homogeneity of the environment for Spanish trotters (with a limited breeding area). The breeding values were adjusted to a normal distribution for all the traits. The evolution of the breeding values related to the year of birth showed a positive progression for all of them.

Session 17 — Theatre 3

Estimation of variance and covariance components of race performance in German thoroughbreds

A. Hahn, E. Bruns and A.R. Sharifi, Institute of Animal Breeding and Genetics, Georg-August-University Goettingen, Albrecht-Thaer-Weg 3, 37075 Goettingen, Germany

Data composed of 210,325 race records from 43.837 2 to 4-years-old German thoroughbreds started in flat races between 1991 and 2005 in Sprint (S) (1000m-1400m), Mile (M) (1401m-1800m), Intermediate (I) (1801m-2100m) and races of Long (L) distances (2100m-2400m). Race performance was defined as rank within a race taking into account the probability of achieving a certain placing for a given number of horses started. Variance and covariance components were estimated by using ASReml 2.0 with age, sex, carried weight, race course and trainer as fixed effects and permanent environmental and additive genetic effects of a horse as random effects, separately for the four race distances and for age of horses. Heritability estimates were low for the race distances considered and ranged from.03 to.09 for 2, 3 and 4-years-old thoroughbreds. Race performances recorded in S and M race distances were highly genetically correlated,.91,.86 to.86 for 2, 3 and 4-years-old horses, respectively. For M and I race distances race performances were highly genetically correlated for 3 and 4-years-old horses (.97,.87) and moderate for M and L race distances for 3 and 4-years-old (.70,.78), respectively. Low genetic correlations were estimated between race performances in S and I and S and L race distances. They decreased from.49 (S and I) to.20 (S and I) for 3-years-old and from.32 (S and I) to.14 (S and I) for 4-years-old, respectively.

Session 17 Theatre 4

Genetic parameters for show jumping ability in young horse competitions in Ireland

K.M. Quinn[1,2], K. Hennessy[3], D.E. Machugh[1], D. Feely[2] and P.O. Brophy[1], [1]University College Dublin, School of Agriculture, Food Science & Veterinary Medicine, Belfield, Dublin 4, Ireland, [2]Irish Horse Board, Block B, Maynooth Business Campus, Maynooth, Co Kildare, Ireland, [3]University College Dublin, Centre for Sports Studies, School of Public Health & Population Science, Belfield, Dublin 4, Ireland

The genetic evaluation of show jumping horses in Ireland aims to identify horses of superior genetic merit for high level show jumping ability. In 1997, competitions confined to 4-year old, 5-year old or 6 and 7-year old horses were introduced. Young stallions seeking approval for breeding within the Irish Sport Horse population are routinely performance tested in open competition and many compete in young horse competitions. Genetic parameters were estimated for results obtained from young horse competitions. The data available consisted of approximately 45,000 performances from 3,159 horses recorded up to 2003 in both young horse and high level show jumping competitions. Heritabilities for a single performance in young horse competitions ranged from 0.04-0.11. Genetic correlations between age groups were high (0.83-0.89). Genetic correlations between young horse competitions results and high level show jumping ability varied from 0.50-0.83. Young horse competitions can be regarded as providing a realistic performance test for stallions.

Session 17 Theatre 5

Sport status and the genetic evaluation for show jumping in Belgian sport horses

S. Janssens, N. Buys and W. Vandepitte, K.U.Leuven, Biosystemen, Division of Gene Technology, Kasteelpark Arenberg 30, 3001 Heverlee, Belgium

Breeding values for show jumping horses in Belgium are based on performances in competition. However, some horses never enter jumping competitions which may point at preselection and this may bias ebv's or make selection inefficient. The objective was to investigate sport status in the framework of show jumping for Belgian warmbloods. Pedigree data from the studbooks BWP and sBs were integrated and made uniform (unique horse numbers). Performance data on show jumping were obtained from 2 riding organizations in Belgium and were treated as 2 traits: JN is performance at national level (KBRSF) and JR is at recreational level (LRV). Sport status (ST) was defined as a binary variable (at least one record in competition *vs* no performance present in the data). 74859 horses and 723908 performances were used in the computations. Genetic parameters were obtained in a Bayesian framework using the Gibbs sampler. Posterior modes of heritabilities were 0.094 (JR), 0.077 (JN) and 0.72 (ST). Genetic correlations were 0.70 (JR,JN), 0.12 (JR, ST) and 0.45 (JN,ST). The results indicate that status is highly heritable and moderately positively correlated with performance at national but not at recreational level. Correlations between ebv's obtained in models with and without ST were high: 0.90 (all horses) and 0.95 (stallions). Preselection exists for national competitions but inclusion of JR may prevent bias in ebv's.

Session 17 Theatre 6

The influence of foreign stallions on the Swedish Warmblood breed

E. Thorén Hellsten, A. Näsholm, E. Strandberg, H. Jorjani and J. Philipsson, Swedish Univ. of Agricultural Sciences, Dep. of Anim. Breeding and Genetics, Box 7023, 750 07 Uppsala, Sweden

The purpose of this study was to survey the use of foreign stallions in the Swedish Warmblood (SWB) breed and to investigate how these individuals have influenced the SWB population. Since the early 1980-ies the genetic progress has been 0.9 and 0.55 genetic SD for jumping and dressage, respectively. An open studbook is practised and foreign stallions have been used over the years to improve different characters of the breed. Data consisted of 202,808 horses included in the SWB routine genetic evaluation for 2006. Studbook of origin (STB) was determined for stallions with at least 5 progeny tested in Sweden. Those 757 stallions had together 116,505 progeny registered in SWB. The proportion of progeny sired by foreign stallions born each year has increased since the early 1980-ies. For 2006 this proportion was about 80%. Of the foreign stallions, those from Holstein had the largest no. of progenies, followed by Hanoverian and Thoroughbred stallions. An analysis of variance showed a significant effect of age group of stallion on the breeding values for dressage and jumping. STB of sire had significant effects for show jumping but not for dressage. The most favourable STB for show jumping was Holstein.

Session 17 Theatre 7

Exploring the possibility to include competition traits in the genetic evaluation of Icelandic horses

E. Albertsdóttir[1], S. Eriksson[2], A. Näsholm[2], E. Strandberg[2] and T. Árnason[1], [1]The Agricultural University of Iceland, Hvanneyri, 311 Borgarnes, Iceland, [2]The Swedish University of Agricultural Science, Box 7023, 750 07 Uppsala, Sweden

The possibility of including competition traits in the genetic evaluation of Icelandic horses, which is currently based on breeding field-test data, was explored. Linear animals models were used to analyse 18 982 records of 3790 horses competing in Iceland and Sweden 1998–2004. The traits included were seven original competition traits: two measures of four-gait, five-gait and tölt, and one pace trait. Additionally, three new combined competition traits were formed and analysed. The estimated heritabilities were low to moderate (0.18–0.35) for all competition traits. Genetic correlations estimated among competition traits were generally strong and favourable. Genetic correlations were estimated between breeding field-test traits and combined competition traits, along with one original competition trait. The breeding field-test data included 16 401 individual records of Icelandic horses evaluated in 11 countries during 1990–2005. High genetic correlations were generally estimated between field-test riding ability traits and competition traits. Moderate genetic correlations were estimated between some field-test conformation traits and most of the competition traits. It was concluded that the combined competition traits and one original competition trait could be added to the current genetic evaluation.

Session 17 — Theatre 8

Analysis of genetic progress in the Hungarian Sport Horse population

J. Posta, I. Komlósi and S. Mihók, University of Debrecen, Institute of Animal Science, Böszörményi str. 138., H-4032 Debrecen, Hungary

The aim of the study was to analyse the genetic response in performance tests of the Hungarian Sport Horse mares based on the test results of 435 three-year-old, and 240 four-year-old mares from 1993 to 2004. Conformational traits, free jumping performance and movement analyses were scored on the tests. Breeding value estimation was based on BLUP AM. Test year, age and owner were included in the animal model as fixed effects. The breeding values showed significant positive genetic progress for every trait. Higher genetic progress was achieved for type, impulsion and elasticity of movement and canter traits. The stallions' breeding values rarely exceeded the average with two standard deviations.

Session 17 — Theatre 9

Analysing the effective population size in the partially closed and fragmented breeding population of the Trakehner Horse breed

R. Teegen, C. Edel and G. Thaller, Institute of Animal Breeding and Husbandry, Christian-Albrechts-University, Hermann-Rodewald-Straße 6, D-24118 Kiel, Germany

The objective of this study was to examine the population structure of the Trakehner breed. The analysis was based on 12504 pedigree records consisting of the actual population and all their ancestors back to 1950. Ancestors with birth years before 1950 were treated as base-animals. The average generation interval calculated was 10 years. The effective population size was estimated from the increase in 1) average inbreeding coefficients and 2) average coancestries based on a) the Numerator-Relationship-Matrix with no correction of missing ancestries and b) the Uncertain-Parentage-Matrix with probable parentages. There were no major differences found between a) and b) with respect to the rate of increase in inbreeding although the global level by using b) was higher. Estimates varied drastically between 1) and 2) indicating a mating scheme with a strong emphasis on the avoidance of inbreeding. From the early 1990s onward a strong increase in the rate of inbreeding was observable that corresponds with an increased variance of family size of sires and might be a consequence of a growing use of artificial insemination. Analysing coancestries within and between the ten centrally managed breeding regions in Germany further revealed a genetically fragmented population with the main partition corresponding to the former division of Germany into East and West.

Accounting for migration rates to compute effective population size in three Arab derived Spanish horse breeds

I. Cervantes[1], J.P. Gutiérrez[2], F. Goyache[3], E. Bartolomé[1], A. Molina[1] and M. Valera[4], [1]University Córdoba, C.U. Rabanales, 14071 Córdoba, Spain, [2]UCM Madrid, Faculty Veterinary, 28040 Madrid, Spain, [3]SERIDA, Somió, 33203 Gijón, Spain, [4]University Sevilla, Ctra. Utrera km1, 41013, Spain

Arab horse has been used in Spain to build 3 breeds: Anglo-Arab (Arab x Thoroughbred), Hispano-Arab (Arab x Spanish Purebred) and Spanish Sport Horse (composite with influence of Arab, Thoroughbred and Spanish Purebred among other breeds). We present different estimates of inbreeding and effective population size (Ne) for the three Arab-derived Spanish horse breeds when gene flow from the paternal breeds is considered or simulating that no migration is allowed in the last two generations. Data consisted of 8,595 Anglo-Arab, 3,484 Hispano-Arab and 7,117 Spanish Sport Horse genealogical records. Additionally, 1,848 Thoroughbred, 2,061 Spanish Purebred and 1,915 Arab horse genealogical records related to the migrants from the parental breeds were available. Inbreeding coefficients and Ne were computed using the program ENDOG v4.0. Ne was computed from increase of inbreeding rates. Realised (average) inbreeding for Anglo-Arabs, Hispano-Arabs and Spanish Sport horses were 2.35%, 4.08% and 0.26%, respectively. Ne values were 71.1, 55.4 and 110.3, respectively for the three populations. If migrants are considered as no inbred, the average F reach values of 0.77%, 0,32% and 0.16%. On opposite, Ne values fitted for effect of migration decreased to 24.8, 48.02 and 97.3, respectively.

Multiple trait selection for radiographic health of the limbs, conformation and performance in Warmblood riding horses

K.F. Stock and O. Distl, University of Veterinary Medicine Hannover, Institute for Animal Breeding and Genetics, Bünteweg 17p, 30559 Hannover, Germany

Results of standardized radiological examinations of 5,157 Hanoverian Warmblood horses, conformation evaluations from studbook inspections of 20,637 mares, and performance evaluations from mare performance tests and auction horse inspections of 16,116 horses were used for multivariate genetic analyses. All horses were born between 1992 and 2001. Genetic parameters were estimated with REML and relative breeding values (RBV) were predicted with BLUP in linear animal models for four radiographic health traits, three conformation traits and five performance traits. Heritability estimates for osseous fragments in fetlock joints (OFF), osseous fragments in hock joints (OFH), deforming arthropathy in hock joints (DAH) and distinct radiographic findings in the navicular bones (DNB) were $h^2 = 0.14$-0.34 after transformation to the liability scale. Front limb conformation, hind limb conformation, withers height, walk, trot, canter, rideability and free jumping showed heritabilities of $h^2 = 0.10$-0.50 and additive genetic correlations with OFF, OFH, DAH and DNB of $r_g = -0.51$ to $+0.51$. Comparison of different modes of RBV-based single-trait and multiple-trait selection revealed that it is possible to simultaneously select for radiographic health of the limbs, quality of gaits, rideability, free jumping and limb conformation, with relative decrease of prevalences of radiographic findings of 13-17% after one generation.

Session 17 Theatre 12

Prediction of genotype probabilities at eight coat colour loci in the Icelandic horse in mate selection

T.H. Árnason, Agricultural University of Iceland, Hvanneyri, IS-311 Borgarnes, Iceland

The vast variation in coat colours in the Icelandic horse has been shown to be largely governed by segregation of alleles in eight biallelic individual loci. The global database for the Icelandic horse (WorldFengur) contains more than 230 thousand horses. Of those 60% have registered coat colour phenotypes by a four digit code. The Genotype Elimination algorithm of Lange was used to list out revised phenotypic records indicating those genotypes in each loci which were compatible with the pedigree data. The computational algorithm of Kerr and Kinghorn was used for the calculation of genotype probabilities for each individual horse at each loci. The genotype frequencies in the population were estimated from the data and were used as priors when the conditional probability that a horse has a certain genotype given all the data was calculated. The following gene frequencies of the dominant allele were estimated: E=0.392; A=0.195; C=0.962; D=0.052; G=0.038; Z=0.017; To=0.047; R=0.002. The resulting genotype probability estimates were listed into SQL tables which can be accessed by a Java servlet on the web. The use of the mate selection servlet for predicting possible coat colour genotypes and indicating the probabilities of corresponding phenotypes in the resulting offspring of any specific mating will be illustrated.

Session 17 Theatre 13

Phenotypic study on longevity in Italian Heavy Draught mares

R. Mantovani[1], B. Contiero[1], A. Sartori[1], C. Stoppa[2] and G. Pigozzi[2], [1]University of Padua, Department of Animal Science, AGRIPOLIS, 35020 Legnaro (PD), Italy, [2]Italian Heavy Draught Horse Breeders Association, Via Francia, 3, 37135 Verona, Italy

This Study investigated the phenotypic relationship between maternal ability, inbreeding, morphology and other sources of variation on longevity of Italian Heavy Draught Horse (IHDH) mares, born between 1987 and 2000. A data set obtained from the IHDH Breeders Association, including 3,386 mares was used. Data analysed consisted of all parities up to the 15th. Longevity was defined as length of productive life (LPL) and calculated as days form first foaling to culling. A proportional hazard model was applied using the PHREG proc. of SAS, including the sire origin (SO, i.e. IHDH or French Breton), age at first parity (AF, early or late age), 3 dummy variables depending of the four morphological score (MS), individual inbreeding coefficient and maternal ability (MA) as no. of foals born per year of reproductive life used as time dependent covariate. Data were stratified accounting for birth years (14 levels). Among different effect analysed, AF and MA had the highest magnitude on LPL ($P < 0.001$), while SO and F were not significant. MS resulted close to be significant (P=0.07). According to relative risk estimated, an early age at first parity and an higher MA resulted more favourable (20% and 30% higher, respect.) for a longer LPL. Among different MS, good and very good scores were related with a longer LPL. Further knowledge on genetic aspects of LPL in IHDH mares could be therefore investigated.

Session 17 Theatre 14

Estimates of repeatability for conformation traits in Arabian horses

R. Kuokkanen and M. Ojala, University of Helsinki, Department of Animal Science, P.O. Box 28, SF-00014 Helsinki, Finland

The aim of the study was to estimate repeatabilities for conformation traits (type, head and neck, body and upper line, and legs) evaluated subjectively in Arabian horse shows, held once a year in August in southern Finland. Data consisted show results from 20 shows between years 1986-2006. On average there were 37 horses in a show. Each horse was evaluated independently by three judges in one show. Thus, all horses had at least three records from each trait. On average, a horse had entered in 2.3 shows. Data included a total of 327 horses with 2356 records. The pedigree extended for five generations, including a total of 1473 individuals. Information about sex and age of a horse, year of birth, year of show, and the judge was available for statistical analyses. Genetic parameters were estimated by repeatability animal model and REML method using VCE5-program. Estimates of repeatability for type, head and neck, and total points were approximately 0.40, for body and upperline about 0.30, and for legs about 0.15.

Session 17 Theatre 15

Estimates of genetic parameters for body measures and subjectively scored traits in the Finnhorse

M. Suontama[1], M.T. Saastamoinen[2] and M. Ojala[1], [1]University of Helsinki, Animal Science, P.O Box 28, 00014 Helsinki, Finland, [2]Agri-Food Research Finland, Equine Research, Varsanojantie 63, 32100 Ypäjä, Finland

The aim of this study was to estimate genetic parameters for body measures (height at withers, height at croup, length of body, circumference of girth, circumference of cannon bone) and subjectively scored traits (character, body, leg stances, quality of legs, hooves, movements) in the Finnhorse. Genetic parameters were estimated in the Finnhorse trotter population from the studbook inspection data. Data included observations from 6381horses. Univariate and bivariate animal models including sex, age and year of judgement as fixed effects, and animal as a random effect, were applied. Data were analyzed also with a model combining year and sex effects into a year-sex subclass effect. The data were analyzed separately within the two sexes as well. Genetic parameters were estimated with REML-method using VCE-programs. Estimates of heritability for body measures were in the range of 0.53 to 0.80 ± 0.02 to 0.06 and for subjectively scored traits from 0.06 to 0.25 ± 0.02 to 0.10. Genetic correlations between body measures were high and positive, varying from 0.72 to 0.99 ± 0.01 to 0.07. Genetic correlations between subjectively scored traits varied from -0.33 to 0.58, most of them being positive. Genetic correlations between body measures and subjectively scored traits were low to moderate, -0.44 to 0.10, most of them being negative.

Session 17 Poster 16

Jumping parameters on different distances of the obstacle combination in free jumping tests (preliminary study)

D. Lewczuk, Institute of Genetics and Animal Breeding Jastrzebiec, Animal Breeding, ul.Postępu 1, 05-552 Wola Kosowska, Poland

The aim of the work was to investigate influence of the distance between obstacles combination on linear and temporal parameters of horse's jump. Group of stallions on performance test stations (14 Małopolski horses) were filmed during their regular work in training centre one week before performance test. The linear and temporal parameters of the jump were obtained by video image analysis (25 frames per second). The influence of distance was investigated for two distances - 6,8m and 7m between the last doublebarre obstacle and previous obstacle of combinations. Horses jumped the combination two times on every distance. The doublebarre obstacle was of the height of 100cm second part and 85cm first part and the wide of 80cm. Data from 56 jumps were analysed by analysis of variance (random effect of the horse, fixed effects of the distance and successive number of the jump). The distance between obstacles influenced the jumping parameters mainly for the parameters describing movements of the back part of horses' body. Longer distance lowered the elevation of the croup and lifting of the hind legs above the obstacle (significant at $p < 0,05$). Traits described as "work of coup" and "bowl of the upper line" were also affected by the distance. Temporal parameters were less affected by the distance in the combination. Only the elevation time of the jump was significantly lower by the longer distance.

Session 17 Poster 17

Genetic parameters for young eventing competition in Spain: correlation between dressage, jumping, cross and conformation

I. Cervantes[1], E. Bartolomé[1], M.D. Gómez[1], C. Medina[1], M.A. González[1] and M. Valera[2], [1]University Córdoba, C.U. Rabanales, 14071 Córdoba, Spain, [2]University Sevilla, Ctra. Utrera km1, 41013 Sevilla, Spain

The young eventing competition combines dressage, jumping, cross exercises and a conformation evaluation. The aim of this study was to estimate the correlation between the four type of proofs of eventing in order to determine the best criterion to improve the horse performance in this equestrian sport. The data sets used in the present study took into account 35 young horse competitions (4-6 years old) held between 2004 and 2006. The total number of records were 1097 entries of 236 horses. The participants were Spanish Sport Horses (142), Anglo-Arabs (46) among others sport breeds. Each competition included an average of 31 participants. A multivariate BLUP animal model was designed to evaluate each exercise using VCE program. The heritabilities ranged between 0.10 and 0.21. Genetic correlation between dressage and cross or eventing were positive but low (0.08). This value was higher between jumping and cross (0.29). The jumping ability had the best correlation with conformation evaluation (0.24). These results indicated that dressage performance and jumping-cross abilities are different criterion to attain the genetic improvement. This makes the selection for eventing more difficult than for other type of disciplines.

Session 17 Poster 18

Morphological data analysis of Spanish Arab Horse aimed to define a line type trait system

I. Cervantes[1], M.D. Gómez[1], E. Bartolomé[1], J.P. Gutiérrez[2], A. Molina[1] and M. Valera[3], [1]University Córdoba, C.U. Rabanales, 14071 Córdoba, Spain, [2]UCM Madrid, Faculty Veterinary, 28040 Madrid, Spain, [3]University Sevilla, Ctra. Utrera km1, 41013 Sevilla, Spain

Body measurements from Spanish Arab Horses have been analysed in order to define a linear type traits system. It would be useful to improve the traits related to the performance in the competitions. Data consisted of 37 body measurements from 135 Spanish Arab Horse (52 males, 62 females and 21 geldings) ranging from 2-10 years old. The body measurements included 16 lengths, 9 angles, 6 heights, 4 widths and 2 perimeters variables. Gender and age effect were analysed using a general linear model. Gender was significant factor for most of the traits. The 61.1% of total phenotypic correlations were significant, ranging their absolute from 0.49 to 0.92. The heights were the most correlated with the other variables and the angles the fewest. The angles obtained the highest coefficient of variation (30.8%) and the heights the lowest (2.2%). This is probably due to the difference between animals bred to improve functionality or morphological traits. Factor analysis showed some associations among variables. The first factor grouped proportional traits, the second and third factors described functionality traits (angles). This analysis is expected to be useful to describe the different morphological types in Spanish Arab horse population (morphological lines, endurance and race type).

Session 17 Poster 19

Selection by phenotypic traits of potential founders of the new Lithuanian Heavy Draught horse line

R. Sveistiene, Institute of Animal Science of Lithuanian Veterinary Academy, Animal breeding and genetics, R.Zebenkos 12, Baisogala, LT-82317, Lithuania

All available genetic potential of horses should be used to widen the heterozygosis of the Lithuanian Heavy Draught horse breed. Therefore, the aim of our investigation was to evaluate the stallions chosen for the development of the new line in order to stop the disappearing of the genealogical structure of the Lithuanian Heavy Draught breed. According to the expedition plans, the data on horse parentage have been collected, typical horses found and assessed by measuring and complex evaluating. The genealogical analysis of Gandras 0697 line progeny indicated that there might be found five generations and, thus, the genealogical group of these horses may be looked upon as a separate and individual line. The genealogical analysis of the progeny shows that 30% of breeding horses were obtained by inbreeding, the average inbreeding coefficient being 6.6%. The stallions of Gandras 0697 line are of a desirable type and body conformation. The selected typical stallions will be included in the general programme for Lithuanian Heavy Draught horse breeding. It is suggested to breed the horses by circular mating scheme, thus preserving their genealogical structure.

Session 17 Poster 20

Evaluation of performance traits in the genetic resource of the Old Kladrub horse

L. Andrejsová[1], I. Majzlík[1], V. Jakubec[1] and J. Volenec[2], [1]Faculty of Agrobiology, Food and Natural Resources, Genetics and Breeding, Kamýcká 129, 165 21 Prague, Czech Republic, [2]National Stud at Kladruby, Kladruby nad Labem, 533 14, Czech Republic

Performance evaluations of 372 horses of the Old Kladrub breed for 15 traits within the period of 1995–2004 were used. For the analysis a linear model with fixed effects and the least squares method were used. These effects and interactions were analysed: variety (gray and black), stud, sex, year at birth, age at classification, sire line, interaction variety × stud and sire line × stud. The subjects of the analysis were the traits: type and sex expression, body lines, fundament, general harmony, dressage, general impression, walk, trot, canter, marathon, dressage test, obstacle driving test, first, second and third pull. The difference between the means of traits was significant or highly-significant within the: varieties for 5 traits, studs for 4 traits, sex for 11 traits, year at birth and age at classification for 5 traits and sire lines for 9 traits. The interaction variety × stud was only significant for 2 traits and the interaction sire line x stud was significant and highly significant for 8 traits. The traits can be chosen as selection criteria for the breeding of varieties and sire lines when the differences between trait means of varieties and sire lines were significant and highly-significant and if interactions do not exist.

Session 17 Poster 21

Linear type trait analysis in the varieties and studs of the Old Kladrub horse

V. Jakubec[1], I. Majzlík[1], J. Volenec[2] and L. Vostrý[1], [1]Faculty of Agrobiology, Food and Natural Resources, Genetics and Breeding, Kamýcká 129, 165 00 Praha 6 - Suchdol, Czech Republic, [2]National Stud at Kladruby, Kladruby nad Labem, 533 14 Kladruby nad Labem, Czech Republic

Linear type evalutions of 494 horses of the Old Kladrub breed for 32 traits were used to analyse the effect of variety, stud, sex, year of birth and age at classification. The linear model with fixed effects was used. The specific properties and variation of the Old Kladrub horse in its current state were characterized by the overall mean, standard deviation, coefficient of variation and number of utilized scores. The highest coefficient of variation showed the forelimbs-side view (40.14 %), chest girth (36.25 %) and height at withers (30.97 %). 28 traits from 32 were within the span from 7 to 9 utilized scores. Significant differences between both varieties were found in 13 of 32 traits. Significant differences were found in 12 of 32 traits between the Kladruby stud and private studs. In only 7 traits a significant interaction variety x stud was recorded. Significant differences between stallions and mares were recorded in a large number of front and body traits (in 11 of 18 traits) and in 2 rear traits. Despite of a remarkable number of significant differences between the years of birth (in 18 of 32 traits) and age at classification (in 13 of 32 traits) both factors are not important. The linear type classification, performance recording and selection are carried out in the population of four year old horses born in the given year.

Session 17 — Poster 22

Melanoma in grey old Kladruber Horse

B. Hofmanova and I. Majzlik, Czech University of Agriculture, Department of Genetics and Breeding, Kamycka 129, 165 21 Prague 6 -Suchdol, Czech Republic

The aim of this study was to verify the possible occurence of melanoma and vitiligo in gray variety of Old Kladruber horse in connection with typical breed trait „greying"with the age as revealed in other grey horses and grey breeds resp. This preeliminary study after detailed inspection of the fenotype of 148 horses of the two greatest studs confirmes an occurence of melanoma which is corresponding with literature information.The occurence of melanoma was detected by adspection and palpation in 5 grades (acc. to Sőlkner *et al.*, 2004).The incidence of melanoma in grey Kladruber horse is related strictly to the age – tumor occur first mostly at the age of 5-6 years. Up to 15 years of age is the melanoma detected in 20% of horses, which is less then in other breeds. Horses older than 15 years showed melanoma in 82% which is consistent with literature. Total occurence of melanoma seems to be in this breed substantially lower. Melanoma grade 3 reached 5 horses only, grade 2 is detectable at the age of 17 and more, grade 5 was not revealed. The occurence of vitiligo is substantially higher and growing with the age-the highest level was noticed at the age 22 and more with the range 50 – 100%. The occurence of melanoma in relation to line or family origin of the horse is in this study not statistically significant.

Session 18 — Theatre 1

Genomic selection: a break through for application of marker assisted selection to traits of low heritability, promise and concerns

W.M. Muir, Purdue University, Animal Sciences, West Lafayette, Indiana, 47907, USA

Genomic selection (GMAS), a form of maker assessed selection, uses all markers, equally spaced, and spanning the genome, for prediction of breeding values. Previous simulations showed great promise for GMAS, but was based on a trait with high heritability and other simplistic assumptions. In this presentation GMAS is compared to BLUP under a range of heritabilities, with differing number of training generations (TG), population sizes, and more realistic assumptions regarding the genetic architecture. TG were the number of generations in which both genotypic and phenotypic information were collected. The combined information was used to estimate effects of all marker using a simple single-marker, mixed-model analysis. Thus during the TG, GMAS uses both genomic and phenotypic information for prediction of breeding values, while in the post TG, BLUP uses only ancestor information, while GMAS uses only genomic information. Results showed that during the TG, for traits of high (.5) or low heritability (.1), the accuracy of selection increased between 10% and 30%. In the post TG, for traits of high heritability, a TG of 3 was needed to increase accuracy of GMAS over BLUP. However, GMAS continued to yield a higher accuracy that of BLUP for over 7 generations post TG. For traits of low heritability, with TG of 3, GMAS accuracy with 45% higher than BLUP. Questions remain as to marker density and effective population size needed to generate population wide linkage disequilibrium needed for GMAS.

Session 18 Theatre 2

A comparison of different regression methods for genomic-assisted prediction of genetic values in dairy cattle

J. Sölkner[1,2], B. Tier[2,3], R. Crump[2,3], G. Moser[2], P. Thomson[2] and H. Raadsma[2], [1]University of Natural Resources and Applied Life Sciences, Gregor Mendel Str. 33, A-1180 Vienna, Austria, [2]University of Sydney, Dairy CRC, Camden, NSW 2570, Australia, [3]University of New England, AGBU, Armidale, NSW 2351, Australia

The availability of large arrays of single nucleotide polymorphisms (SNP) is changing the approach of predicting breeding values from molecular information. A pool of 1546 Holstein Friesian bulls, mostly of Australian origin, with highly accurate estimates of breeding values (EBV) was genotyped for 15380 SNP. Methods of regressing EBV, considered as proxies for true breeding values, on SNP genotypes coded as 0, 1 (heterozygous) and 2 were compared. The traits considered were total merit, protein yield, overall type, fertility and somatic cell count. The methods applied were ordinary least squares regression (OLSR) with LAR variable selection, OLSR using a genetic algorithm for variable selection and modified prediction (OLSR-GA) and partial least squares regression (PLSR). To avoid overfitting due to the large number of regressors, cross-validation techniques were applied and the predictive capacity was evaluated from 5 repeated runs separating 200 bulls as test data not involved in the estimation. Correlations of true and predicted values for these test data sets were in the range of 0.65-0.8 for most traits, including fertility, a low heritability trait. OLSR-GA and PLSR performed significantly better than OLSR.

Session 18 Theatre 3

Using partial least square regression (PLSR) and principal component regression (PCR) in prediction of genomic selection breeding values

T.R. Solberg[1], A.K. Sonesson[2], J.A. Woolliams[3] and T.H.E. Meuwissen[1], [1]University of Life Sciences, Dept. of Animal and Aquacultural Sciences, P.O.Box 5003, N-1432, Aas, Norway, [2]AKVAFORSK (Institute of Aquaculture Research Ltd.), P.O.Box 5010, N-1432 Aas, Norway, [3]Roslin Institute (Edinburgh), Roslin, Midlothian, EH25 9PS, United Kingdom

Use of dense marker information combined with genomic selection has been suggested in animal breeding to predict breeding values for animals. When dense marker maps are available the number of predictors (markers) is large compared to the number of observations, and traditional regression methods are no longer feasible because of multicollinearity. This is a typical case where the number of effects to be estimated is larger than the number of records. Partial least square regression (PLSR) and principal component regression (PCR) are designed for such situations, and they try to explain the relationship between predictors and a response by means of a reduced number of principal components. Here we used PLSR and PCR to predict genome wide breeding values on offspring based on single SNP marker effects, where the density ranged from 1 to 0.125 cM spacing (resulting in a total of 1010 and 8080 markers, respectively). Using PLSR and PCR, the selection accuracy varied from 0.467 to 0.556 and from 0.570 to 0.649, respectively. In an earlier study, we used Bayesian methods which gave higher accuracies of selection (varying from 0.663 to 0.820). However, the PLSR and PCR were computationally much faster and simpler.

Session 18 Theatre 4

Validation of genomic selection in an outbred mouse population

A. Legarra, E. Manfredi, C. Robert-Granié and J.M. Elsen, INRA, SAGA, BP52627, 31326 Castanet Tolosan, France

Genomic selection was tested in an outbred mice population (1928 weights at 6 weeks, 10946 SNP, http://gscan.well.ox.ac.uk/), composed of 172 full-sib families. We used cross-validation. The data y was split at random 100 times in an estimation set (y_1) and a validation set (y_2). Breeding values $g=(g_1, g_2)$ were estimated using y_1 only (y_1 = sex + Zg + e), i.e., breeding values in y_2 (g_2) were predicted from y_1. Three models were used: (1)classical polygenic BLUP; (2)a SNP mixed model, and (3)both. Correlation between y_2 (corrected by sex) and estimates of g_2 was the goodness of fit criterion. Moreover, it is proportional to the expected genetic gain selecting animals in y_2 using information in y_1. We split y in two ways: by excluding whole families to form y_2 (i.e., population LD is used) ("sampling families"); or by splitting each family into two (i.e., LD due to close relationships is also used) ("splitting families"). The results "sampling families" show a correlation of 0 if classical BLUP is used, and 0.20 with any model including SNP. The results "splitting families" show a correlation of 0.59 for classical BLUP, 0.48 for SNP information only and 0.60 for BLUP+SNP. SNP information recovers a good part of the family information and perhaps of the population LD but its predictive ability is poorer than classical BLUP for practical purposes, contrary to Meuwissen *et al.* (2001) simulations. It is unknown if this result is due to an incorrect genetic model or if more sophisticated methods might do better.

Session 18 Theatre 5

Detection of SNPs associated with chick mortality in broilers: a machine learning approach

N. Long[1], D. Gianola[1], K.A. Weigel[1], G.J.M. Rosa[1] and S. Avendano[2], [1]University of Wisconsin, 1675 Observatory Dr., Madison, WI 53706, USA, [2]Aviagen Ltd., Newbridge, Midlothian, EH28 8SZ, United Kingdom

Genome-wide studies with single nucleotide polymorphisms (SNPs) can identify variants related to complex traits. Efficient methods of selecting influential SNP markers are needed, for their incorporation into statistical models for predicting phenotypes. A two-step method was developed, consisting of filtering (information gain) and wrapping (naive Bayesian classification). Filtering reduces the number of SNPs, to facilitate wrapping step. A discretization approach of continuous phenotypic values was used, to enable SNP selection in a classification framework. Methods were applied to chick mortality rates on progeny from 201 sires in a broiler line, each typed for over 5000 SNPs. To mimic case-control studies, sires were clustered into low and high mortality groups using arbitrarily chosen cut points. Eleven different case-control' samples were formed, and the two-step SNP selection procedure was applied to each. The 11 sets of chosen SNPs were evaluated with a linear model using cross-validation predicted residual sum of squares (PRESS) as end-point. Classification accuracy was improved from around 50% using all SNPs to above 90% with feature selection. Results were consistent over the 11 case-control samples. The case-control group with lowest PRESS selected 17 SNPs accounting for 36% of the variation in mortality rates across all sire groups.

Session 18 — Theatre 6

Use of SNP for marker assisted selection in French dairy cattle

F. Guillaume[1,2], S. Fritz[3], D. Boichard[1] and T. Druet[1], [1]INRA, UR337, Domaine de vilvert, 78350 Jouy en Josas, France, [2]Institut Elevage, 149 rue de Bercy, 75595 Paris, France, [3]UNCEIA, 149 rue de Bercy, 75595 Paris, France

A Marker Assisted Selection program based on linkage equilibrium is applied since 2001 in France. This study aims at evaluating the benefits from the replacement of the microsatellite markers by some 300 SNP. A simulation study was realized to compare the efficiency of both MAS programs. For all dairy traits, MAS proved to predict more precisely breeding values of young bulls before testing than classical selection. With microsatellite, the gain of reliability were equal to +0.05, +0.06,+0.04, +0.11 and +0.08 for milk, fat and protein yields and fat and protein contents, respectively. With SNP markers, the gain was even better: +0.09, +0.11, +0.06, +0.16 and +0.11, respectively. However, thanks to a higher marker density, SNP markers offer also the possibility to trace small chromosomal regions over more generations. The study of our Holstein pedigree showed that more than 75 % of the gametes of young bulls came from 22 founder alleles. Using haplotypes of 10 SNP in 1 cM allowed us to group correctly these gametes. The number of gametic effects to estimate was strongly reduced and therefore the precision of the estimation was improved. MAS with SNP is expected to perform even better because the linkage disequilibrium can be used to trace small chromosomes fragments conserved over much more generations.

Session 18 — Theatre 7

Empirical and theoretical considerations on the impact of genetic interactions on response to selection

A. Le Rouzic[1], P.B. Siegel[2] and O. Carlborg[1], [1]Uppsala University, Linnaeus Centre for Bioinformatics, BMC Box 598, SE-751 24 Uppsala, Bahamas, [2]Virginia Polytechnic Institute and State University, Department of Animal and Poultry Sciences, Blacksburg, VA 2406-0306, USA

The genetic architecture of a quantitative trait influences responses to selection (natural and/ or artificial). Animal and plant breeders, as well as quantitative geneticists, have been aware that gene interactions influence quantitative traits, however, theories used to predict selection responses still focus on additive genetic variation. Molecular genetics allows for further studies of the importance of genetic architecture on responses to selection. We, through individual-based simulations, analyzed the dynamic properties of a four-locus gene network influencing response to bi-directional selection for juvenile body weight in chickens. Our results show how epistasis can modify the selection response, leading to a progressive release of genetic variation, as well as different final outcomes of selection depending on the initial allele-frequencies in the population. Strong genetic interactions may also mislead Quantitative Trait Loci detection experiments based on crosses between selected lines by leading to fixation of the same alleles in lines undergoing different selection pressures.

Session 18 Theatre 8

Modelling genetic epistasis between selected candidate genes for milk production traits in Jersey cattle

A. Gontarek[1], J. Komisarek[2] and J. Szyda[1], [1]University of Life Sciences, Institute of Animal Genetics, Kożuchowska 7, Wrocław, 51-631, Poland, [2]Agricultural University of Poznan, Department of Cattle Breeding and Milk Production, Wojska Polskiego 71 A, Poznań, 60-625, Poland

The aim of the study was to estimate the epistatic relationships between polymorphisms within DGAT1, LEP, GH, GHR, and BTN genes. The animal material comprised 100 Jersey cows from a single herd. The phenotypic records of milk, fat and protein yields, fat and protein contents, and somatic cell scores were obtained through the routine milk recording programme and available for multiple lactations. A previous study revealed significant main effects of some of the polymorphisms. Here we report results of fitting a mixed model with random additive polygenic and permanent environmental effects and fixed effects of selected polymorphisms and their epistasis parameterised according to the F-infinite metric model defined by Kao and Zeng. The impact of the number of additive, dominance and epistatic genetic effects fitted into the model on the estimates of model parameters and model selection was illustrated. Original BIC and a BIC, modified following Baierl *et al.* in order to give a penalty appropriate for epistatic effects, are used for model selection.

Session 18 Theatre 9

Detection and use of single gene effects in large animal populations

N. Gengler[1], M. Szydlowski[2], S. Abras[2] and R. Renaville[2], [1]Gembloux Agricultural University (FUSAGx), FNRS, Passage des Déportés 2, 5030 Gembloux, Belgium, [2]Gembloux Agricultural University (FUSAGx), Passage des Déportés 2, 5030 Gembloux, Belgium

Unbiased estimation of single gene effects can only be achieved by estimating them simultaneously with other environmental and polygenic effects. As in large animal populations the vast majority of animals are however not genotyped, missing genotypes have to be estimated. Currently used methods are unpractical for large datasets. Recently an alternative method to estimate missing gene content, defined as the number of copies of a particular allele was developed. Unknown gene content is approximated from known genotypes based on the additive relationships between animals. This method was tested for the detection of candidate gene effects for bovine transmembrane GHR on first lactation milk, fat and protein test-day yields in Holsteins where 961 mostly recent sires out of 2,755,041 animals were genotyped. The GHR gene was estimated to show moderate to small effects of 295 g/day for milk, -8.14 g/day for fat yield and -1.83 g/day for protein yield (phenylalanine replaced by tyrosine, frequency 23.3%). The accuracy of the procedure was estimated by doing 15 simulations using gene dropping and adjustment of the observed 12,858,741 records using the estimated values. The new method proved functional and accurate as relative bias in the estimation of allele frequency was very low (0.2%) as were the biases for moderate effects (milk: 3.7%; fat: 3.3%). Bias was larger for small effect on protein (55.3%).

Session 18 Theatre 10

Model comparison criteria in a global analysis of a microarray experiment

C. Díaz[1], N. Moreno-Sánchez[1], J. Rueda[2], A. Reverter[3], Y.H. Wang[3] and M.J. Carabaño[1], [1]INIA, Dept. Mejora Genética Animal, Ctra. de la Coruña Km. 7,5, 28040 Madrid, Spain, [2]Facultad de Biología, UCM, Dept. Genética, Ciudad Universitaria s/n, 28040 Madrid, Spain, [3]CSIRO Livestock Industries and CRC for Cattle and Beef Quality, Queensland Bioscience Precint, 306 Carmody Rd., St Lucia 4067, QLD, Australia

Analysis of data from cDNA microarray experiments is an area of intense research. Options include models at a gene level or at global level. The later joining information from all profiled genes. In general, a joint analysis is expected to be more powerful than gene specific analyses. Global analysis of microarray data requires definition of a model for data normalization and analysis jointly. The objective was to assess alternative models for data normalization in an experiment to identify differentially expressed genes between two muscles in Avileña Negra Ibérica calves. Thirteen models to analyze expression intensities were comparedusing Bayes Factors (BF) and cross-validation predictive densities (PD). Models included array or array-block effect to account for the spatial arrangement of spots, dye channel, muscle effect and their interactions as systematic effects. Additionally, all models included gene, array by gene, dye by gene, muscle by gene and gene by animal effects as random factors. Both, the BF and PD indicated that the model including array-block, dye, muscle, array-dye as systematic effects and all gene related effects as random was the best model for normalization and analysis of this data.

Session 18 Theatre 11

A Bayesian mixed-model approach for the analysis of microarray gene expression data assuming skewed Student-t distributions for random effects

J. Casellas and L. Varona, IRTA-Lleida, Millora i Genètica Animal, Alcalde Rovira Roure 191, 25198 Lleida, Spain

The analysis of large data sets from microarray studies has been an area of intensive statistical development for the last years. Mixed models have been recently adapted to microarray gene expression data and random effects are commonly stated as normally distributed, although this assumption may be in some cases biased. The differential expression between two treatments or tissues could take a wide range of skewed distributions, whereas the Gaussian assumption forces the analysis to a stringent scenario. The objective of this research is to present a Bayesian mixed-model approach allowing for skewed and heavy-tailed distributions for the random effects associated with genes or treatments. The model included the systematic effect of each array and two random sources of variation, the gene and the treatment nested within gene. The Bayesian likelihood for the microarray data was assumed as multivariate normal with heterogeneous residual variances, and the a priori distribution for the effect of each array was stated as flat. Moreover, the random effects of each gene as well as the differential expression between treatments nested by gene were modeled under two independent skewed Student-t distributions. The model was developed using Gibbs sampling, with a Metropolis-Hastings step for the nonstandard conditional posterior distributions. This procedure has been applied on free access microarray data with appealing results.

Session 18 Poster 12

A differential evolution Markov chain algorithm to map epistatic QTL

M.J.M. Rutten, M.C.A.M. Bink and C.J.F. Ter Braak, Biometris, Wageningen UR, Bornsesteeg 47, 6708 PD Wageningen, Netherlands

Genetical Genomics aims to unravel metabolic, regulatory and developmental pathways by application of QTL analysis on high throughput genomics data. The number of markers and traits surpass greatly the analyzing capacity of the commonly used QTL methodologies. Furthermore, the current QTL methodology often is less suited to handle models including multiple QTL with putative epistatic interactions. We propose a novel and flexible Markov chain Monte Carlo version of the genetic algorithm Differential Evolution, called DE-MC, to map multiple main– and epistatic QTL. The DE-MC is a population MCMC algorithm, in which multiple chains are run in parallel. DE-MC chooses an appropriate direction and scale of the jumping distribution and thereby solving a common problem in MCMC methods. In DE-MC the jumps are simply a fixed multiple of the differences of two randomly chosen parameter vectors that are currently in the population. We aim to increase the sampling efficiency of the algorithm so that large genetical genomics datasets can be analyzed by fitting complex models. The potential of the proposed method will be illustrated by an analysis of simulated data.

Session 18 Poster 13

Modelling epistasis between quantitative trait loci on swine chromosome six

J. Szyda[1], M. Szydłowski[2] and E. Grindflek[3,4], [1]University of Life Sciences, Institute of Animal Genetics, Kożuchowska 7, Wrocław, 51-631, Poland, [2]Agricultural University of Poznan, Department of Genetics and Animal Breeding, Wołyńska 33, Poznań, 60-637, Poland, [3]The Norwegian Pig Breeders Association, Hamar, 2304, Norway, [4]Norwegian University of Life Sciences, CIGENE and Department of Animal and Aquacultural Sciences, Ås, 1432, Norway

There is growing evidence that the genetic architecture of quantitative characters is not only based on additive genetic effects, but also on epistasis. Our long term study devoted to detection of quantitative trait loci (QTL) revealed multiple linked QTL affecting the amount of intramuscular fat on swine chromosome 6 and also indicated a possible effect of epistasis. Here we describe a further step of the analysis, where staitistical models included polygenic effect, multiple random QTL and epistasis described as a covariance between QTL effects. The data consisted of 305 F2 individuals resulting from a backcross involving Duroc boars and Norwegian Landrace sows, for whom intramuscular fat content (IMF) and genotypes of 24 microsatellite markers spanning 147.7 cM on chromosome 6 were recorded. Multilocus Monte Carlo Markov Chain algorithm was used for the estimation of IBD matrices at 1 cM steps within the 60 cM region of chromosome 6. Several diagnostics were involved to monitor the convergence of the algorithm. The (co)variances and other parameters of the models were estimated by the REML method.

Estimation of Quantitative Trait Loci parameters for somatic cell score in the German Holstein dairy population

C. Baes[1], M. Mayer[1], A. Tuchscherer[1], F. Reinhardt[2] and N. Reinsch[1], [1]Research Institute for the Biology of Farm Animals (FBN), Wilhelm-Stahl-Allee 2, 18196 Dummerstorf, Germany, [2]United Information Systems Animal Production (VIT), Heideweg 1, 27283 Verden, Germany

The mixed inheritance model, with random quantitative trait locus (QTL) and fixed additive effects, is a suitable tool for statistically describing genetic variation in quantitative traits. Mixed QTL models account for additive polygenic relationships between animals as well as gametic relationships at specific positions on the genome, resulting in more exact parameter estimation. The objective of this study was to estimate the proportion of total genetic variance attributed to a QTL on *Bos Taurus* autosome 18 for somatic cell score in the German Holstein dairy population using a mixed inheritance model. Genotype information was provided by the German MAS-program for 6,050 bulls and 470 bull dams. The pedigree contained 12,008 animals, including nongenotyped parents of genotyped animals. The inverse of the QTL-relationship matrix was derived using a partial pedigree approach. A random QTL model was applied to incorporate marker information into parameter estimation. The ratio of QTL to polygenic variance was estimated at 0.06 for somatic cell score, with a likelihood ratio test statistic of 14.62. These results indicate that the highly significant QTL in the chromosomal area studied is responsible for approximately 6% of the genetic variance in somatic cell score in the German Holstein population.

The use of the ant colony algorithm for analysis of high-dimension gene expression data sets

K.R. Robbins, W. Zhang, J.K. Bertrand and R. Rekaya, University of Georgia, Animal And Dairy Science, Department of Animal and Dairy Science, The University of Georgia, 30602, USA

Gene expression profiling is becoming a common technique for the study of gene expression variation; however, due to the high dimensions and complex structure of expression data sets, traditional statistical models may not be adequate for the analysis of such data. To address issues associated with commonly used methods for the identification of predictive gene sets, the ant colony algorithm (ACA) is proposed for use on data sets with large numbers of features (genes) and complex structures. The ACA is an optimization algorithm capable of modeling complex data structures without the need for explicit parameterization. The incorporation of prior information and communication between simulated ants allow the ACA to search the sample space more efficiently than other optimization methods. When applied to a high-dimensional microarray data set, the ACA was able to identify small subsets of highly predictive and biologically relevant genes without the need for simplifying assumptions. Using genes selected by the ACA to train a latent variable model yielded increases in prediction accuracy of 16.6% and 6.5% over genes sets selected by test statistics and other optimization models. Furthermore, the ACA was able to converge to good solutions without the need for significant truncation of the data, as required by the other optimization algorithms. The ACA was also able to achieve higher prediction accuracies using fewer genes than other methods.

Session 18 Poster 16

Comparison of Single Nucleotide Polymorphism and microsatellite polymorphism for QTL mapping

G.C.B. Schopen, H. Bovenhuis, M.H.P.W. Visker and J.A.M. van Arendonk, WUR, Animal Breeding and Genomics Centre, Marijkeweg 40, 6709PG Wageningen, Netherlands

The objective of this study was to compare the information content on single nucleotide polymorphisms (SNPs) and microsatellite polymorphism in full-sib chicken data and half-sib cattle data. For this study, real data was available for 42 SNPs and 24 microsatellite markers of one chromosome of chicken, and 37 SNPs and 63 microsatellites of one chromosome of cattle. The two marker types were compared for their information content, which is the variance of the probability that an offspring inherited a certain allele from its parent. The analysis of the chicken data showed that the 42 SNPs provide an average information content of 0.36 and the 24 microsatellites provide an average information content of 0.29. For the cattle data, the 37 SNPs provide an average information content of 0.47 and the 63 microsatellites provide an average information content of 0.73. This study, therefore, suggests that to obtain an equal amount of information about the inheritance of chromosomal segments from parents to offspring, 1 to 2 SNPs are needed for each microsatellite marker within chicken. For cattle, 3 to 4 SNPs are needed for each microsatellite marker.

Session 18 Poster 17

A method to approximate parents' contribution to gene content at marker locus

M. Szydlowski[1], N. Gengler[1] and M.E. Goddard[2], [1]Gembloux Agricultural University (FUSAGx), Animal Science Unit, Passage des Déportés 2, 5030 Gembloux, Belgium, [2]University of Melbourne, Royal Pde, Parkville, 3010, Australia

Gene content is the number of copies of an allele in a genotype of an animal. It can be used to study additive gene action of a candidate gene. A new method to approximate paternal and maternal contributions to gene content for an untyped animal being a member of large complex pedigree with incomplete molecular data was proposed. The proposed method was derived from gametic model and mixed model equations. In the model the sire and dam contribution to gene content at single biallelic locus are treated as random additive effects. A genotype elimination algorithm was used before computation to construct (co)variance matrix of parental contribution based on pedigree and molecular data available for the considered locus and linked markers, and to extend the vector of observations by concluding the only possible gene content for as many gametes as possible. The proposed method was compared to MCMC approach on a large complex pedigree and simulated data for hypothetical SNP. An average absolute difference between the results obtained by the use of MCMC approach and the proposed method was 0.037. The potential of the method was further evaluated for the model under exact (co)variance matrix, which was obtained by the use of MCMC approach. It was concluded that it can provide good approximation to parental contribution of gene content at single marker locus when MCMC approach is impractical.

Session 19 Theatre 1

What makes a good poster?

B. Malmfors[1], P. Garnsworthy[2] and M. Grossman[3], [1]Swed. Univ. Agric. Sci., Dept of Animal Breeding and Genetics, PO Box 7023, 75007 Uppsala, Sweden, [2]Univ Nottingham, Loughborough, LE12 5RD, United Kingdom, [3]Univ Illinois, Urbana, IL, 61801-4733, USA

A good poster attracts viewers, raises interest, and emphasizes the most important points, or "take-home" messages. A poster should be informative, but too much information and design, distract from your messages. The ABC of a poster is: Attractive and Audience-adapted, Brief, and Clear in conveying messages. Additional information can be provided in a handout. A good poster has a brief, informative title; clearly stated objectives; and clear conclusions easily found. Contents can be arranged in columns or rows, or in some other structure, e.g. circular. Text should be brief, be in a font large enough to be read from a distance, and have good contrast between text and background. Results are best presented in tables or figures, preferably with a take-home message attached. Photographs can enhance the poster. Colours are useful to harmonize a poster and to highlight, separate, or associate information. Using too many colours distracts or gives an uncoordinated effect. In a good poster there is a balance between text and illustrations; and there is some empty space. This guideline is an abstract of information provided at the EAAP workshops on Writing and Presenting Scientific Papers, and in Malmfors, B., Garnsworthy, P. and Grossman, M. 2004. Writing and Presenting Scientific Papers, 2nd ed. Nottingham University.

Session 19 Theatre 2

Genetic response in piglet survival in a selection experiment carried out under outdoor conditions

R. Roehe[1], N.P. Shrestha[2], P.W. Knap[3], K.M. Smurthwaite[4], S. Jarvis[1], A.B. Lawrence[1] and S.A. Edwards[2], [1]Scottish Agricultural College, Edinburgh, E26 0PH, United Kingdom, [2]University of Newcastle, Newcastle upon Tyne, NE1 7RU, United Kingdom, [3]PIC International Group, Schleswig, 24837, Germany, [4]Grampian Country Food, Turriff, AB53 4NH, United Kingdom

Selection for piglet survival (SV) may have low response due to low heritability, genotype by environmental interactions (GxE), etc. This selection experiment was carried out to predict selection response of SV under outdoor conditions when sires and maternal grandsires have been selected for SV under indoor conditions. Data were recorded on 6,589 piglets from first-parity litters of 497 dams. The dams' sires (MGS) had been selected for a high or average breeding value (EBV) for maternal genetic effects of SV during the nursing period (SVNP). Sires of piglets were selected for direct genetic effects EBV groups of SVNP only. Phenotypic SVNP means of the high and average MGS EBV groups were 91.1 and 88.4%, respectively. Considering all fixed and random effects, the difference (selection response: 3.03%, se=1.18%) was significant, and slightly higher than the difference in maternal EBV (2.67%) of the MGS groups at time of selection. Birth weight (BW) was identified to be one main factor for the selection response (adjustment for BW reduced the response to 1.39%). This selection experiment shows that selection for SVNP under indoor conditions will improve SVNP under outdoor conditions, indicating the absence of GxE.

Session 19 Theatre 3

Correlated responses for pre- and postweaning growth and backfat thickness to six generations of selection for ovulation rate or prenatal survival in French Large White pigs

A. Rosendo[1], L. Canario[1], T. Druet[1], J. Gogué[2] and J.P. Bidanel[1], [1]INRA, UR337 SGQA, F-78350 Jouy-en-Josas, France, [2]INRA, UE332 Domaine de Galle, F-18520 Avord, France

Correlative effects of 6 generations of selection for either ovulation rate (OR) or prenatal survival (PS) on growth rate and backfat thickness were estimated. Genetic parameters for piglet weight at birth (BW), at 3 weeks of age (W3W) and at weaning (WW), average daily gains from birth to weaning (ADGBW), from weaning to 10 weeks of age (ADGWT) and during performance test (ADGT), age (AGET) and backfat thickness at the end of test (ABT), were estimated using REML methodology applied to a multivariate animal model. Estimates of direct and maternal heritabilities were, respectively, 0.10, 0.12, 0.20, 0.24, 0.41 and 0.17, 0.33, 0.32, 0.41, 0.21 (SE = 0.03 to 0.04) for BW, W3W, WW, ADGBW and ADGWT. Genetic correlations between direct and maternal effects were moderate at birth (-0.21 ± 0.18), but much larger after birth (-0.59 to -0.74). Maternal effects were not considered for on test performance traits. Direct heritabilities were 0.34, 0.46 and 0.21 (SE = 0.03 to 0.05) for ADGT, AGET and ABT, respectively. Genetic correlations of OR and PS with performance traits were low (below 0.30) except maternal genetic correlations of PS with preweaning growth traits (-0.34 to -0.65). Estimated genetic trends were low and non significant, except negative maternal trends for BW and favourable direct trends for ADGT and AGET in both lines.

Session 19 Theatre 4

Inbreeding depression in Irish Holstein-Friesians

S. Mc Parland[1,2], J.F. Kearney[3], M. Rath[2] and D.P. Berry[1], [1]Moorepark Dairy Production Research Centre, Fermoy, Co. Cork, Ireland, [2]University College Dublin, Belfield, Dublin 4, Ireland, [3]Irish Cattle Breeding Federation, Bandon, Co. Cork, Ireland

Inbreeding within the Irish Holstein-Friesian population is increasing at a rate of 0.1% per annum. The objective of this study was to quantify the effect of inbreeding depression on milk production and fertility. Production and calving records from the years 2003 to 2005 inclusive were extracted from the national database and analysed using a sire model, with fixed effects including herd-year-season of calving, age nested within parity, parity, lactation length and inbreeding coefficient fitted as a continuous variable. A random permanent environmental effect was also included the model. Only cows with 3 complete generations of pedigree information were included in the analyses. Inbreeding had a significant deleterious effect ($P < 0.01$) on all milk and fertility traits analysed although the effect was sometimes non-linear (milk yield) or varied significantly by parity (protein yield and somatic cell score). A primiparous cow with 12.5% inbreeding had a reduction in milk, fat and protein yield of 61.8kg, 5.3kg and 1.2kg, respectively with an increase in somatic cell score of 0.03 units. Age at first calving and calving interval from first to second parity increased linearly by 0.2 and 0.7 days respectively per 1% increase in inbreeding, while survival from first to second parity decreased by 0.3% per 1% increase in inbreeding.

Session 19 Theatre 5

Inbreeding control in commercial pig breeding

D. Olsen, H. Tajet and B. Holm, NORSVIN, P.Box. 504, No-2304 Hamar, Norway

In most breeding programmes expected breeding values aggregating the economically most important traits have been used as a major selection criterion. Inbred offspring is the result of mating related parents. Hence, inbreeding will increase in all closed breeding populations. The question is not how to avoid inbreeding, but at which level it should be controlled and how this can be achieved maximising genetic gain simultaneously. Historically, several approaches have been used to control inbreeding. In pig breeding keeping a sufficient number of sires per generation and maintaining several sire family lines over several generations have been some of these approaches. By using such restrictions, annual ΔF in Norsvin Landrace and Duroc the 10 last years have been 0.50 % and 0.68 %, respectively. However, more advanced tools, e.g. Gencont, have been developed based on optimum contribution theory (OC). In a commercial breeding programme Gencont can be used to maximise genetic gain with restriction on a chosen increase in inbreeding. Continuous and multiple selections steps were a challenge implementing OC in practical selection. Norsvin now utilizes OC on all selection steps on boars, running Gencont weekly. A new selection criterion on females combining the total EBV and the relationship coefficients between the animal itself and the rest of the population has also been developed. By utilizing these new techniques to control inbreeding, Norsvin aims at the same ΔF as previously, but with an even higher annual genetic progress.

Session 19 Theatre 6

Control of the coancestry in breeding programs for aquaculture species

J. Fernández, L. Vega and M.A. Toro, INIA, Mejora Genética Animal, Ctra. Coruña Km 7,5, 28040 Madrid, Spain

The high reproductive potential of aquaculture species make possible to exert high selection intensities. Moreover, it allows for the development of large families, leading to more precise genetic evaluations even for low heritability traits. However, the same properties could imply a fast increase in the inbreeding levels and the loss of genetic diversity. Accounting for inbreeding in a selection scheme will help to avoid the loss of fitness due to inbreeding depression in the short term, and would guarantee a standing response in the long term. There is a consensus on the optimal way of controlling inbreeding in a selection scheme, optimising the contributions of evaluated individuals in generation t to the evaluated individuals in generation $t + 1$. However this strategy can not always be implemented in aquaculture. Here, we consider a different strategy: optimising contributions from selected individuals in generation t to those in $t + 1$ (it can be called Modified Within-Family Selection). We found several attractive solutions that maintain acceptable levels of response without an excess of inbreeding. For example, 38 – 80% reductions on the levels of coancestry can be achieved with only 2 – 18% decrease in response to selection, compared to pure within-family selection based on a familiar index. Obviously, when practical conditions allows for a more precise control, optimum strategy should be implemented and, thus, reductions in inbreeding levels could be obtained without loss of response.

Session 19 Theatre 7

A method to maintain population allele frequencies in conservation programmes

M. Saura[1], A. Pérez-Figueroa[1], J. Fernández[2], M.A. Toro[2] and A. Caballero[1], [1]Universidad de Vigo, Bioquímica, Genética e Inmunología, Facultad de Biología, Universidad de Vigo, 36310, Spain, [2]Instituto Nacional de Investigación y Tecnología Agraria y Alimentaria INIA, Mejora Genética Animal, Crta. Coruña Km 7.5, 28040 Madrid, Spain

Optimisation of contributions of parents by minimising the average coancestry of the progeny has been shown to be the best strategy for maintaining genetic diversity in conservation programmes. This strategy, however, homogenizes the allele frequencies at each locus, changing the distribution of allele frequencies present in the original population. Because one of objectives of a conservation programme is to preserve the genetic composition of the original endangered breed, we propose a method aimed to maintain the allele frequency distribution at each locus. Contributions of parents are obtained so as to minimize the allele frequency changes in a set of molecular markers in a population of reduced size. Computer simulations are used to assess the effectiveness of the method in comparison with that from contributions of minimum coancestry. Different population parameters are investigated: population sizes, number of markers, degree of linkage, presence or absence of artificial selection on a quantitative trait, etc. The results indicate that the proposed method is effective in maintaining the original distribution of allele frequencies, particularly under strong selection and linkage, and that also maintains low levels of average coancestry in the population.

Session 19 Theatre 8

Categorical expression of social competition

I. Misztal and R. Rekaya, University of Georgia, Athens, GA 30602, USA

A model by Muir and Schinckel used to model social competition among animals assumes that the animal competitive effects are expressed on a continuous scale. A model is proposed where these effects could be expressed in a few discrete categories (strongly dominant,.., independent,.., passive) or as binary (dominant, passive). Let y_{ij} be a record generated under a set of environmental effects i, and let d_j and c_j be the direct and competitive effects of animal j, respectively. Further, let $\alpha(k,x_i)$ be the effect of animal with dominance status k on its mates in the same pen in a set of environmental effects represented by x_i. Let p_j be a social dominance category of animal j. The model could be represented as:

$y_{ij}= other + d_j+\sum \alpha(p_k,x_i) +e_{ijk}$

where other are effects other than animals, and the summation are over all the remaining animals in the pen. Additionally, the dominance category of an animal can be described through a liability model:

$l_{ij}= other +c_j+e_{ij}$

where l_{ijk}, is an unobserved liability. If the dominance status of all animals is known, implementation of the proposed models can be achieved through a modified linear-threshold model. If the dominance status is not known, an additional step is needed where the dominance status is inferred using the observed data via a Bayesian MCMC approach. The social dominance model that assumes a categorical expression can allow for a more realistic expression of social dominance for animals housed in pens.

Data transformation for rank reduction in multi-trait MACE models

J. Tarres[1,2], Z. Liu[1], V. Ducrocq[2], F. Reinhardt[1] and R. Reents[1], [1]VIT, Heideweg 1, 27283 Verden, Germany, [2]INRA, Station de génétique quantitative et appliquée, Domaine de Vilvert, 78350 Jouy-en-Josas, France

Milk production traits are evaluated in Germany with a multiple lactation random regression test day model (three coefficients for each of the first three lactations). In order to reduce the number of within country traits for international bull comparison, different sub-models of a multi-trait MACE model were implemented based on German data. The first submodel, a multiple lactation MACE model, analysed daughter yield deviations on a lactation basis. This reduction works well except for the youngest bulls with daughters with short lactations. Alternatively, a principal component approach can be used as dimension reduction technique. A second submodel applied a rank reduction based on the largest eigenvalues of the genetic correlation matrix and their associated eigenvectors. This submodel showed that for international genetic evaluation purposes the German breeding values on production can be reduced from 9 to 5 or 6 traits by applying data transformation without loosing much accuracy in any of the random regression coefficients. However, as the combined lactation proofs in Germany depend mainly on the first coefficient of each lactation, a rank reduction accounting for the weight of each trait in the combined lactation allowed to perform rank reductions to 3 or 4 traits keeping proofs correlations with the full model on a combined lactation basis between 0.995 and 0.998.

Usage of xylose as non-enzymatic browning agent for reducing ruminal protein degradation of soybean meal

A. Can and A. Yilmaz, Harran University, Animal Science, Harran University Agriculture Faculty, 63100 Sanliurfa, Turkey

An in situ trials with three ruminally cannulated Akkaraman rams were conducted to evaluate non-enzymatic browning on ruminal protein degradation of soybean meal (SBM). A combination of xylose level (XL; 1 and 3% SBM; dry matter, DM, basis), heating temperature (HT; 120 and 150 °C) and heating length (HL; 30 and 60 min) were compared using a completely randomized design (CRD). In this study, all of the xylose-treated SBM had lower crude protein (CP) degradation during 2, 16, 24 and 48 h of in situ incubation, higher escape protein (EP) and acid detergent insoluble nitrogen (ADIN) values and ADIN corrected escape protein (CEP) than untreated SBM ($P < 0.01$). Increasing XL and HL decreased in situ protein degradation during 16 and 48 h incubations and increased EP and CEP ($P < 0.01$). Data from this study showed that controlled non-enzymatic browning was an effective method for protecting SBM protein from ruminal degradation.

Session 20 Theatre 2

Effect of water quality, grain type and micro-organism on lactic acid production of fermented feeds

A.T. Niba, J.D. Beal, A.C. Kudi and P.H. Brooks, University of Plymouth, School of Biological Sciences, Drake Circus, Plymouth, PL4 8AA, United Kingdom

The present study evaluated *in vitro* the effect water quality on lactic acid production on 4 cereals grains and 3 lactic acid bacteria species, conducted as a 3 factor, factorial design, with 3 replicates per treatment. The factors were: grain type (barley, maize, wheat or sorghum), LAB inoculants (*Lactobacillus farciminis* (F), *L. plantarum* (S) or *Pedococcus acidilacti*) and Calcium carbonate concentration of the water used for fermentation (T0 (0g/l), T25 (0.01g/l), T50 (0.02g/l), T75 (0.03g/l) or T100 (0.04g/l). Grains were hammer-milled and irradiated with 25 kGy from 60Co in 100 g sachets and fermented at a feed to water ratio of 1:1.4.Results show highly significant ($P < 0.001$) differences in the interaction between all the factors in the lactic acid after 24 hour fermentation. For barley, lactic acid production tended to increase in with increasing water mineral content. With maize, sorghum and wheat the trend was reversed for the three LAB species with the exception of L. plantarum for maize and wheat and P. acidilacti for wheat. The highest overall mean concentration of lactic acid, 317.78 (SEM 5.36), was recorded with T0. This was 42.58(SED=8.004), 51.58 (SED=8.087), 61.64 (SED=8.004) and 91.47(SED=8.004) mM higher ($P < 0.001$) than the concentrations of T75, T100, T25 and T50 respectively. The results show that microbial fermentation of these cereals is affected by the mineral concentration of the water used in fermentation.

Session 20 Theatre 3

Milk consumption in suckling llamas (*Lama glama*) measured by an isotope dilution technique

A. Riek, M. Gerken and E. Moors, Institute of Animal Breeding and Genetics, University of Goettingen, Albrecht-Thaer-Weg 3, 37075 Goettingen, Germany

The objective of the study was to estimate daily milk intake in llama crias and relate nutrient intakes at peak lactation to growth data. Milk intake in 11 suckling llamas was estimated from water kinetics using deuterium oxide (D_2O) at days 17, 66 and 128 post partum. The body water pool tended to decrease with age. Daily milk intakes averaged 2.6, 2.3 and 2.0 kg at 17, 66 and 128 days post partum. Milk intake decreased with age when expressed as daily amount, percentage of body weight or per kg metabolic size, but the influence of age was eliminated when expressed per g daily gain. As llamas only have one young per parturition, milk intake was equivalent to the daily milk output of the dam, which ranged from 27.6 to 96.9 g per kg maternal body weight$^{0.75}$. Compared with different ruminant species, the milk production potential in llamas appears to lie between wild and domestic ruminants used for meat production. Nutrients (dry matter, fat, protein, lactose) and energy intakes from the milk calculated by combining milk intake and milk composition data decreased with age. Maintenance requirement for suckling llamas at peak lactation (17 days post partum) was 312 kJ ME per kg body weight$^{0.83}$. Combined with milk composition data, the present milk intake estimations at different stages of the lactation can be used to establish recommendations for nutrient and energy requirements of suckling llamas.

Session 20 Theatre 4

Effect of selenium source and dose rate on selenium content and speciation in milk and cheese

R.H. Phipps[1], A.K. Jones[1], A.S. Grandison[1], D.T. Juniper[1] and G. Bertin[2], [1]University of Reading, School of Agriculture, Policy and Development, Earley Gate, Reading, RG6 6AR, United Kingdom, [2]Alltech France, Regulatory Affairs, Levallois-Perret, 92300, France

Forty Holstein cows were used in a 16-wk continuous design study to determine effects of selenium (Se) source: selenized yeast (SY:Sel-Plex®) from *Saccharomyces cerevisiae* CNCM I-3060 and sodium selenite (Na_2SeO_3) and inclusion rate of SY on Se concentration and speciation in milk and cheese. Cows received *ad libitum* a TMR with 1:1 forage to concentrate ratio. The four diets (T1-T4) differed only in source and dose of Se additive. Estimated total dietary Se for T1 (no supplement), T2 (Na_2SeO_3) and T3 and T4 (SY) was 0.16, 0.30, 0.30 and 0.45 mg/kg DM. Milk samples analysed at 28-d intervals all showed linear effects ($P < 0.001$) of SY on milk Se values. At day 112 the milk Se values for T1-T4 were 24, 38, 57, 72 ± 3.7 ng/g fresh material, and indicate improved Se bioavailability (T2: 38 *vs* T3: 57 ng/g fresh material) from SY. In milk, Se source had no marked effect on selenocysteine (SeCys), while it had a marked effect on selenomethionine (SeMet). At day 112 the SeMet content of milk from T3 was 111 *vs* 36 ng Se/g for T2, and the level increased further to 157 ng Se/g for T4. At day 112 milk from T1, T2 T3 was made into cheese and replacing Na_2Se0_3 (T2) with SY (T3) increased Se, SeMet and SeCys content from 180 to 340 ng Se/g, 57 to 153 ng Se/g and 52 to 92 ng Se/g.

Session 20 Theatre 5

Effect of methionine supplementation on performance and carcass characteristics of awassi lambs

B.S. Obeidat, A.Y. Abdullah, M.S. Awawdeh and R.I. Qudsieh, Jordan University of Science and Technology, Irbid, 22110, Jordan

The objective of this study was to evaluate the effects of ruminally-protected methionine supplementation (0, 7, or 14 g/head/day) on performance and carcass characteristics of Awassi ram lambs. Twenty four Awassi lambs were randomly assigned to 3 treatment diets (8 lambs/treatment) and housed in individual pens. Lambs were given an adaptation period of 7 days before the intensive feeding period that lasted for 86 days. On day 74 of the trial, a digestibility experiment was performed. At the end of the trial, all lambs were slaughtered to evaluate carcass characteristics. Increasing the level of methionine supplementation did not improve performance or feed conversion ratio. Nutrient intakes and digestibilities were not influenced by methionine supplementation. There were no differences in final weight, hot and cold carcass weights, dressing-out percentage, or any of the measured non-carcass components. Tissues and fat depth measurements together with all meat quality attributes were not affected by methionine supplementation, the only parameters affected ($P < 0.05$) were redness (a*) and the hue angle. These results suggest that methionine supplementation is not likely to produce any production benefits in nutrient digestibilities, performance, or carcass characteristics of lambs fed a high performance diet.

The growth performance, caecal fermentative activity and digestibility of nutrients in growing rabbits fed the diet containing chicory roots (*Cichorium intybus L*)

Z. Volek and M. Marounek, Institute of Animal Science, Pratelstvi 815, CZ-10401, Prague Uhrineves, Czech Republic

The effect of a diet containing chicory roots on the growth performance, fermentative activity and digestibility of nutrients was studied on 86 rabbits (Zika-Hybrid), weaned at 35 days of age. Sixty rabbits (30 per group) were used for the growth performance trial, while 26 rabbits (13 per group) were used for the digestibility of diets (between 45 and 49 days of age), and fermentative activity evaluation (49. day of age). The control and chicory diet were formulated. The chicory diet contained 10% of chicory roots at the expense of oats. The feed intake tended to be lower (35.-56. day of age) in rabbits fed the chicory diet than in control rabbits, while no differences were observed for the whole period (35.-77. day of age). No effect on growth rate was observed. Caecum relative weight and its content were significantly higher, and the caecal pH and ammonia concentration lower in rabbits fed the chicory diet. Acetate molar proportion was significantly higher and that of propionate and butyrate lower in rabbits with chicory diet. Digestibility of cellulose was significantly higher and that of crude protein, fat, starch and hemicelluloses lower in rabbits fed the chicory diet. Results of this study suggest the good utilisation of the chicory diet in rabbits. However, other experiments regarding an optimal level of chicory roots in the diet, in relationship to digestive health, are needed. (MZE-0002701403)

Animal nutrition and food quality

J. Gundel[1], T. Pálfi[2], T. Ács[1], I. Erdélyi[2] and A. Hermán[1], [1]Research Institute for Animal Breeding and Nutrition, Pig Nutrition, Gesztenyés u. 1., 2053 Herceghalom, Hungary, [2]University of Debrecen, Centre of Agricultural Sciences, Böszörményi út 138., 4032 Debrecen, Hungary

For the past one and a half – two decades due to the development of nutrition science and to the creation and strengthening of consumers' health consciousness nutrition biological value of food products especially that of meat and meat products, milk and dairy products and eggs has become more significant (quantity and rate of fat, fatty acids, cholesterol, vitamin and mineral and other bioactive agents). In addition to this, from the production side sustainable development, animal protection aspects, striving for quality consciousness, national traditions and origin protection have come to the fore and become more important. All these trends result in quality improvement of food products. In this present paper the authors tries to present the possible wordings of quality and the correlation between feeding and animal product quality. In several cases the author underlines the identity or the clash of interests. How to feed our animals to achieve good quality? The answer is still missing to the question. This is mainly due to the lack of vertical integration in the field of production of animal products. Finally the author states that all the representatives of human and animal nutrition science have to inform the unprofessional society about the required quality of food products according to mutually accepted principles free from interests and trends.

Session 20 Theatre 8

Enhancement of pig antioxidant status with vitamin E in the case of moderate trichothecene intoxication

T. Frankič, J. Salobir and V. Rezar, University of Ljubljana, Biotechnical faculty, Department of Animal Science, Chair of Nutrition, Groblje 3, 1230 Domžale, Slovenia

The objective of the present study was to establish the effect of deoxynivalenol (DON) and T-2 toxin on lipid peroxidation and lymphocyte DNA fragmentation in young growing pigs, and furthermore, to evaluate the potential of vitamin E in prevention of toxin mediated changes. Young male pigs (n=48) were randomly assigned to five experimental groups: control, T-2 (3 mg/kg T-2 toxin), T-2+E (3 mg/kg T-2 toxin+100 mg/kg vit. E), DON (4 mg/kg DON) and DON+E (4 mg/kg DON+100 mg/kg vit. E. Parameters as malondialdehyde (MDA), total antioxidant status (TAS) of plasma and erythrocyte glutathione peroxidase (GPx) were determined. DNA damage in lymphocytes was measured by Comet Assay. The data were analysed by the General Linear Models procedure of the SAS/STAT module. Production parameters of toxin treated groups were significantly impaired in comparison to the control. DON and T-2 toxin increased the amount of DNA damage in lymphocytes by 28 % and 27 %, respectively. The levels of TAS were lowered only by addition of DON. Plasma and urinary MDA and GPx levels did not differ among groups. Supplementation with vitamin E only partially protected lymphocyte DNA from toxin impact. Enhancement of the antioxidant status with vitamin E in the case of moderate trichothecene intoxication may benefit the DNA integrity of the immune cells and thus improve the immune response of young growing pigs in the case of infection.

Session 20 Theatre 9

Some performance aspects of doe rabbits fed diets supplemented with fenugreek and aniseed

M. Sayed, A. Azoz, A. El-Maqs and A. Abdel-Khalek, Animal Production Research Institute, Dokki, 12311, Egypt

A seven-treatment experiment was carried out to study the effect of dietary supplementation with 0.0, 0.5 or 1.0 % fenugreek or aniseed or their mixture (1:1) at 0.5 or 1.0 % on some performance aspects of Bouscat doe-rabbits and their offspring (up to 8th week of age) through three parities. Results obtained reveal that: -Improved litter size ($P < 0.05$) and litter weight ($P < 0.01$) at 8 weeks of age were detected due to the tried additives. - Milk yield significantly were affected by additives evaluated in the third and fourth weeks of lactation. -At 7-14 days of pregnancy, total protein ($P < 0.01$), albumin ($P < 0.05$), globulin ($P < 0.05$), GPT ($P < 0.01$) and GOT ($P < 0.01$), all were significantly differed between studied treatments. At 21-28 days of pregnancy, only GOT and GPT ($P < 0.01$) were affected by treatments. -Progesterone (P4; $P < 0.05$) and estradiol (E2;$P < 0.01$), prior to pregnancy, triiodothyronine (T3; $P < 0.05$), and E2 ($P < 0.01$), at 7-14 days of pregnancy, and again E2 ($P < 0.01$), and P4 ($P < 0.01$) at 21-28 days of age were significantly affected by studied treatments. It is recommended to improve litter size and litter weight at 8 week after parturition to supplement the diet with a mixture of fenugreek and aniseed at the level of 0.5%.

Session 20 Poster 10

An *in vivo* 48-hour model to study feed preferences in weaned pigs

G. Tedó[1], X. Puigvert[2], X. Manteca[3] and E. Roura[1], [1]Lucta SA, R & D Feed Additives, Ctra. Masnou-Granollers, km. 12,4, 08170 Montornés, Spain, [2]Univ. de Girona, EQATA, Avda. Lluís Santaló, 17071 Girona, Spain, [3]Univ. Aut. de Barcelona, Cell Biology, Physiology & Inmunology, Fac. Veterinària, 08193 Bellaterra, Spain

Double choice models to assess feed ingredient preferences in pigs are useful to evaluate palatability when testing periods not shorter than 4 days are used. The aim of this study was to improve the double choice model by studying the effect of a shorter testing period (24, 36 or 48 h), and the necessity of having an adaptation period (4 or 6 h) before starting the double choice test. The ideal number of 2-pig replicates per treatment (6 or 8) was also tested. The double choice consisted of simultaneously feeding a control and a test feed. Six trials including a preliminary 16-h fasting period were conducted. Trials 1, 2, 3 and 4 consisted of an adaptation period of 4 or 6 h followed by a 36-h test with 6 replicates. In trials 5 and 6, a 4-h adaptation period was followed by a 48-h test with 8 replicates.. Results showed that an adaptation period of 4 h was sufficient. Six replicates were enough to detect significant preferences ($P < 0.05$). In all cases, the 48-h model was validated by finding no differences for the double control choice test. In conclusion a new preference model has been developed based on a 48-h test with a preliminary 16-h fasting and a 4-h adaptation periods and including a minimum of 6 replicates of 2 pigs per treatment.

Session 20 Poster 11

Assessment of nutritive value of Bt-maize using rats and rabbits

M. Chrenková[1], L. Chrastinová[1], Z. Ceresnáková[1], J. Rafay[1], G. Flachowsky[2] and S. Mihina[1], [1]Slovak Agricultural Research Centre, Hlohovská 2, 94992 Nitra, Slovakia (Slovak Republic), [2]Institute of Animal Nutrition, Bundesallee 50, 38116 Braunschweig, Germany

Bt-maize is characterized by the introduction of gene for Bt-toxin which protects maize against the European maize borer. The Bt-maize was used in one experiment with rats and one with rabbits. The complete feed mixture for pigs containing 70 % maize Prelud (*Zea Mays* L. Line CG 00256-176) and feed mixture with 70 % of isogenic maize as a control were tested on rats. The mixtures were isoenergetic and isonitrogenous. Crude protein digestibility was 88 or 86 %, PER value 1.65 or 1.75 and feed conversion on the level 3.4. We found no pathological changes on inner organs in experimental animals. The performance of feed conversion in feeding mixtures with the same content of Bt-maize and isogenic maize (11 %) intended for broiler rabbits was studied. Feed conversion was the same for both feed mixtures (3.6 or 3.5) and daily gain was 40 g or 38 g. Rabbits were slaughtered in 73 days of age when they achieved the live weight 2.5 kg. The musculus longissimus dorsi was analysed for individual nutrients. All qualitative parameters of meat did not differ significantly. The results showed the suitability of Bt-maize in feed mixture for production of rabbit meat of good quality.

Session 20 Poster 12

Effect of selenium source and dose on glutathione peroxidase activity in the whole blood and tissues of finishing cattle

D.T. Juniper[1], R.H. Phipps[1] and G. Bertin[2], [1]University of Reading, School of Agriculture, Policy and Development, Earley Gate, Reading, RG6 6AR, United Kingdom, [2]Alltech France, Regulatory Affairs Department, Levallois-Perret, 92300, France

Thirty-two castrated Limousine cross cattle (489 ± 42.9 kg) were enrolled on a 16-wk continuous design study investigating effects of selenium source (selenized yeast (SY) [*Saccharomyces cerevisiae* CNCM I-3060: Sel-Plex®] and sodium selenite [Na_2SeO_3]) and inclusion rate of SY on total Se and glutathione peroxidase (GSH-Px) activity in whole blood and skeletal tissue. Cattle received *ad libitum* access to one of four maize silage TMRs that differed in either source or dose of Se additive. Estimated dietary Se values for BM1 (background), BM2 (Na_2SeO_3) and BM3 and BM4 (SY) were 0.15, 0.3, 0.3 and 0.5 mg/kg DM, respectively. Blood samples were taken prior to slaughter and samples of the *L.Thoracis* taken post-mortem for assessment of Se content and GSH-Px activity. There were significant treatment ($P < 0.001$) and linear dose effects ($P < 0.001$) on whole blood total Se and GSH-Px activity with ascending inclusion of SY. There were significant treatment (P=0.017) and linear dose effects (P=0.007) on tissue total Se and GSH-Px activity with ascending inclusion of SY. In addition, total Se and GSH-Px activity in blood and tissue were higher in SY than Na_2SeO_3 when fed at comparable doses. Whole blood and tissue Se contents were positively correlated to GSH-Px activity (R=783 [P=0.003] and R=898 [$P < 0.001$], respectively).

Session 20 Poster 13

Nutritional characterisation and effect of using distillers dried grains-based diets for fattening pigs (50-100 kg)

T. Panaite and M. Iliescu, National Research-Development Institute for Animal Biology and Nutrition, Physiology of nutrition, Calea Bucuresti nr. 1, 077015, Balotesti, Romania

Two tests (digestibility and production) were conducted to investigate the possibility of using distillers dried grains (DDC) as energy and protein source in the compound feeds for fattening pigs (50-100 kg). The experiment used hybrid Landrace x LS-Periş 345, castrated male half-brothers assigned to two groups. The experimental compound feed included 30% DDC. Control slaughtering was performed at the beginning and end of the experiment, biological samples were collected and analysed chemically and calorimetrically. We monitored the compound feed ingesta, the feed conversion ratio, body weight and the meat to fat ratio of the gain. After 42 days 2 pigs per group were used in the digestibility trial (5 days) on week VI and then again on week VII. Based on the daily records and on the chemical analyses conducted on average weekly samples we calculated the digestibility coefficients of the dietary nutrients. The experimental diet produced the best organic matter digestibility (89.7%) and the best gross energy digestibility (88.9%). It also produced better performance ($P < 0.05$) for gain and intake (0.723 kg/day average gain and 3,81 FCR).

Session 20 Poster 14

Safety alternative additives to antibiotics in rabbit nutrition: prebiotics and probiotics

A. Abdel-Khalek, Animal Production Research Institute, Dokki, 12311, Giza, Egypt

Rabbit meat production is a very important activity in the Mediterranean area, amounting for about 50% of total world production and consumption. Rabbit meat is interesting for its dietetic and nutritional characteristics and bio-security for the consumer. In addition, the consumer as a product of high quality already accepts it. With intensification of production systems and exposure to stressors, the growth promoters, antibiotics, have been introduced to rabbits for long time. Recently, evidences on the increased health risk to people who consumed the products they contain urged the EU (January, 1st 2006) and other countries to ban the usage of sub-therapeutic levels of antibiotics as antimicrobial growth promoters, which in turn, forced the producers to look for other natural, safe, reliable, and economic additives serve the same goals achieved by antibiotics. This paper reviews the effects probiotics and prebiotics as alternatives to antibiotics in health status and growth performance of rabbit.

Session 20 Poster 15

Use of poultry litter, corn, rice and sugar cane by products in confined lamb nutrition, blood metabolites

G.E. Nouel B.[1], P. Hevia O.[2], R.J. Sánchez B.[1], M.A. Espejo Diaz[1] and M.J. Velàsquez[1], [1]Universidad Centroccidental Lisandro Alvarado, UIPA-Agronomìa, La Colina, Tarabana, Cabudare, Estado Lara, 3023, Venezuela, [2]Universidad Simòn Bolìvar, Ciencias Bàsicas, Lab. Nutriciòn, Sartenejas, Baruta, Estado Miranda, 1000, Venezuela

Eight ruminally fistulated growing lambs (14,176 ± 2,03 kg metabolic weight), were used in a replicated 2 x 4 factorial treatment arrangement in latin square, to determine response variables included blood metabolites of lambs in eight rations with increasing levels of poultry litter (PL) and rice polishing (RP) or corn homini (CH), and sugar cane bagasse (SB). Rations were PL (11,4; 20,4; 29,4 y 38,4%), RP or CH (48, 39, 30 y 21%), SB (40%), salt: sulfur flower (4:1, 0,6 %). Lambs were housed in metabolism pens. Physiological variables for the rations with PL and CH the values were normal for pH (7,41 to 7,39), glucose (62,6 to 48,3 mg/dl), transaminasas (15,7 to 15,6 UI/dl) and cholesterol (57,2 to 69,9 mg/dl), while the ones that had PL and RP were altered being increased abnormally promoting a tendency to liver damage by the high levels of pH (7,56 to 7,77), cholesterol Mg/dl) and some high values of transaminasas (16,7 and 18,0 UI/dl) in the levels of 11,4 and 20,4% of CH. The results indicate that the corn by products used permitted to maintain physiological values for animals in growth phase evaluated, without risks for the animal health, which did not occur with the rations based on RP.

Session 20 Poster 16

Effects of selenium-rich yeast supplementation on the luteal function of postpartum cows

H. Kamada[1], Y. Ueda[1], T. Mitani[1], M. Miyaji[1], K. Nakada[2], T. Yasui[3] and M. Murai[1], [1]National Agriculture Research Center for Hokkaido Region, Hitsujigaka-1, Toyohira, Sapporo, Hokkaido, 062-8555, Japan, [2]Rakunou Gakuen University, Midorimachi-582, Bunkyoudai, Ebetsau, Hokkaido, 069-8501, Japan, [3]Bussan Biotech Co., Ltd., Shibadaimon-2-5-8, Minato, Tokyo, 105-0012, Japan

Early recovery of postnatal ovary function is necessary for increased conception rate. It is reported that the postnatal progesterone level is positively correlated to the conception rate. In this experiment, the effects of selenium-rich yeast on the luteal function of postpartum cows were investigated. Selenium yeast containing 300ppm selenium was fed 10g per day and 20g per day in prenatal and postnatal periods to cows (n=7), respectively. The control group (n=7) was fed the same amount of normal yeast. Blood sampling was carried out every two days after delivery. The plasma progesterone concentrations were measured by EIA. The selenium concentration of blood plasma in the selenium yeast treatment group was significantly higher than in the normal yeast treatment group. Consequently, selenium yeast supplementation elevated the plasma progesterone concentration earlier than in the control. Our previous *in vitro* experiments using cultured luteal cells showed that LH stimulated luteal cells produce a large amount of progesterone accompanying peroxide accumulation, and that selenium addition to cells decreases peroxide. The present results suggest that the anti-oxidant effect of selenium may elevate luteal function in postnatal cows.

Session 20 Poster 17

Isolation of lactic acid bacteria from chickens that demonstrate probiotic properties of autoaggregation and coaggregation with *S. enteritidis*

S. Savvidou, J. Beal and P. Brooks, University of Plymouth, Biological Sciences, Drake Circus, PL4 8AA, Plymouth, United Kingdom

In this study, a total of 53 lactic acid bacteria (LAB) were isolated from the contents and mucosa of the crop, caecum and small intestine, of three organically farmed chickens. Isolates were examined for autoaggregation and coaggregation with *Salmonella enteritidis*. Autoaggregation and coaggregation tests were assessed according to methods of Kmet *et al.* (1999) and Drago *et al.* (1997). Suspensions were observed by scanning electron microscope. LAB strains were identified using API CHL kit (Biomerieux, UK). From 23 LAB that were aggregative, 11 bacteria showed rapid and 12 normal autoaggregation.These were further tested for their ability to coaggregate with *S. enteritidis*. One LAB strain showed maximum aggregation, two showed marked aggregation, six showed good aggregation, nine partial aggregation and three showed no or almost no aggregation. The strain that showed rapid autoaggregation and maximum co- aggregation with *S. enteriridis* was identified as *Lactobacillus plantarum*. This organism could be used for further screening for its potential use as probiotic in chicken nutrition.

In vitro gas production of fresh alfalfa under different pH

R.A. Palladino[1], G. Jaurena[1], M. Wawrzkiewicz[1], J.L. Danelon[1], M. Gallardo[2] and M. Gaggiotti[2], [1]FAUBA, Av. San Martin 4453, (C1417DSQ) Buenos Aires, Argentina, [2]INTA, Ruta 34 km 227, (2300) Rafaela, Santa Fe, Argentina

The aim of this work was to define the range of tolerance of fresh alfalfa digestion to ruminal pH reductions by the *in vitro* gas production (GP) technique. Fresh alfalfa samples (c.a. 3.5 g wet matter) were incubated at four pH (6.8, 6.3, 5.8 and 5.3) in three independent periods using a phosphate-citrate buffer free of indirect GP. Rumen liquor was collected from two rumen cannulated cows grazing alfalfa. The GP was measured at 1, 2, 4, 6, 9, 12, 16, 20, 24, 30, 36 and 48 h and GP hourly rates were estimated from raw data, and subsequently peak rate (ml/g OM h), time to peak rate (h) and lag time (h) variables were calculated and statistically analysed according to a complete block design (blocking by incubation) and linear orthogonal contrasts were performed. Cumulative *in vitro* GP increased linearly ($P < 0.05$) with pH at 6 and 48 h ($P < 0.05$) and GP at 48 h decreased below pH 5.7 – 5.8 (T 5.8). As pH decreased, both lag time and time to peak rate increased linearly ($P < 0.05$), but no differences were observed in peak rates among treatments ($P > 0.05$). The highest GP was observed at pH 6.2 – 6.3 (T 6.8). These results suggest that digestion of fresh alfalfa would be limited by a ruminal pH lower than that commonly reported in the literature (pH 6.2).

In vitro gas production and substrate digestion relationship for a buffer free of indirect gas production

G. Jaurena, M. Wawrzkiewicz, R.A. Palladino and J.L. Danelón, FAUBA, Av. San Martín 4453, (C1417DSE) Buenos Aires, Argentina

A constant gas production (GP) to dry matter (DM) digested relationship is a basic assumption of the GP technique. The aim of this study was to verify if this relationship was constant along the time of incubation (H) for fresh alfalfa incubated at different pH and using a non-conventional buffer. Fresh alfalfa samples were incubated at 4 pH (T6.8, T6.3, T5.8 and T5.3) in 3 periods using a phosphate-citrate buffer instead of the conventional carbonate-bicarbonate buffer. The GP was measured at 1, 2, 4, 6, 9, 12, 16, 20, 24, 30, 36 and 48 h and the DM disappeared (DMD) was estimated by collecting the digestion residues at 6, 12, 24 and 48 h. Data were analysed by linear regression and ANOVA. The best adjustment for regression of GP on DMD was obtained by splitting the data set by H (H=6 and H > 6) so that the intercept (α) was respectively -38 ml ($P < 0.01$) and -12 ml ($P > 0.05$). The slope of GP on DMD was 0.18 ml/g DMD ($P < 0.01$) for both H. The GP/DMD ratio was affected by pH (ml/g DMD, 0.15a, 0.13ab, 0.12b and 0.12b respectively for T6.8, T6.3, T5.8 and T5.3, different letters differ $P < 0.05$) and by H (H=6, 0.06; H>6, 0.16 ml/g DMD; $P < 0.01$). These results indicate that the GP was proportional to DMD and remained constant along the incubation for the new buffer, but suggest that part of the material measured as digested was lost during filtrate at early times of incubation (as indicated by $\alpha<0$ for H=6), and that this effect was higher at low pH.

Session 20 — Poster 20

Effects of ß-D-glucanase and ß-D-mannanase addition alone or in combination to diets based on barley-soybean containing two metabolizable energy levels on performance of broiler chicks

K. Karimi, A.A. Sadeghi, F. Forodi and P. Shawrang, Department of Animal Science, Science and Research Branch, Islamic Azad University, P.O. Box 14515.4933, Tehran, Iran

This study was completed to determine the effects of addition of β-mananase and β-glucanase individually and in combination to barley-soybean based broiler diets containing metabolizable energy (ME) according to Ross recommendation or 100 kcal lower, on feed intake, body weight gain, feed to gain ratio and abdominal fat content. In this experiment, four hundred and eighty day-old chicks from Ross 508 strain were used in 2×4 factorial arrangement of treatments with three replicates and twenty chicks per pen. The birds had free access to feed and water from 0 to 42 d of ages. Data were analyzed as a CRD design using GLM procedure of SAS. Results indicated that the interactions between ME levels and enzyme treatments on feed intake, body weight gain and feed to gain ratio were significant. Higher ME level (100 kcal/kg) had no effect on performance traits ($P > 0.05$). Abdominal fat content was differ among treatments ($P < 0.05$) and it was highest in broiler chicks fed diet containing β-glucanase individually. Addition of β-mananase and β-glucanase in combination to barley-soybean based diet had better effects on body weight gain and feed to gain ratio. The present results not only confirm that there are no negative interactions, but also suggest that it may be advantageous to include both enzymes in barley based broiler diets.

Session 20 — Poster 21

Use of staining activity for detecting xylanase activity in the feed and digesta of broiler chicks

A.A. Sadeghi and P. Shawrang, Department of Animal Science, Science and Research Branch, Islamic Azad University, P.O. Box 14515.4933, Tehran, Iran

Sixty day-old broilers were selected and randomly allocated to diets based corn-soybean without or with xylanase supplementation. At 3 and 6 weeks of age, fifteen chicks from each treatment were slaughtered and the crop, gizzard, duodenum, jejunum and ileum contents were emptied and collected for electrophoresis and activity staining. Xylanase activity in native-PAGE gel was detected by overlaying the gel with 2% agar dissolved in 100 mM MES buffer, pH 6.5, containing 1.5% Remazol Brilliant Blue-Xylan. The color developed in the gel as a single band reflected the xylanase activity. Data were analyzed as a CRD design using GLM procedure of SAS. Exogenous xylanase protein was detected by SDS-PAGE in the xylanase diet as well as in all of the digesta. Xylanase activity was detected by the activity stain native-PAGE assay in the xylanase diets and the digesta collected from the crop, gizzard and duodenum with the exception of jejunum and ileum. Xylanase activity in the digesta from the crop was higher than that from proventriculus or that from gizzard ($P < 0.05$). Differences in pH and the degree of endogenous protease action resulted in varying activity of xylanase in different segments of the gastrointestinal tract. The result of the study suggests that the activity stain assays allow the detection of low levels of exogenous xylanase activity in the diet as well as in the digesta of the broiler chicks.

Session 20 Poster 22

The study of different levels of RDP in the ration of lactating cows and their effects on estradiol and progesterone levels in the blood

A. Moharrery, Shahrekord University, Animal Science, Shahrekord University, Iran, 115, Iran

Ruminally degradable protein or ruminally undegradable protein in excess of requirement can contribute to reduced fertility in lactating cows. Dietary protein nutrition or utilization and the associated effects on ovarian or uterine physiology have been monitored with urea nitrogen in blood or milk. Twenty-one multiparous Holsteincows in the late of lactation period were used in complete randomized design to determine the effect of excessive intake of ruminally degradable crude protein on the some reproductive hormones (estradiol and progesterone levels in the blood). Experimental periods were 6 wk in length, with d 1 to 14 used for adjustment and wk 2 and wk 6 used for a sampling (blood, and milk). Three concentration of a rumen-degradable protein (RDP) supplement according to National Research Council recommendations (9.3, 11.4, and 14% of dry matter intake) were treatments. No significant effect of concentration of RDP supplement was detected on difference levels of both hormones form wk2 to wk 6. Higher undegradable protein or escape protein showed highest value for increasing progesterone levels but simultaneously lowest value for estradiol. In this regard no significant correlation was observed between estradiol with blood urea nitrogen (BUN) (P=0,1990) and progesterone with BUN (P=0,8919), but significant correlation (0,51) was observed between estradiol and progesterone (P=0,0108).

Session 20 Poster 23

A comparison of three proteolytic enzymes for predicting in sacco protein degradation constants of protein meals

C. Guedes, A.L.G. Lourenço, M.A.M. Rodrigues, S. Silva and A.A. Dias-da-silva, CECAV-UTAD, Animal Scence, PO Box 1013, 5001-801 Vila Real, Portugal

The objective of this study was to estimate the *in sacco* protein degradation constants using *in vitro* incubation with proteolytic enzymes - protease from *Streptomyces griseus*, bromelain and ficin. Twelve samples of 7 protein meals (3 soybean meals, 2 groundnut meals, 1 cottonseed meal, 1 sunflower meal, 2 coconut meals, 2 palm kernel meals and 1 rapeseed meal) were used. Three mature fistulated rams fed a diet of meadow hay, soybean meal and sugar beet pulp (49:36:15 on DM basis) were used for *in sacco* incubations. Samples were incubated in the rumen for 2, 4, 8, 16, 24, 48, 72 and 96 h and with each enzyme for 1, 2, 4, 6, 8, 24 and 48 h. Degradation constants were estimated from the exponential equation $p=a-b[1-\exp(-ct)]$. Correlation and regression analysis were used to estimate the *in sacco* degradation constants using proteolytic enzymes. No significant correlations were found between *in sacco* degradation constants and the enzymatic hydrolysis observed after 1, 2, 4, 6, 8, 24 and 48 h of incubation *in vitro* irrespective of protease source. Degradation constants obtained using ficin as protease source explained 85.0, 73.5, 63.0 and 64.8% of the variation in predicting a, b, c and a+b from *in sacco* incubations, respectively. The other proteolytic sources gave much lower and non-significant correlations ($P > 0.05$). We concluded that incubation with ficin has potential to predict accurately *in sacco* protein degradation constants of protein meals.

Session 20 Poster 24

Apparent ileal amino acid digestibility in growing pigs fed low protein diets added with pancreatin

R. Gomez, M. Cervantes, W. Sauer, N. Torrentera, A. Morales and A. Araiza, UABC-ICA, Mexicali, 21100, Mexico

Secretion of pancreatic proteases appears to be reduced when pigs are fed low protein diets. The experiment was conducted to evaluate the effect of adding a mixture of pancreatic proteases (pancreatin) to a low protein, sorghum-soybean meal-based diet on the apparent ileal digestibility (AID) of amino acids (AA). Six barrows adapted with cannulas at the terminal ileum were used according to a repeated Latin square design. The treatments were: 1) basal, 10.5 CP diet, 2) basal diet plus 0.5 g pancreatin/kg feed, and 3) basal diet plus 1.0 g pancreatin/kg feed. The basal diet was formulated with sorghum, soybean-meal, vitamins and minerals. All diets contained 0.2% chromic oxide as digestibility marker. Feed was mixed with water, and offered twice a day in equal amounts, at 0700 and 1900 h. Ileal digesta were collected continuously from 0700 to 1900 h on d 6 and 7. The AID values (%) of CP and AA in T1, T2, and T3 were: CP, 81.7, 81.2, 81.1; arg, 80.9, 80.3, 80.2; his, 80.4, 81.0, 80.0; ile, 82.0, 81.5, 81.5; leu, 84.5, 84.0, 84.0; lys, 78.0, 76.6, 76.2; met, 81.2, 82.3, 81.5; phe, 83.4, 82.7, 82.6; thr, 77.7, 76.6, 75.8; val, 81.2, 80.5, 80.2, respectively. There was effect of pancreatin supplementation to sorghum-soybean meal diets on the AID of AA ($P > 0.10$). These results indicate that the addition of pancreatic protease to sorghum-based diets for growing pigs, does not improve the apparent ileal digestibility of AA.

Session 20 Poster 25

Lysine requirement of growing pigs fed wheat-soybean meal diets

L. Buenabad, M. Cervantes, W. Sauer, S. Espinoza, N. Torrentera, A. Morales and A. Araiza, UABC-ICA, Mexicali, 21100, Mexico

Two experiments were conducted to assess the lysine requirement of growing pigs fed wheat-soybean meal diets as compared to corn-soybean meal diets. In Exp. 1, eight barrows (43 ± 3.6 kg BW), fitted with a simple T-cannula were fed used. Diet 1 was a wheat-soybean meal diet. Diet 2 was a corn-soybean meal diet. Both diets were formulated to contain similar crude protein (14.2%) and lys (0.91%) contents. The AID of ile ($P = 0.042$), lys ($P = 0.013$), phe ($P = 0.014$), val ($P = 0.035$), and thr ($P = 0.012$) were higher in the wheat-based diet. There were no differences ($P > 0.11$) in the AID of arg, his, leu, and met between the wheat- and the corn-based diets. In Experiment 2, 25 pigs (20.8 ± 1.8 kg BW) were fed five diets. Diet 1 was a wheat-soybean meal basal diet. Diets 2, 3, and 4 were the basal but added with 0.13, 0.26, and 0.39% L-Lysine HCl, respectively. Diet 5 was a corn-soybean meal diet. Feed: gain ratio was higher ($P = 0.013$) for pigs fed the basal diet than for the other diets. There were no differences ($P > 0.119$) between treatments in ADG. There was a linear increase in ADG ($P = 0.032$), feed/gain ratio ($P = 0.004$), and muscle growth ($P = 0.011$) as lysine content increased in the wheat-soybean meal diet. These data indicate that the requirement of total and apparent ileal digestible lysine for growing pigs is not affected by the cereal type included in the diet.

Session 20 Poster 26

Whole-crop maize treated with urea: effects on chemical composition and apparent digestibility

C. Guedes, M.A.M. Rodrigues, A.L.G. Lourenço, S. Silva, L.M. Ferreira and A. Mascarenhas-Ferreira, CECAV-UTAD, Animal Science, PO BOX 1013, 5001-801 Vila Real, Portugal

The objective of this study was to evaluate the preservative and upgrading potential of urea added to whole-crop maize. Whole-crop maize harvested at milk-dough stage was ensiled with 3 levels of urea - 0 (S0), 45 (S45) and 60 (S60) g kg^{-1} DM - and stored for 60 days. Apparent digestibility of organic matter (OMD) was determined using 4 mature rams. Un-hydrolysed urea, pH, total volatile fatty acids (VFA), chemical composition and *in vitro* organic matter digestibility (IVOMD) were determined. Data were analysed using urea level as the main factor. Urea breakdown was not affected by urea level ($P > 0.05$) and more than 50% of added urea was hydrolysed. Microbial activity measured by pH (>8.0), VFA production (<20 mM) and total non-structural carbohydrates concentration (>150 $gkgDM^{-1}$) was significantly reduced by urea treatment. Urea treatment significantly ($P < 0.001$) increased water soluble and ammonium N; more than 50% of the added nitrogen was retained. Application of urea at a rate of 45 and 60 gkg^{-1} DM significantly ($P < 0.01$) decreased the neutral detergent fibre content. The *in vivo* and *in vitro* OM digestibility increased significantly ($P < 0.01$) after urea addition. The OMD increased ($P < 0.05$) from 615 g kg^{-1} on S0 samples to 644 and 689 g kg^{-1} on S45 and S60 samples, respectively. Whole-crop maize can be effectively preserved and upgraded by ensiling with 45 and 60 g urea kg^{-1} DM.

Session 20 Poster 27

The effect of Akomed R and weaning age on performance, small intestine lipase aktivity and blood picture in broiler rabbits

L. Zita[1], E. Tůmová[1], V. Skřivanová[2] and M. Marounek[2,3], [1]Czech University of Life Sciences Prague, Department of Animal Husbandry, Kamýcká 129, 165 21 Prague 6 - Suchdol, Czech Republic, [2]Research Institute of Animal Production, Přátelství 815, 104 00 Prague 10 - Uhříněves, Czech Republic, [3]Czech Academy of Sciences, Institute of Animal Physiology and Genetics, Vídeňská 1083, 142 20 Prague 4 - Krč, Czech Republic

The objective of this study was to evaluate the effect of the weaning age and commercially available source of oil Akomed R® (AarhusKarlshamn AB, Sweeden, containing medium-chain acids – caprylic, capric, lauric) and lipase (Iontex, Czech Republic) on growth, feed conversion, slaughter parameters, small intestine lipase activity and blood picture in broiler rabbits. Rabbits were weaned at 25 and 35 days, were split into 2 groups which were fed with a control and experimental feed mixture (with 1% of Akomed R® and 0.5% lipase). There was no significant effect of supplements on weight gain and feed conversion. In slaughter parameters there were no significant differences in dressing percentage but Akomed R® and lipase significantly ($P \leq 0.05$) increased thigh share and reduced renal fat. Rabbits weaned at 25 days had generally higher intestinal lipolytic activity. The lipase and oil supplement did not significantly affect the lipolytic activity in small intestine. All measurements of blood picture were in a physiological range.

Session 20 Poster 28

Laboratory evaluation of vegetable wastes as alternative feed sources for livestock farmers in South Africa

B.D. Nkosi[1], M.M. Ratsaka[1], K.-J. Leeuw[1], D. Palic[1] and I.B. Groenewald[2], [1]ARC-LBD: Animal Production, Animal Nutrition, Bag X2, 0062, Irene, South Africa, [2]UFS, Centre for Sustainable Agriculture, Box 938, 9300, Bloemfontein, South Africa

South African fresh produce markets and food industries are dumping tons of vegetable wastes which may be used in livestock feeding. Incorporating such wastes as animal diets may reduce environmental pollution, but their use is limited by their low dry matter contents, unknown nutritive value and pathogenic agencies, which require further treatments before use. Cull beetroot, cabbage and pumpkin from fresh produce markets and selected wastes of frozen broccoli, steamed broccoli, steamed carrot, butternut peels and spinach from food companies were collected, sun-dried and analysed for DM, CP, CF, ME and IVOMD. DM ranged from 7.28 -17.25%, with the highest being butternut peels while the lowest being from frozen broccoli. CP was 14.94%, 18.8%, 18.68%, 38.31%, 22.59%, 13.75%, 5.48% and 20.69% for beetroot, cabbage, pumpkin, frozen broccoli, steamed broccoli, steamed carrot, butternut and spinach respectively. Pumpkin showed a CF of 20.35% which was the highest, while beetroot had the lowest (8.15%). Pumpkin had the highest ME (18.58%) while spinach has the lowest (8.6%). Higher IVOMD were observed from steamed carrot (85.37%) and the lowest from spinach (70.24%). The results show the vegetable wastes to have potentials to be included in livestock rations. However, palatability and digestibility tests still need to be done.

Session 20 Poster 29

Influence of limestone particle size on egg production and eggshell characteristics during early lay

F.H. De Witt, H.J. van der Merwe, M.D. Fair and J.P.C. Greyling, University of the Free State, Animal, Wildlife and Grassland Sciences, PO Box 339, 9300 Bloemfontein, South Africa

A study was conducted to determine the effect of limestone particle size on egg production and eggshell characteristics during the early laying period. Samples consisting of small (<1.0 mm), medium (1.0-2.0 mm) and large (2.0-3.8 mm) particle sizes were obtained from a specific calcitic South African limestone source that is extensively used in poultry diets. Calcium content of the limestone (360 g Ca/kg) and limestone inclusion level (95.8 g/kg) in the isocaloric and isonitrogenous basal diet was analogous for all particle sizes. Ninety-nine, individual caged Lohmann Silver pullets, 17 weeks of age, were randomly allocated to the three treatments with thirty-three birds per treatments for the determination of egg production (EP), egg output (EO), egg weight (EW), egg content (EC), feed intake (FI), feed efficiency (FE), body weight (BW), shell weight (SW), percentage eggshell (PES), eggshell calcium content (ESCa), shell ash (SA), egg surface area (ESA), shell weight per unit surface area (SWUSA) and eggshell thickness (ST). Data recorded at 24, 28 and 32 weeks of age were pooled to calculate parameter means for early lay. Limestone particle size had no significant effect on any of the parameters during the experimental period. These results suggested that large particles limestone is not necessarily essential to provide sufficient Ca^{2+} to laying hens during early lay.

Session 20 Poster 30

Effect of Allzyme® SSF on growth performance of broilers receiving diets containing high amounts of distillers dried grains with solubles

J.L. Pierce, T. Ao, B.L. Shafer, A.J. Pescatore, A.H. Cantor and M.J. Ford, Alltech-University of Kentucky Nutrition Research Alliance, 3031 Catnip Hill Pike, Nicholasville, KY 40356, USA

Allzyme® SSF is an enzyme complex produced by solid state fermentation. An experiment was conducted to evaluate the effects of Allzyme® SSF on the growth performance of male broiler chicks when fed diets containing 26% dried distillers grains with solubles (DDGS). A 21-day growth assay was conducted with 144 Cobb broilers allotted to four dietary treatments in a randomized complete block design. The treatments were 1) corn-soy (CS) reference diet, 1.24% Lys, 22% CP and 3150 kcal/kg ME, 2) positive control diet containing 25% DDGS with 1.24% Lys, 22% CP and 3150 kcal/kg ME, 3) and 4) 26% DDGS with 1.11 % Lys, 21% CP and 2835 kcal/kg ME without and with 200 g/tonne Allzyme® SSF, respectively. Feeding 25% DDGS significantly decreased weight gain (737 *vs* 691g, $P<.01$) and gain:feed (0.766 *vs* 0.676, $P < 0.01$) comparing with corn-soy reference diet. Reducing the dietary energy and crude protein concentration also reduced gain (691 *vs* 626 g, $P < 0.01$) and gain: feed (0.676 *vs* 0.579, $P < 0.01$). The addition of Allzyme® SSF tended to increase gain (626 *vs* 648 g) and significantly increased gain:feed (0.579 *vs* 0.610, $P < 0.01$). These results indicate that Allzyme® SSF improves growth and efficiency when high levels of DDGS are used in broiler diets.

Session 20 Poster 31

The requirement of Zn provided as organic Zn for broiler chicks fed corn-soy based diet with or without supplementation of phytase

T. Ao, J.L. Pierce, A.J. Pescatore, A.H. Cantor, M.J. Ford and B.L. Shafer, Alltech-University of Kentucky Nutrition Research Alliance, 3031 Catnip Hill Pike, Nicholasville, KY 40356, USA

A study was conducted to investigate the requirement of Zn when provided as Bioplex Zn® (a chelated Zn proteinate) for broiler chicks fed a practical corn soybean meal diet (25 mg/kg Zn content) with or without supplementation of phytase. A total of 864 day-old broiler chicks was randomly assigned to each of twelve dietary treatments with six replicate cages of 12 chicks in 3wk study. Treatment structure consisted of 2 x 6 factorial arrangement with two levels of phytase (0 or 500 U/kg) and six levels of Bioplex Zn® providing 0, 2, 4, 8, 16 and 32 mg Zn/kg. Dietary inclusion of phytase increased ($P < 0.01$) the feed intake, weight gain, plasma Zn and tibia Zn content. Dietary supplementation of Bioplex Zn® linearly ($P < 0.01$) increased feed intake, weight gain, plasma Zn concentration, liver Zn concentration and tibia Zn content. Significant interactive effects ($P < 0.05$) of phytase and BioplexZn® on feed intake and weight gain were found. When the supplemental level of Zn was below 8 mg/kg, the dietary inclusion of phytase increased ($P < 0.05$) feed intake and weight gain of the chicks. One slope, straight broken-line analysis of weight gain regressed on the supplemental Zn level provided as BioplexZn® indicated that 12 mg/kg supplemental Zn without phytase and 7.4 mg/kg supplemental Zn with phytase were required for the maximal weight gain of chicks.

Session 20 Poster 32

Investigation of antagonism and absorption of zinc and copper when different forms of minerals were fed to chicks

T. Ao, J.L. Pierce, R. Power, A.J. Pescatore, A.H. Cantor, M.J. Ford and B.L. Shafer, Alltech-University of Kentucky Nutrition Research Alliance, 3031 Catnip Hill Pike, Nicholasville, KY 40356, USA

The aim of this study was to investigate the antagonism of Zn and Cu when organic or inorganic forms of these minerals were fed to chicks. A corn-soybean meal diet was used as a basal diet. Bioplex Zn® and Bioplex Cu® were used as the organic sources. Reagent grade sulfate salts provided the inorganic sources of Zn and Cu. Supplements provided 20 ppm Zn and 8 ppm Cu. Ten groups of six day-old broilers were assigned to each of seven treatments during the 3 wk trial. Treatments consisted of feeding chicks basal diet alone or basal diet plus Zn or Cu or both from organic or inorganic source. The Zn and Cu uptake of mucus and the mucosa in duodenum was observed. Weight gain and feed intake were increased by Cu ($P < 0.01$) and were further increased by Zn or Zn + Cu ($P < 0.01$). Gain to feed ratio was decreased ($P < 0.01$) by Zn + Cu provided as inorganic forms but not as the organic forms. Zinc supplementation increased ($P < 0.01$) tibia and plasma Zn concentrations. Tibia Zn and Cu levels were higher ($P < 0.01$) for the organic Zn + Cu treatment than for the inorganic Zn + Cu treatment. The Cu content in the mucosa of chicks fed both organic Zn and Cu was significantly higher ($P < 0.01$) than that of chicks given no supplementation or both inorganic Zn and Cu. This suggests that the antagonism between Zn and Cu can be avoided through using proteinated forms of these minerals.

Session 20 Poster 33

Fat supplementation and placental transfer of polyunsaturated fatty acids (PUFA) in goats

C. Duvaux-Ponter[1], Y. Schawlb[2], K. Rigalma[1] and A.A. Ponter[2], [1]AgroParisTech, 16 rue C Bernard, 75005 Paris, France, [2]ENVA, 7 av Gen de Gaulle, 94704 Maisons-Alfort, France

During foetal development the brain's requirements for PUFA is high. In ruminants which consume very little arachidonic (C20:4n-6) and docosahexaenoic (C22:6n-3) acids, supply may be too low for correct brain growth. The C20:4n-6 and C22:6n-3 precursors, linoleic (C18:2n-6) and linolenic (C18:3n-3) acids respectively, may not be absorbed in sufficient quantities due to rumen bio-hydrogenation. We studied the effect of the type of fat supplement on PUFA transfer from dam to foetus. From 1.5 months pre-term, 2x15 goats received a diet containing a concentrate high in linseed (L, rich in C18:3n-3, 23%; low in C18:1n-9 (oleic acid), 7%) or rapeseed (rich in C18:1n-9, 25%; low in C18:3n-3, 5%). At birth, goat and kid jugular plasma, cotyledon and umbilical blood were sampled to measure FA profiles. Data was analysed by ANOVA. The proportions of C18:3n-3, C20:5n-3 and C22:6n-3 increased and those of C20:4n-6 decreased with the L diet ($P < 0.05$). The diet effect on the FA proportions was modest compared to the differences in diet composition, probably due to rumen bio-hydrogenation. The transfer of C18:3n-3 and C18:2n-6 from dam to foetus appears to be very low while C20:4n-6 and C22:6n-3 appear to be concentrated in the foetus compared to the dam. There may be either a selective placental transport system for PUFA, or C18:2n-6 and C18:3n-3 may be elongated and desaturated in the placenta to C20:4n-6 and C22:6n-3.

Session 20 Poster 34

Effect of broiler strain and dietary protein on the performance and litter quality of broilers

B. Dastar[1], M. Shams Shargh[1] and M. Mohajer[2], [1]Agricultural Science & Natural Resources of Gorgan University, Animal Science, Gorgan, 1, Iran, [2]Jehad-e-Keshavarzi, Gorgan, Golestan Province, 2, Iran

This experiment was conducted during hot summer season. A standard protein diet (protein recommendation by NRC, 1994) for starter (CP=21.6%) and finisher (CP=18.75%) periods and also a low protein diet (90% protein recommendation by NRC, 1994) for starter (CP=19.4%) and finisher (CP=16.87%) periods were fed to Ross 308 and Cobb 500 broilers for 42 day. All diets had 3000 Kcal ME/Kg. Birds were placed in floor pens. The resultant data were analyzed in a completely randomized design with a 2*2 factorial arrangement consisted of 2 broiler strains (Ross308 *vs* Cobb500) and 2 dietary protein levels (NRC *vs* 0.9NRC). Five replicate groups of 40 unsexed chicks were allocated to each treatment. Results indicated that there was significant differences in the performance of birds ($P < 0.05$). At 42 day of age Ross308 birds had higher body weight gain (1.85 *vs* 1.79 Kg) and lower feed conversion (1.99 g feed: g gain *vs* 2.10 g:g). Broilers were fed low protein diet had lower weight gain (1.77 *vs* 1.88Kg) and higher feed conversion (2.10 g feed: g gain *vs* 1.99). Reducing dietary protein level decreased protein intake (653 *vs* 735 g) and litter nitrogen percent (2.86 *vs* 3.27%). Results of this experiment showed that Ross308 broilers had higher performance than Cobb500. Reducing dietary protein level decreased broiler performance, but improved litter quality because of reducing litter nitrogen percent.

Session 20 Poster 35

Effect of protein source and mechanical extracting of fat on nutrient digestibility of sheep diets

K. Zagorakis[1], D. Liamadis[1] and C. Milis[2], [1]Aristotle University of Thessaloniki, Animal Nutrition, Thessaloniki, 54006 Thessaloniki, Greece, [2]Ministry of Rural Development and Foods, Laboratory for feeds analysis, Thermi Thessaloniki, 57001 Thermi, Greece

An *in vivo* digestibility trial using a latin square 4x4 experimental design with castrated rams was conducted to evaluate the effects of isonitrogenous replacing of Soy Bean Meal (SBM) with Full Fat Soya (FFS), Whole Cottonseed (WCS) and Cottonseed Meal (CSM), on nutrients' digestibility. Another goal was to examine whether mechanical extracting of ether extract (EE) affects nutrient digestibility of mature sheep diets. Rams were fed four isonitrogenous rations containing alfalfa hay, grounded corn grain plus one of the feeds under study, differing in main protein source and/or EE's content. The SBM and FFS rations had higher Dry Matter (DM) and Organic Matter (OM), digestibility, in comparison with WCS and CSM rations. Crude Protein (CP) digestibility of SBM diet was the highest compared to all other rations (83.5; 79.8; 78.9; and 77.5 %, for SBM, FFS, WCS, and CSM, respectively). Mechanical extracting of EE (SBM and CSM) negatively affected (10%) rations' EE digestibility. Nevertheless, the higher content in EE of FFS and WCS diets did not negatively affect crude fiber (CF) digestibility. CSM diet showed the lowest CF and NFE digestibility which in turn negatively affected DM and OM digestibility of this ration.

Session 20 Poster 36

Nutritive effects of mulberry leave enrichment with L-aspargine on silk production of silkworm *Bombyx mori* L. in north of Iran

R. Rajabi Kanafi[1], R. Ebadi[1], S.Z. Mir Hoseini[2], M. Fazilati[3] and A.R. Seidavi[4], [1]isfahan university of technology, plant protection, isfaha-iran, 0311, Iran, [2]guilan university of technology, animal science, rasht-guilan, 0131, Iran, [3]isfahan university of technology, food science, isfahan-iran, 0311, Iran, [4]azad islamic university-rasht branch, animal science, rasht-guilan, 0131, Iran

The effect of mulberry leaves enrichment with L-aspargine in 0.01, 0.1, 0.2 and 0.5%concentrations on cocoon characteristics of *Bombyx mori* L., Hybrid 103×104 was studied. L-aspargine diluted to mentioned concentrations with distilled water. Silkworm larvae were fed on fresh mulberry leaves of shin inche nevise variety enriched with L-aspargine, once a day. Normal leaves used for Control treatment. All economic parameters were determined by using standard technique in sericulture. Maximum female cocoon weight (1.468gr) was recorded in 0.5% concentration while maximum male cocoon weight (1.181gr) was recorded in 0.2% concentration. Male and female showed maximum pupal weight (1.343 and 0.904gr) in 0.1 % and 0.2% concentrations respectively. Maximum amount of female shell weight obtained in 0.01% concentration while maximum male shell weight recorded in 0.5% treatment to be 0.279gr. Female had maximum shell ratio (19.89%) in 0.01% treatment but male had maximum shell ratio (25.85%) in control. The overall results showed that L-aspargine could not increase silk production of silkworm *Bombyx mori* L.

Session 20 Poster 37

Effect of whole cottonseed by products on nutrient digestibility of sheep rations based on corn silage

D. Liamadis[1], M. Dasilas[1] and C. Milis[2], [1]Aristotle University of Thessaloniki, Animal Nutrition, Thessaloniki, 54006 Thessaloniki, Greece, [2]Ministry of Rural Development and Foods, Laboratory for feeds Analysis, Thermi - Thessaloniki, 57001 Thermi - P.O. Box 60511, Greece

An *in vivo* digestion trial with castrated rams was conducted by the use of four protein sources, whole cottonseed (WCS; control), cottonseed meal (CSM), cottonseed cake (CSC) and cottonseed pellets (PCS), in order to evaluate the effects of processing WCS on nutrient digestibility of sheep rations. Four diets were formulated that contained corn silage plus one of the investigated feedstuffs (WCS, CSM, CSC, and PCS, for rations A, B, C and D, respectively). Ration C presented a tendency for higher crude protein (CP) digestibility in comparison with ration A and D (64.6; 69.3; 71.1; and 64.6 %). Rations A and D had higher ether extract (EE) digestibility compared to rations C and B (94.2; 85.4; 89.6; and 93.7 %), whilst EE's digestibility was higher in ration C in comparison with ration B. Ration A had the lowest NFE digestibility (68.8; 72.3; 73.4; and 72.2 %) probably due to higher particle size. Ration B had higher NDF digestibility compared to rations A and C (57.2; 61.3; 57.6; and 59.5 %), and also had the higher hemicelluloses digestibility in comparison with all other rations (65.1; 71.9; 68.1; and 67.9 %).

Session 20 Poster 38

The effect of artificial inoculation with Fusarium strains on quality of corn silage

L. Křížová[1], S. Pavlok[1], F. Kocourek[2] and J. Nedělník[3], [1]Research Institute for Cattle Breeding, Ltd., Vídeňská 699, 691 23, Pohořelice, Czech Republic, [2]Crop Research Institute, Drnovská 507, 161 06, Praha 6 - Ruzyně, Czech Republic, [3]Research Institute for Fodder Crops, Ltd., Zahradní 1, 664 41 Troubsko u Brna, Czech Republic

The objective of the study was to determine the influence of Fusarium contamination on the ensiling process. Silages were prepared from the conventional corn hybrid Monumental that was either untreated (C) or artificially inoculated with a suspension of Fusarium strains (I). The inoculation was made in the growing crop in milk stage of maturity after harming the cops and stalks with wire brush. The corn crops were grown, harvested and ensiled under identical conditions. Entire corn plants were harvested at the soft dough stage of maturity, and ensiled in microsilage tubes (approximately 6.5 kg per tube, 3 tubes per treatment) and fermented at 25 °C (± 1°C) for 15 weeks. The dry matter and CP content of silage C was higher than that of silage I (345.4 and 66.7 g/kg *vs* 329.0 and 64.4 g/kg, $P > 0.05$). Silage I showed a higher degree of proteolysis measured as N-NH 3 (% of total N) than silage C (7.8 *vs* 6.8, $P > 0.05$) and had a lower content of lactic acid than silage C (4.7 *vs* 7.8 g/kg, $P > 0.05$). This study was supported by NAZV 1B53043

Session 20 Poster 39

The effect of various crude protein levels on milk yield and physiological parameters in dairy cows

V. Kudrna, P. Lang and K. Poláková, Institute of Animal Science, Nutrition and feeding technique of farm animals, Pratelstvi 815, 10401, Prague 10 - Uhrineves, Czech Republic

The aim of this experiment was to compare the effect of two different crude protein (CP) levels in total mixed rations (TMR) fed *ad libitum* to dairy cows (18.2 % CP - P; 16.1 % CP - K; 16.1 % CP+16 g/head/day protected methionine – M) on dry matter intake (DMI), daily milk yield (DMY) and physiological parameters. A total of 27 dairy cows were used. Protected methionine was supplied individually in the milking room. Cows were approx. 34 days after calving. The highest average daily DMI (22628 g/head) was found in ration P, followed by ration K (22459 g/head) and ration M (20631 g/head). The highest average DMY (39.9 kg/head) was found in ration P, while there were no differences in DMY between ration M and K (36.0 *vs* 35.8 kg/head, respectively). Significant differences were found in milk protein and urea contents. Cows fed ration P and M had consistently higher level of rumen NH^3 than cows fed ration K. Plasma concentration of glucose was significantly lower on ration P compared with other rations. A CP content of 18.2 % in the TMR substantially increased average DMY and FCM production, but resulted in the lowest milk protein production. Supplementation with protected methionine lowered DMI. Some of blood plasma and rumen fluid parameters were worsened. The project No. 1G 46086 was supported by NAZV, Ministry of Agriculture, Czech Republic.

Session 20 Poster 40

Effect of escape protein level on finishing performance of Awassi lambs

A. Can, N. Denek and S. Tufenk, Harran University, Animal Science, Harran University Agricultura Faculty, 63100 Sanliurfa, Turkey

Two lamb finishing trials were conducted to evaluate escape protein (EP) level on finishing performance of fat tailed male and female Awassi lambs under neutral and high ambient temperatures. Experiment was arranged as a 2 x 2 x 2, factorial in which were concentrate feed EP level (low 3.4% *vs* high 5.0%), gender (male *vs* female), and temperature (neutral 18.7 °C *vs* high 30.4 °C). Two concentrate feeds were formulated with 21.25 % and 31.44% EP of total CP (16% of DM) for low and high EP levels, respectively. Increasing EP level from 3.4 to 5.0 % did not affect average daily gain (ADG), DM intake, and feed efficiency ($P > 0.05$). Male lambs had higher ADG, DM intake and better feed efficiency ($P < 0.05$). All the lambs had similar ADG, feed consumption, and feed efficiency values in both neutral and high ambient temperatures ($P > 0.05$). As a result, barley, cotton seed meal, and urea can meet the EP requirements of Awassi lambs without fish meal (FM) addition to the high crude protein (CP) containing lambs diet.

Session 20 Poster 41

The nutritional quality of alternative types of total mixed rations differing in linseed contents measured by the *in vivo* method

V. Koukolová, P. Homolka and O. Tománková, Institute of Animal Science, Přátelství 815, 10400 Praha, Czech Republic

The objectives of this study were to investigate relationships among *in vivo* digestibilities of dry matter (DM), organic matter (OM), and gross energy (GE), including their relationships with the chemical components of two total mixed rations (TMR) differing in linseed supplementation. The linseed TMR was a compound of maize silage (46 %), alfalfa silage (20 %), extracted grains (9 %), linseed (5 %), alfalfa hay (2 %), soybean meal (1.3 %), oats (14 %), vitamins, and minerals. The control TMR had a similar proportion of feedstuffs to the linseed TMR, but the linseed supplementation was compensated for by a small part (1.5 %) of Ca-salt palm oil oleic acid. The *in vivo* digestibilities of DM, OM and GE were determined in metabolic trials using six wethers of the Romanovské breed. The chemical compositions of fat, crude fibre, ash, neutral-detergent fibre, acid-detergent fibre and acid-detergent lignin were 3.6, 19.8, 8.5, 38.6, 19.1, 4.6 % in control TMR and 4.3, 17.5, 6.5, 37, 21.8, 4.8 % in linseed TMR, respectively. The GE values of the control TMR and linseed TMR were 19 and 19.5 J/g, respectively. The *in vivo* digestibilities of DM, OM, and GE averaged 64.6, 66.9 and 64.6 % for control TMR and 60.2, 63.5 and 61.5 % for linseed TMR, respectively. Significant differences ($P < 0.05$) between the control TMR and linseed TMR in the *in vivo* digestibilities of DM and GE were observed. The study was supported by the project MZE0002701403.

Session 20 — Poster 42

The nutritional value of four genotypes canola forage

A.R. Safaei, Y. Rouzbehan and M. Aghaalikhani, Tarbiat Modares University, Animal Science, Tehran, Iran

The nutritive value of four genotypes canola forage Olimp, Midas, Global and Hybrid (Cobra × Regent) were evaluated by chemical composition, *in vitro* (gas production), in situ (dry matter and nitrogen) degradation and palatability (a short term intake rate procedure with four Shall breed sheep) methods. Organic matter (OM), crude protein (CP), neutral detergent fiber (NDF), acid detergent lignin (ADL) content of the four genotypes ranged from 805 to 813, 218 to 247, 298 to 338 and 37 to 53 g/kg, respectively (DM basis). *In vitro* dry matter and organic matter digestibility of the forages ranged from 62.9 to 79.1 and 53.7 to 67.5%, respectively. Calculated metabolism energy content of the genotypes ranged from 7.8 to 10.1 MJ kg-1 DM. Dry matter and CP ruminal degradability of the forages ranged from 67.1 to 68.0% and from 89.6 to 91.0%, respectively. Palatability for Olimp, Midas, Global and Hybrid genotypes forage were 15.4, 13.7, 11.2 and 10.9 (g DM/min) respectively. On the basis of the nutritional value of the forages, there were significant differences ($P < 0.05$) among them, and the ranking order of these species was Olimp, Midas, Global and Hybrid (Cobra × Regent).

Session 20 — Poster 43

Effects of deoxynivalenol contaminated feed on some parameters in piglets

D. Dinu[1], I. Taranu[2], D. Marin[2], M. Costache[1] and A. Dinischiotu[1], [1]University of Bucharest, Department of Biochemistry and Molecular Biology, 91-95 Spl. Independentei, Bucharest, Sector 5, 050095, Romania, [2]Institute of Biology and Animal Nutrition, Animal Biology Department, 1 Calea Bucuresti, Balotesti, Ilfov, 077015, Romania

The effects of 0.5 ppm and 1.5 ppm DON contaminated feed on hepatic and renal function on piglets were examined. Thirty healthy local hybrid male piglets aging 35 days were monitored daily and had free access to water and feed. The basal diet contained corn, wheat and proteic supplements. The animals were randomly divided into three groups of ten. The first group continued to receive the standard diet, the other ones were feeded with basal diet contaminated with 0.5 ppm DON and 1.5 ppm DON, respectively. After 14 days all animals were slaughtered and blood was collected. Serum glutamate dehydrogenase specific activity was diminished by 23.6% in 0.5 ppm DON and by 31.1% in 1.5 ppm DON administration ($p < 0.05$), suggesting that the gluconeogenetic pathway was affected by this mycotoxin. The significant increase of alanine aminotransferase and gamma-glutamyltranspeptidase specific activities noticed in serum of DON treated piglets compared with the control group ($p < 0.05$) suggested alterations of the liver cells. The increase in serum urea levels observed at both DON concentrations versus control group ($p < 0.05$) suggested a progressive kidney damage induced by this mycotoxin. The DON related affection of liver and kidney was obvious.

Session 21 — Theatre 1

Application of a synthetic maternal pheromone reduces post-mixing aggression and lesions in weaned pigs

J.H. Guy, S.E. Burns, J.M. Barker and S.A. Edwards, Newcastle University, School of Agriculture, Food and Rural Development, Kings Road, NE1 7RU, United Kingdom

Although mixing pigs is not recommended, mixing of pigs from different litters often occurs at weaning in order to form even-sized groups and the resulting aggression can lead to skin lesions and reduced growth performance. The aim of this experiment was to determine the effect of application of a synthetic maternal pheromone to pens of weaned pigs on levels of aggression, skin lesions, general behaviour, feed intake and growth rate. Litters of pigs were weaned at approximately 28 days of age into a commercial weaner building to give 8 rooms, each of 4 pens containing approximately 20 animals. Pigs were allocated on a per room basis to either Pheromone pens (pen sides and feeder treated with 1.0 ml per pig of synthetic pheromone; Suilence, Ceva Sante Animale, France) or control pens (nothing applied) and fed *ad libitum* for 28 days. Skin lesions were scored before and after mixing and behaviour observed during 2 days after mixing by both direct means and video recording. Pigs in the Pheromone treatment had significantly fewer lesions 24 hrs after mixing (17.7 *vs* 27.6 control, $P < 0.001$) and displayed significantly less fighting over this period ($P < 0.05$). Feed intake and growth performance were unaffected by treatment. In conclusion, application of a synthetic pheromone reduced levels of aggression and lesions sustained by weaned pigs after mixing which would lead to improved welfare.

Session 21 — Theatre 2

Behavioural synchronisation in a needs adequate pig facility

K.H. de Greef, S.A. Alders and H.M. Vermeer, Animal Sciences Group of Wageningen UR, PO Box 65, 8200 AB Lelystad, Netherlands

Behavioural synchronisation substantially affects the facility requirement of group housed pigs. Images from a needs-adequate facility were analysed for degree of synchronisation with two straight forward facility-aspects: m^2 space allowance and number of feeders. *Ad libitum* fed pigs were housed in groups of 12, 24 or 48 with on average 2.2 m^2 and 0.5 feeders per pig, calculated to be exceeding requirements. To date, the three available batches have not been analysed fully [planning summer 2007], but the global pattern of results is clear enough to share preliminary results. Resting behaviour was the only behaviour showing full synchronisation (21% of obs), resulting in the maximal theoretical area requirement. When lying, pigs showed location-synchronicity: they rather lay clustered than spaced. Clusters averaged 3.1 pigs; 52% lying single. During light hours, feeders were in use for 53% of time, in which 49% of the meals was performed alone. 93% of all meals was in groups of 3 or less pigs. When eating, pigs were rather minimising their mutual distance (location synchronicity) than spacing (43% versus 12% of obs.). Results demonstrate that pigs not only synchronise the two studied behaviours in time, but also in location. Behavioural optima for both needs exceed the economic optima. The within-group variability involved produces no clear cut requirements, but rather 'diminishing returns' figures. Data allow transparent choices between animal needs, societal requirements and economy.

Session 21 Theatre 3

Effect of winter accommodation on dairy cow behavioural synchrony

K. O'Driscoll[1,2], L. Boyle[1], P. French[1] and A. Hanlon[2], [1]Moorepark Dairy Production Research Centre, Fermoy, Co. Cork, Ireland, [2]University College Dublin, Agriculture, Food Science and Veterinary Medicine, Belfield, Dublin 4, Ireland

The study aimed to investigate differences in behaviour synchrony between three wood-chip out-wintering pads (OWP) and cubicle housing. Treatments were; indoor cubicles (IC), a covered OWP, an uncovered OWP and an OWP with a self feed silage pit. There were 10±6.7 animals per group, replicated 6 times. Observations were carried out every 30min between 06:00 and 01:30 three times in Jan and Feb 06. The outcomes for measurement were eligible cows lying (ECL), cow comfort index (CCI) and proportion animals feeding (AF). Data were analysed using SAS V9.1. Autocorrelation was calculated using the Durban Watson statistic, and compared across treatments (PROC GLM). A centred moving average was computed and used to obtain the residual and analysed using a least squared GLM. Low autocorrelations were recorded in IC ($P < 0.01$) indicating stronger temporal behavioural synchrony in the 3 OWP designs. Overall, the highest proportion of ECL, CCI and AF occurred in IC ($P < 0.05$). However, high proportions (>90%) ECL were recorded on OWPs in the early morning while the range of ECL in IC (22%-87%) was much lower. The low overall proportions for ECL and CCI in the OWPs were caused by the cows standing a lot without feeding during daylight hours. However, these cows performed more synchronised lying at night which suggests that OWPs promote a more natural circadian behaviour pattern.

Session 21 Theatre 4

Effect of switching milking frequencies on indicators of discomfort in dairy cows

L. Boyle, K. O'Driscoll, G. Olmos, P. Gazzola, D. Gleeson and B. O'Brien, Teagasc, Dairy Production Research Centre, Moorepark, Fermoy, Co. Cork, Ireland

Changing the frequency of milking dairy cows during lactation offers benefits to producers with seasonal calving herds. The aim of this study was to compare the effects of switching from once (1x) to twice (2x) daily milking and from 2x to 1x daily milking together with milking 2x daily throughout lactation on indicators of cow discomfort. Spring calving Holstein-Friesian cows (n=42) were blocked according to calving date, previous milk yield and parity and randomly assigned to three treatments: i) 2x full lactation (2x); ii) 2x switched to 1x (2x1x) and iii) 1x switched to 2x (1x2x). Measurements were taken on days -2, -1, +1, +2 and + 7 relative to the switch day (day 0 =110 [19.7 s.d.] days in milk [DIM]). Five aspects of locomotory ability were scored from 1 (normal) to 5 (severely abnormal) prior to the morning milking. Milk leakage was also recorded at this time. Udder firmness was scored from 0=loose to 3=hard in the parlour. Step/kicking behaviour performed by cows during udder preparation and cluster attachment was also scored. On day +1, six 2x1x cows, and no cows in the other two treatments, were leaking milk ($P < 0.01$). Furthermore, there were significantly more 2x1x cows with scores of 3 for udder firmness compared to 1x2x cows ($P < 0.05$). Treatment had no effect on locomotory ability or on cow behaviour in the parlour ($P > 0.05$). Switching from 2x to 1x daily milking at 110 DIM had minor transient effects which did not cause discomfort to the cows.

Session 21 — Theatre 5

Estimation of body condition score in dairy cattle using digital images

J.M. Bewley[1], A.M. Peacock[2], O. Lewis[2], M.P. Coffey[3], D.J. Roberts[3] and M.M. Schutz[1], [1]Purdue University, West Lafayette, IN, USA, [2]IceRobotics Ltd., Roslin, Scotland, United Kingdom, [3]Scottish Agricultural College, Dumfries, Scotland, United Kingdom

Body condition scoring, as an indirect measure of levels of subcutaneous fat in dairy cattle, has been widely adopted as a research tool and field-assessment tool for management purposes. However, the subjectivity and time commitment of current systems have interfered with adoption and routine execution. Automating body condition scoring would eliminate these barriers to adoption. The feasibility of utilizing digital images to determine body condition score was assessed for 6 weeks for lactating dairy cows within the SAC Crichton Royal Farm. Scores were obtained using the primary systems utilized within the United Kingdom (developed by Mulvany) and the United States (developed by Edmonson and Ferguson). Up to twenty-three anatomical points were manually identified for images captured automatically as cows passed through a weigh station. Points around the hooks were easier to identify than points around the ribs, thurls, or pins. All identifiable points were utilized to define and formulate measures describing the cow's contour. Angles around the hooks and pins were significant predictors of body condition score ($p < 0.05$). A higher percentage of variation was explained for US body condition score than UK body condition score. Future efforts will attempt to automate this process using thermal imaging technology to better discern anatomical landmarks.

Session 21 — Theatre 6

The effects of supplementing gestating ewe diets with DHA from algal biomass on responses of their lambs to natural parasitological challenge

E. Scott-Baird, S.A. Edwards, C. Leifert and G. Butler, University of Newcastle upon Tyne, Nafferton Ecological Farming Group, Nafferton Farm, Stocksfield, Northumberland, NE43 7XD, United Kingdom

A rapidly rising level of anthelmintic resistance and increasing public concern over drug residues in animal products make it necessary to develop more sustainable nematode control programmes. The fatty acid, docosahexaenoic acid (DHA), produced in high concentrations by some marine algae, has been shown to modify immune response and decrease the production of proinflammatory cytokines. This action might reduce gut inflammation and associated hypophagia in response to parasite infestation. 48 twin bearing mule ewes were allocated to a 2x2 factorial design comparing presence or absence of an herbal anthelmintic product and an algal biomass (AB) supplement supplying 6g DHA/day in the last 6 weeks of gestation. The effect of the herbal product was taken into account in evaluating the effect of AB supplementation. Residual effects of supplementation were seen on DHA content of milk (at 3 weeks) and lamb plasma (at 6 weeks) of lactation. There were no significant differences in faecal egg counts from ewes or their lambs; however lambs from DHA supplemented ewes showed reduced lymphocyte counts at times following exposure to new nematode challenges. This was associated with significantly greater lamb liveweight and condition score at 4 weeks post weaning. Additional study is required to further investigate the underlying mechanisms of action.

Session 21 Theatre 7

Can kefir reduce coccidial oocysts output in goat kids?

G. Daş, C. Ataşoğlu, H.I. Ülkü, C. Tölü, T. Savaş and I.Y. Yurtman, Çanakkale Onsekiz Mart University, Department of Animal Science, Faculty of Agriculture, Campus of Terzioglu, 17020 Çanakkale, Turkey

Investigations on the effects of kefir usage following weaning on coccidial oocyst output of dairy goat kids together with some growth and feed consumption parameters were subjects of this study. Twenty twin kids were randomly allocated to two groups with equal distributions of genders. Kids from the first group (KEF) were received 20 ml/day kefir daily for 6 weeks, while other kids remained in the sham treated control group (CONT). Faecal samples were taken from each kid three times per week and were examined with a modified McMaster counting technique to quantify oocyst output. There were no differences between groups in terms of feed consumption and growth rate values of kids ($P > 0.05$). Kids from KEF group had a 30% lower oocyst output than those from CONT group during the study. The highest and the lowest oocyst output values were found lower in KEF group than in KON group. Although there was no significant difference between Logarithmic transformed oocyst output (Log-OpG) of groups during the whole study period ($P=0.09$) CONT group had relatively higher level of Log-OpG than KEF group (7.14 *vs* 6.45). CONT group had higher level of Log-OpG than the KEF group for the first two ($P < 0.05$) and the third week ($P=0.068$), but the differences were found to be similar in the last three weeks ($P > 0.05$). The results of this study suggest that kefir may have a potential for controlling coccidiosis.

Session 21 Theatre 8

Assessment of genotoxicity and cytotoxicity of AFB1 in peripheral blood mononuclear cells of goats

B. Ronchi, A. Vitali, P.P. Danieli, U. Bernabucci, N. Lacetera and A. Nardone, University of Tuscia, Department Animal Production, Via S. Camillo De Lellis, 01100 Viterbo, Italy

Aflatoxins are known as natural substances highly toxic and carcinogenic for humans and animals. Nevertheless, limited knowledge is available on effects of this metabolites of fungi on immunology of small ruminants. For this reason, a study was carried out to ascertain the effects of Aflatoxin B_1 (AFB_1) on peripheral blood mononuclear cells (PBMC) of goats. Eight non-lactating and non-pregnant Saanen goats were exposed to different levels of AFB_1 (0, 25, 50 and 100 μgAFB_1 kg^{-1} body weight) given orally in a single dose. Individual blood samples were collected at 0, 24, 72, 120 and 192 h after AFB_1 administration. PBMC were isolated and submitted to a multi-level assay scheme to evaluate the effects of AFB_1 on cell proliferation (DNA synthesis test), DNA damage (Comet Assay genotoxicity test) and cell viability (MTT test). DNA synthesis and cells viability were severely depressed in a dose and time dependent manner ($P < 0.01$). For all AFB_1 levels tested, tail moment was higher ($P < 0.01$) after 24 h from the administration. Only for the highest dose of AFB_1, tail moment was higher ($P < 0.01$) also 72 h after the administration. These findings confirm the high cytoxicity and genotoxicity of AFB_1, and suggest that exposure to AFB_1 may compromise the immune response of goats.

Session 21 Poster 9

Associations between vaginal fluid diagnosis at the day of insemination using a novel intravaginal device (Metricheck) and Non-Return Rates in dairy cattle

D. Völker[1], S. König[1], U. Janowitz[2], J. Potthast[2], J. Spicker[2] and M. Gauly[1], [1]Institute of Animal Breeding and Genetics, Albrecht Thaer Weg 3, 37075 Goettingen, Germany, [2]RUW e.G., Schiffahrter Damm 235,, 48147 Münster, Germany

The diagnosis of endometritis is undertaken using a variety of techniques including vaginoscopy, manual examination, cytology and ultrasonography. In this study a novel test device (Metricheck, Simcrotech, Hamilton, New Zealand) to win vaginal fluid was used in Holstein Friesian cows (n= 989) from 99 herds (average herd sizes: 83 cows, average milk yield: 8855 kg) at the day of insemination. Any material retrieved was scored on a 1 to 5 scale (1 = clear mucus, 2 = flecks of purulent material within otherwise clear mucus, 3 = mucopurulent but < 50 % purulent material, 4 = mucopurulent with > 50 % purulent material and 5 = mucopurulent with > 50 % purulent material and with an odour). The relationship between the test score and the non-return rate 90 (NRR 90) was estimated. All inseminations were done by two technicians. 70.5 % (n = 697) of all animals showed clear mucus, whereas 29.5 % were scored between 2 and 5 (19.7 % - 2, 7.6. % - 3, 1.7 % - 4, 0.5 % - 5). Animals with clear mucus had in average an 7.68 % higher NRR 90 (55.71 %) when compared with animals scored between 2 and 5 (48.03 %). The statsistical significance was: p = 0.044. It is concluded that examination with the metricheck device at the day of insemination is sensitive in detecting endometritis. Scores between 2 to 5 were associated with poorer NRR 90.

Session 21 Poster 10

Growth performance and behaviour of pigs raised outdoors in different size groups

V. Juškienė and R. Juška, Institute of Animal Science of LVA, Animal Hygiene and Ecology, R. Žebenkos 12, LT-82317, Baisogala, Lithuania

Sixty Lithuanian white x Large white x Swedish Yorkshire crossbred pigs were used in the study to determine growth performance and behaviour of pigs raised outdoors in the groups of different size. The trial was carried out in June-October. Group 1 size was 15 animals per enclosure and Group 2 size was twice bigger comprising 30 pigs per enclosure. The enclosures were set up outdoors and the space allowance was 56.7 m^2/pig in both groups. Every enclosure was fitted with three-wall shades with the area of 0.5m^2 per animal. Both in the growing and finishing phases of the trial and during the whole experiment, the pigs of Group 1 gained higher weights, respectively, by 32, 44 and 37 g than the pigs in Group 2, however, the weight difference was statistically insignificant. Behaviour studies indicated that the time spent for resting was almost similar in both groups and accounted for 65.5-67.2% of the whole time. The pigs in Group 1 were a little more active and spent on the average 11.9 minutes more for walking, yet the differences were insignificant. The pigs in Group 1 spent more time for comfort behaviour, and those in Group 2 spent 16.7% more time at nuzzling, but the differences were statistically insignificant.

Session 21 Poster 11

The effect of dairy cows housing during first week of live on their avoidance behaviour in adult age

P. Kisac, J. Broucek, M. Uhrincat, S. Mihina, A. Hanus and S. Marencak, Slovak Agriculture Research Center, Hlohovska 2, 949 92 Nitra, Slovakia (Slovak Republic)

Early weaning from mother can decrease fear of new environment. Animals with varied experience are more prepared to deal with situations which they can meet latter. The aim of experiment was to find out the effect of housing during first week of life on avoidance behaviour in adult age. Holstein heifers were kept in individual hutch (IH, n=19) and in the pen with mother (PM, n=32). Since 2nd week of live were animals of both groups kept in IH. In aversive alley (16.3x1.86 m) was during three days (12 tests) observed aversive behaviour of cows in seventh month of first lactation. During testing on the first day a manipulator (standing in the end of alley) brushed animals. On the second day the animals received negative impulse in the end of alley – single noise 128 dB.s^{-1} and on the third day the animals received electric shock 1.5 kV.s^{-1}. The results showed, that animals weaned from mother at the second day of live needed longer time for crossing the alley (83.5 s IH; *vs* 47.3 s PM; p=0.0011); they returned more to start (0.3 times IH *vs* 0.1 times PM; p=0,0016); they stood more in last part of alley (13.3 s IH *vs* 8.8 s PM). They stood more (1.6 parts IH *vs* 0.9 part PM; p=0.0003). It is possible, that animals reared without mother are able to remember negative treatment and they anticipate negative impuls in the end of aversive alley. Way of rearing affects aversive behaviour of animals in adult age.

Session 21 Poster 12

Interbreed differences of metabolic profile parameters in cattle of holstein-friesian and slovak spotted breed

J. Buleca[1], J. Szarek[2], M. Húska[1], E. Dudriková[1], M. Tučková[1], A. Ondrejková[1], E. Beličková[1] and S. Mardzinová[1], [1]University of Veterinary Medicine, Komenského 73, 041 81 Košice, Slovakia (Slovak Republic), [2]Agricultural University, Department of Cattle Breeding, Al. Miczkiewicza 24/28, 3059 Krakow, Poland

In our work biochemical parameters of two groups of holstein-friesian and slovak spotted cattle (n=30+30) were evaluated. Routine hematological parameters were completed by analyses of selected parameters of energy profile and methods of spectrophotometry. Numerically similar values of biochemical parameters of blood serum: total lipids (4.05 and 3.96 g/l), triglycerides (0.52 and 0.48 g/l), glucose (3.76 and 3.90 g/l), total cholesterol (2.9 and 3.15 g/l) and HDL (1.36 and 1.47 mmol/l) were noticed. Together individual variance of values in both experimental groups was evident and confirms relatively higher values of variance coefficients. Obtained results could serve as source of actualization of reference interval of physiologic values in mentioned cattle category. This work was supported by the Slovak Research and Development Agency under the contract No.APVV-20-063205.

Session 21 Poster 13

Simple monitor for daily behaviors with a wireless temperature data logger on a collar of unrestricted cows

M. Okamoto and S.B. Park, Rakuno Gakuen University, Department of dairy science, Bunkyodai Midorimachi 582, Ebetsu, Hokkaido, 069-8501, Japan

It is very important to monitor behaviors including eating, ruminating and resting of cows for successful production and herd health. There has not been a simple monitoring system for daily behaviors except estrus monitor. The objective of this study was to evaluate a simple temperature monitor put on a collar around the neck of unrestricted cow as a monitor of behaviors such as eating and rumination. A temperature sensor was attached at the fore and rear edge of the collar around the neck, respectively. The fore edge of the collar touched on the jaw because of keeping head down during eating. The rear edge touched on the breast because of keeping head up during rumination. The temperature difference between fore edge and rear edge of the collar (fore minus rear) was positive value during eating, and decreased to negative value during rumination. The discrimination between eating and rumination coincided with actual behavior in healthy cows in a dairy barn and in a pasture. The ratio of eating time to rumination time changed with grass availability in the pasture. Cows on heat became restless, and recovered to normal behavior after duration of estrus. A cow also became restless from several hours before calving. Licking resulted in increase of fore edge temperature. A hypomagnesemia cow tended to keep standing, and the temperature difference between fore edge and rear edge of the collar remained in a small range around zero.

Session 21 Poster 14

Investigations on dairy cows' bone structure

I. Zitare[1], M. Pilmane[2] and A. Jemeljanovs[1], [1]Latvia University of Agriculture, Research Institute of Biotechnology and Veterinary Medicine Sigra, Instituta street 1, LV 2150, Sigulda, Latvia, [2]Riga Stradins University, Institute of Anatomy and Anthropology, Dzirciema street 16, LV 1007, Riga, Latvia

Humerus bone in 5-6 years lactating cows were examined after animals compulsory slaughtering. The Cutting-Grinding Technique for Hard Tissue was used for dissection of bone. Also mineral density test was used for cows' bones. Growth factors – BMP2/4, FGFR, were used to detect cell growth and cellular differentiation by immunohistochemistry (IMH). TUNEL method was performed to detect cell death and for matrix degradation we used MMP2 and MMP9 IMH detection. Bone showed thin trabecules with variable number of osteocytes from 20.30±3.79 to 54.30±5.66 in mm^2. Osteones also presented different diameter – from 0.0668±0.0183 to 0.1596±0.0285mm. Intensive proliferation of connective tissue and small capillaries were seen in osteon channels. Regions with granular, optically intensively stained basophilic substance took places here and there in bone with density from 2206.45±714 to 3017.94±744g/cm^2. Fragments of articular cartilage seemed not changed in routine histological sections. Few BMP2/4-containing cells were detected in all chondrocytes of articular cartilage in all animals and in main part of bone in cows. Numerous to abundance of chondrocytes expressed FGFR1 in articular cartilage, but only few osteocytes of spongy bone contained these receptors. Apoptosis affected mainly chondrocytes.

Session 21 — Poster 15

The relationship between type traits of Holstein heifers and their subsequent production breeding values and production life

M. Stipkova, J. Bouska and E. Nemcova, Institute of Animal Science, Pratelstvi 815, 104 00 Prague, Czech Republic

The relationships between different type traits of Holstein heifers scored at 14 months and their breeding values (EBV) for production traits calculated at the end of the third lactation were analysed. A total of 2978 heifers calving from 1998 to 2003 were included into the analysis. The level of the rearing period until 14 months of age is characterised by the average daily live weight gain 790 g, live weight 385 kg, and height at sacrum 131.3 cm. Pearson's correlation coefficients (r) between type traits and EBVs for milk and protein yields were estimated. The coefficients between milk yield and angularity, chest width, and body depth were r=0.39, r=0.66, and r=0.76, respectively. For complex characteristics, the coefficients of correlation were r=0.47 and r=0.65 for dairy character and total score for type, respectively. The heifers were classified into 5 groups according to the score of dairy character. The average EBV for milk yield was 216 and 156 for the group with the highest and lowest dairy character score, respectively. Length of production life was most associated with feet and legs traits. The groups of heifers with the highest score for claw conformation and hock joint quality had longer production life by 2.5 and 3.5 months, respectively, compared to the groups with the lowest score. The financial support from the project NAZV 1B44035 is acknowledged.

Session 21 — Poster 16

The relationship between somatic cell count in milk and linear type traits in Holstein cows

E. Nemcova, M. Stipkova, L. Zavadilova, J. Bouska and M. Vacek, Institute of Animal Science, Pratelstvi 815, 104 00 Praha, Czech Republic

Test-day records of somatic cell count (SCC) and six linear type traits (fore udder attachment, udder depth, central ligament, rear udder height, front teat placement, teat length) of 22 613 first lactation Holstein cows from 117 herds were included into this study. SCC was log-transformed into somatic cell score (SCS). For each analyzed linear trait, the cows were assigned into three levels according to the linear type score: level 1 (score 1 and 2); level 2 (score 5 or 6); level 3 (score 8 and 9). A linear model was used to estimate the effect of different type traits on SCS. The highest values of SCS were found for the first levels. The differences between the first and second level were on average 0.33, 0.54, 0.28, and 0.36 for fore udder attachment, udder depth, central ligament, and rear udder height, respectively. Cows with deep udders, weak central ligaments and fore attachments, and low rear udder heights showed highest SCS. Low SCS appeared to be associated with intermediate distance between front teats and longer teats. The financial support from the project NAZV 1G46086 is acknowledged.

Session 21 Poster 17

Possibilities of reducing milk somatic cell count

H. Kiiman, E. Pärna, T. Kaart and O. Saveli, Estonian University of Life Sciences, Kreutzwaldi 64, 51014 Tartu, Estonia

Somatic cell count (SCC) in milk is a measure of udder health and milk quality. Milk SCC data are used as a proxy for udder health at the cow level. Herds with high SCC have to adopt a short-term goal of reducing SCC as quickly as possible. Estonia faces problems with SCC affecting milk quality and udder health. In 2005, the culling rate of cows due to mastitis was 26%, while milk SCC ranged from 347,000 to 446,000 ml on our dairy farms. The SCC recording in Estonia dates back to 1979 and since 1987 the herds have participated in a milk recording scheme on a monthly basis. Since 2004 the required milk SCC is in compliance with the proposed EU legal limit. The objective of this study was to estimate heritability of SCC for different breeds and lactations. The investigation involved 84, 79 and 56 1st, 2nd and 3rd lactation cows of Estonian Red breed as well as 255, 319 and 284 1st, 2nd and 3rd lactation Estonian Holstein cows, respectively. The milk SCC heritabilities in the 1st, 2nd and 3rd lactation of the daughters of Estonian Red were 0.10; 0.14; 0.18 and those of Estonian Holstein bulls 0.09; 0.13; 0.16, respectively. Since the heritabilities were not high we decided to investigate the impact of milking procedures on milk SCC. Monitoring of the working operations of milkers, was carried out. The duration of each element of the working process was recorded. The milk SCC was essentially associated with adequate pre-milking cow preparation, delay in application of the milking unit, and over-milking ($P < 0.001$).

Session 21 Poster 18

Bacterial counts of common environmental mastitis pathogens in relation to bedding materials

V. Krömker[1], N. Schwarzer[2], E. Moors[2] and M. Gauly[2], [1]University of Applied Sciences and Arts, Heisterbergallee 12, 30453 Hannover, Germany, [2]Institute of Animal Breeding and Genetics, Albrecht Thaer Weg 3, 37075 Goettingen, Germany

The aim of the study was to compare under in-vitro conditions different bedding materials (25 g/material) in the bacterial counts of common environmental mastitis pathogens after the contamination with milk (1 ml) or liquid manure (1.5 ml). The bedding materials used were: 1. saw dust, 2. saw dust + 10 % quick lime, 3. chaffed wheat straw, 4. chaffed wheat straw + 10 % quick lime and 5. sand. The experiments were five times repeated. The following parameters were measured on day 0, 1, 2, 4 and 7 after contamination: total bacterial count, no. of *Enterococcus spp.*, *Streptococcus uberis*, coliform bacterias and pH-value. Significantly highest bacterial counts were found in saw dust with an average pH-value of 6.26. The addition of 10 % quick lime did significantly increase the pH to an average of 12.25 and decrease bacterial counts to the lowest level of all beddings. Wheat straw (pH = 6.07) did not differ significantly in total bacterial counts from sand (pH = 6.41). However both beddings showed lower counts when compared with pure saw dust. Bacterial counts were not reduced in wheat straw when 10 % quick lime was added. But the average pH-value was significantly higher (9.74).

Session 21 Poster 19

Isolation of *Escherichia coli* O 157:H7 from dairy raw milk samples

E. Beličková, B. Holečová, J. Buleca, A. Ondrejková, R. Ondrejka, Z. Beníšek and M. Prokeš, University of Veterinary Medicine, Komenského 73, 041 81 Košice, Slovakia (Slovak Republic)

Because of an increasing occurrence of alimentary diseases that are caused also by pathogen *E. coli* O 157:H7 we aimed at the monitoring of its presence in one of the basic foods of animal origin for human nutrition–in raw cows milk. Monitoring was carried out in East-Slovak region, where in 8 districts we have collected 196 samples of cow's milk from various dairy cows. The method of immuno-magnetic separation has been used to defect *E. coli* O 157:H7. Out of total number of examined milk samples strains *E. coli* O 157:H7 were isolated only in two dairy cows in Trebišov district. Although complex evaluation revealed only low incidence (1.02%) in respect of health safety of the pathogen it is necessary to continually and in territorial range to observe its occurrence also in basic raw material raw milk as early as in the stage of primary milk production, to determine source and way of contamination and not only propose but also realize effective measures for its eradication with subsequent checking of their effectiveness.

Session 21 Poster 20

Effect of microorganisms on D-amino acid content of milk

G. Pohn[1], C.S. Albert[2], S.Z. Salamon[2], Z.S. Csapó[1] and J. Csapó[1,2], [1]University of Kaposvár, Faculty of Animal Science, Chemistry-Biochemistry, Kaposvár, Guba S. u. 40., H-7400, Hungary, [2]Sapientis Hungarian University of Transylvania, Food Science, Csíkszereda, Szabadság tér 1., RO-4100, Romania

It was established that certain microbe species causing mastitis (*Streptococcus dysgalactiae*, *Streptococcus uberis*, *Escherichia coli*, *Staphylococcus aureus*, *Pasteurella multocida*, *Corynebacterium bovis*, *Arcanobacter pyogenes* and *Pseudomonas aeruginosa*) contributed to D-aspartic acid, D-glutamic acid and D-alanine contents of milk to a different extent, however, examination of amino acids was only partially suitable for identification of pathogen microbe species causing mastitis. Out of D-amino acids of peptidoglycan D-glutamic acid contents provides the possibility of identifying the microbes. Based on D-aspartic acid contents only Mastitest-negative milk sample and the species *Staphylococcus aureus* can be identified. On the basis of D-alanine contents microbes examined by us with the exception of the species *Escherichia coli*, *Streptococcus aureus* and *Pseudomonas aeruginosa* can be identified. Free amino acid contents of milk derived from mastitic udder with mastitis caused by the individual bacterial species do not differ significantly from each other. The species *Streptococcus uberis* produces the least glycine, for the *Escherichia coli* is typical the very high phenylalanine contents. Milk derived from mastitic udder with mastitis caused by *Pseudomonas aeruginosa* contains the most of free lysine.

Session 21 Poster 21

Effect of total germ number of raw milk on free amino acid and free D-amino acid contents of various dairy products

C.S. Albert[1], S.Z. Salamon[1], G. Pohn[2], Z.S. Csapó[2] and J. Csapó[1,2], [1]Sapientia Hungarian University of Transylvania, Food Science, Csíkszereda, Szabadság tér 1., RO-4100, Romania, [2]University of Kaposvár, Faculty of Animal Science, Chemical-Biochemical, Kaposvár, Guba S. u. 40., H-7400, Hungary

In the course of our researches we have examined free amino acid and free D-amino acid contents of milk samples with different germ numbers and composition of dairy products produced from them. Total germ number of milk samples examined varied from $1.25 \cdot 10^6$ to $2.95 \cdot 10^6$. It was established that with an increase in germ number concentration of both free D-amino acids and free L-amino acids increased, however, increase in D-amino acid contents was bigger considering its proportion. There was a particularly significant growth in the germ number range of $1.5 \cdot 10^6$ to $2.9 \cdot 10^6$. In the course of analysis of curds and cheese samples produced using different technologies we have come to the conclusion that for fresh dairy products and for those matured over a short time there was a close relation between total germ number and free D-amino acid and free L-amino acid contents, ratio of the enantiomers was not affected by the total germ number, however. For dairy products, however, where amino acid production capability of the microbial cultures considerably exceeds production of microorganisms originally present in the milk raw material, free amino acid contents of the milk product (both D- and L-enantiomers) seem to be independent of the composition of milk raw material.

Session 21 Poster 22

Sustainable cattle breeding supported by health reports

C. Egger-Danner[1], B. Fuerst-Waltl[2], R. Janacek[3], M. Mayerhofer[1], W. Obritzhauser[4], F. Reith[5], F. Tiefenthaller[6], A. Wagner[1], P. Winter[7], M. Wöckinger[6], K. Wurm[8] and K. Zottl[9], [1]ZAR-ZuchtData, Dresdnerstr.89, 1200 Vienna, Austria, [2]Univ. Nat. Res. and Appl. Life Sci., Gregor Mendel-Str.33, 1180 Vienna, Austria, [3]TGD, Landhauspl.1, 3109 St. Pölten, Austria, [4]ÖTK, Biberstr.22, 1010 Vienna, Austria, [5]LKV Stmk, Am Tieberhof 6, 8200 Gleisdorf, Austria, [6]LWK OÖ, A. d. Gugl 3, 4021 Linz, Austria, [7]Univ. Vet. Med., Veterinärpl.1, 1210 Vienna, Austria, [8]LWK Stmk, Hamerlingg.3, 8010 Graz, Austria, [9]LKV NÖ, P. W. Deibl-Str.4, 3910 Zwettl, Austria

Increasing herd sizes and pressure on producer prices result in high demands for successful herd management. To support management decisions of cattle breeders and herd health management of their veterinarians, health reports are provided for the first time within a health monitoring project in Austria. The health report will be available approx. 10 times a year. It combines performance recording data and data from health monitoring to enable early detection of health problems and their therapy. The report is divided into two sections. First, an overview is provided including relevant information in the fields Fertility, Udder health, Metabolism and Digestive tract, Feet and Legs and Miscellaneous (e.g. Disposals). Subsequently, individual results are shown within each listed field. Generally, first diagnoses of three preceding months are taken into account. It is planned to include claw trimming results and to elaborate an annual report.

Session 21 Poster 23

Isolation of bacterial pathogens from teat duct and milk samples of ewes following machine-milking

I.A. Skoufos, A.S. Tzora, G.E. Maglaras, D.V. Vassos, K.G. Fotou and C.G. Voidarou, Technological Educational Institute of Epirus, Animal Production, Kostakioi, Artas, 47100, Arta, Greece

The objective of this investigation was to determine whether there were differences in bacterial populations in the teat duct and milk of ewes before and after machine-milking. In total, 308 paired-samples of teat duct material and milk, were collected from three flocks of 30 Friesarta-breed ewes before and 50-70 min. after machine-milking. Samples were processed bacteriologically. We compared changes in bacterial isolation following milking, for duct and milk samples. Bacteria were isolated from 18 (6%) duct and 19 (6%) milk samples collected before the milking procedure; respective figures after it, were 81 (26%) and 33 (11%). In 77 (25%) cases, bacteriological findings in the two duct samples of each pair were different; in 7 cases bacteria were isolated only before, whilst in 70 cases bacteria were isolated only after milking ($P < 0.005$); respective results for milk samples were 26 (8%): 6 and 20 cases (P=0.693). The majority of bacterial isolates were staphylococci, accounting for 63% of 99 isolates. Neither changes in milk, nor mammary abnormalities were detected in ewes sampled during the study. The milking procedure predisposes to entrance of bacteria into the teat duct; however, increased bacterial isolation from the teat did not result to increased mammary infections. The present study found only a very small and non-significant increased risk of intramammary infection associated with machine-milking.

Session 21 Poster 24

Malabsorption of vitamin A in calves during *Cryptosporidium parvum* – infection

P. Klein, Institute of Animal Science, Department of Nutrition, Pratelstvi 815, CZ-10401, Praha, Czech Republic

Cryptosporidiosis followed by impairment of intestinal functions and diarrhea is recognized as one of the major infections in calves. Infection may affect also absorption of other substances such as vitamins. The aim of this study was to determine the effects of cryptosporidiosis on absorption of vitamin A (VA) during the clinical course of disease. The tolerance tests with 7,500 IU of VA.kg^{-1} (as retinyl-palmitate in vegetable oil) were performed in 10 newborn calves experimentally infected by 10^7 of *C. parvum* oocysts. Nine uninfected animals of the same age served as control. The tests were performed in 1-week intervals from day of challenge up to the recovery of animals. During the tests gain of plasmatic concentration of VA was measured for 8 hrs in 1-hr-intervals, the absorption curve was constructed, and $AUC_{(0-8)}$ was calculated for each animal. The failure of intestinal absorption in infected calves was observed in day 7 *p.i.* (AUC: 140.7±20.6 *vs* 230.3±26.5 mg.min.L^{-1} in control; $p < 0.001$). In contrast, the differences in day 14 *p.i.* were non-significant and the results obtained in day 21 *p.i.* suggest on full recovery of the infected intestine. The study confirmed that cryptosporidiosis highly affects intestinal absorption of VA in calves, namelly in the acute stadium of disease followed by diarrhea and oocysts output. Injection application of VA thus should be a necessary part of care about sick calves (Project MZe-0002701403).

Session 21 Poster 25

Colour of mucous membrane as an indicator of endoparasite infections in sheep

E. Moors, C. Fasshauer and M. Gauly, Institute of Animal Breeding and Genetics, Albrecht Thaer Weg 3, 37075 Goettingen, Germany

In relation to the pathology of endoparasites different indirect indicators can be used to quantify the total worm burden. In case of parasites, which lead to blood looses (e.g. *Haemonchus contortus*) packed cell volume (PCV) seems to be an useful indicator. Based on that methods were developed earlier to score the colour of mucous membranes (e.g. FAMACHA) as a sign of anemia. The aim of the present study was to examine the correlation between PCV, FAMACHA score, colour saturation (Chroma) and colour angle (Hue-angle, HA, Minolta Chroma Meter CR-200b). The parameters were taken from German Black Head Mutton (GB, n=118) and Leine sheep (LE, n=114) at an age of five months. Furthermore individual faecal samples were collected at that time. There were no significant differences between the two breeds in PCV (p=0.768) and log FEC (p=0.934). FAMACHA score was significant higher in GB than in LE lambs (p=0.008). Chroma and log HA were significant higher in LE lambs compared to the GB animals ($p < 0.05$). There were significant ($p < 0.05$) phenotypic correlation coefficients (r_p) between PCV and colour traits (FAMACHA, Chroma and HA) with higher rp in the GB lambs. It is suggested that colour measurements of mucous membranes have to be adapted to different sheeep breeds. Irrespective of the breed the FAMACHA system is more adequate to detect animals with a lower PCV (r_p $p < 0.01$), whereas Chroma and HA are more adequate when PCV is higher (r_p $p < 0.001$).

Session 21 Poster 26

The role of preparation technique, culture media and incubation time for embryonation of *Heterakis gallinarum* eggs

U. Püllen, S. Cheat, E. Moors and M. Gauly, Institute of Animal Breeding and Genetics, Albrecht Thaer Weg 3, 37075 Goettingen, Germany

In a first experiment intact female *H. gallinarum* worms were kept for development of the unembryonated eggs in four different media and incubated under constant temperature (20 – 22°C) for 2, 4, 6 or 8 weeks. Afterwards the body wall of the worms were ruptured and the number of embryonated and unembryonated eggs were determined. Based on that the percentage of embryonated eggs were calculated. After 8 weeks of incubation 27.6 % of the eggs cultivated in 0.5 % formalin developed into third larval stage. For worms cultivated in 0.1 N sulphuric acid, 26.7 % and in 0.1% potassium dichromate culture 29.4 % developed, respectively ($p > 0.05$). In contrast only 18.6 % cultivated in 2 % formalin were embryonated ($p < 0.05$). In a second experiment *H. gallinarum* eggs were harvested directly from worms uteri and cultivated afterwards in different media (2 % formalin, 0.1 N sulphuric acid, 0.1% potassium dichromate) at 20 to 22° C for 6 weeks. Compared with the results of experiment one after 6 weeks of incubation a significantly ($p < 0.05$) higher percentage of eggs kept in 2.0 % formalin and 0.1% potassium dichromate media developed into the third stage larvae, whereas no significant differences were found for 0.1 N sulphuric acid at that time. The results suggest that preparation technique, media and time of incubation has an essential influence on the development rate of *H. gallinarum* eggs.

Session 21 Poster 27

Effect of weaning strategy on immune function of beef suckler cows and their calves

E. Lynch[1,2], M. Mc Gee[1], S. Doyle[2] and B. Earley[1], [1]Teagasc, Grange Beef Research Centre, Dunsany, Co. Meath, Ireland, [2]NUI, Maynooth, Co. Kildare, Ireland

The effect of weaning strategy on immune function was determined using 36 previously grazed, spring-calving, beef × Holstein-Friesian cows and their calves. Following abrupt weaning, calves (276, s.d. 37.0 kg) were either 1) housed (WH) immediately in a slatted floor shed and offered grass silage *ad libitum* and concentrates or 2) remained at pasture (WP). Post-weaning, all cows remained at pasture (separate from the calves). Calves from WP and cows were housed 35 days later. Blood samples were obtained from the cows and calves at weaning (day 0) and days (d) 2, 7, 14, 21, 28, 35, 37, 42, 48 and 56. At d2 post-weaning, there was no change ($P > 0.05$) in white blood cell (WBC) number in WH calves while WBC number increased ($P \leq 0.05$) in WP calves. Neutrophil percentage (%) was unchanged ($P > 0.05$) in WH and WP calves between weaning and d 2. By d 7, WBC number and neutrophil % increased ($P \leq 0.05$) in WP whereas in WH calves neutrophil % remained unchanged ($P > 0.05$) and WBC number decreased ($P < 0.01$). By 2 days post-housing (d 37), neutrophil % increased ($P < 0.05$) and subsequently decreased by d 42 ($P=0.06$) in WP calves while WBC number decreased ($P < 0.01$) by d 42. In cows, WBC number ($P=0.06$) and neutrophil ($P < 0.001$) % increased by d 2 and decreased ($P < 0.05$ and $P < 0.001$, respectively) by d 7. Post-housing, neutrophil % decreased ($P < 0.001$) by d 7 (d 42) with no change ($P > 0.05$) in WBC number. In conclusion, weaning and housing altered immune function in cows and calves.

Session 21 Poster 28

Effect of residual feed intake on immune function of beef suckler cows at weaning and housing

M. Mcgee, B. Earley and M.J. Drennan, Teagasc, Grange Beef Research Centre, Dunsany, Co. Meath, Ireland

The effect of residual feed intake (RFI), an alternative measure of feed efficiency, on immune function at weaning and housing was determined using 36 spring-calving, beef × Holstein-Friesian cows,comprising 2 genotypes (G). Expected grass silage intake (GSI) was calculated for each G separately by regressing average daily GSI, recorded over a 53-day period in late pregnancy, on average daily liveweight gain (ADG) and mid-test liveweight$^{0.75}$. The RFI for each animal was then calculated as actual GSI minus the GSI predicted from the regression analysis. Within each G, the cows were ranked on RFI and 0.5 were classified as having low or high RFI. At pasture, the low and high RFI cows and their calves were rotationally grazed, together. Following abrupt weaning of calves, cows remained at pasture for a further 35 days (d) until housing in a slatted floor shed with *ad libitum* access to grass silage. Blood samples were obtained from the cows at weaning (d 0) and d 2, 7, 14, 21, 28, 35 (housing), 37, 42, 48 and 56. Cow RFI had no effect ($P > 0.05$) on white blood cell number. Neutrophil and lymphocyte % did not differ ($P > 0.05$) pre-weaning or housing but neutrophil % was higher 2 d post-weaning ($P=0.08$) and housing ($P < 0.05$) but not subsequently, and lymphocyte % was lower ($P < 0.05$) at those times but not subsequently, in high than low RFI cows. In conclusion, the results indicate that high RFI cows are more susceptible to stress at weaning and housing than low RFI cows.

Session 21 Poster 29

Influence of group housing floor pens and cage on behaviour of fattening rabbits

S. Silva, J. Mourão, A.L.G. Lourenço, C. Domingues and V. Pinheiro, CECAV-UTAD, Animal Science, PO BOX 1013, 5001-801 Vila Real, Portugal

The rabbit welfare can be enhanced by housed animals in groups and enriched their environment. This study reports the effect of different housing on rabbit's behaviour. A total of 238 New Zealand White rabbits were randomly distributed to 21 pens (10 rabbits/pen) and 7 cages (4 rabbits/cage). The animals were kept at an equivalent density and were assigned to one of the four following treatments: Tl, T2 and T3- for pens with sawdust, with 50 % sawdust and 50% wire net, and with wire net floor, respectively; T4- for wired cages. A pelleted dict was offered *ad libitum*. Lights were on from 08:00 to 20:00 h. Room temperature and relative humidity were 23 ± 2°C and 65%, respectively. Observations, for each pens and cages were observed with a video camera during sequences of 2 min each starting at 06:00, 08:00, 12:00, 16:00, 20:00, 24:00, 04:00 h to cover 24 h behavioural time budget. Stereotypic, eat/drink, locomotor, maintenance, and resting behaviours were studied. A General Linear Model was used. Sawdust penned rabbits engaged more time (from 20 to 40%) in locomotor behaviour ($P < 0.05$) than caged rabbits. Caged rabbits showed more stereotypic behaviours ($P < 0.05$) (digging, floor chewing, bar biting), than rabbits on pens. There are no housing effect ($P > 0.05$) on resting, maintenance and eat/drink behaviours. The results suggest that sawdust is the most effective material floor in reducing abnormal behaviour and improving the locomotor behaviour of rabbits.

Session 21 Poster 30

Contens of phthalic acid esters in feeding additives and their packages

J. Harazim[1], A. Jarosova[2], L. Kratka[2], D. Kolencikova[2] and P. Suchy[3], [1]Central Institute for Supervising and Testing, Hroznova, Brno, Czech Republic, [2]Mendel University of Agriculture, Zemedelska, Brno, Czech Republic, [3]University of Veterinary, Palackeho, Brno, Czech Republic

Samples of feeding additives (n=15) and others feeds (n=3) and their packages (n=18) were taken from the industrial producers in the Czech Republic in accordance with the National Law of Feeding Stuffs. Packages of feeds consist of plastic material only, or plastic material and paper or aluminium (n=15) and paper and aluminium or aluminium only (n=3). The concentrations of phthalic acid esters as di-n-butyl phthalate (DBP) and as di-2-ethylhexyl phthalate (DEHP) were measured in samples of feeds and their packages. The highest/lowest concentration as a sum of DBP and DEHP of feed was detected in additive Niacin Amid (6,01 mg. kg^{-1}) / in additive Vitamin A – No. 2 (0,11 mg. kg^{-1}). The average content as a sum of DBP and DEHP was 1,80 mg. kg^{-1}. The highest/lowest concentration as a sum of DBP and DEHP of packages was detected in additive Vitamin E – No. 1 (526,80 mg. kg^{-1}) / in additive Vitamin E No. 3 (6,21 mg. kg^{-1}). The average content as a sum of DBP and DEHP was 73,06 mg. kg^{-1}. There was no evident relationship between the amount of DBP and DEHP and between feed and packages ($P < 0,05$). The work was supported by the Czech National Agency for Agricultural Research. Project No.: QG60066/2005.

Session 21 Poster 31

Applicability of gonadotropin hormones on regulation gonodal functions in pigs

A. Siukscius and R. Nainiene, Institute of Animal Science of LVA, Department of Animal Reproduction, R.Zebenkos 12, Baisogala, LT-82317, Radviliskis distr., Lithuania

Various new preparations are being developed throughout the world for the replacement of the PMS in the synchronisation scheme for pig oestrus and ovulation. The follicle stimulating preparation folitropin (FSH) that has a similar effect to that of PMS. The replacement of blood serum of pregnant mares by a single injection of folitropin (FSH) (300 units) in the scheme of oestrus and ovulation synchronisation indicated that in 65 hours after the administration of the preparations for ovulation synchronisation, the level of follicle ovulation amounted to only 53.3%. Also, the conception rate in this group amounted to only 53.3% and the litter size was by 27.1% lower compared with the gilts administered PMS ($P > 0.5$). The replacement of PMS by twice injection of folitropin (FSH) of 300 units following every 12 hours indicated that the concentration of estradiol 17b amounted to 28.5±1.48 pmol/l of blood plasma and was by 1.9 times higher in comparison with the PMS treated gilts ($P < 0.001$). The conception rate and fertility data were also similar. Therefore, it can be concluded that twice injection of FSH is applicable for estrus stimulation. A single injection of FSH for oestrus stimulation resulted in a 24.5% decrease of the follicle number in the ovaries of a pig and the number of ovulated follicles was by 21.5% lower in comparison with the control PMS treated group of pigs ($P < 0.05$).

Session 22 Theatre 1

Maternal nutrition from day 30 to day 80 of pregnancy in singleton bearing ewes increases the lamb birth weight

M. Kuran, U. Sen, E. Sirin, Y. Aksoy, K. Kilinc and Z. Ulutas, Gaziosmanpasa University, Department of Animal Science, 60250, Tokat, Turkey

A total of 31 Karayaka breed mature ewes were allocated randomly into three feeding groups to investigate whether undernutrition or overnutrition from day 30 to day 80 of pregnancy in singleton bearing ewes has any effect on the birth weight of the lambs. Ewes in control group (C; n=9) were offered a diet to meet their daily requirement while ewes in overnutrition group (ON; n=7) consumed the quantity of 175% and those in undernutrition group (UN; n=15) consumed the quantity of 50% of their daily requirements. Live weights of the ewes changed significantly in all treatment groups during the feeding period ($P < 0.01$). Maternal plasma ammonia concentrations were higher ($P < 0.05$) in ewes of ON group than those in C and UN groups at one hour after the feeding. ON increased lamb birth weights significantly ($P < 0.05$). Lamb birth weights were 4.61±0.42, 3.92±0.04 and 3.87±0.13 kg for ewes in ON, C and UN groups, respectively. There were no significant differences between treatment groups in terms of maternal plasma glucose and blood urea nitrogen concentrations, gestation length, the total placenta weights and the numbers and weights of placentomes. These results indicate that maternal nutrition during mid-pregnancy in singleton bearing ewes can affect the fetal growth which may result in altered birth weights of the offsprings. This study was financed by TUBITAK (TBAG-U/148).

Leptin, insulin-like growth factor-I and luteinizing hormone secretion and oestrous behaviour in fat-tailed Tuj lambs following ovulation induction using progestagen sponges plus PMSG in non breeding season

S. Yildiz[1], O. Ucar[1], M. Cenesiz[1], M. Kaya[2], F. Onder[1], M. Uzun[1], D. Blache[3], M. Blackberry[3] and G.B. Martin[3], [1]Kafkas University, Faculty of Veterinary Medicine, Department of Physiology, Pasacayiri Kampusu, Kars, Turkey, [2]Kafkas University, Faculty of Veterinary Medicine, Department of Reproduction and Artificial Insemination, Pasacayiri Kampusu, Kars, Turkey, [3]University of Western Australia, Animal Biology, Crawley, Perth, Australia

Aim of this study was to evaluate the effects of progestagen sponges and PMSG on the characteristics of LH surge and oestrous behaviour in fat-tailed Tuj ewe-lambs during non-breeding season and to correlate these findings with plasma leptin and insulin-like growth factor-I. Progestagen sponges were inserted into and left in the vagina for 14 days (n=36). For determination of LH surge blood samples were collected with 2 h intervals for 96 h (sponge removal + PMSG injection at 0 h). Oestrous behaviour was also assessed visually at 2 h intervals for 96 h by leaving rams in the pens. Mean bodyweight and body condition scores were 32.3±0.6 kg and 2.84±0.08 units (5 point-scale), respectively. Mean plasma leptin and IGF-I concentrations were 0.90±0.04 ng/ml and 20.4±1.2 ng/ml, respectively. Oestrus or LH surges were succesfully induced in all Tuj ewe-lambs during the first year of their life.

Effect of dietary polyunsaturated fatty acids on the expression of genes involved in prostaglandin biosynthesis in the bovine uterus

S. Waters[1], S. Childs[1] and D. Kenny[2], [1]Teagasc, Athenry, Co. Galway, Ireland, [2]UCD, Dublin, 4, Ireland

Nutrition plays a critical role in regulating cow fertility and there is emerging evidence that dietary omega-3 polyunsaturated fatty acids (w-3 PUFA) may act as potent regulators of the reproductive process. w-3 PUFAs, eicosapentaenoic acid (EPA) and docosahexaenoic acid (DHA) have been shown to play pivotal roles in the suppression of uterine derived prostaglandin $F2_a$ (PGF_{2a}) synthesis, though the biochemical mechanisms are unclear. Beef heifers were fed a high or low w-3 PUFA diet, generating, in turn, combined uterine endometrial concentrations of EPA and DHA which were more than two-fold higher, and EPA concentrations alone that were more than fourfold higher in the high and low supplemented animals, respectively. Total RNA was isolated from endometrial tissue and real-time RT PCR was carried out to measure the relative expression of 10 genes known to be involved in the prostaglandin biosynthetic pathway. Expression of mRNA for prostaglandin E synthase (PGES) and the peroxisome proliferator-activated receptors, PPAR a and d was increased in animals fed the high compared with low PUFA diet. mRNA expression of phospholipase A_2 (PLA_2) tended to be down-regulated. In conclusion, differential regulation of bovine endometrial gene expression in response to w-3 PUFA supplementation suggests a possible mechanism by which PUFAs may influence uterine function and in turn embryo survival.

Investigation on alteration in and relationship between thyroid hormones and testosterone in ram plasma during and out of the breeding season

A. Alaw[1], A. Seidavi[1], R. Al-Rekabi[2] and H. Deldar[3], [1]Islamic Azad University, Rasht Branch, Animal Science, Rasht, 41857, Iran, [2]Baghdad University, Biology, Baghdad, 41667, Iraq, [3]Tehran University, Animal Scince, Karaj, 31485, Iran

This experiment was conducted to investigate the effects thyroid hormones on reproductive performance in Varamin rams and determine the relationship between thyroid hormones and testosterone during and out of the breeding season. Blood samples were collected weekly during two seasons. Thyroid hormones concentration in plasma was measured by RIA. It was showed that month had significant effect on testosterone and thyroid hormones ($P \leq 0.05$). Plasma testosterone was lowest and highest in October (breeding season) and June (non-breeding season), respectively. Thyroxine (T4) concentration declined at the beginning of the breeding season and was lowest in the breeding season (17.2 ng/ml). The highest plasma level of T3 was observed in June (non-breeding season) and December (end of breeding season). Season had significant effect on testosterone and thyroid hormones concentration ($P \leq 0.05$). There was a negative and significant correlation between testosterone and thyroid hormones in the trial period ($P \leq 0.05$). Correlation between thyroid hormones (T3 and T4) concentration was significant ($P \leq 0.05$). Overall, it was concluded that factors decreasing thyroid hormones concentration will lead to increased testosterone concentration.

Session 22 Theatre 5

A simple modelling approach of regulations in energy partitioning for the lactating female: application to the dairy goat

L. Puillet, O. Martin, M. Tichit and D. Sauvant, INRA, PHASE & SAD, 16 rue Claude Bernard, 75231 Paris cedex 5, France

Predicting kinetics of production, intake and live weight changes in ruminant is controlled by homeorhetic regulations (HR). Aimed at investigating the effect of long term regulations on energy partitioning, a dynamic model of a lactating dairy goat including simple HR principles was developed. The model predicts the dynamics of raw milk yield (RMY), dry matter intake (DMI), body weight (BW) and reserves anabolism/catabolism through the whole animal productive life. Production potential and weight at maturity are driving inputs. The model incorporates three sub-systems. The reproduction events sub-system coordinates the major events during goat's life (service, kidding and drying off). The regulating sub-system controls dynamics of body reserves (mobilisation and reconstitution) and milk production through the pregnancy/lactation cycles with a simple four-compartment structure. The operating sub-system represents the major flows of energy through the goat: maintenance, growth, pregnancy, mobilisation, reconstitution, RMY and DMI. External validation indicates that simulated kinetics of milk production and composition are fairly close to the mean data of the French Milk Control (FMC). However, some kinetics extracted after a cluster analysis of FMC cannot be well simulated. Results are discussed in the perspective of integrating this animal model into a herd simulator.

IGF-1 expression in leukocytes at two ages in growing cattle

Z. Saprõkina and A. Karus, Institute of Veterinary Medicine and Animal Sciences / Estonian University of Life Sciences, Chemistry Union, Kreutzwaldi 64, EE51014, Tartu, Estonia

It has been shown that the total increase in serum IGF-1 level during the first postnatal year can be positively associated with milk protein and milk fat production in first lactation. This suggests that measures of serum IGF-1 in the first postnatal year has potential as productivity prognosis in dairy cattle. The aim of the present study was to describe changes in blood serum IGF-1 levels and in the expression of IGF-1 and IGF-1R in leukocytes of 4-6 weeks and 13 months old Holstein heifer calves. IGFBP-3 mRNA was measured only in 13 months old cattle. The mRNA was investigated by one step RT-PCR. Serum IGF-1 was measured by RIA. Changes in expression levels were observed by relative quantification (GAPDH and UBQ were used as housekeeping genes). Results of our experiment show that leukocytes may contribute to blood serum IGF-1 content. Our study confirms the responsiveness of IGF-1 in blood to age-changes in Holstein female calves. Inter-age association was observed between IGF-1 expression and serum IGF-1. In 4-6 weeks old calves serum IGF-1 concentration was lower than around puberty.

Growth performance of New-Zealand white rabbits fed diets containing different levels of untreated or fungal treated sugar beet pulp

A.Y. El-Badawi[1], A.A. Hassan[2] and A.A. Abedo[1], [1]National Research Center, Animal Production, Tahrir st., Dokki, Giza, 12622, Egypt, [2]Animal Prodution Research Institute, Animal Prodution, Dokki, 12622, Egypt

Fifty New-Zealand White (NZW) growing rabbits aged eight weeks and weighing 877±34 g were randomly divided into five similar groups (10 animals each). Untreated sugar beet pulp (USBP) or fungal treated sugar beet pulp (TSBP) with *Trichoderma reesei* were introduced in rabbit's diets at 25 or 50 % in replacement of the traditional concentrate feed mixture (CFM). All experimental diets were manipulated to be iso-caloritic and iso-nitrogenous. The results showed that the rabbits fed either 25 or 50 % TSBP were significantly ($P < 0.05$) better than those fed control and 25 or 50 % USBP in daily weight gain, nutrients digestibility and dietary nitrogen utilization. Results of carcass characteristics showed higher dressing percentage and significantly ($P < 0.05$) higher yield of edible giblets for rabbits fed 25 and 50 % TSBP. Chemical composition of lean meat showed more ($P < 0.05$) DM and ash content and lower fat content for rabbits fed either USBP or TSBP compared with the control group. However, no significant difference was detected for the protein content among groups.

Session 22 — Theatre 8

Effect of feeding biologically treated corn stalks on plasma prolactin levels in growing rabbits

A. El Shahat and A. Morad, NRC, Animal Production, Cairo, 12622 Dokki, Egypt

The experimental animals of five weeks old were randomly distributed into three nutritional groups. The animals of the first group (control) were given a commercial diet (e.g. diets without supplementation) whereas those of the second and third groups were fed on rations, with either 10% or 30% biologically treated corn stalks. At the end of the experimental period which lasted for twelve weeks, blood samples were collected in heparinized tubes to estimate plasma prolactin concentration levels in growing rabbits. Enzyme Immunoassay technique (ELISA) was used to measure levels of prolactin in plasma. It was generally observed that the mean values of plasma prolactin levels of the animals of the first group (control) were highest; 52.90 ng/ml, followed by those of the second group (10%); 46.66 ng/ml, and then those of the third group (30%); 45.70 ng/ml. However, the statistical analysis revealed that the differences were non-significant. On the basis of the present results, it could be stated that the biological treatments may decrease plasma prolactin levels in growing rabbits and that feeding the biologically treated low quality roughages can be used successfully and safely without adversely affecting physiological state, growth performance and feed intake of the rabbits.

Session 22 — Poster 9

Evaluation of diet effect on blood redox status and on milk quality and production in lactating cows

R. De Rosa[1], L. Ferrara[1], P. Abrescia[2], A. Carbone[2], R. Baculo[1] and M.S. Spagnuolo[1], [1]ISPAAM-CNR, via Argine 1085, 80147 Napoli, Italy, [2]Università Federico II, via Mezzocannone 8, 80134 Napoli, Italy

Reactive oxygen species (ROS) can cause tissue oxidative stress. During lactation, the increase in metabolic processes is associated with increased ROS production. Retinol (Ret) and α-tocopherol (Toc), major liposoluble antioxidants, scavenge free radicals. Milk is the most important source, for calves, of these antioxidants, that also save food quality and prevent lipid oxidation. Our aim was to check whether diet affects blood and milk redox status and milk quality and production. Twenty Holstein Friesian cows were divided into two groups (A and B). The group A was fed a diet with a higher amount of starch and organic nitrogen than group B. Milk and plasma samples were collected weekly and analysed for Ret, Toc and Nitro-tyrosine (N-tyr) concentrations. Milk yield was recorded. Milk content of lactose and fat, and somatic cells concentration (SCC) were measured. Plasma levels of Ret and N-tyr did not differ between the groups. Milk and plasma levels of Toc were higher ($P < 0.01$) in group B. Milk content of Ret did not differ between the groups, while milk yield, and lactose and fat amount were higher ($P < 0.01$) in group A. SCC was higher ($P=0.03$) in milk from group B. Our results suggest that the two diets do not influence plasma concentration of Ret and N-tyr. Moreover diet A seems to positively affect milk production and quality by improving its organoleptic properties and safety.

Session 22 Poster 10

Gene expression of the constitutive nitric oxide synthase (NOS-3) in bovine vas deferens

H. Hassanpour[1], G.H. Niknakht Brojeni[2] and A. Mohammad Zade[3], [1]Veterinary Medicine Faculty, Sharekord University, Basic Sciences, Sharekord, 15, Iran, [2]Veterinary Medicine Faculty, Tehran University, Immunology, Tehran, 22, Iran, [3]Veterinary Medicine Faculty, Azad University, Garmsar Branch, Basic Sciences, Garmsar, 46, Iran

The nitric oxide synthases (NOSs) are a group of enzymes that catalyse the nicotinamide adenine dinucleotide phosphate (NADPH)-dependent oxidation of L-arginine to nitric oxide (NO) and L-citrulline. To date three distinict isoforms of NOS have been purified and characterized: NOS-1, NOS-2 and NOS-3. NO is an important cellular messenger molecule that is implicated in a wide range of physiological and pathophysiological actions in the reproductive system. In this study, vas deferens from Holstein adult bulls was prepared in the abattoir and transported in liquid nitrogen to the laboratory. Total RNA of vas deferens tissues was extracted by single stage method of acid ganidinium-thiocyanate-phenol-chloroform. Then the gene expression of nitric oxide synthase type 3 was evaluated by use of RT-PCR technique. This investigation showed that the gene of NOS-3 is activated in vas deferens and its mRNA is found in this site of the bovine reproductive system, and this could be related to the synthesis and physiological function of nitric oxide in vas deferens.

Session 22 Poster 11

Dairy cow energy status in early pregnancy does not affect the reproductive performance of primiparous female progeny

D.P. Berry[1], P. Lonergan[2], S.T. Butler[1] and A.C.O. Evans[2], [1]Moorepark Dairy Production Research Center, Fermoy, Co. Cork, Ireland, [2]School of Agriculture, Food Science & Veterinary Medicine, University College Dublin, Belfield, Dublin 4, Ireland

The concept of fetal origins of adult disease suggests that environmental influences early in fetal life are reflected in impaired growth and development, leading to increased risks of disease in adulthood. The objective was to investigate if parity or metabolic status of the dam in early pregnancy (approximated by average fat to protein ratio, net energy of milk output, and milk fat and protein composition in the first 6 weeks and 6 to 12 weeks post-conception) affected the subsequent reproductive performance of female offspring (measured by age at first calving and calving interval from first to second calving). Test-day milk production records and calving dates were available for Holstein-Friesian dairy cows calving between 1995 and 2005. Data, consisting of up to 22,237 dam-offspring pairs, were analysed using mixed models with sire and maternal grandsire of the offspring included as random effects. Fixed effects in the model pertaining to the offspring were herd-year of birth, day of the year at birth, and Holstein percent while fixed effects pertaining to the dam were parity, age at calving nested within parity and indicators of metabolic status. Neither parity of the dam nor milk output and composition in early pregnancy significantly affected reproductive performance of primiparous female progeny.

Session 22 Poster 12

Different changes in postprandial plasma ghrelin and GH levels in wethers fed concentrate or timothy hay

T. Takahashi, Y. Kobayashi, S. Hasegawa, K. Katoh and Y. Obara, Graduate School of Agricultural Science, Tohoku University, Department of Animal Physiology, Amemiyamachi, Aoba-ku, Sendai, 981-8555, Japan

Effects of feed compositions on the somatotropic hormones are poorly understood in the ruminant. The aim of the present study was to investigate whether changes in feed composition influence postprandial plasma concentrations of ghrelin and growth hormone (GH) in wethers. After 4 animals were fed concentrate (Group C) or timothy hay (Group R) for 14 d, blood sampling was performed to analyze the hormonal levels. The circadian changes in plasma ghrelin levels were close to those in GH levels in Group C, while this was not the case in Group R. In addition, the basal ghrelin levels in both groups were rapidly and significantly ($P < 0.05$, paired *t*-test) decreased after eating although the change in Group C was greater than that in Group R. However, there was no significant change in AUC or incremental area for ghrelin. The basal GH levels were also rapidly and significantly decreased after feeding in both groups. The decrease in GH levels in Group C was significantly greater than that in Group R, resulting in a significant difference between the two groups for AUC of GH. Plasma insulin levels in both groups increased after eating although the change in Group C was greater than that in Group R. In conclusion, feed composition significantly changes postprandial plasma levels of ghrelin and GH in wethers although the mechanism remains to be clarified.

Session 22 Poster 13

Effect of maternal undernutrition on the hypothalamic-pituitary-gonadal axis function in sheep offspring

G. Papadomichelakis, B. Kotsampasi, S. Chadio, S. Deligeorgis, D. Kalogiannis, I. Menegatos and G. Zervas, Agricultural University of Athens, Animal Science, 75 Iera Odos Str, 11855, Athens, Greece

Fetal programming hypothesis implies that an insult acting during embryonic life may lead to permanent alterations, including reproductive system development that could affect reproductive potential in adulthood. Thus, a study was conducted to examine the effects of maternal nutrient restriction during gestation on the hypothalamo-pituitary-gonadal axis function in sheep at different ages postnatal. Pregnant ewes were fed a 50% nutrient restricted diet from day 0-30 (R1, n=7), or from day 31-100 of gestation (R2, n=7) or a control diet (C, n=8). After birth the onset of endocrine puberty was determined in female offspring. At 2, 5.5 and 10 months of age lambs were challenged with gonadotrophin releasing hormone. At 10 months lambs were slaughtered and testes and ovaries were collected. Birth weight was not affected by maternal undernutrition. At the age of 10 months a higher ($P < 0.05$) FSH response was observed in R1 female and R2 male lambs. A higher ($P < 0.05$) number of small (2-3mm) follicles was observed in R1 and a lower ($P < 0.05$) number of large (8-11mm) corpora lutea in R2 lambs. In conclusion maternal undernutrition can affect differentially hypothalamo-pituitary axis and gonadal function not only with respect to the time and the duration that it is imposed, but also in a sex specific manner.

Session 22 Poster 14

Estral and ovulatory responses following short-term protocols for induction/synchronization of estrus in Altamurana ewes

G. Martemucci, A.G. D'Alessandro and N. Paradiso, University of Bari, PRO.GE.S.A., Via Amendola 165/A, 70126 Bari, Italy

In autumn, four short-induction/synchronization protocols were compared with a traditional protocol of long-term (14 days). Seventy-five multiparous and dry Altamurana breed ewes were subdivided into 5 experimental treatment groups (N=15): A) FGA (Fluorogestone acetate, 40 mg in vaginal sponges, 14d) + PMSG (400 IU at sponge removal), as control; B) FGA (5d) + PGF2α (Cloprostenol, 100 μg, 5th d) + PMSG (200 IU, 5th d); C) PGF2α (100 μg, d0) + FGA (40 mg, 5d) + PMSG (200 IU, 5th d); D) PGF2α (100 μg, d0) + FGA (40 mg, 5d) + GnRH (Fertagyl, 100 μg, 7th d); E) GnRH (100 μg d0) + PGF2α (100 μg, 5th d) + PMSG (200 IU, 5th d). The animals were observed for estrus onset by teaser rams every six hours. Ovulation time was monitored by laparoscopy (N=10). Percentage of ewes in estrus was not influenced by treatments (93.3, 84.6, 78.6, 71.4, 69.2% in A, B, E, C, D groups, respectively). The interval from sponge removal to onset of estrus was lower ($P < 0.05$) in group A (28.3 h) than in groups B (43.6 h) and D (37.3 h). Within the short-term protocols, onset of estrus was significantly earlier in group E compared with group B (39.9 *vs* 43.6 h; $P < 0.05$). No differences were found in ovulation time between conventional protocol (group A, 58.0 ± 4.4 h after sponge removal) and the short-term treatments. We can conclude that the short-term protocols seem efficient to induce/synchronize estrus and ovulation.

Session 22 Poster 15

An update on anestrous synchronization in sheep

H. Solgi[1], M. Reza Zadeh Valejerdi[2], P. Eftekhari Yazdi[1], A. Shahverdi[1], M. Daneshzadeh[1], A. Dalman[1], F. Hasani[1] and V. Siavashi[1], [1]Royan, Embryology lab NT (nucluer tansfer), Theran, 19395-4644, Iran, [2]Tarbat Modares University, Anatomy, Tehran, 14115-111, Iran

The objective of the current study was to determine the effects of hormonal treatments on ovarian follicular development and ovulation rate in anestrous Iranian native sheep. In treatment group 1, twenty ewes were given a controlled internal drug release (CIDR) device (at day -11). After removal of CIDR at day 0 (follicular phase), 750 unit/dose PMSG (pregnant mare serum gonadotrophin) was injected. In treatment group 2, twenty ewes received the CIDR without PMSG as control. Ovaries were studied by laparotomy at day 1. Follicles ≥1 mm were counted and ovulation rate was recorded. In groups 1 and 2, number of visible follicles was different ($P < 0.01$), but ovulation rate was not different ($P > 0.05$). The ovulation rate was 100% and 94% in group 1 and 2, respectively. Thus, CIDR given alone did not affect follicular development, follicle count and uterus quality in anestrous ewes. In addition, CIDR treatment protocols used in ovine IVF (*in vitro* fertilization) programs should be carefully designed to minimize adverse effects on fertilization rates.

Postfasting leptin levels in Awassi, Friesian × Awassi and Friesian × Merino ewes

S. Yildiz[1,2], G.B. Martin[1], R. Bencini[1], M. Blackberry[1], G. Pedrana[1], S. Agra[1] and D. Blache[1], [1]University of Western Australia, Department of Animal Biology, Crawley, Perth, Australia, [2]Present address: Kafkas University, Faculty of Veterinary Medicine, Department of Physiology, Pasacayiri Kampusu, Kars, Turkey

Aim of the current study was to evaluate the effects of fasting on plasma leptin concentrations in three breed groups of sheep. Awassi (n=10), Friesian × Awassi (n=5) and Friesian × Merino (n=5) ewes were fed once daily at 09.00 h. They were fed on the first and third day but not on the second day. Blood samples were collected twice daily at 08.00 and 16.00 h for leptin analysis. There were no difference in mean body condition scores (3.14 ± 0.17 units) and leptin concentrations (2.05 ± 0.19 ng/ml) between the groups studied ($P > 0.05$). On the first day, leptin concentrations slightly increased from 08.00 h to 16.00 h but dramatically decreased when the feed was not offered on the next day. Afterwards, leptin concentrations started to rise following re-feeding. A strong positive relationship was observed between body condition score and mean plasma leptin concentrations (R^2=0.454, P=0.005). The results suggest that (1) rather than breed of sheep, body condition affects plasma leptin concentrations, and that (2) short-term fasting decreases plasma leptin concentration in dry ewes.

The developmental process of rumen fistulation in Iranian native ruminants

A.R. Safaei, H. Fazaeli, M. Zahedifar, H. Mansouri and S.A. Mirhadi, Animal Science Research Institute, Animal Nutrition, Karaj, Iran

A rumen fistula is the direct route for access to the inside of the rumen. To describe the developmental process for the use of rumen fistulas, a number of 40 male and female cows, 60 male sheep, 30 male goats, 15 male buffaloes and 10 male camels were used in a period of 10 years (1996-2006). A fistula is made of industrial hard plastic (a kind of polyethylene) and consists of: main body, flanges (internal and external), screw clamp and screw cap, where all parts have been constructed according to FAO recommendations with fundamental modifications. These modifications were made according to physiological conditions and the volume of the rumen. Results showed that the main internal diameter of the fistula body and the height of fistula were largest for cows and were smallest for sheep and goat, because of a smaller rumen to skin height and volume of rumen. Goats were the most sensitive to fistulation and camels the most difficult to operate because of the water reserve bag in the rumen, bigger distance between rumen and skin, and due to caudal spleen. When analysed at slaughter, it was observed that only around the fistula hole in a radius of 10 cm there was no papillae and the main reason was the internal flange, but otherwise rumen papillae were healthy. Overall, it is concluded that the two-step method of FAO fistulation with a few modifications is superior to other methods.

Session 22 Poster 18

Assessment of rabbit semen quality using resazurin reduction test

K.A. El-Battawy, National Research Centre, Animal Reproduction and A.I., Tahrir street, 12622 Dokki, Cairo, Egypt

Resazurin (7-hydroxy-3H-phenoxazin-3-one-10-oxide) is a non-nutrient, non-toxic redox dye that is used as indicator of dehydrogenase activity. When resazurin is activated by metabolically active viable cells as spermatozoa, it is reduced to resorufin and to colorless hydroresorufin upon further reduction. In this study, the resazurin reduction test (RRT) was used to assess the color change of resazurin reduction in butanol extracted samples to evaluate rabbit semen quality. 100 samples of rabbit semen were used and the absorption was read at 580 nm and 615 nm, respectively. Dividing the absorption at 580 nm by the absorption at 615 nm, various ratios were obtained to assess semen quality. Results showed that the ratios of RRT were decreased as the incbation time of the diluted semen increased. In addition, the results indicated the presence of a high correlation with sperm motility percentage ($r=0.901$, $P < 0.001$) and acrosomal integrity ($r=0.813$, $P < 0.001$).

Session 23 Theatre 1

Farm meat marketing in cattle suckler breeding: economic results and impacts on breeding system management

S. Ingrand[1], P. Veysset[2] and M. Limon[1], [1]INRA, SAD, UMR Metafort, equipe TSE, 63122 St Genes Champanelle, France, [2]INRA, SAE2, Laboratoire Economie de l Elevage, 63122 St Genes Champanelle, France

We evaluate the impacts of direct selling, both on the breeding system management and on the economical results. By making surveys in 20 farms of the Massif Central Suckler Farming Area, we showed that the farm marketing allows a better valorisation of the animals than the traditional channels, in spite of the additional costs. The study highlightes explanatory criteria of the variability of the economic performances: the method of carving management and the mode of production (Conventional or Organic Farming). The degree of engagement of the breeder in the farm marketing cannot be assessed only iwith the number of animals sold, because the breeder also makes choices according to investment and working duration. The study underlines the existence of various strategies of orientation of the animals, according to the category they belong to, together with their quality. It appears that the farm marketing allows to valorise non "standard" animals (steers, heavy calves, Organic Farming production) while satisfying the customers. The practice of the farm marketing does not decrease the labour productivity and can be seen as an alternative to the enlarging of the structures because it makes it possible to create more added values and to maintain more workers with a constant herd's size.

Session 23 Theatre 2

Animal mobility contributes to optimize the forage systems

P.P.M.R. Pierret Roux, ENESAD, DPA, BP 87999, 21079 Dijon Cedex, France

In Burgundy, the annual flow of store cattle exported amounts to 280000 head including 2/3 to Italy and especially the males of charolais breed at an average of 9 to 18 months. We make the assumption that the movement of the cattle and export contribute to optimize the forage systems in Burgundy, in others part of France and in Italy by specializing the areas, the farm area accordind to their respective potential. Three study levels are developed:

- at the regional level, the increase in males exports, counted between the French National Agricultural census of 1988 and 2000, made it possible to release from forage areas and to increase the suckler herd and develop the beef breeding cow fattening for french market. We will describe the opportunities created by this double mouvement: release forage areas in Burgundy and using new forage areas in the import country.
- at the level of charolais systems breeding, the diversity of the products (store cattle at different age: 9 to 18 months) allows a differentiated land use, we describe the breeding systems plasticity thus obtained.
- at the breeding level, we analyze how the production is organized and how the animals according to the quality of the pastures are directed (case study).

This trade sustainability founded on the areas specialization, depends on moderated transport cost, welfare standards and sanitary risk control.

Session 23 Theatre 3

An integrated approach to sheep breeding

C.F. Nakielny[1], J.A. Roden[2] and D.A. Jones[1], [1]Innovis Genetics Ltd, Peithyll Centre, Capel Dewi, Aberystwyth, Ceredigion, SY23 3HU, United Kingdom, [2]Institute of Rural Sciences, University of Wales Aberystwyth, Llanbadarn Campus, Aberystwyth, SY23 3AL, United Kingdom

Innovis Genetics Ltd, a new sheep breeding company, was formed in 2005 to exploit new developments in sheep genetics to breed sheep which meet market requirements. Innovis Genetics has developed a new integrated approach to sheep breeding which provides an alternative crossbred ewe to the Mule type common in UK lowland sheep farming. The approach utilises the Inverdale® fecundity gene, a naturally occurring sex-linked gene first identified in NZ Romneys in the 1980's. Following introgression into a nucleus flock of predominantly Texel genetics there is an ongoing breeding programme in the UK using additional gene marker technology and quantitative genetic evaluation. Innovis Genetics also developed a contract production system that leases Inverdale® carrying rams to hill farmers. The female progeny of the hill ewes are then purchased by Innovis Genetics at a pre-agreed price for subsequent sale to lowland producers. This system formalises the use of hill ewes to produce a robust and prolific crossbred ewe with known health status and performance potential. Technical advice is offered to all producers to optimise their performance and profitability. This integrated breeding system provides a unique opportunity within the sheep industry for clear and consistent data collected in a commercial production environment to inform the selection decisions at a nucleus level.

Session 23 Theatre 4

Dual purpose-breed, a more sustainable choice

C. Gaillard[1] and F. Casabianca[2], [1]ENESAD, BP 87999, 21079 Dijon, France, [2]INRA, Quartier Grossetti, 20250 Corte, France

Dairy cattle can be distinguished according to breeding orientation, in specialized one and dual-purpose one, balanced between milk and meat. Dual-purpose breed can constitute an interesting alternative to dairy specialization in the context of sustainable agriculture. We seek to understand how breeding managers build dual-purpose breed within an animal population. Breeding schemes have been studied. They concerned six dual-purpose breeds as classified by experts, Normande, Simmental, Montbéliarde, Abondance, Bleue du Nord and Vosgienne. Surveys carried out near managers of these schemes identified mobilized criteria and implementation methods.The results allowed to better define the concept of dual-purpose breed and to understand the complexity of its implementation. Dairy cattle selection tended to decrease the beef production.The methods used to control this degradation according to breeds, have showed two main strategies in the choice of candidate breeding animals. The first is "*dominant dairy abilities for the sire and complementarity in matings*" through the choice of a female with a good meat value. The second is "*balanced abilities*" through the choice of complete candidate breeding animals on the various required criteria (dairy ability/ morphology/beef production/functional criteria). Preserving the meat ability in a dairy selection led in maintaining a form of rusticity for these dual-purpose cows, through the animal autonomy and the safeguarding of a balance in the body reserve mobilisation.

Session 23 Theatre 5

Investigation on the effect of energy level changing on white meat production costs at broilers farms

J. Azizi, Islamic Azad University Rasht branch, Rasht, 4185743999, Iran

This experiment was conducted for investigation on effect of energy level changing on white meat production costs at broilers farms. Seven combinations for high and low energy diets application were used at starter period (1-14days), grower period (14-35 days) and finisher period (35-56 days) respectively: 1)high energy, high energy, high energy., 2) high energy, low energy, low energy., 3) low energy, high energy, low energy., 4) low energy, low energy, high energy., 5) low energy, high energy, high energy and 7) high energy, high energy, low energy. At end of rearing period, cost and income parameters recorded individually and analyzed. Statistical analysis of obtained data showed that feed efficiency and economical efficiency of rearing system at this case (treatment 5) is higher than other treatments ($P < 0.05$). It is predicted compensatory mechanism is caused improvement of production and economical efficiency at these rearing systems. On the basis of these results, it is concluded that feed cost per one Kg meat weight at rearing system using low energy and inexpensive at total period were caused to minimum feed cost per one Kg white meat. Cause of this matter is expensive high energy ingredients in comparison with low energy ingredients. Therefore with attention to alteration of ingredients price in market, criterion of diet energy level must be reviewed and selected with on the based of ingredients and finished production price.

Session 23 Theatre 6

Methods of milk preservation in Egypt and developing countries

K. Soryal, H. El Shaer and S. Aboul Ezz, Desert Res. Center, Animal Prod. Division, Matarya, 31111 Cairo, Egypt

Post harvest loss of milk is a major concept particularly in small scale dairy production and processing systems in developing countries. Small holder's dairy farms could reduce their losses if they use effective preservation methods. Milk production, processing and marketing operations could be enhanced. This article will discuss several methods of milk preservation used in the rural areas i.e. refrigeration, heat treatment and chemical additives. Among chemical additives, activation of lactoperoxidase system (LPs) may reduce the use of non-approved methods. The use of LPs extends the milk collection distances and improves milk quality due to its bacteriostatic and bacteriocidal effects. Goat and camel milks react with the (LPs) in a different way compared to other milks, since somatic cell counts is higher in normal goat milk and camel milk has a higher content of lactoperoxidase and other natural milk enzymes. The advantages and precautions of (LPs) activation will be discussed. Companies should take care of preparing forms of sachets containing small amounts of chemical used in the (LPs) activation to favor the small holder's uses to the least amount of 5 kg. milk. The cost of additives is significantly lower compared to other preservation methods. The Codex guidelines concerning the use of milk preservatives are needed to match with the recent approaches of milk preservation methods especially the (LPs) activation.

Session 23 Poster 7

Dynamics of pastures and fodder crops for Mirandesa cattle breed – ii Mineral composition

L. Galvão[1], O.C. Moreira[2], F. Sousa[1], R. Valentim[1], T. Correia[1], J.R. Ribeiro[2] and V. Alves[3], [1]IPB-ESA, Apartado 1172, 5301-855 Bragança, Portugal, [2]INIA-EZN, Apartado 17, 2005-048, Portugal, [3]UTAD-CECAV, Apartado 1013, 5001-911 Vila Real, Portugal

Mirandesa cattle are a local breed from the Northeast region of Portugal, playing an important role on the maintenance of the rural spaces, contributing to the fixation of the populations and to the environmental preservation. The aim of this study is to characterise the feed resources (mineral composition available along the year in this farming system. The animals graze natural pastures in Spring and beginning of Summer, being after fed with hays (of natural pasture or oat) and straws (oat, barley or wheat). Samples of feeds were taken from three different farms in two consecutive years and analysed for crude protein (CP), cell wall components, minerals and *in vitro* organic matter digestibility (OMD). Data were evaluated using the ANOVA statistical approach. Results of organic composition of feeds were presented by Galvão *et al.* (EAAP, 2005). For mineral composition, seasonal variations were observed in natural pastures with decreases of K and P (from Spring to Summer 2 to 0.97 and 0.33 to 0.18% DM, respectively). Farm variations were observed for Ca, K and Mg. Regarding hays composition, differences were observed for Ca and Mg contents which was lower for oat hay, compared with that from natural pasture (0.16 *vs* 0.32 and 0.08 *vs* 0.15% DM, respectively). The composition of straws varied between farms and type of straw.

Session 23 Poster 8

Animal waste valorisation (Azores - Portugal)

L. Falcão, F. Rodrigues and F. Moreira Da Silva, University of the Azores, DCA - Agricultural and Environmental Sciences, Angra do Heroísmo, 9700 Angra do Heroismo, Portugal

The biogas technology introduction in animal's farms was successful in Portugal mainland and less in the AzoresIslands, while the integration of the technology has been less successful in different socio-ecological situations in both places. For such purpose, quantitative changes in physical-chemical properties of poultry, rabbit and quail wastes were studied to understand the technologies process and to evaluate the suitability of the compost product as a soil amendment. After an economic valorisation of poultry manures simulation it was observed that there are large differences between profits of poultry farms that produce exclusively meat or eggs if they intent to capture and treat biogas for energy waste valorisation. Growth and concentration of the animal industry create opportunities for the proper disposal of the large quantities of manures generated at rabbit, bobwhite quail and poultry farms in Portugal. Pollutants from unmanaged livestock wastes can degrade the environment, and methane emitted from the biodegradability of the manure may contribute to global climate change as green house gas. Results of the present study clearly demonstrated differences between animal's farms, animal's species, as well as between Portugal Mainland and the AzoresIslands managements.

Session 23 Poster 9

Estimation of the genotype × environment interaction for the weaning weight of beef cattle breeds in the Czech Republic

L. Vostrý[1], V. Jakubec[1], J. Přibyl[2], I. Majzlik[1], Z. Veselá[2] and M. Bjelka[3], [1]Czech University of Agriculture, Kamycka 129, 16000 Prague, Czech Republic, [2]Institute of Animal Science, Pratelstvi 815, 10400 Prague, Czech Republic, [3]Research Institute for Cattle Breeding, Vyzkumniku 267, 78813 Vikyrovice, Czech Republic

The genotype×environment interaction (G×E) was estimated for the weaning weight of four beef cattle breeds: Angus, Beef Simmental, Hereford and Charolais (n=19,760) in the CzechRepublic. The environment was defined in regions by five criteria: Altitude (A), Crop-plat Growing Regions (CGR), Economic Value of Soil (EVS), Less Favored Areas (LFA) and Performance Level in a Herds (PL). The existence of G×E was examined by the mixed model (fixed and random effects) with and without interaction. The suitability of various types of environment for the estimation of G×E was tested by means of the residual error variance (REV) and Akaik's information criterion (AIC). The LFA and PL were the best from all criteria of environment. The incorporation of G×E into the models for both types of environment had: a. No impact on the estimated residual error variance for models with G×E (LFA- 957, PL- 954) and without G×E (LFA- 958, PL- 954). b. An evident impact on the AIC for models G×E (LFA- 195102, PL -193343) and without G×E (LFA-195166, PL- 193393). The estimated G×E were for both types of environment highly significant ($P < 0.0001$). G×E should be considered into evaluation of the weaning weight of beef cattle.

Session 23 Poster 10

Effects of Japanese quail parents age and egg weight on hatchability and chick quality

T.M. El-Sheikh, Sohag University, Animal Production, Faculty of Agriculture, 82786, Egypt

A 2400 eggs were used in this investigation during summer season. These eggs were obtained from the Japanese quails at the age of 8 weeks (1st group), at the age of 16 weeks (2nd group) and at the age of 24 weeks (3^{ed} group). The eggs were grouped according to their weights as follows; 8.50 -10.50, 10.51-11.50 and 11.51-13.50 g. The percent of Pre-incubation, early, mid and late embryonic mortality were 3.18, 5.81, 7.35 and 9.76%, respectively for 1st group, 3.35, 5.56, 7.62 and 10.41 for 2nd group and 1.85,3.02, 3.99 and 5.21% for 3^{ed} group. The % of Pre-incubation, early, mid and late embryonic mortality were 1.96, 4.55, 5.85 and 7.61%, respectively for the smaller egg weights, 3.12, 4.79, 6.86 and 8.98 for mid egg weight and 3.31,5.04, 6.25 and 8.79% for the largest egg weight. Malformation and malposition of the embryonic death and piped eggs were affected by breeder age and egg weight. Fertility was decreased as the parents age increase while the opposite trend was found with hatchability. The breeder age and egg weight had significant effect on fertility, hatchability and chick quality ($P < 0.05$). Abnormal chicks, dead in shell and naval wet were increased with older parents and smaller egg weights. Chick weight was significantly increased as egg weight and parent age increases. The incubation period was shorter with increasing egg weight and breeder age.

Session 23 Poster 11

Profitable Estonian dairy herd – reproduction considerations

M. Voore and O. Saveli, Estonian University of Life Sciences, Institute of Veterinary Medicine and Animal Sciences, Kreutzwaldi 62, 51014 Tartu, Estonia

In dairy cattle management, production per day of calving interval is a trait of economic importance due to of its effect on production costs and main incomes in dairy production. In this research the profitability of an Estonian test farm was analysed in relation to the length and relative economic efficiency of calving interval. Five test groups were formed from different Estonian dairy cow breeds. Data of 112 cows in the first and of 58 cows in the second calving interval were analysed. Milk production, and costs of feed, insemination and veterinary treatment per cow were calculated. Labour costs and other costs per farm were calculated and divided between the test groups according to number of cow places. The results revealed that in economic milk production, the genetic capability of different breeds should be considered. Milk production of the Holstein groups was high, yet the turnover rate of the herd was high as well. The production of Estonian Red cows was lower but their longevity was better. In this study, the most profitable test group according to highest income from milk production was the Holstein group with an average breeding value. Insemination and veterinary costs had a significant effect on milk production costs with prolonged calving interval, although feed costs remained almost the same. Increase in the length of a calving interval had a considerable negative effect on milk production costs.

Session 23 Poster 12

Efficacy of homemade lick supplements for cattle in rural areas of Namibia

I.B. Groenewald[1] and Z.K. Katjiteo[2], [1]Centre for Sustainable Agriculture, University of the Free State (68), P.O. Box 339, Bloemfontein, 9300, South Africa, [2]Ministry of Agriculture and Rural Development, Private Bag X 5556, Oshakati, 9000, Namibia

The objective of this study was firstly to explore the efficacy of using locally based raw materials in the formulation and production of lick supplements. Conventional raw materials were partially replaced by locally produced products of plant origin like marula fruit shells, coconut meal and acacia pod meals. Secondly, the objective was to measure the effect thereof on animal production of indigenous sanga cattle during the dry season at Ongongo Agricultural College in Northern Namibia. Heifers between 2 and 3 years of age, with an average live mass of 270kg, were earmarked and randomly divided into four treatment groups i.e. control group (no supplement), two groups receiving different homemade lick supplements and a group receiving a commercial lick supplement. The experimental period was 196 days. Changes in live mass and body condition scores (0 to 5) were monitored at fortnightly intervals. Voluntary intake of these lick supplements were measured and translated in economic terms. All three treatment groups, entailing lick supplementation, had a positive effect on live mass changes as well as body condition scoring. The positive effect of these parameters on the possible reproductive performance of heifers is highlighted. Such response justifies the additional costs incurred.

Session 23 Poster 13

Influence of the geographical area and the season on the milk urea content in Wallonia

S. Meura, J.F. Cabaraux, L. Istasse, J.L. Hornick and I. Dufrasne, Liege University, Animal Production - Nutrition Unit, Bd of Colonster, 4000 Liege, Belgium

Milk urea concentration is routinely determined in commercial dairy farms along with the official milk analyses carried on milk samples for the dairies by the “Comité du lait”. The milk urea content does not affect milk price but can be useful for the farmers for diet calculation, milk urea content being related to the energy-protein ratio in the diet. The aim of this paper was to study the evolution of the milk urea content according to the months and the areas. The survey was carried in 2005. On average, there were 12 samples per month and per milk producer, so for Wallonia, there were 563862 data. Milk urea concentration changed according to the months owing to the diet. The summer diets, mainly composed by grass in many farms, were characterized by higher nitrogen contents with as results an increase in milk urea concentration. By contrast, during the winter period indoor, the diet was generally more diversified and balanced, so that, the milk urea concentration decreased. The geographic areas also influenced milk urea concentration due to dietary difference. For example, in the Hesbaye area, mean milk urea content was low due to feeding of cereals and by – products of sugar beet locally produced. In Ardennes the mean milk urea content was higher, grasslands being dominant. The urea content in milk can be a useful measurement for the diet calculations in order to decrease nitrogen waste and to reduce the feed costs.

Session 23 Poster 14

Milk urea or urine nitrogen: indicators to quantify nitrogen rejections by grazing dairy cows according to fertilisation types

S. Meura[1], R. Lambert[2], J.F. Cabaraux[1], I. Istasse[1], J.L. Hornick[1] and I. Dufrasne[1], [1]Liege University, Animal Production - Nutrition Unit, Bd of Colonster, 4000 Liège, Belgium, [2]Catholic University of Louvain, Ecology Unit, Pl. Croix du Sud, 5 Bte 1, 1348 Louvain-La-Neuve, Belgium

Nitrogen (N) rejections are a large problem in grazing cattle because the amounts of N rejected are high since grass N content exceeds the animal requirements. The present study aims to quantify N rejections from urine with dairy cows in a rotational system. Paddocks were fertilized with compost, slurry or mineral N and grazed during 7 days by 35 dairy cows in late lactation. Milk urea content was determined in tank milk samples every day. Urine samples were taken on each cow on days 3 and 5 after the entry in the paddock. Urine N and creatinin were determined to quantify urine N excretion (UNE). These observed values were compared to calculated UNE obtained by the difference from N intake and milk N and faeces N. The mean observed UNE excretion at 312 g N day^{-1} was similar to the mean calculated UNE at 330 g N day^{-1}. The correlation between calculated UNE and observed UNE was significant ($P < 0.001$; $r^2 = 0.22$). The correlation between observed UNE and milk urea content tended to be significant ($P < 0.10$; $r^2 = 0.60$). So, it appears that UNE prediction can be more precise with tank milk urea than with calculated UNE from N intake. This research has to be continued and repeated during more grazing periods and with cows at different lactation periods to validate the results.

Session 23 Poster 15

Grass nitrogen nutrition index and nitrate residues in pastures grazed by dairy cows and fertilised with mineral fertiliser, pig slurry or cattle compost

S. Meura, J.F. Cabaraux, L. Istasse, J.L. Hornick and I. Dufrasne, Liege University, Animal Production - Nutrition Unit, Bd of Colonster, B43, 4000 Liege, Belgium

A code of good practices was established by each European member state according to the nitrate directive. In Belgium, the nitrogen (N) inputs from slurry or compost are limited to 230 kg N/ha in pastures. This trial aims to measure N balance and soil nitrates in pastures fertilised with mineral N fertiliser (min N), pig slurry (S) or cattle compost (C). The experiment was carried out during 5 years on permanent pastures grazed by dairy cows. The fertilisation allowed similar efficient N levels. Total N inputs by fertilisation were different at 169, 170 and 102 kg N/ha in C, S and min N plots respectively. The use of S and C as compared with min N fertiliser increased N balance and reduced apparent N efficiency. The N nutrition index, the number of grazing days and the milk yields per ha were not different. The use of pig slurry and cattle compost as compared with mineral N fertiliser did not influence sward characteristics, except legumes proportion. The nitrate residues were not different between fertilisation systems: 19.8, 19.7 and 29.0 kg N-NO3 per ha in C, S and min N respectively. The low nitrate contents suggested a low nitrate leaching with the three types of fertilisation. These results prove that when slurry or compost is used with good practice, they do not decrease animal production and they do not increase nitrate leaching risks.

Effects of twice-daily milking of cows at very unequal intervals upon milk production

B. Rémond[1], D. Pomiès[2] and C. Julien[2], [1]ENITA, Unité Elevage et production des ruminants, BP 35, 63370 Lempdes, France, [2]INRA, Unité de recherche sur les herbivores, Theix, 63122 St Genès-Champanelle, France

This work aimed to study, through 3 trials, the effects upon milk production of cows of very unequal milking intervals (MI). In trials 1 (20 cows) and 2 (28 cows) three groups milked twice-daily (2M groups; intervals between the 2 milkings of the daytime: 11h, 7h and 3h in trial 1; 11h, 5h and 2.5h in trial 2), and one group milked once-daily (1M group) were compared after the peak of lactation, for 3 weeks. In trial 3 (35 cows), two 2M groups (daytime MI of 10h and 5h) and one 1M group were compared from the second week of lactation for 24 weeks. In trials 1 and 2, milk yield decreased in an accelerated manner by 5%, 12% and 28% for daytime MI of 5h, 3h and for 1M group, respectively. In trial 3, the decrease of milk yield was 10% for a daytime MI of 5h, and 40% for the 1M group. The composition of the daily milk (content in fat, protein, lactose, Ca, P, SCC) was little different between the 2M groups, but the fat content in the milk of milking augmented as the MI decreased (60 g/kg after a MI of 3h). SCC showed a similar evolution. Protein content did not change clearly. After resumption of twice-daily milking with a 10h or 11h interval, productions (milk, fat, proteins) were no more different between 2M groups (no carry-over effect). The unequal milking regime does not decrease the quantity of work but it can give more flexibility in the professional and personal life.

Effect of milking frequency of heifers on milk production in the subsequent lactation

B. O'Brien, D. Gleeson, J. Mee and L. Boyle, TEAGASC, Moorepark Dairy Production Research Centre, Fermoy,, CO. Cork, Ireland

This study examined the effect of once a day (OAD) milking of heifers on milk production in the 2nd lactation when twice a day (TAD) milking was applied. In year 1, thirty-two spring-calving heifers were assigned to treatments from calving; TAD milking on a high or low nutritional level (NL); OAD milking on a high or low NL. NL was defined by concentrate (420 kg or 135 kg) and post-grazing sward height (75 or 55 mm). In year 2, nine 2nd lactation cows were selected from each milking frequency (MF) group (balanced across NL) and milked TAD from calving. Milk yield was measured daily while milk composition was measured weekly. Mixed models with block included as a random effect were used to determine the effect of treatment on milk production. In year 1, milk yield was reduced with OAD (2573 kg) compared to TAD (3760 kg) milking ($P < 0.001$). Milk protein content was higher with OAD (3.56%) compared to TAD (3.34%) milking ($P < 0.01$). Fat content was not affected by MF (TAD=4.01%, OAD=4.22%). Also, lactation length was not affected by MF (TAD=248 d, OAD=239 d). In year 2, milk yield and protein content were not affected by the MF applied in year 1 (TAD year1=6639 kg, OAD year 1=6441 kg) and (TAD year1=3.48 g/100g, OAD year 1=3.45 g/100g), respectively. Fat content and lactation length were also not affected by MF. Thus, milking heifers OAD in their 1st lactation does not adversely impact on milk production in the subsequent lactation when changed to TAD milking.

Influence of different milking regimes on milk secretion

J. Hamann, M. Schridde, F. Reinecke and R. Redetzky, Institute of food quality and food security, University of veterinary medicine, Bischofsholer Damm 15, 30173, Germany

Milk secretion rate and milk yield are affected by the degree of udder evacuation, milking interval and milking frequency. The effects of two different milking regimes (CON: conventional system; VMS: automated system) on milk secretion and milk yield were compared. 66 cows were randomly distributed to either a conventional (CON) or a voluntary milking system (VMS). CON lasted for 64 days and milking frequency was twice a day; VMS lasted for 252 days (2.7 milkings daily). In the CON group quarter foremilk samples were taken twice a day (at 6.00 and 16:00 h), whereas the VMS group was sampled continuously for 24 h. Parameters were: cytobacteriological findings, NAGase, electric conductivity, secretion rate per quarter and yield per cow. A quarter was "normal" if the foremilk cell count was $<$ 100.000 cells/ml and culturing was negative. SAS were applied for the statistical analysis. The milk secretion activity per cow was significantly increased in the VMS group ($p < 0.0001$) compared to the CON group. The average daily milk yield per cow was significantly ($p < 0.05$) higher (5%) in the VMS group than in the CON group. Increasing milking frequency to 2.7 times daily will increase the secretion rate, mainly due to the short milking intervals ($<$ 8 h). It can be concluded that optimizing the milking interval may result in higher milk yields.

Relationships between water intake, feed intake and milk yield

E. Kramer[1], E. Stamer[2] and J. Krieter[1], [1]Institute of Animal Breeding and Husbandry, Christian-Albrechts-University, Olshausenstr.40, D-24098 Kiel, Germany, [2]TiDa Tier und Daten GmbH, Bosseerstr.4c, D-24259 Brux, Germany

The aim of the present study was to investigate the relationship between milk yield, water- and feed intake to answer the question, if the easier collectable water intake can pose as an information trait for feed intake. Data recording was performed on the dairy research farm Futterkamp of the Landwirtschaftskammer Schleswig-Holstein. A dataset of about 21,000 observations from 178 Holstein cows was used. Average milk yield, water- and feed intake were 34.9, 82.4 and 19.8 kg, respectively. Estimation of variance components have been accomplished with applying a linear mixed model. Lactation number ($p \leq 0.03$), test day ($p \leq 0.001$), and the parameters of the function of lactation day ($p \leq 0.001$) were included as fixed effects in the model. Repeatabilities were assesed to 0.34 for feed intake, 0.41 for water intake and 0.76 for milk yield. Correlations between milk yield and water- and feed intake were 0.73 and 0.59, respectively. The correlation between water- and feed intake was 0.73. Hence, estimation of feed intake could be achieved by using water intake as an information trait. Improvements for dairy management could be given with a more exactly concentrate partitioning for dairy cows and thus descending feeding costs. With an enlargement of the dataset, it is imaginable to estimate genetic parameters to include water intake as information trait for feed intake in breeding programs.

Session 24 Theatre 5

The link between energy balance pattern and fertility in dairy cows

G.E. Pollott, SAC, Sustainable Livestock Systems Research Group, Bush Estate, Penicuik, Midlothian EH26 0PH, United Kingdom

Energy balance and fertility are two major traits underlying robustness in dairy cows. Several authors have highlighted the relationship between the two traits when cows are in early lactation. However the relationships are not strong and are influenced by a number of other factors, including the pattern of energy balance. In this work energy balance profiles of 363 lactations from cows were categorised into four groups in relation to the first 100 days of lactation; 1) no decline in energy balance and no negative energy balance, 2) no decline in energy balance but starting the lactation in negative energy balance, 3) a decline in energy balance but no negative energy balance and a decline in energy balance totally in negative energy balance. The cows were fully recorded for fertility characteristics and were also monitored for milk progesterone for the first 140-d of lactation. This enabled both the underlying fertility and the farm-observed fertility to be monitored. The fertility was analysed within category to investigate the factors driving the onset of cycling, first heat and early cycle characteristics. The commencement of luteal activity, day of first heat and day of first service all varied by category of energy balance pattern group. Where no negative energy balance was found a consistent trend in the onset of breeding was observed. Lactations experiencing negative energy balance showed a close relationship between energy balance curve characteristics and fertility.

Session 24 Theatre 6

About the influence of production intensity on the specific energy requirement for milk production

C.G. Rus and R. Brunsch, Leibniz-Institute for Agricultural Engineering Potsdam-Bornim, Livestock Management, Max-Eyth-Allee 100, 14469 Potsdam, Germany

Energy efficiency is an important factor of sustainability. The aim of the study is a comparison of different levels of milk production in relation to their energy efficiency, expressed in MJ NEL/kg milk for the total life span of a cow. North-Eastern Germany is the region of the study. The averaged milk yield is about 8,500 kg/ cow and year with a rate of replacement of about 40%. Calculations have shown that the total fodder energy demand per kg milk (from new born calf to slaughter cow) in a herd with 8,000 kg milk yield and 40% replacement rate is nearly the same as in a herd with 6,000 kg milk and 25% replacement. The age at first calving influences the absolute energy demand per kg milk but has only minor influence to the general result. The presentation provides results and discusses the importance of intensity factors like the intensity of rearing (age at first calving), life time of cows, and milk yield of the herd based on data from the region of study. The presentation likes to encourage a discussion about criteria of sustainability of dairy farming systems.

Session 24 Theatre 7

Electrical conductivity of milk as an indicator of mastitis

C. Henze, W. Junge and J. Krieter, Institute of Animal Breeding and Husbandry, Christian-Albrechts-University, Olshausenstraße 40, 24098 Kiel, Germany

The automated detection of mastitis either in automated milking systems or in modern milking parlours is still a problem, because of low substantial parameters. In this study the usage of routine measuring of the electrical conductivity in the premilking of each udder-quarter was tested. During one year the measurements were done twice weekly including the electrical conductivity of each premilking and the somatic cell count (SCC) of a milkingsample of each cow. In total data of 9011 milkings were available. Additional data like activity, concentrate intake and visits at the feedingtable were available. Mastitis was determined according to two different definitions: SCC over 100,000 and/or a medical treatment (1), SCC over 400,000 and/or a medical treatment (2). With mixed models the variance and covariance components had been analysed. The models included fixed effects (lactationnumber, lactationsegment, daily activity, visits at the feedingtable and for electrical conductivity the variance between the quarter-measurements and the difference between highest and lowest value) and random effects(date and animal). Significant influences ($p < 0.05$) came from the traits of electrical conductivity and the visits at the feedingtable. With regard of the traits of electrical conductivity at one milking, sensitivities between 73 % to 80 % and specifities between 50 % and 64 % have been reached (depending on the definition of mastitis).

Session 24 Theatre 8

Automatic detection of mastitis and oestrus in dairy cattle

R.M.G. Roelofs[1], R.M. de Mol[2], K. Odinga[3], A.H. Ipema[2] and E.P.C. Koenen[1], [1]NRS BV, P.O. Box 454, 6800 AL Arnhem, Netherlands, [2]ASG-WUR, P.O. Box 65, 8200 AB Lelystad, Netherlands, [3]Nedap Agri, P.O. Box 104, 7140 AC Groenlo, Netherlands

In modern dairy farming there is a need for increasing automation with increasing number of cows per person. The main condition with automation, especially in relation to food safety and milk quality, is that techniques are reliable. Sensor technologies like conductivity meters and pedometers have already been available for some years, but have not widely been adopted because of the many false positive alerts. In this study a new two-phase processing of sensor data was tested on two Dutch dairy farms. Data on milk yield, temperature and conductivity and cow's activity were regularly sent automatically from the process computers to the central database. First a time-series model analysed the sensor measurements and generated alerts. Secondly a fuzzy-logic model estimated the reliability of these alerts by using cow-calendar data. The reliable alerts were then reported to the farmers on a web-based herd management system. In this study mastitis detection results were better compared to milking-parlour software as it produced much less false positives. Oestrus detection results differed between farms. On one farm the alerts were frequently too late whereas on the other farm they trusted solely on the automatic alerts. It was concluded that central processing of sensor data using fuzzy logic can contribute to a more sustainable udder health and fertility in dairy cattle.

Session 24 Theatre 9

Evaluation of mid-infrared spectroscopy as a technique for predicting coagulation properties of milk

M. De Marchi, R. Dal Zotto, P. Carnier, C. Cassandro, L. Gallo and G. Bittante, University of Padova, Dept. Animal Science, Viale Università 16, 35020 Legnaro, Italy

This study aimed to investigate the application of mid-infrared spectroscopy (MIRS) as a tool for predicting milk coagulation properties (MCP). Experimental data were from 79 Holstein Friesian cows; 6 milk samples were collected from each cow, and half of them were added with Azidiol preservative. Samples were analyzed at 3 different times: at collection (T0), after 4 (T4) or 8 d (T8) of storage. Milk coagulation time (RCT,min) and curd firmness (a_{30},mm) were measured using Computerized Renneting Meter (CRM) on T0 samples with or without preservative. MIRS analyses on T0, T4 and T8 samples with or without preservative were performed using MilkoScan FT120. Prediction equations were estimated by PLSR (WINISI II version 1.02). The reference average RCT and a_{30} values were 15.2 min and 36.1 mm, respectively, with a variation coefficient of 0.26 and 0.23 respectively. Correlation coefficients between CRM and MIRS analysis were 0.78 and 0.67 for RCT and $a_{30,}$ respectively. MCP at different times predicted by MIRS were comparable, while the coefficients of determination in calibration and in cross validation were better in the milk sample added with preservative. In conclusion, MIRS allowed for a rapid and rather accurate analysis of the MCP of cow milk and could be proposed as a tool for recording milk coagulation properties on a large scale, such as within milk recording schemes.

Session 24 Theatre 10

Predicting approaching calving of the dairy cow by a behaviour sensor

E. Maltz and A. Antler, Agriculture Research Organization, The Volcani Center, Institute of Agricultural Engineering, P.O. Box 6, 50250 Bet Dagan, Israel

A basic algorithm, based on once-daily observations of activity and lying behaviour of 15 cows that calved, was applied to data obtained from 12 dry cows before calving. These cows were kept in a dry-cow barn equipped to enable automatic downloading of behavioural data collected by a behaviour sensor fitted to each cow. On average the cows' steady behaviour changed significantly within 24 h prior to calving, with increases in daily steps and restlessness, and decreased lying time. The prediction of individual calvings was improved by incorporation of qualitative limits into the algorithm: 10 of the 12 calvings could be detected a day in advance, and 9 false positive alarms were reduced to 4 by incorporation of a restlessness variable, i.e., less than 5% of all the cow-days measured.

Economic weights of Holstein cattle in Slovak dairy production system

Z. Krupova, J. Huba, J. Dano, E. Krupa and M. Oravcova, Slovak Agricultural Research Centre-Research Institute for Animal Production, Department of Animal Breeding, Hlohovska 2, Nitra, 94992, Slovakia (Slovak Republic)

The bio-economic model was used for calculation of economic weights (EWs) for milk production traits (305-day milk yield, fat content, protein content), for four growth, seven functional and three carcass traits of Holstein cattle raised in Slovakia in 2005. Milk production is with quota limited in Slovakia, but the quotas limits are not filled up if the whole dairy population is taken into account and deletion of quota from 2015 is considerate. Calculation of EWs was simulated for two alternatives: A - milk production with quota, B - without market quota. Both alternatives include three market payment systems of milk. Base price for milk value was corrected according to content of fat, protein and somatic cells. Marginal EWs for milk production traits and for some functional traits were influenced by quotas system. Mainly EWs for fat content got different values (-103.8 EUR and -21.9 EUR/% per cow and year in system A and B). The EWs for all growth and carcass traits were similar in both systems. Values for fat and protein content were also influenced by payment system. Standardised EWs for lifetime of cows and daily gain of calves during rearing and feedlot in system A shared 20 %, 20 % and 18 % of the value for milk yield. In system B values reached 18 %, 14 % and 13 % of the value for milk yield. The relative EWs for milk fat content shared about –31 % and –4 % for system A and B.

The effect of age at first calving and gestation length on calving difficulty in Holsteins

M. Fiedlerová, D. Řehák, M. Štípková, E. Němcová and J. Volek, Intitute of Animal Science, Přátelství 815, 10400 Prague, Czech Republic

In the Czech Republic, calving difficulty (CD) is scored on three-point scale, e.g. normal calving, hard pull, complicated calving. Totally 2.5 million records have been collected in Holstein population since 1992. Classes 1, 2, and 3 represented 93.1, 6.0, and 0.9% of raw data, respectively. The objective of this study was to analyse the effect of age at first calving and gestation length on the course of calving. Age at first calving and gestation length were analysed separately. The first parity records (n=806 463) were detached from the basic data set and were divided by age at the calving into groups of one-month intervals; the average age at first calving was 841±103 days (e.g. 27.6 months). When heifers calved at the extremely low (less then 21 months) or extremely high (more then 40 months) age, their CD fluctuated. When they calved between 21 and 27 months of age CD declined; from 27 to 40 months CD was balanced. Gestation length was 279.5±7.2 days on average (n=1 757 364) and was categorized to one-day classes. The relationship between gestation length and CD was not linear. Cows with short or prolonged gestation tended to have more difficult delivery. Calving difficulty should be adjusted for both age at the first calving and gestation length in subsequent analyses including a genetic evaluation. An exclusion of records with the extreme age at first calving is recommended. Study supported by MZE 0002701402.

Session 24 Poster 13

Factors influencing on parturition time in range Holstein cattle in a dairy herd in Tehran suburb

A. Ebrahimi, F. Gharagozloo and M. Vojdgani, University of Tehran, Department of Clinical Science, Faculty of Veterinary, Tehran, 14155/6453, Iran

Cows having difficulty in parturition and newborn calves are more likely to be left unassisted at night. The objective of this study was to evaluate the influences of age of dam, season, sex and calving abnormality like stillbirth or abortion on the time of parturition. Cross-sectional surveys of 6606 calving records in 1989 to 2005 were analyzed by the least-squares and chi-square. Cases were divided in four groups based on time of parturition in 24 hours in each six hours period of 0-6, 7-12, 13-18, 19-24 with the frequency of 1762, 1787, 2031 and 1027 calving records respectively. Calving were divided in four groups based on season with the frequency of 24.4% in spring, 26.2% in summer, 26% in autumn and 23.4% in winter. The frequency of stillbirth in seasons was 3.9%, 4.3%, 8% and 5.5% respectively ($p=0.06$). Abortion in seasons was12.7%, 9.8%, 10.5% and 12.8% of total calving of each group respectively ($p=0.008$). The incidence of abnormal parturitions in day was more than night significantly ($p < 0.001$). The frequency of stillbirth in night is slightly but not significantly higher than day calving (4.7% *vs* 3.9%).There is no significant difference between the calving time and fetal sex(48.2% female *vs* 51.8% male). More frequency of calving in August and September but less in February and March.

Session 24 Poster 14

Effects of postinsemination administration of GnRH or CIDR on pregnancy in dairy cows during heat stress

A.A. Moghaddam, M. Kamyab Teimouri and M. Kazemi, Free (Azad) University, Animal Science Department, Agriculture Faculty, Felestin Square, P.O.Box 366, 39187 Saveh, Iran

The objective of this study was to evaluate protocols for improving pregnancy rates in dairy cows during heat stress. Estrus cycles of Holstein cows were synchronized using two i.m. injections of prostaglandin (PG) 12 d apart, and cows were artificially inseminated 12 h after detected estrus. After AI (Day 0), the cows were randomly divided into Control (n=30), GnRH (n=30), CIDR (n=30) and CIDR/GnRH (n=30) groups. Control group did not receive any treatment. On Day 8 postinsemination, Group GnRH cows received GnRH agonist, while Group CIDR cows treated with CIDR. Group GnRH/ CIDR cows were given CIDR and GnRH on Day 8 postinsemination. CIDR was removed at 21 days after estrus. Cows were palpated per rectum at 45 days after AI to determine pregnancies. The pregnancy rate was significantly higher ($P < 0.05$) in group CIDR cows (60%) as compared to groups Control (30%), GnRH (40%) and CIDR/GnRH (44.4%) cows. Results indicate that CIDR administration on Day 8 postinsemination increased pregnancy rate of heat stressed dairy cows.

Session 24 — Poster 15

Etiology and efficacy of different treatment methods in inactive ovary of cows from Terhran dairy farms.

M. Mohammadsadegh and B. Khabbaz, Islamic Azad University, Garmsar Branch, Veterinary Medicine, Enngelab Street,Garmsar, Smnan Province, Iran

In order to study the etiology and efficacy of different treatment of inactive ovary, 64 non pregnant cows which had not shown any signs of estrous after more than 60 days of parturition were selected in some Tehran dairy farms. Rectal palpation and evaluation of follicle and corpora lutea on the ovaries were repeated after 10 days and the ovaries without any follicle and corpora lutea were encountered as inactive ovaries. Anestrous cows were treated by ov synch (n=32), CIDR +ov synch (n = 12), PMSG (3000 IU) + ov synch (n = 20). Some reproductive indices such as: treatment to pregnancy interval (retrospective open days) and numbers of service for pregnancy were analyzed in different treatment groups. The rates of uterine infection, lameness, pneumonia, and mastitis were analyzed in anestrous and normal cows. The results show that numbers of service for pregnancy, treatment to pregnancy interval were significantly lower in PMSG treated animals ($P < 0.05$). Lameness and pneumonia have significant correlation with inactive ovary however; uterine infection and mastitis have not such correlation. It is concluded that PMSG therapy is an important method to remedy anestrous and pneumonia and lameness are two important factors in producing inactive ovaries in present condition of this study.

Session 24 — Poster 16

Bovine embryo freezability after biopsy

R. Nainiene, J. Kutra and A. Siukscius, Institute of Animal Science of LVA, Department of Animal Reproduction, R. Zebenkos 12, Baisogala, LT-82317, Lithuania

The aim of the study was to determine the effect of biopsy on bovine embryo freezability and viability depending on the method and magnitude of biopsy. Late morulae or blastocysts of excellent or good quality were selected for biopsy. Biopsy was carried: 1) by suction (with micropipette); 2) by cutting (with a microrazor blade). The biopsied embryos were cryopreserved with 1.4 M glycerol by controlled freezing regime. Embryo viability after biopsy, freezing and thawing were evaluated: 1) by cultivation in *vitro* in modified TCM-199 for 48 hours; 2) by the pregnancy rate after embryo transfer to recipient cows. *In vitro* embryo cultivation indicated that cutting is more harmful than suction and reduced embryo development results by 35.7%. Blastula as well as morula were more susceptible to cutting than to suction: after 48-hour cultivation, respectively, 28.6 and 54.5% blastula and 0 and 44.4% morula have developed. The transfer of 18 postthaw embryos after biopsy by suction resulted in 7 pregnancies (38.9%). This pregnancy rate did not differ significantly from the pregnancy rate after transfers of intact embryos, when 17 embryo transfers resulted in 8 pregnancies (47.1%) The transfers of 8 postthaw embryos after biopsy by cutting did not result in pregnancies. It is possible to freeze bovine embryos after biopsy by suction and get only slightly lower pregnancy rates than those of intact embryos. The establishment of embryo cryobanks could satisfy specific genotype needs of breeding programs.

Session 24 Poster 17

In vitro fertilizing capacity comparison of bull semen from an Azorean rare breed "Ramo Grande"

P. Santos, A. Chaveiro, A. Marques, G. Antunes and J. Moreira Da Silva, University of the Azores, Department of Animal Science, Universidade dos Açores, Departamento de Ciências Agrárias, 9701-851 Terra Chã, Portugal

The present study was conducted to assess fertility potential of 5 bulls from an Azorean breed called Ramo Grande (RG), by evaluating the effect of sire on the kinetics of *in vitro* cleavage rates and embryo development after IVF. For such purpose, 4 replicates were carried out, with a total of 1106 oocytes matured and randomly divided in 6 groups. Groups 1 to 5 were fertilized with different RG bulls and in Group 6, oocytes were fertilized with a commercial bull (CB). Cleavage rates from CB (86.0%) were statistically ($P < 0.05$) higher than in 4 RG bulls, and between the 5 RG bulls there were also differences ($P < 0.05$), (83.7% *vs* 49.5%, for higher and lower cleavage rates). The proportion of embryos developing to blastocyst stage from 1 of the RG bulls was statistically ($P < 0.05$) lower than CB (10.2% *vs* 20.2%). The proportion of blastocysts that continued the development to hatched blastocysts stage was also different among bulls ($P < 0.05$), both between RG bulls and all 6 bulls, being higher ($P < 0.05$) in 1 RG bull than CB (87.5% *vs* 72.2%). In conclusion, results clearly showed a sire effect in *in vitro* fertilizing ability and subsequent embryo development competence, being possible to notice different ability between RG sires to produce embryos with good development competence.

Session 24 Poster 18

Study of cervical mucus crystallization, sperm survival in cervical mucus and reproductive results of Holstein cows

A. Jezkova[1], L. Stadnik[1] and F. Louda[2], [1]Czech University of Agriculture Prague, Kamycka 129, Prague 6–Suchdol, 165 21, Czech Republic, [2]Research Institute for Cattle Breeding, Vyzkumniku 267, 788 13 Vikyrovice, 788 13, Czech Republic

The objective of this study was to determine the relationship between calving year and season, parity, number of AI, milk production, reproductive diseases occurrence, sperm motility (SM) during 30, 60 and 90 minutes of the cervical mucus survival (CMS) test, cervical mucus crystallization (CMC) and their influence on days to insemination (interval), open days (SP), inseminations number for pregnancy (index), and pregnancy rates (PR) in Holstein cows (n=391). Healthy cows without had better results of interval, SP ($P < 0.01$), index and PR. The higher results of PR were discovered in relation to ferny-like crystallization of cervical mucus ($P < 0.001$). CMC affected results of cervical mucus survival test, the highest motility of sperms after the 60 and 90 minutes was assumed in the case of club moss – ferny and ferny-like crystallization with statistical significance ($P < 0.05$). The lowest motility of sperms at all times (30, 60, and 90 minutes) was in mucus with nothing or dotted crystallization with statistically significance ($P < 0.05$, respectively $P < 0.01$). Presumption of relevance time of estrus, ovulation prediction, best conception rate in cows with fern-like patterns crystallization of their cervical mucus and higher motility and longer survival ability of sperms CMS test was confirmed.

Sperm survival in cervical mucus and reproduction results in synchronized Holstein cows

L. Stádník, A. Ježková, M. Vacek and F. Louda, Czech University of Agriculture, Department of Animal Science, Kamycka 129, 165 21, Prague 6 - Suchdol, Czech Republic

The objective of this study was to determine the relationship between calving year, calving season, parity, number of AI, days in milk, milk production in the 1st 100 lactation days, disease occurrence (retained placenta, endometritis, or cysts) and their influence on sperm motility (SM) during 30, 60 and 90 minutes of the cervical mucus survival test (n=284) or pregnancy rates (PR) in Ovsynch-treated Holstein cows (n=238). Significant differences of SM in the 30th minute and PR ($P < 0.05$) were determined in relation to the calving year. Cows calved from July to September had the lowest SM in the 30th minute ($P < 0.05$) and also the lowest PR. The best SM was defined in the 1st lactation after 60 and 90 minutes ($P < 0.05$ to $P < 0.01$). No statistical significance was detected for the effect of the number of inseminations. Cows inseminated before the 72nd lactation day showed lower SM and the lowest PR ($P < 0.05$ to $P < 0.01$). The lowest milk production during the first 100 lactation days significantly correlates with the lowest SM during the survival test ($P < 0.05$ to $P < 0.01$). The lowest PR was detected in cows with milk production over 3831 kg ($\geq + 0.25\ s_d$). Health traits affected SM and PR significantly ($P < 0.05$ to 0.01). The best SM at the beginning of the survival test signified the best results during the entire test ($P < 0.05$ to 0.001) and also higher PR, but with no statistical significance.

Effect of urea on bovine oocyte's maturation

M.J. Brilhante, P. Santos and F. Moreira Da Silva, University of the Azores, DCA - Animal Reproduction, Angra do Heroísmo, 9700, Portugal

In the present study the effect of the urea in the *in vitro* maturation of bovine cumulus-oocyte-complexes (COCs) were evaluated. Experiments were performed using oocytes aspirated from ovaries of slaughtered dairy and beef cows within 6 hours after slaughter, in a total of 7 replicates. COCs were aspirated from follicles of 85 ovaries and classed on basis of their morphological appearance, selected those of good quality and matured for 24 hours at 38.5°C in a humidified atmosphere of 5% CO_2 having been divided in two groups. In group I COC's (n = 172) were maturated without any urea addition while in group 2 the COC's (n = 48) were maturated with urea (concentration = 64.24 mg/dl). After maturation (22-24 hours), COCs were denuded by vortexing and individually fixed in acetic acid:ethanol (1:3) for about two days and nuclear maturation was characterized after aceto-orcein staining. Further, nuclear oocytes were observed under a phase-contrast microscopy (400 ×) and nuclear development was divided into germinal vesicle, metaphase I, telophase I and metaphase II. Only COCs in the metaphase II stage were considered as matured. Results clearly demonstrated that the presence of urea in the maturation media has a statistic negative effect ($p < 0.05$) on the metabolism of meiosis of bovine oocytes. On average, for the control 83.70% of the COCs achieved the maturated stage, while when urea was present, only 70.80% of the COCs reached the stage of metaphase II.

Session 24 Poster 21

Effect of urea on apoptose regulation of the granulosa cells in bovine: a flow cytometer study

T. Pereirinha, A. Chaveiro, P. Santos, A. Marques, G. Antunes and F. Moreira Da Silva, University of the Azores, DCA -Animal Reproduction, Angra do Heroísmo, 9700, Portugal

The protein excess ingested for the cows is associated with the decrease of the fertility, due an increase of urea and ammonia in the blood, modifying the metabolism of the animals and consequently reducing the conception rate as high levels of urea are present in the follicular fluid, affecting oocyte's quality and the granulosa cells. For such purpose the aim of the present study was to evaluate the effect of urea in the apoptosis and necrosis of the granulosa cells in bovines. Cells were obtained using ovaries (n=80) of slaughtered cows within 6 hours after slaughter, in a total of 8 replicates. After follicular punction, granulosa cells were diluted to a final concentration of 5 x 10^6 cells /ml, divided into three groups and the cellular suspension was cultivated in bottles for 48 hours at 38.5ºC in a humidified atmosphere of 5% CO_2 with: 0; 64.24 and 128.24 mg urea/dl cells. After 48 hours culture, to 100 µl of the suspension 5 µl of Annexin V-FITC and 5 µl of propidium iodide were added, the suspension was incubated in the dark for 15 minutes and measured by flow cytometry. Results clearly demonstrated that when granulosa cells are incubated with urea, a significant increase in the necrotic as well as apoptotic cells is observed. It can be speculated that urea toxicity can induce apoptosis of granulosa cells leading to their death.

Session 24 Poster 22

Influence of apoptosis in the early arrestment of pre-implanted bovine embryos

G. Antunes, P. Santos, A. Marques, A. Chaveiro and F. Moreira Da Silva, University of the Azores, DCA- Animal Reproduction, Angra do Heroísmo, 9700, Portugal

Early arrest of embryos represents the main cause of implantation failure following *in vitro* fertilization. Most arrested embryos displayed nuclei with chromatin condensation and fragmentation, highly suggestive of apoptosis. Although apoptosis occurs during normal preimplantation development of bovine embryos, the proportions of cells with the classic features of apoptosis is larger in arrested embryos than in the ones that reach the blastocyst sage. The aim of this study was to evaluate the potential relationship between early cattle embryo arrestment in different stages of development (2-8 cells; 9-16 cells and morula) and cellular apoptosis of these embryos. For that, arrested bovine embryos and normal developed blastocysts were assisted for apoptosis, biochemical (TUNEL) and morphological (cytoplasm fragmentation) to establish the apoptotic cell rate (ACR) for each group. Results demonstrated that, arrested 2-8 cells embryos had an apoptotic cell rate lower than all the other groups. This fact might be correlated to the embryonic genome activation, which only occurs after the 8-cell state. ACR in the 9-16 cells and morula groups were significantly higher then in blastocysts, suggesting that although apoptosis is a normal feather in pre-implanted embryos, there is a point when ACR contributes to the embryo arrestment. No correlation was observed between the level of cell's apoptosis and cytoplasmic fragmentation.

Session 24 Poster 23

Milk production, milk components content and reproduction of cows in conventional *vs* ecological system of farming

L. Stádník[1], F. Louda[1,2], A. Ježková[1] and M. Bjelka[2], [1]Czech University of Agriculture, Department of Animal Science, Kamycka 129, 165 21, Prague 6 - Suchdol, Czech Republic, [2]Research Institute for Cattle Breeding, Ltd., Výzkumníků 267, 788 13 Vikýřovice, Czech Republic

Research and acquisition of basic data was performed in agricultural enterprise which farms in altitude 550–750 m, on 1432 ha of agriculture land. Periods of farming in 5 conventional (C) years and subsequent 5 ecological (E) years were monitored. Reproduction result, milk production in lactation, content of milk components of 167-174 Czech Pied cows was measured in both periods. Significant differences was ascertained in this traits with using ANOVA, SAS STAT. Total milk production was the higher by 5.62-18.94% in summer period of C farming in relation to winter period. A trend of higher total milk production by 7.66-16.00% was observed in summer period of E farming in relation to winter period. The higher daily milk production in C farming was affected by intensive fertilizing of pasture with commercial fertilizers. Content of milk components varied 4.25% - 4.88% of fat, 3.20% - 3.33% of protein, and 4.73 - 4.80% of lactose in period of C farming and 4.15% - 4.32% of fat, 3.27% - 3.34% of protein, and 4.48% - 4.82% of lactose in period of E farming. The average of milk production was 4.722 kg milk in period C and decline by 418 kg per lactation was measured in period E ($P \leq 0.05$). Differences in reproduction (service period and calving interval) result were not significant.

Session 24 Poster 24

Effects of propylene glycol on milk production, milk composition, blood metabolites and nutrient digestibility of lactating dairy cows

A.H. Toghdory, Member of young researchers club, Islamic Azad University, Gorgan branch, Islamic Azad University of Gorgan, 49147-39975, Iran

Eight multiparous Holstein cows with an average milk production of 32.1 kg/d and body weight of 624.3 kg were used to evaluate the effect of propylene glycol (PG) on animal performance. The experimental design was a 4×4 Latin Square with 21 day periods. Experimental treatments were: 1) No PG (NPG), 2) 250 g/d PG per cow (PG 250), 3) 500 g/d PG per cow (PG 500) and 4) 750 g/d PG per cow (PG 750). Propylene glycol was blended with 0.4 kg of ground corn and fed as a part of TMR. Diets contained 17.4% crude protein, 21% ADF, 34% NDF and 41% NFC. Animals were milked three times per day. Milk yield was measured at each milking during the last 7 days of each period. Individual milk samples for component analysis were collected on the last 2 days of each period and pooled from 3 consecutive samples. Dry matter intake was not different between PG groups ($P > 0.05$). Supplementing cows with 250 and 500 g/d PG did not have any effect on animal performance. Supplementing cows with 750 g/d PG significantly increased plasma glucose ($P < 0.05$). Apparent digestibility of dry matter and organic matte was not affected by PG administration ($P > 0.05$). In conclusion results of this experiment indicate that administration of PG as a part of TMR don't have any significant effect on dairy cows performance in early lactation.

Session 24 Poster 25

Productive performance of lactating cows fed rations supplemented with fibrozyme

M.S. Saleh, Fac. of Agric, Kafr El-Sheikh, Dept. Anim. Prod., Kafr El-Sheikh, 33516, Kafr El-Sheikh, Egypt

Twenty-four lactating Friesian cows averaged 482 kg LBW from 7 to 18 weeks postpartum were used in this study. Cows were divided into four similar groups. Cow groups were fed one the following experimental rations:- LR0: 40% roughage (R) + 60% concentrate mixture, CM); LRF: LR0 + Fibrozyme; HR0: 60% R + 40% CM; HRF: HR0+ Fibrozyme. Roughages were consists of corn silage, berseem hay and rice straw by 50, 25 and 25 %, respectively. Fibrozyme was supplemented (10 g/h/d) to CM just prior to the feeding. Results indicated that, most of nutrient digestibility improved ($P < 0.05$) by Fibrozyme supplementation. Concentration of NH_3-N decreased ($P < 0.05$) for HR0 compared to LR0. Total VFA and propionic acid concentrations increased ($P < 0.05$) for treated rations. Higher acetic acid, but lower ($P < 0.05$) propionic acid percentage was observed for HR0 than LR0. Fibrozyme supplementation improved ($P < 0.05$) milk yield by 12.96 for cows fed LRF compared to those fed LR0 and improved by 13.37 HRF compared to HR0. Most of milk component yields were higher ($P < 0.05$) for cows fed both LRF and HRF than those fed LR0 and HR0. Feed cost decreased by 8.00 and 8.14% for cows fed both LRF and HRF compared to those fed LR0 and HR0, respectively. Moreover, it decreased by 14.00% for cows fed HR0 compared to those fed LR0. Economic return increased by 24.32 and 22.21% for cows fed LRF and HRF compared to those fed LR0 and HR0, respectively. Moreover, it increased by 15.34% for cows fed HR0 compared to those fed LR0.

Session 24 Poster 26

Relationship between concentration of selected biochemical indicators of blood measured in heifers and the latter's milk performance

J.M. Oprzadek, E. Dymnicki and A. Oprzadek, Institute of Genetics and Animal Breeding, animal breeding, Jastrzebiec ul. Postepu1, 05-552 Wolka Kosowska, Poland

The aim of the study was to determine the relationship between concentrations of selected biochemical indicators of blood (thyroxin, triiodothyronine, insulin, alanine aminotransferase, urea, glucose and cholesterol) measured in heifers and the latter's milk performance in 100 and 305-day lactations (milk, fat, protein yields, as well as fat and protein contents). The blood samples were taken from 109 heifers 7-8 months old. The first blood samples were collected under standard feeding conditions when the heifers were 200 days old. Cholesterol concentration in heifers' blood had no effect on any milk performance traits. There was some relationship found between concentrations of T_3 and T_4, insulin, glucose and urea as well as GPT activity and some milk performance traits in 200 and 250 days old heifers. Nonetheless, it cannot be stated that these are good indicators for prediction of future milk performance.

Session 24 Poster 27

New insight in mechanism of action of the California Mastitis Test (CMT)

J. Hamann, D. Kleinschmidt, F. Reinecke and R. Redetzky, Institute of food quality and food security, University of veterinary medicine, Bischofsholer Damm 15, 30173 Hannover, Germany

The California Mastitis Test is a simple, accurate, cow-side tool to determine indirectly the somatic cell count in a milk sample., mainly from individual quarters. The test includes the addition of an anionic surfactant to milk that interacts with the DNA in somatic cells to form a gel. This study details the influences of different cell types (polymorphonuclear leucocytes (PMN); lymphocytes), different status of PMN (vital; necrotic) and ultrasonic treatment of milk on the reaction of CMT. Blood samples of six cows were used to get PMN and Lymphocytes. These cells were diluted (with PBS buffer) in five steps in a range between 20,000 to 3,000,000 cells/ml. Then the CMT was performed with the different cell concentrations. In 191 quarter milk samples the cell vitality of PMN was measured by means of flow cytometry (FACS). Ultrasonic treatment of 20 foremilk samples was tested. PMN and lymphocytes showed identical results in the CMT reaction. The comparison of vital and necrotic PMN did not result in significant different CMT reactions. Ultrasonic treatment changed the DNA structure of cells. It is concluded that only undamaged cellular DNA can react with the CMT reagent. The CMT has a sufficient potential for early detection of subclinical mastitis.

Session 24 Poster 28

Genetic trend of milk yield of bulls in Vojvodina

S. Trivunović, M. Ćinkulov, M. Plavšić and D. Glamočić, Faculty of Agriculture, Animal Science, Trg D. Obradovica 8, 21000 Novi Sad, Serbia

The analysis of genetic trend for milk yield and milk fat in progeny test of bulls used for the artificial insemination was done on the basis of milking results of Holstein-Frisian cows in Vojvodina. The genetic trend for milk yield was calculated as the regression of breeding value and the year of birth of bull and cow. The genetic parameters were calculated as well as the effect of the systematic environmental factors of the farm, year, season, duration of lactation, number of lactation, age at calving and duration of days open on milk production traits. According to the obtained results the most accurate model for estimated breeding value was chosen. The genetic variances were calculated by the REML methodology. The estimated of breeding value was calculated Animal Models with and without repetition, AM for one and more traits and AM with and without genetic groups. The average milk yield of 5741 kg in the standard lactation, with 3,47 % and 197 kg milk fat showed that the milking in this population was lower than in the countries from INTERBULL organization. All studied traits were significantly affected ($P < 0,01$) by all environmental factors.It implies that it is necessary to include all of these effects into the models for the genetic evaluation. The genetic trend estimated by AM for several traits of bulls and cows was negative.

Session 25 — Theatre 1

Predicting composition of lamb carcases

D.L. Hopkins[1] and E.N. Ponnampalam[2], [1]NSW DPI, PO Box 129, Cowra, N.S.W., 2794, Australia, [2]DPI, 600 Sneydes Road, Werribee, Vic. 3030, Australia

In the assessment of breeding stock in Australia, ultrasonic fat depth over the 12th rib is measured along with the depth of the loin muscle (LL). These measures are used to produce breeding values which are combined into various indices to facilitate selection of breeding stock. There has been some discussion about the use of multiple measurement sites on the live animal on the basis that this will improve the discriminatory power of selection for carcase lean. Data from 311 lambs was used to examine this issue. At slaughter hot carcase weights were recorded and the GR (total tissue depth over the 12th rib, 110 mm from the midline) measured using a GR knife. The right sides were scanned by dual energy X-ray absorptiometry using a Hologic QDR 4500A fan beam X-ray bone densitometer and predictions of fat, ash and protein (lean) derived. After scanning the fat depth over the LL at the 12th rib was measured (Fat C), as was the depth of the LL. The muscle depth of the rump (m. *gluteus medius*) and of the subcutaneous fat was measured on the cranial end of the hindleg 30 mm distal to the lumbar/sacral junction. Models were developed to predict either the percentage of fat or lean in the carcase. GR was the best single predictor (R^2 = 52.3) and measures in the rump region not as useful (R^2 = 40.0). The use of multiple measures at different anatomical positions was of marginal value, although GR and FatC together had the highest prediction accuracy.

Session 25 — Theatre 2

The effect of body weight and aging on meat quality of Awassi ram lambs

A.Y. Abdullah and R.I. Qudsieh, Jordan University of Science and Technology, Irbid, 22110, Jordan

Thirty Awassi ram-lambs were used to study the effects of slaughter weight and aging time on meat quality. Lambs were slaughtered at three different live weights (20, 30, and 40 kg). Four selected muscles of each lamb carcass were aged for either 24 hours or 7 days in chiller at 4°C post-slaughter. Dressing-out% was higher ($P < 0.001$) for lambs slaughtered at 30 kg compared to other weights. Total lean and bone decreased ($P < 0.05$) as a percent of cold carcass weight while total fat% increased ($P < 0.05$) with increasing weight. Muscle-to-bone ratio increased ($P < 0.001$) while muscle-to-fat ratio decreased ($P < 0.001$) with increasing body weight. Shear force values increased ($P < 0.01$) with increasing weight and decreased ($P < 0.01$) by increasing aging. Color of the muscles became darker ($P < 0.001$) with increasing weight but was not affected by aging time. Cooking loss was reduced ($P < 0.001$) by increased aging in *M.Longissimus* and by increasing ($P < 0.001$) weight in *M.Semitendinosus*. Aging time had no effect on water holding capacity of all muscles. In conclusion, aging improved meat quality in ram lambs slaughtered at different body weights while slaughtering ram lambs at weights up to 30 kg resulted in higher dressing-out%, better carcass composition and meat quality than ram-lambs slaughtered at heavier weights.

Sensory quality evaluation of Serrana Kids meat: effect of sex and carcass weight

S. Rodrigues and A. Teixeira, Escola Superior Agrária - Instituto Politécnico de Bragança, Dep. Zootecnia, Campus Sta Apolónia Apt 1172, 3501-855 Bragança, Portugal

The main purpose of this work is the characterization of Serrana kids carcass and meat, which is a Protected Origin Denomination product. The effects of sex and carcass weight were studied. Parameters of toughness, juiciness, flavour intensity, flavour quality, odour intensity, fibre presence, sweet intensity and overall acceptability were evaluated on sixty males and females allocated to 3 carcass weight groups: 4, 6 and 8 kg. Sensory quality of meat was evaluated by a trained taste panel of 11 experts, in five sessions. Meat was previously cooked in a conventional oven until inner temperature reached 70/80°C. Then it was cut in sample pieces of 2*2*0.5 cm and given to the panel members to be evaluated following a standard methodology. Sex effect was detected by experts. Males presented higher juiciness, flavour quality and general acceptability than females. Cabrito Transmontano DOP includes animals from 4 to 9 kg carcass weight. Still, differences among them can be important, since taste panel found differences between animals from distinct weight. Light carcasses were considered more tender and with less flavour and odour intensity than heavy carcasses. This can be an indication to breeders that they should produce light carcasses at lower production costs. In fact, this may lead to higher profitability since lighter animals have a higher market price.

Dose response of cinnamaldehyde on lamb performance and carcass characteristics

A.V. Chaves[1], K. Stanford[2], L.L. Gibson[3], T.A. Mcallister[1], F. van Herk[1] and C. Benchaar[4], [1]Agriculture and Agri-Food Canada (AAFC), P.O. Box 3000, T1J4B1, Lethbridge, AB, Canada, [2]Alberta Agriculture and Food, 5401-1st Ave. S., T1J4V6, Lethbridge, AB, Canada, [3]AAFC, 6000 C&E Trail, T4L1W1, Lacombe, AB, Canada, [4]AAFC, P.O. Box 90 STN Lennoxville, J1M1Z3, Sherbrooke, QC, Canada

The objective of this study was to determine the effect of different doses of cinnamaldehyde (CIN) on feed intake, gain, feed efficiency, and carcass yield of lambs fed a concentrate-based diets. Forty-eight healthy lambs were stratified by live weight (LW) and randomized among treatments (n=4) at weaning (LW=20.4±1.12 kg). Animals had *ab libitum* access to pelleted diet (14% CP; 31% NDF) and water over a 13-week period. There were 4 treatments with 12 animals each: 1) Control (no CIN); 2) CIN 0.01%; 3) CIN 0.02%; 4) CIN 0.04% (% dry matter basis). Feed deliveries were recorded daily, and refusals were weighed weekly on an individual basis, for determination of intake. Animals were weighed on a weekly basis and slaughtered after reaching 40 kg LW. Feeding CIN diets did not affect intake (928±50.6g/day) or the average daily gain (ADG; 220.9±18.05g/day; $P > 0.05$) of lambs fed supplemented diets as compared to the control. There were no trends for linear or quadratic responses for DMI or ADG when CIN was fed. Feed conversion (FC) was also similar among treatments (4.2 g of DM/g of gain; $P > 0.05$). Saleable meat yield (as proximal cuts) from the carcasses did not differ among treatments (15.5±0.81; $P > 0.05$).

Session 25 Theatre 5

Genetic parameters for *M. longissimus* depth, fat depth and carcass fleshiness and fatness in Danish Texel and Shropshire

J. Maxa[1,2], E. Norberg[1], P. Berg[1] and J. Pedersen[3], [1]University of Aarhus, Research Centre Foulum, Department of Genetics and Biotechnology, P.O. Box 50, 8830 Tjele, Denmark, [2]University of Göttingen, Institute of Animal Breeding and Genetics, Albrecht-Thaer-Weg 3, 37075 Göttingen, Germany, [3]Danish Agricultural Advisory Service, Udkærsvej 15, Skejby, 8820 Aarhus N, Denmark

Genetic parameters for muscle depth (MD) and fat depth (FD) and carcass fleshiness (FORM) and carcass fatness (FAT) were estimated for Danish Texel and Shropshire, the most common sheep breeds in Denmark. Data used in this study were collected from 1990 to 2005 by the Danish Agricultural Advisory Service. A multivariate animal model was used for estimation of (co)variance components for muscle depth (MD), fat depth (FD), carcass fleshiness (FORM) and carcass fatness (FAT). Heritabilities for MD were similar 0.28 and 0.29 for both breeds. Different heritabilities were found for FD; 0.39 for Texel and 0.12 for Shropshire. Carcass fleshiness was highly heritable, 0.45 for Texel and 0.36 for Shropshire. The heritability for FAT was 0.11 for Texel and 0.19 for Shropshire. Genetic correlations between MD and FORM, and FD and FAT were positive and favourable which support the use of ultrasound measurements on live animals as good predictors for final carcass classification.

Session 25 Theatre 6

Short- and long-term lactational effects of pre-pubertal nutrition differ according to breed in dairy sheep

A. Zidi, G. Caja, M. Ayadi, X. Such, V. Castillo, C. Flores, A.A.K. Salama and E. Albanell, Universitat Autònoma de Barcelona, Ciència Animal i dels Aliments, Campus universitari, 08193 Bellaterra, Spain

A total of 57 ewe-lambs of Manchega (MN) and Lacaune (LC) dairy breeds were used to evaluate the effects of pre-pubertal nutrition on lactational performances during whole lifespan. Treatments consisted of *ad libitum* or restricted feeding to achieve the maximum growth rate (MN, 254 g/d, n = 18; LC, 293 g/d, n = 12) or 65% (MN, 164 g/d, n = 17; LC, 189 g/d, n = 10) during wk 7-22 of age. Computerized axial tomography (CAT) of udder was done at wk 16 and 36 for parenchyma and fat pad evaluation. Puberty was reached earlier in ad lib. than in restricted fed ewe-lambs. CAT images at wk 16 showed greater fat pad in ad lib. fed ewe-lambs of both breeds ($P < 0.05$). At wk 16, parenchyma percentage was lower in ad lib. than in restricted MN ($P < 0.05$), but no differences were detected for LC. No differences in CAT values were reported at wk 36. Reproduction traits at 1st lambing did not vary by treatment and breed. At 1st lactation, restricted ewe-lambs yielded more milk than ad lib. in the case of MN (+25%; $P < 0.05$), but it was the opposite in LC (-10%; $P < 0.05$). Treatment effects on average milk yield persisted for 2 to 4th lactations, the difference being significant in MN (+30%; $P < 0.001$) but not in LC (-9%, $P > 0.05$). No significant effects in milk component contents were detected by effect of pre-pubertal feeding treatment, the milk of MN having greater fat and protein than LC ($P < 0.001$).

Effect of somatic cell count on longevity in dairy ewes using survival analysis

V. Riggio, D.O. Maizon, M. Tolone and B. Portolano, Università degli Studi di Palermo, S.En. Fi.Mi.Zo. - Sect. Animal Production, Viale delle Scienze, 90128 Palermo, Italy

Mastitis incidence has an important effect on culling decision. Somatic cell count (SCC) has been recognized as a very useful indicator of mastitis. The aim of this study was, therefore, to evaluate the effect of SCC on culling decisions in Valle del Belice ewes. The data used for the analysis accounted for 4,430 lactations of 2,071 ewes. Data for SCC were collected by the University of Palermo between 1998 and 2006 in 11 flocks. Longevity was defined on a lactation basis, from one lambing to the next instead of from the first lambing to culling. SCC information were divided in 5 classes and included in the analysis as a time dependent covariate changing at each stage of lactation, which changed at 60, 120, 180, and 270 dim. A Cox model was used for the analysis. The percentage of right-censored data was 59.4%. A ewe with a test-day in the highest class of phenotypic level for SCC had a significant 1.21 times greater hazard risk of being culled when compared to the risk for the class with the lowest SCC level (SCC < 600,000); in general all classes had greater hazard risks than the one in the lowest level. As a conclusion, the phenotypic level of SCC played a role in culling decisions in Valle del Belice ewes.

Multiple trait genetic evaluation of ewe traits in Icelandic sheep

T.H. Árnason[1] and J.V. Jónmundsson[2], [1]Agricultural University of Iceland, Hvanneyri, IS-311 Borgarnes, Iceland, [2]Farmers Association of Iceland, Bændahöllin v/Hagatorg, IS- 107 Reyjavík, Iceland

The fertility of ewes were defined as 4 traits measured as number of lambs born in age 1 to 4 years. The maternal ability related to the same age interval was measured by special production indices which have been developed previously. The genetic parameters for these traits were estimated by series of bivariate REML analyses using animal models. The material used for the genetic analysis contained 540,518 records on 193,213 ewes. The heritability for number of lambs born were h^2 = 0.17, 0.13, 0.11, 0.10 for the four respective age classes. Corresponding estimates for ewe production index were h^2 = 0.16, 0.17, 0.17, 0.15. The genetic correlations between lamb born at different ages were ranging from 0.63 to 0.98. For the ewe production index the corresponding range was 0.82 to 0.99. The genetic correlations between fertility and production index were generally low. It was concluded that the multiple trait genetic evaluations needed to include two sets of 4 correlated traits. The material used for estimating the breeding values by the MT-BLUP Animal Model consisted of 1,4 million individuals in the pedigree file. 749,429 ewes had records for number of lambs born and 701,818 ewes had production index (at least one year). All possible missing patterns were present in the data. In the iteration process expected values for missing traits were generated and solutions were obtained on a canonical transformed scale.

Session 25 Theatre 9

Bottleneck detection in Sicilian goat breeds based on molecular information

M. Siwek[1], R. Finocchiaro[1], J.B.C.H.M. van Kaam[2], A. Zumbo[3] and B. Portolano[1], [1]University of Palermo, S.En.Fi.Mi.Zo., Viale della Scienze, 90128 Palermo, Italy, [2]Istituto Zooprofilattico Sperimentale della Sicilia "A. Mirri", Via Gino Marinuzzi 3, 90129 Palermo, Italy, [3]University of Messina, Mo.Bi.Fi.Pa., Polo Universitario dell Annunziata, 98168 Messina, Italy

Due to Italian sanitary regulations, which prohibited rearing animals within the urban centers, goat populations declined, leading to potential loss of unique genetic variation. Loss of genetic variation and further lack of variability blocks the possibility of selection. In Sicily four goat populations: Girgentana, Derivata di Siria, Maltese, and Messinese are reared. The information available on genetic structure and diversity of these breeds is very limited. Based on 20 microsatellites markers the current analysis of genetic bottlenecks was performed. Blood samples were obtained from animals of both sexes, from all 4 populations reared on different farms. Roughly: about 40 samples per breed were sampled on 5 farms. Genomic DNA was isolated from blood with the GFX DNA extraction kit (Amersham). The PCR product was visualized with an ABI3130 Genetic analyzer. Alleles were scored with GeneMapper software (Applied Biosystems). Heterozygosity excess (generated by a bottleneck) was calculated using Bottleneck software assuming the Infinite Allele Model and Stepwise Mutation Model. Results indicate recent bottlenecks in all breeds except Maltese, what gives an important insight into goat breeding system.

Session 25 Theatre 10

Atypical scrapie jeopardises genetic-based scrapie eradication in sheep

G. Lühken[1], A. Buschmann[2], H. Brandt[1], M.H. Groschup[2] and G. Erhardt[1], [1]University of Giessen, Department of Animal Breeding and Genetics, Ludwigstrasse 21b, 35390 Giessen, Germany, [2]Friedrich-Loeffler-Institute (FLI), Institute for Novel and Emerging Infectious Diseases, Boddenblick 5a, 17493 Greifswald-Insel Riems, Germany

In this study genetic and epidemiological aspects of German atypical and classical scrapie outbreaks were compared. Scrapie cases were grouped as atypical or classical scrapie-positive based on several analyses. Scrapie-positive sheep and flock mates were genotyped for *PRNP* codons 136, 141, 154 and 171 by direct sequencing and RFLP analysis. Classical scrapie was diagnosed in 121 sheep from 17 flocks, whereas atypical scrapie was observed in 116 sheep from 103 flocks. There were significant epidemiological differences between classical and atypical scrapie outbreaks, e. g. regarding the number of scrapie-positive sheep per flock. Furthermore, differences were evident in the distribution of *PRNP* haplotypes. While more than 90 percent of classical scrapie-positive sheep carried the *PRNP* genotype ALRQ/ALRQ, sheep carrying the *PRNP* haplotypes ALHQ and/or AFRQ had the highest risk of being positive for atypical scrapie. Ten percent of atypical scrapie-positive sheep carried the genotype ALRR/ALRR which is selected for in actual European scrapie eradication and resistance breeding programs. The results of this study stress the need to reconsider actual scrapie eradication strategies within the European Union.

Assessing global warming and weather effects on milk production traits in Valle del Belice ewes

J.B.C.H.M. van Kaam[1], R. Finocchiaro[2], B. Portolano[2] and S. Caracappa[1], [1]Istituto Zooprofilattico Sperimentale della Sicilia "A. Mirri", Via Gino Marinuzzi 3, 90129 Palermo, Italy, [2]University of Palermo, Department S.En.Fi.Mi.Zo., Viale della Scienze, 90128 Palermo, Italy

Data on 83898 test-day milk records between 1994 and 2006 and belonging to 8638 lactations of 5562 ewes in 17 Valle del Belice flocks were used to assess: (1) The effects of weather conditions of the preceding day and (2) The effect of global warming. Precipitation, solar radiation, sun hours, air pressure, wind-speed, wind-direction and day and night temperature-humidity indexes (THI) were used. For each of the weather parameters in combination with each production trait a fixed effect model was applied, which included a flock × year of test-day × season of test-day interaction and a litter size × days in milk class × parity class interaction and a weather parameter. The results indicated effects of all weather parameters on milk yield. The lowest R^2 was 0.530. THI-day was the parameter which resulted in the highest R^2 (0.575). The lowest R^2 values were for somatic cell score. The effect of 1°C of global warming was predicted by: (1) Shifting the frequency distribution of days per degree Celsius with one degree (2) Calculating the weighted average production across all temperatures before and after the shift and comparing the results. Daily milk yield, fat+protein and somatic cell count would decrease with 0.5%, 1.1% and 0.4% respectively. In the last 13 year Sicilian maximum temperatures increased 0.4°C.

Heritability estimates for milk traits in Slovenian dairy sheep by random regression model

A. Komprej, G. Gorjanc, S. Malovrh, M. Kovač and D. Kompan, University of Ljubljana, Biotechnical Faculty, Department of Animal Science, Groblje 3, 1230 Domžale, Slovenia

The estimation of genetic parameters for milk traits in Slovenian dairy sheep was performed by random regression model. For the period 1994-2002, 38983 test-day records from 3068 ewes were used. The fixed part of the model contained breed and season of lambing as classes, and days after lambing, parity and litter size as covariates. Random part of the model consisted of random regression coefficients for the common flock environment effect, additive genetic effect, permanent environment effect over lactations, permanent environment effect within lactation, and residual. Legendre polynomials from the 1st to the 4th power were used to model random effects on standardized time scale (days after lambing), where polynomials of the 3rd power were found to be sufficient. Heritability estimates varied from 0.08 to 0.17 for daily milk yield, from 0.08 to 0.14 for fat content and from 0.15 to 0.28 for protein content. The analysis of eigenvalues showed that 76 to 87 % of additive genetic variance was explained by the first eigenvalue and that 13 to 24 % was additive genetic variance in the shape of the lactation curve.

Session 25 Poster 13

Comparison of carcass quality in different German sheep breeds

U. Baulain[1], W. Brade[2], A. Schoen[2] and S. Korn[3], [1]Institute for Animal Breeding, Federal Agricultural Research Centre (FAL), Hoeltystr. 10, 31535 Neustadt, Germany, [2]Chamber of Agriculture Lower Saxony, Johannssenstr. 10, 30159 Hannover, Germany, [3]Nuertingen University, Neckarsteige 6-10, 72622 Nuertingen, Germany

A consequence of the new German animal breeding law is that performance testing is not a public duty anymore, but has to be conducted by the breeding organisations themselves. In sheep breeding carcass quality is primarily determined in stationary progeny testing while performance testing of rams is not prevalent. Ultrasound is not routinely applied and measurements of muscle and fat are not standardized between and within breeds. But increasing demands on carcass quality as well as demands for a cost-efficient performance testing require an extended usage of ultrasound in the near future. A total of 152 carcasses of Suffolk, Blackheaded Mutton, Mutton Merino, Bleu du Maine and Leine sheep from stationary progeny testing were examined to study anatomical differences and conformation of legs with loin as a primal cut. After taking regular measurements as required for the test, legs with loin were separated from the carcass between 5th and 6th thoracic vertebra and carcass composition was measured by Magnetic Resonance Imaging (MRI). Significant breed differences were found in muscle and fat scores as well as in MRI tissue volumes. Best carcass quality was observed in Bleu du Maine and lowest in Merino. Corresponding lean to fat ratios determined by MRI were 0.26 and 0.31, respectively.

Session 25 Poster 14

Genetic and environmental factors which influence reproductive traits of german fawn goats in Serbia

M. Cinkulov, S. Trivunovic, I. Pihler and M. Krajinovic, Faculty of Agriculture, Department of Animal Science, Trg Dositeja Obradovica 8, 21000 Novi Sad, Serbia

The objective of this study was to evaluate genetic and environmental factors which influenced the reproduction traits of German Fawn on one goat farm in Serbia between 2003 and 2007. The studied traits were: length of gestation period (GP), number of born kids (BK), birth weight (BW) and weaning weight (WW). The statistical model included fixed effects of month and year of kidding, type of parturition, sex of kid (male, female) and number of born kids. The average values for studied traits were: 152 days (GP), 2.2 (BK), 3.38 kg (BW) and 18.39 kg (WW), while the heritabilities were 0.102, 0.084, 0.17 and 0.11 respectively. All included fixed effects significantly influenced ($P < 0.05$) the BK, BW and WW, while just month of kidding and number of born kids affected the GP. The results showed adequate management in the researched flock, and good genetic base for the further selection.

Growth performance and carcass traits of growing Chios and Farafra lambs fed diets containing different hay levels

H. Hamdon[1], F. Abo Ammo[2], M. Abd El Ati[3], M. Zenhom[2] and F. Allam[3], [1]Environmental Studies & Research Institute, Animal Production, 1Environmental Studies & Research Institute, Minufiya University, 32897, Egypt, [2]Animal Production Research Institute, Sheep & Goat Research Division, Animal Production Research Institute, -, Egypt, [3]Faculty of Agriculture, Animal Production Dept., Faculty of Agriculture, Assiut University, Egypt

Thirty lambs aged 3 months with average body weight 18.93 kg were divided to 3 groups fed 0%, 6% and 15% hay. Farafra final body weight was significantly higher 41.76 than Chios lambs 36.57kg. Total daily gain was 218 g/day for Farafra and 168 g/day for Chios lambs. The highest feed conversion efficiency value was of 15% hay fed group (5.88 DM). Feed cost/kg gain when using diets containing hay was reduced by about 15.5 and 20.2% for 6% hay and 15% hay, respectively. Eye muscle area was significantly ($P < 0.01$) higher in Farafra than Chios lambs (11.91 *vs* 11.42 cm2). Protein was higher (73.77%) in Farafra meat than Chios (71.86%) on dry basis. In conclusion, using concentrates with alfalfa hay for early lambs fattening improved growth performance.

Influence of dairy production level on reproductive activity induced by the male effect in Sarda sheep breed

V. Carcangiu, G.M. Vacca, M.C. Mura, M. Pazzola, M.L. Dettori, S. Luridiana and P.P. Bini, Dipartimento di Biologia Animale, Via Vienna 2, 07100 Sassari, Italy

The aim of this research was to evaluate the influence of production level on reproductive activity induced by the male effect in Sarda sheep breed. 120 pluriparous ewes, with a mean age of 4.2 ± 2 years, lambing in the month of November, were utilized. Milk production of each ewe was registered monthly. On the basis of dairy production level, animal were separated in three groups, each one made up of 40 ewes. Group A, low production (<600 g/day); Group B, intermediate production (600-1000 g/day); Group C, high production (>1000 g/day). From May 1st, the day rams were joined with the flock, to June 30th, blood samples were collected fortnightly to obtain progesterone level. Data were submitted to chi-squared test. Thirty days after the males joining, Group A showed more ewes in sexual activity (70%) than Group B (45%) and Group C (25%), with statistical difference among groups ($P < 0,001$). At the end of the trial there was no difference regarding animal in sexual activity, between Group B (70%) and Group C (68%), while these two groups differed ($P < 0,01$) from Group A (90%). Results pointed out that milk productive level affects strongly reproductive activity.

Session 25 Poster 17

Evalution of protein characteristics of poultry byproduct meal with CNCPS model and its different levels effect on Baluchi lambs performance

T. Ghoorchi, S. Hasani, M. Roodbari, B. Dastar and M. Birjandi, Gorgan University of Agricultural Science & Natural Resources, Animal Sceince, College of Agriculture, Gorgan University of Agricultural Science & Natural Resources, Gorgan, Iran, 49138-15739, Iran

An experiment was conducted to evaluate protein characteristics of poultry by - product meal (PBM) with CNCPS(Cornell Net Carbohydrate and Protein System) model and its effect on Baluchi lambs fattening performance in a completely randomized design with 4 treatment(0, 2.5, 5 and 7.5 percent PBM) and 6 replications(Lambs)per treatment. The lambs were fed for 90 days. Average dry matter intake of 7.5 percent group was significantly lower then other groups($P < 0.05$). PBM level had no significant effect on average daily gain and feed conversion($P > 0.05$). Urea nitrogen of plasma of control treatment(35 mg/dl) was significantly higher than tratments ($P < 0.05$), but blood plasma glucose of different treatments were not significantly different ($P > 0.05$). The results of CNCPS analysis showed that PBM contained high level of bypass protein. The value of A,B1,B2,B3 and C protein portion of PBM were 11.8, 8.23, 21.22, 1.19 and 3.81(percent), respectively. Generally, the results of economic evaluations showed that using of PBM, 7.5 percent level was the best level from the economic point of view.

Session 25 Poster 18

The effect of different levels of monensin on finishing performance and blood metabolits in Moghani lambs

T. Ghoorchi, M. Keyvannloo Shahrestanaki, S. Hasani and Y. Jafari Ahangri, Gorgan University. of Agriculture Sciences and Natural Resources, Animal Science, Gorgan University.of Agriculture Sciences and Natural Resources, Gorgan, 49138-15739, Iran

This investigation was carried out to study of the effect of monensin on performance and blood metabolites of Moghani finishing lambs. Forthy male Moghani lambs with mean body weight of 30.89 kg and 5-6 months of age were in a 84-day feeding experiment. The trial was carried out using Completely Randomized Design with 4 treatment. The applied treatments were: 1) Control without monensin; 2)15 mg monensin per kg of dry matter; 3)30mg monensin per kg of dry matter and 4)45 mg monensin per kg of dry matter. For measuring blood metabolites and urine pH, 3 Lambs were randomly selected in each treatment and blood samples were collected. The results indicated that effect of monensin on averge daily gain and final weight was significant($P < 0.05$). Average daily gain and final weight were highest in 30 mg monensin treatment(193.69 gr, 47.24 kg respectively) and lowest in control group(135.83 gr, 42.41 kg respectively). Monensin had signifcantly effect on blood glucose and phosphere but had no significant influence on urea, calcium and total protein. Monensin had no significant effect on urine pH($P > 0.05$). According to this experiment,due to desirable effects of monensin on lamb performance, use of monensin was recomemended.

Session 25 Poster 19

Production efficiency and feeding behavior of Awassi lambs and Baladi kids fed on a high concentrate diet

S.G. Haddad and B.S. Obeidat, Jordan University of science and technology, Animal Production, p o box 3030, Irbid 22110, Jordan

Fifteen Awassi lambs and 15 Baladi kids (males, averaging 14.3 kg)were used to study the differences in feeding behavior and performance of sheep and goats fed a concentrate finishing diet (cp=16kg/100 kg DM, ME=2.85 Mcal/kg DM) in a complete randomized design experiment lasting 60days. Dry matter and OM intakes were significantly ($P < 0.05$) higher in lambs. Kids had higher ($P < 0.05$) apparent OM, crude protein and gross energy digestibilities. No significant ($P > 0.05$) differences were observed in apparent NDF digestibility, eating, chewing and ruminating times. However, eating and ruminating times (as min/kg NDF intake) was significantly ($P < 0.05$) higher in kids. Final body weight and average daily weight gain were significantly ($P < 0.05$) higher in lambs while kids had significantly ($P < 0.05$) lower feed to gain ratio. Feed cost/kg weight gain for kids was better than for lambs. Results demonstrated that Awassi lambs consumed more feed and grow faster than Baladi kids. However, kids were more efficient feed converter than lambs.

Session 25 Poster 20

Oleic acid effect on ovine preadipocyte differentiation gene expression

A. Arana, B. Soret, P. Martínez, I. Encío, J.A. Mendizabal and L. Alfonso, Universidad Pública de Navarra, Producción Agraria, Campus de Arrosadía s.n., 31006 Pamplona, Spain

The addition of oleic acid to the culture medium during the ovine preadipocyte differentiation process induced an increase on differentiated cells. For studying if the increase on cell differentiation caused by oleic acid was accompanied by changes on the expression of genes involved in ovine preadipocyte differentiation, primary subcutaneous preadipocyte cultures differentiated by adding serum-free differentiation induction media containing 1.6 mg/ml insulin, 2nM tri-iodothyronine, 10 nM dexamethasone, 10 mM rosiglitazone and 200 mM oleic acid through the differentiation period (7 days) were studied. mRNA expression levels of the transcription factors PRARg, ADD1, C/EBPb and C/EBPV and the lipogenic enzymes lipoprotein lipase (LPL), which uptakes fatty acids from blood, and acetyl CoA carboxylase (ACC), which regulates de novo fatty acid synthesis, were analyzed by quantitative real time PCR. Addition of oleic acid decreased mRNA expression level of ADD1 and ACC on the day 7 of the differentiation process ($p < 0.05$). These results showed that, although the addition of oleic acid increased preadipocyte differentiation, there was a decrease in the expression of the ACC gene, enzyme involved in de novo fatty acid synthesis. Nevertheless other studies are necessary to gain a better understanding of the role of fatty acids incorporated from the blood on the preadipocyte differentiation and the lipogenic enzymes gene expression.

Session 25 Poster 21

Sexual seasonality of Alpine and Creole goats maintained without reproduction

L. Bodin[1], S. Dion[1], B. Malpaux[2], F. Bouvier[3], H. Caillat[1], G. Baril[2], B. Leboeuf[4] and E. Manfredi[1], [1]INRA, UR631 SAGA, F-31326 Castanet-Tolosan, France, [2]INRA-CNRS, UMR6175 PRC, F-37380 Nouzilly, France, [3]INRA, UE332 La Sapinière, F-18390 Osmoy, France, [4]INRA, UE88 UEICP, F-86480 Rouillé, France

Seasonality of goats was studied on 3 cohorts of Alpine goats (n=78, 59, 41) issued from 13 bucks of the French selection program and 1 cohort (n=16) of Creole goats issued from embryo transfer. Blood samples collected each week from about 7 months of age and during the first sexual season without reproduction were used to determine the ovarian activity through progesterone levels. Females were weighed each month and body condition scores were recorded each 3 months. For all traits analyzed, they were no cohort effects for both breeds. For Alpine goats, the first sexual season started at the beginning of November independently of the small age variability. At mid-November 100% of the females presented ovarian activity and the end of the sexual season ranged from February 23rd to Marsh 14th according to cohorts. Creole goats started their first sexual season 4 weeks later than Alpine goats but presented ovarian activity until mid of April. The second sexual season of Alpine goats started at the same period than their first sexual season but ended 2 weeks later. In contrast, the second sexual season of Creole goats started 2 weeks before the date of their first sexual season but ended at the same period. Environmental and genetic effects of the features of sexual season will be presented.

Session 25 Poster 22

Radiocesium (Cs-137) in soil-plant-animal continuum on some areas in Bosnia and Herzegovina and Croatia

L. Saracevic[1], S. Muratovic[2], Z. Steiner[3], E. Dzomba[2], N. Gradascevic[1], Z. Antunovic[3], S. Cengic-Dzomba[2] and Z.Z. Steiner[3], [1]Faculty of Veterinary Medicine, Zmaja od Bosne 90, 71000 Sarajevo, Bosnia and Herzegowina, [2]Faculty of Agriculture and Food Sciences, Zmaja od Bosne 8, 71000 Sarajevo, Bosnia and Herzegowina, [3]Faculty of Agricullture, Trg svetog trojstva 3, 31000 Osijek, Croatia (Hrvatska)

Radiocesium (Cs-137) is one of the most important artificial radionuclide in environment. Cs-137 was spewed into atmosphere by atomic bomb testing and similar activities. The reactor incident in Chernobyl was mayor radioactive source in eastern part of Europe. Mayor route of Cs-137 entrancing into human food chain is soil-plant-animal continuum. Therefore, it has been investigated Cs -137 radiation on some important sheep raising areas in Bosnia and Herzegovina and Croatia. In period 2002-2004, samples of soil, grasslands, wool, milk of sheep and produced cheese were taken from five localities in Bosnia and Herzegovina, each year. Simultaneously, soil plants and wool were sampled from four localities in Croatia. The highest radiocesium activities were measured in soil samples ranging from 5,88 to 274,12 Bq kg^{-1}. Cs-137 activities in other substrates ranged from 0,32 to 51,26 Bq kg^{-1}; 0,54 – 12,97 Bq kg^{-1}; 0,50-5,08 Bq kg^{-1}, and 2,42-3,37 Bq kg^{-1} for hay, wool, milk and cheese, respectively. Generally, the highest values in almost all samples measured in Central Bosnia region and they increased with altitude what is connected to inhomogenous distribution of rainfalls.

Session 25 Poster 23

Correlation between intramuscular fat and body fat depots as an indicator of ewe body fat reserves

S. Silva, A.L.G. Lourenço, C. Guedes, V. Santos, J. Azevedo and A.A. Dias-da-silva, CECAV-UTAD, Animal science, PO Box 1013, 5001-801 Vila Real, Portugal

It is well established that intramuscular fat (IF) tissue is an important determinant of meat quality. However the use of this fat depot as indicator of body fat reserves was neglected. The information available about the relationship between IF and other fat depots allows pondering the IF as a potential indicator of body fat reserves. Therefore the aim of our study was to establish the correlations between IF and the internal and carcass fat depots. A group of 47 female sheep (42±7 kg LW) was used. After slaughter internal fat depots- mesenteric, omental, and kidney and pelvic- were carefully obtained and weighed. The carcass was entirely dissected and carcass fat depots- subcutaneous fat and intermuscular fat were obtained. For IF determination a sample of longissimus muscle was used. The correlations between fat depots and IF were performed with SAS software. High correlations (r between 0.702 and 0.935; $P < 0.001$) were found for all fat depots. The correlations between IF and internal fat and carcass fat depots were 0.892 and 0.929 ($P < 0.001$), respectively. The knowledge of these relationships allows the exploitation of *in vivo* techniques, such as RTU, to monitor all fat body reserves through IF.

Session 25 Poster 24

Divergent selection for reproduction: genetic parameters and genetic change

S.W.P. Cloete[1], J.J. Olivier[2] and J.B. Van Wyk[3], [1]University of Stellenbosch, P/Bag X1, 7602 Matieland, South Africa, [2]ARC: Livestock Business Division, P/Bag X5013, 7599 Stellenbosch, South Africa, [3]University of the Free State, PO Box 339, 9300 Bloemfontein, South Africa

Genetic trends and parameters were derived for reproduction, fleece and live weight of 3 394–3 485 mature ewes of a Merino population divergently selected for 20 years (1986 – 2005) either for (H line) or against (L line) ewe multiple rearing ability. Repeatability model h^2 estimates were 0.11 for number of lambs born, 0.05 for number of lambs weaned, 0.06 for weight of lamb weaned (TWW), 0.52 for live weight (MLW) and 0.59 for greasy fleece weight (GFW). Data of 1 235–2 691 hoggets were used to estimate genetic trends and parameters for live weight and wool traits. Estimates of h^2 were 0.38 for live weight (LW), 0.29 for clean fleece weight, 0.39 for staple length, 0.52 for fibre diameter (FD), 0.52 for CV and 0.23 for staple strength (SS). Genetic correlations (r_g) of reproduction with MLW were positive and medium to high. SS was positively related to CFW (r_g=0.64) and FD (r_g=0.52). The r_g of SS with CV was favourable, but not significant (-0.21). Genetic trends for reproduction in the H and L lines were divergent. Breeding values in the H line increased with 0.33 kg (or 1.7%) per annum for TWW, while it decreased with 0.17 kg (or 0.9%) in the L line. H line hoggets became heavier, with an opposite trend in the L line. Genetic progress in lamb output was attainable, without unfavourable correlated genetic changes.

Session 25 Poster 25

The potential of the Damascus goat breed for meat production

A. Koumas and C. Papachristoforou, Agricultural Research Institute, P.O.Box 22016, Lefkosia, Cyprus

Several characteristics associated with meat production of the Damascus goat, were evaluated over the last 20 years in medium and high input systems. The breed is characterized by its large body size (males 80 to 90, females 65 to75 kg), and good reproductive performance. In a herd comprising of 28% yearlings and 72% adults, the litter size at birth was 2.0, kids born live 1.88, and kids weaned 1.75 per goat kidding. Kids are weaned at 7 weeks and then fattened until 4 to 5 months of age. From a total of 2711 records (21.0% singles, 63.1% twins, 15.9% triplets and quadruplets), mean birth weight of males was 4.8 and of females 4.3 kg, while at weaning, males reached 16.9 and females 14.8 kg. The average daily gain (ADG) from birth to weaning was 248 and 214 g for males and females, respectively. For the same dataset, males weighed 35.1 and females 28.3 kg at 120 days of age. In several growth trials, the ADG from 50 to 120 days, varied from 231 to 313 for males, and 207 to 239 g for females; the feed to gain ratio ranged from 3.5 to 4.3 in both sexes. Regarding carcass characteristics of male kids, the dressing percentage (on empty fasted body weight) at 110 days of age was 54.7, and at 140 days, 54.9. Meat, fat and bone content of carcass at the two ages respectively, were 65.6, 11.1 and 23.3, and 65.6, 12.8 and 21.6%. Taking into consideration published information on meat production performance of other goat breeds, it seems that the Damascus goat compares favourably with them.

Session 25 Poster 26

A model for predicting the retention of electronic ruminal boluses according to their physical features in goats

S. Carné, G. Caja, J.J. Ghirardi and A.A.K. Salama, Universitat Autònoma de Barcelona, Ciència Animal i dels Aliments, Campus universitari, 08193 Bellaterra, Spain

A total of 1,725 electronic boluses were applied to Murciano-Granadina (dairy) and Blanca de Rasquera (meat) Spanish goat breeds. Boluses consisted of 15 capsules containing a 32 mm half-duplex transponder, and varying in physical features: length (37 to 84 mm), outer diameter (9 to 22 mm), weight (5 to 110 g), volume (2.6 to 26 mL) and specific gravity (1 to 4.3). Boluses were read at 7 d, 1 mo, and thereafter every 2 mo by using a hand-held transceiver. Results correspond to readings made on 2 to 24 mo post-application. No problems occurred when applying large-dimensioned bolus (26 mL) on adult goats. Retention rate ranged 0 to 100% depending on bolus type, but only the largest one (26 mL,110 g, and SG = 4.23) showed a satisfactory retention rate (> 99%). A logistic model was build-up ($R^2 = 0.956$; $P < 0.001$) to predict the retention of boluses by using their volume and weight as independent covariates. The regression coefficients obtained from the model were: 0.809, -0.879 and 0.283, the last two referred to bolus volume and weight, respectively. Estimated minimum weight and specific gravity for a standard-sized bolus (22 mL) to reach >99% retention rate were 95 g and 4.32, respectively. Further research is required to improve the prediction accuracy of the model and for developing bolus designs allowing a reliable retention rate in goats.

Session 25 Poster 27

Effect of slaughter weight on slaughter characteristics in Croatian Cres lambs

B. Mioc[1], V. Pavic[1], I. Vnucec[1], F. Poljak[2], Z. Prpic[1] and Z. Barac[2], [1]Faculty of Agriculture, Svetosimunska 25, 10000 Zagreb, Croatia (Hrvatska), [2]Croatian Livestock Center, Ilica 101, 10000 Zagreb, Croatia (Hrvatska)

A total of 60 male lambs were, after weaning, raised on island Cres pastures without any supplementary feeding or treatment until slaughtering which was performed in local abattoir, after 12-h fasting. The lambs were slaughtered at 15-18 kg (group 1), 18.3-21 kg (group 2) or 21.1-24 kg (group 3) of liveweight (mean weights of 16.9, 19.4 and 22.4 kg, respectively). Following slaughter, hot carcasses and ''non-carcass'' components were weighed. After chilling at 4 °C for 24 h, carcasses were again weighed to determine the cold carcass weight. The hot carcass weight increased according to slaughter weight (9.3 kg, 10.4 kg and 11.7 kg respectively), but differences were significant ($P < 0.05$) only between group 1 and group 3. Although the hot and cold carcass weights increased, the dressing percentage diminished (54.9%, 53.4% and 52.3%, respectively), with differences being significant ($P < 0.05$) between group 1 and group 3. Proportion of non-carcass components increased significantly ($P < 0.05$) and was 39.9%, 41.2% and 42.6%, respectively. Higher slaughter weights decreased the percentage of blood, skin, feet, lungs + traquea, heart and kidneys, but significantly ($P < 0.05$) increased the percentage of intestine, liver and spleen. From these results it is recommended to Cres lambs be slaughtered at 15-21kg liveweight to result in optimal carcass dressing percentage.

Session 25 Poster 28

Influence of somatic cell count on ewe's milk composition, its properties and quality of rennet curd

L. Novotná, J. Kuchtík and P. Zajícová, Mendel University of Agriculture and Forestry in Brno, Department of Animal Breeding, Zemědělská 1, 613 00 Brno, Czech Republic

The evaluation of the influence of somatic cells count on ewe's milk composition, its properties and on the quality of rennet curd was carried using milk samples obtained from 10 ewes, F_{112} crossbreeds of Lacaune, East Friesian and Improved Wallachian. All ewes were on the 1st lactation. The sampling was carried on 47th, 81st, 123rd, 152nd and 189th day of lactation. Milk samples were categorized into 5 different SCC groups: 1st group (£ 10 000 SC/ml), 2nd group (11 000–50 000 SC/ml), 3rd group (51 000–100 000 SC/ml), 4th group (101 000–200 000 SC/ml), 5th group (³ 201 000 SC/ml). For the evaluation of the quality of rennet curd was used following scale: 1st class – very good and hard curd, 2nd class – good curd, 3rd class – bad curd, 4th class – very bad curd, 5th class – invisible flocculation of casein. Recorded data were statistically analyzed using the classical least squares method (SAS; PROC GLM variant ss4). The results of this study indicate that no statistical significant effect of SCC was found only on rennetability and quality of rennet curd. The concentration of total solids, fat, protein, casein and titratable acidity were lowest in milks with the highest SCC levels from 5th group (³ 200 000 cells/ml). Also in milk from the same SCC group were highest whey protein content, pH and rennetability.

Session 25 Poster 29

Modelling lactation curve in dairy sheep

V.A.P. Cadavez[1], S. Malovrh[2] and M. Kovač[2], [1]Escola Superior Agrária de Bragança, Zootecnia, Campus de Santa Apolónia, 5301-855 Bragança, Portugal, [2]University of Ljubljana, Zootechnical, Groblje 3, SI-1230 Domzale, Slovenia

The objectives of this study were to investigate the environmental effects on daily milk yield (DMY), and to compare the goodness-of-fit of four mathematical functions for modelling lactation curve on dairy sheep. Data comprising 74,771 test-day records of DMY from 15,624 ewes explored under extensive production system, collected during six consecutive years (2000 to 2006) by the milking recording program of the Churra da Terra Quente dairy sheep breed, were used. Test-day records registered before 30 and after 150 days in milk (DIM) were discharged. A high variability in DMY can be observed between flocks and years. DMY increased ($P < 0.05$) until the third parity, and decreased ($P < 0.05$) after the sixth parity. Ewes with lambing on March presented the highest ($P < 0.05$) and those with lambing in November the lowest ($P < 0.05$) DMY. All models presented similar fitting quality, with coefficient of determination between 50.7 and 51.2%. However, all models underestimate the DMY for high production levels; this trend was more pronounced for Wilmink model. More research is needed to find functions that can model accurately the lactation curve of dairy ewes explored under extensive production systems.

Session 25 Poster 30

Development of claws in two different German sheep breeds

U. Bauer, E. Moors and M. Gauly, Institute of Animal Breeding and Genetics, Albrecht Thaer Weg 3, 37075 Goettingen, Germany

The claw structure, growth rates and elements of size seem to play an important role in the clinical response of sheep to various claw diseases like foot rot. The aim of the present study was to characterise different claw parameters in two German sheep breeds, which may be correlated with the occurrence of foot diseases. The measured parameters could be a future basis for selection for claw health in sheep. Therefore different claw parameters of Leineschaf (n = 52) and German Blackhead Mutton sheep (n = 50) were measured at the left fore- and hindleg in 4 weeks intervals beginning at an age of two weeks until slaughtering (24 weeks). Claw hardness, dorsal border length, diagonal length and dorsal angle were measured. In addition chemical claw parameters (water content, water absorption capacity) were analysed in randomly selected lambs (Leineschaf n = 15, German Blackhead Mutton sheep n = 15) after slaughtering. German Blackhead Mutton sheep claws were in average significantly harder, shorter and lower when compared with Leineschaf claws. Water content and water absorption capacity was not significantly different between the breeds. However the claws of Leineschaf showed significantly higher growth rates. This may lead to an increase in the expenses of claw trimming, especially as long as the animals are kept in barn.

Session 25 Poster 31

Oleic acid influences preadipocyte sheep differentiation

B. Soret, P. Martinez, A. Arrazola and A. Arana, Universidad Publica de Navarra, Campus Arrosadia, 31006, Spain

There is a growing interest in fat supplementation to animal diets in order to altere fatty acid profile or adipose tissue development. We analysed the effect of oleic acid (OA) on the differentiation of sheep preadipocytes *in vitro*. Primary preadipocytes from the stromovascular fraction of subcutaneous and omental adipose tissues were cultured and differentiated by adding induction media (1.6 microg/ml insulin, 2nM tri-iodothyronine, 10 nM dexamethasone, 10 microM rosiglitazone and 200 microM oleic acid) over 7 days. Differentiation was assessed by RedO oil staining. The number of differentiated cells was analyzed by flow cytometry and the differences by Anova. Addition of OA increased the number of differentiated cells in omental (p=0.05) and subcutaneous (p=0.06) adipocytes. The RedO oil staining showed higher amount of lipid droplets on the cells challenged with OA. The higher number of differentiated cells and the bigger amount of neutral lipids forming droplets in the adipocytes could be related to a higher OA cell intake but the mechanisms by which this induction is achieved by an individual dietary fatty acid remain unknown.

Session 25 Poster 32

The effect of different dietary energy levels on fat accumulation and distribution in young replacement Dorper rams

J.P.C. Greyling, L.M.J. Schwalbach, N. Bester, H.J. Van Der Merwe and M.D. Fair, University of the Free State, Department of Animal, Wildlife and Grassland Sciences, P.O. Box 339, Bloemfontein 9300,, South Africa

A study was conducted to evaluate the effect of different dietary energy levels on the body and carcass development and especially fat accumulation and distribution in growing replacement Dorper ram lambs. Thirty six (11 to 12 month old) Dorper rams with a mean initial body weight of 42.0 ± 0.52 kg were randomly allocated to 3 groups (n=12 per group). Each group was randomly assigned to one of 3 experimental diets (treatments), formulated on a crude protein, degradable protein, calcium and phosphorus equivalent basis. A medium energy (Me) diet was formulated according to the NRC(1985) for young growing ram lambs to serve as the control diet. The metabolisable energy (ME) content of this diet (Me) was increased by 15% for the high energy (He) and decreased by 15% for the low energy (Le) diets respectively. The actual ME levels for the Le, Me and He as determined by means of a digestibility study were 6.52, 8.09 and 9.39MJ/kg, respectively. The rams received the diets for a 127 day trial period. The Le group showed a lower back-fat, thickness. Fat accumulation occurred mainly in the eye muscle, scrota and around the kidneys of rams fed the He diet. The undesirable scrotal fat accumulation could be detrimental to the reproductive performance of these rams. The results of the present study suggest that 15 MJ ME/ day is needed by replacement Dorper rams lambs at a weight range of 40-65 kg.

Session 25 Poster 33

Effects of TDS on inorganic selenium in drinking water in young rams

J.A. Meyer, N.H. Casey, K.K.O. Holele and R.J. Coertze, University of Pretoria, Pretoria, 0001, South Africa

Selenium (Se) occurs at potentially hazardous concentrations (>0.01 mg / L) in water sources. The trial quantified effects of volume loaded hypertensive total dissolved solids (TDS) (NaCl) in water on Se status, and the effect on blood Se levels [Se], faecal Se excretion and ADG of 24 young S.A. Mutton Merino rams randomly allocated to four treatments: 0 mg Se / L, <200 mg TDS / L; 0 mg Se / L, >3000 mg TDS / L; 0,7 mg Se / L <200 mg TDS / L; 0,7 mg Se / L >3000 mg TDS / L for a 10-week treatment period. Feed and water were *ad libitum.* Dry matter and water intakes (WI) were recorded. The feed was lucerne with negligible [Se] (0,011 mg / kg DM). Lucerne was used so as to increase the WI by resulting in a higher urea content in the urine, that would cause higher urine clearance rate and a resultant increased dose of the treatment. Se affected WI (p < 0,05) overall, and during the 3 to 6-week period (p < 0,0001). TDS and the combination of TDS and Se had no significant impact on WI. TDS in a Se rich water source tended to decrease the rate of accumulation of Se in the blood (p < 0,05). Se supplement resulted in higher blood [Se], higher faecal [Se] and higher ADG (P < 0.05). TDS resulted in lower blood [Se] (P < 0,05), with an improved functioning of homeostatic controls against chronic selenosis. Se treatment groups had higher WI (P < 0,05). An alleviator treatment of 3000 mg TDS / L of drinking water may reduce the risk of selenosis from potentially hazardous water.

Session 25 Poster 34

The efficacy of Neem (*Azadirachta indica*) seed extracts for flea control in goats

L.M.J. Schwalbach, J.P.C. Greyling, M. David and K.C. Lehloenya, University of the Free State, Department of Animal, Wildlife & Grassland Sciences, P.O. Box 339, Bloemfontein, 9300, South Africa

The aim of this trial was to evaluate the efficacy of two Neem seed extracts (water and oil based) for the control of flea infestations in goats. Sixty goat kids were divided into six groups (n=10). Groups T1 and T2 were thoroughly sprayed with 10% and 25% water-based Neem seed extract at a rate of 10ml/kg body weight, Group T3 received a total of 5 ml of a 25% oil-based Neem seed extract (5 doses of 1 ml each at the horn base, axilas and inguinal regions-spot-on). The control groups C1, C2 and C3 were treated (water spray or vegetable oil) in the same way as their respective treatment groups. Treatments were repeated weekly for 14 weeks. Prior to each treatment, the goat kids were cleaned from fleas by thoroughly brushing them with a piece of cotton-wool soaked in ether and combing the animal with a brush. Flea recording was accomplished by counting all the fleas falling onto a white cloth placed under the animal. Packed cell volume (PCV) was determined and the live weights of the goats were recorded every 2 weeks. Both water-based sprayed Neem seed extracts (10% and 25%) treated groups recorded significantly lower flea counts, higher PCV and live weight values than the 25% oil-based Neem seed extract and their respective control groups. It was concluded that flea infestations can be effectively controlled in goats by a weekly application of 10% or 25% water based Neem seed extract sprayed on the animal at the rate of 10ml/kg body weight.

Session 25 Poster 35

Effect of sub-clinical parasitic infections on growth of dairy goat kids

G. Daş[1,2], C. Tölü[1], M. Gauly[2] and T. Savaş[1], [1]Çanakkale Onsekiz Mart University, Department of Animal Science, Faculty of Agriculture, 17020 Çanakkale, Turkey, [2]University of Göttingen, Institute of Animal Breeding and Genetics, Albrecht-Thaer-Weg 3, 37075 Göttingen, Germany

The aim of the present study was to estimate the effect of endoparasite infections on growth parameters of kids. Therefore 49 kids were divided into three groups. Kids of group 1 (n=20) were kept as untreated controls, while kids in group 2 (n=15) received for two consecutive days 10mg/kg BW toltrazuril (every 15 days). Animals of group 3 (n=14) received in monthly intervals 7.5mg/kg BW fenbendazole. Individual faecal samples were taken regularly (n=1748) to determine the number of oocyst per gram of faeces (OpG) and nematode faecal egg counts (FEC). Cases of diarrhoea were recorded. No significant ($P > 0.05$) differences were found between the groups for body weight development and diarrhoea frequency (1.66 %). Log-OpG were not significantly ($P > 0.05$) different between groups 1 and 3. Low FEC were found in kids from groups 1 and 2 ($P > 0.05$) during the entire study. Kids, which were prophylactically treated against coccidial or helminth infections did not show increased growth rates when compared with the untreated controls. It can be concluded that kids may be able to tolerate effects of parasitic infections under low rates of infections. Therefore, an intensive level of anti-parasitic drug usage should be avoided under circumstances where clinical outbreaks or specific risk conditions are not present.

Session 25 Poster 36

Identification of polymorphism of LALBA locus in Sarda breed goat

M.L. Dettori, G.M. Vacca, V. Carcangiu, M. Pazzola, M.C. Mura and S. Luridiana, Dipartimento di Biologia Animale, via Vienna 2, 07100 Sassari, Italy

As other milk proteins a-lactalbumin is considered as a "major gene" for milk yield, since it plays an essential role in the biosynthesis of lactose. Studies carried out on different ruminant species, revealed that this protein is, anyway, rather monomorphic. In order to detect polymorphisms in goat LALBA gene, and to assess their possible correlations with milk yield and chemical composition, a screening was carried out on 50 Sarda breed animals. Milk and blood samples were collected from each animal in the middle of lactation and at the same time milk yield was registered. Percentage values of fat, protein and lactose were determined on milk samples by infrared method. DNA extracted from leukocytes was analysed by PCR-SSCP of the promoter and codifying region of LALBA locus. Several primer pairs were designed on the basis of the available nucleotide sequence, which allowed amplification of two fragments from the promoter region, one containing the first and one with the fourth exon. SSCP revealed the presence of six polymorphic patterns for the PCR-amplified fragments corresponding to the first exon. The fourth exon showed three different patterns, as well as the promoter region spanning from nucleotide -121 to -369. Four different single stranded DNA conformation patterns were identified in the proximal promoter region. Analysis of variance did not identify any significative association between the polymorphic patterns and milk yield and chemical composition.

Session 25 Poster 37

Association of GH-gene polymorphism with milk yield and composition in Sarda breed goat

A.M. Rocchigiani, G.M. Vacca, V. Carcangiu, M.L. Dettori, M. Pazzola and M.C. Mura, Dipartimento di Biologia Animale, via Vienna 2, 07100 Sassari, Italy

Growth Hormone gene (GH gene) in ruminant species is considered a genetic marker to assess dairy production. The aim of this research was to identify GH-gene polymorphism in Sarda breed goats and to evaluate the possible associations of this gene with quantitative and qualitative dairy productions. Twentyfour milking goats, belonging to as many flocks located in different areas of Sardinia, were studied. During the intermediate period of lactation, milk yield was registered and milk and blood samples were collected. Fat, protein and lactose level were determined using the infrared assay. DNA was extracted from blood to investigate the five exons of the GH gene. The analysis was carried out using PCR-SSCP technique, and it revealed few polymorphism in exons 1 and 2 (one and two electrophoretic patterns respectively), five patterns in exons 3 and 4, and six different polymorphic patterns in exon 5. The comparative analysis (ANOVA) of the different polymorphisms with milk fat, protein, lactose and yield, pointed out a relation between some genotypes of exon 5 and milk yield ($P < 0.01$), and some genotypes of exon 3 and milk fat ($P < 0.05$). No significant relation, was found between any exon and milk protein and lactose levels. The result, even though it comes from a preliminary study, attests that also this locus has an high genetic heterogeneity in Sarda goat and it could be utilized as a genetic marker in selective breeding programmes.

Session 25 Poster 38

Perfomance and digestibility of nutrients in lambs fed with diets containing fish residue silage

A.G. Silva Sobrinho[1], S.M. Yamamoto[2], R.M. Vidotti[3], A.C. Homem Júnior[1], R.S.B. Pinheiro[4] and C. Buzzulini[1], [1]FCAV/Unesp, Via de Acesso Prof. Paulo D. Castellane, 14884900, Brazil, [2]UEM, maringá, Avenida Colombo, 5790, 87020-900, Brazil, [3]Fishing Institute, Rodovia Washington Luiz km 445, 15025-970, Brazil, [4]FMVZ/Unesp, Distr. Rubião Jr, s/n, 18.618-000, Brazil

Thirty six 7/8 Ile de France 1/8 Polwarth (18 males and 18 females) lambs, averaging 18 kg, were confined and distributed among the following diets: control diet; FFRS - 8% freshwater fish (*Oreochromis niloticus*) residue silage diet and SFRS - 8% sea fish*(Lophius gastrophisus)* residue silage diet. The fish residue silages have partially substituted the soybean meal and forty percent of corn silage was used as roughage. The dry matter intake and average daily weight gain have not presented differences between diets and sex, with average of 891.83 and 240.26 g/day, respectively. However, lambs that received SFRS presented worse feed:gain ratio (4.04) in relation that received control diet (3.81) and FFRS (3.47). In relation to the biometric measures at 32 kg of corporal weight, males lambs presented larger corporal lenght (60.09 cm) and anterior member height (56.11 cm) than the female lambs, with values of 58.03 and 54.75 cm, respectively. The ether extract total ingestion and digestibility coeficient (90.39%) was higher in lambs that received FFRS, while the neutral detergent fiber digestibility coeficient (59.20%) was smaller in lambs that received control diet.

Session 25 Poster 39

Food intake and digestive efficiency in temperate wool and tropic semi-arid hair lambs fed different concentrate:forage ratio diets

A.M.A. Silva[1], A.G. Silva Sobrinho[2], I.A.C.M. Trindade[2], K.T. Resende[2] and O.A. Bakke[1], [1] UFCG, Patos, PB, Cx.P.: 64, 58.700-970, Brazil, [2]FCAV/Unesp, Via de Acesso Prof.Paulo Donato Castellane s/n, 14884900, Brazil

Twenty-four lambs, averaging 90 days old and a mean body weight of 20 kg, 12 of them wool lambs (WL), F1 Ideal x Ile de France crosses, and 12 others pure Santa Inês hair lambs (HL), were distributed into a four replication 3 x 2 factorial arrangement consisting of three diets and two genotypes. Experimental diets consisted of: D1 = 60% concentrate mix (C) and 40% *Cynodon* sp. c.v. Tifton-85 hay (F), D2 = 40% C and 60% F, and D3 = 20% C and 80% F. Increasing forage levels in diets resulted in linear reductions in dry matter (DM), organic matter (OM), crude protein (CP), total carbohydrates (TCH) and metabolizable energy (ME) intake, and a linear increase in neutral detergent fiber (NDF) ingestion. Tropic semi-arid hair lambs had higher DM, OM, NDF, CP and TCH intake than temperate wool lambs. Although there were no genotype effects in OM and gross energy (GE) coefficient of digestibility, hair lambs showed more efficient digestibility of DM, CP, NDF and TCH. Increases in forage levels of diets corresponded to a negative linear effect in the apparent digestibility of DM, OM, CP, TCH and GE, while apparent digestibility of NDF increased linearly. Total endogenous nitrogen (fecal plus urinary N) for WL and HL hair lambs were, respectively, 182 an 321mg/kg^{0,75} per day.

Session 25 Poster 40

Net and metabolizable protein requirements for body weight gain in hair and wool lambs

A.M.A. Silva[1], A.G. Silva Sobrinho[2], I.A.C.M. Trindade[2], K.T. Resende[2] and O.A. Bakke[1], [1]UFCG/ Patos, Cx.P.: 64, 58.700-970, Brazil, [2]FCAV/Unesp, Via de Acesso Prof. Paulo D. Castellane, 14884900, Brazil

Thirty-four castrated lambs, 17 of them F1 Ideal x Ile de France wool lambs (WL) and the remaining ones were Santa Inês hair lambs (HL), with homogeneous initial body weight (BW) were used. Five animals from each genotype were slaughtered in the beginning of the experiment period and used as reference. Diets (D) were composed of concentrate mix (C) and *Cynodon* sp. c.v. Tifton-85 hay (R), combined in three different ratios: D1 = 60C:40R; D2 = 40C:60R and D3 = 20C:80R. Animals of each group of three lambs, that showed simultaneously an initial BW of 20 kg at the beginning of the dietary regimen, were slaughtered whit 35 kg, what always happened to be the one fed with D1. Net requirements for BW gain in WL, fleece-free, ranged from 101-110 of protein/kg BW, and for hair lamb ranged from 110-118g of protein/kg BW. Net protein requirements for wool production ranged from 634-642 g/kg of produced wool. Hair lambs presented a 7.8-9.5% higher estimated net protein requirements than WL, according to BW and daily weight gain (DG). Total net protein requirements for HL and WL, with 30 kg of initial BW and an approximate 200g mean DG, were 48.5 and 45.4g/day, respectively. Metabolizable protein requirements for HL and WL, with 20 kg of initial BW and an approximate 200g mean DG were 59.4 and 76.5g/day, respectively. Net protein requirements for wool production was 64g/100g of produced wool.

Session 25 Poster 41

Sensory characteristics of lamb meat aged and injected with calcium chloride

A.G. Silva Sobrinho, N.M.B.L. Zeola, P.A. Souza and H.B.A. Souza, FCAV/Unesp, Via de Acesso Prof. Paulo D. Castellane, s/n, 14884900, Brazil

The sensory characteristics of meat from Morada Nova lambs submitted to ageing and injection of calcium chloride were studied. The lambs were slaughtered with 25 kg of body weight. After *rigor mortis*
development, *Biceps femoris*, *Longissimus* and *Triceps brachii* were harvested and submitted to treatments. The analysis of variance indicated that ageing did not influence flavor and tenderness of meats from *Biceps femoris*. Calcium chloride also did not affect these attributes. The interaction between factors was not significant for tenderness; however, it was significant for flavor. Ageing did not affect sensory attributes of meats from *Longissimus* without and with calcium chloride. Neither ageing nor calcium chloride influenced flavor and tenderness of meats from *Triceps brachii*. The interaction between factors was non-significant for flavor and tenderness. Similar results of sensory analysis were observed for the meat of Morada Nova lambs submitted to ageing and injection with calcium chloride. The muscle *Longissimus* showed adequate correlation between shear force and subjective tenderness.

Session 25 Poster 42

Different ways of artisanal lamb rennet utilization in the manufacture of typical Italian cheeses

K. Carbone, B. Ferri, C. Tripaldi and D. Settineri, CRA-Istituto Sperimentale per la Zootecnia, Feeding and Nutrition-Milk Production, via Salaria, 31 Monterotondo (Rome), 00015, Italy

Lamb and kid artisanal rennet is almost exclusively used in the traditional cheese from sheep and goat milk and it is one of the factors responsible of their unique sensory characteristics. The aim of this paper, part of a more extensive trial, is to study the chemical and biochemical characteristics of rennet, that is used in different ways, in the manufacture of typical Italian cheeses. Wole stomachs, full of milk, from suckling lambs (9-10Kg l.w.) are divided in 2 groups and analyzed at 20h from slaughtering. The first, wall plus internal content of abomasum (HR), is minced and homogenized to semi-liquid paste. The second group, only internal content (IR), after sectioning and extracting, is handled as above. Samples from both groups are analyzed for dry matter, ash, fat, total and soluble protein. Lipolytic activity of extracts is tested using tributyrin as substrate. The results show a significant higher fat (33.2% *vs* 23.9% on d.m.; $P \leq 0.001$) and a lower soluble protein content (7.8mg/ml *vs* 10.8mg/ml; $P \leq 0.0001$) in IR group than in HR. The consequence is a higher lipolytic activity of HR (HR>30% *vs* IR). This is due to effect the large amount of fat, in IR, that determines its lower enzymic protein solubility probably caused by the hydrophobic interactions between them. From these first results it comes out that the different way of utilization of rennet paste could affect its lipolytic activity.

Session 25 Poster 43

Study on behavior of Thomson's Gazelle (TG) in Thailand

S. Inthachinda[1], S. Wongklom[2], A. Na-chiangmai[3], J. Noppawongse Na-ayudthaya[3], S. Anothaisinthawee[3], C. Thothong[4] and P. Sungworakarn[1], [1]Nongkwang Livestock Research and Breeding Center, Amphur Photaram, Ratchaburi Province, 70120, Thailand, [2]Tak Livestock Research and Breeding Center, Tak Province, 63000, Thailand, [3]Animal Husbandry Division, Department of Livestock Development, Rachathevi, Bangkok, 10400, Thailand, [4]Peung-Wann Resort, Saiyok, Kanchanaburi, 71150, Thailand

A study on behavior of Thomson's gazelle (TG) was conducted at Nongkwang Livestock Research and Breeding Center in Ratchaburi province and Peung Wan Resort Farm in Kanchanaburi province from June 2005 to July 2006. For grazing behavior, TGs spent most of the time grazing in the paddock. They especially liked to graze on Mauritius grass. The grazing frequency was 26.75±11.00 times per hour and 27.02±10.51 seconds each time. For nursing behavior, TG kids suckled 2-3 times a day between 10.00-11.30 hours and 17.00-18.00 hours. Suckling durations were between 40 and 90 seconds. The weaning age was approximately 6 months. Horns started to grow at 7-8 weeks of age. For wiggle behavior, TGs wiggled 78.13±7.99 times per minute but when they were frightened they wiggled 100±14.14 times per minute after which they started to run away when they wiggled more quickly. For social behavior, it took 5 days for a TG kid to get used to deer kid after introducing. Approximately 2 weeks after that they could graze close to each other.

Session 25 Poster 44

Estimating intramuscular fat in live ewes from ultrasound images using gray level distribution

S. Silva, A.L.G. Lourenço, V. Santos, C. Guedes, A.A. Dias-Da-Silva and J. Azevedo, CECAV-UTAD, Animal Science, PO Box 1013, 5001-801 Vila Real, Portugal

Real time ultrasound (RTU) is used to estimate intramuscular fat (IF) in live animals, which identifies IF as an ultrasonic speckle backscattered from the small and irregularly shaped marbling deposits. The aim of our study was to develop a technique to allow a quantitative description of the IF by means of image analysis. A group of 47 female sheep (42±7 kg LW) were scanned with an Aloka SSD500V using a linear probe of 7.5 MHz. The probe was placed perpendicular to the backbone, between the 3rd and the 4th vertebrae. The RTU images were captured on a video printer, digitized and analysed with NIH 1.57 software. To avoid deviations in the image statistics the same ultrasound machine and video printer settings were used for all ewes. The image parameters included the statistic characteristics of the gray level distribution in a certain region of interest (ROI) of the RTU image. The ROI was an area over the longissimus muscle (LM) of 10 x 10 mm and 2400 pixels with 256 gray levels (0 is for white and 255 for black). The ROI area of the image is broken down into pixels and a 16 gray level histogram was obtained (GL1 to GL16). The IF was determined by chemical analysis using a LM sample. The relationship between gray level and IF was established by regression and the best equation was achieved with the GL8 (r^2= 0.762, $P < 0.001$). This study indicates that it is feasible to use ultrasound to predict *in vivo* IF.

Trend in dairy sheep BLG genotype found with repeatability test-day model

R. Finocchiaro[1], M.T. Sardina[1], E.F. Knol[1,2], J.B.C.H.M. van Kaam[3] and B. Portolano[1], [1]University of Palermo, Department S.En.Fi.Mi.Zo, Viale della Scienze, 90128 Palermo, Italy, [2]IPG, Schoenaker 6, 6641 SZ Beuningen, Netherlands, [3]Istituto Zooprofilattico Sperimentale della Sicilia "A. Mirri", Via Gino Marinuzzi 3, 90129 Palermo, Italy

A total of 19.207 test-days, collected on 4 farms from 1999-2006 and belonging to 1109 Valle del Belice dairy sheep were analyzed with a repeatability model. After strict outlier analysis 17.747 records were retained. Animals were reared in an extensive system resulting in large environmental influences. However, significant genetic variation was detected in production traits (milk production mean=1336 stdg=93 g/d, fat+protein 167 stdg=12 g/d). Heritability was only 3% while the interaction year by month of test-day explained 27% of the variation. A protein and DNA analysis program was initiated in order to facilitate selection to maintain these flocks under these conditions. In total 427 animals were typed for the BLG locus. Using the recorded pedigree this genotype information was spread over the entire population. The trend in frequency for the AA genotype was significantly negative (p-value=.0057), as were trends in milk yield and lactation length. Possibly farmers have a preference for the BB animals. Relations with production traits were indicative for this trend, but not significant. A possible explanation is the relevant, but non significant difference in length of the lactation between AA and BB animals.

An experience in pedigree reconstruction based on likelihood methods using genetic markers

M. Siwek[1], D.O. Maizon[1], R. Finocchiaro[1], J.B.C.H.M. van Kaam[2] and B. Portolano[1], [1]University of Palermo, S.En.Fi.Mi.Zo., Viale della Scienze, 90128 Palermo, Italy, [2]Istituto Zooprofilattico Sperimentale della Sicilia "A. Mirri", Via Gino Marinuzzi 3, 90129 Palermo, Italy

The breeding of sheep and goats in Sicily is characterized by unrecorded, natural mating with multiple sires joining the flocks. This situation leads to offspring with uncertain parentage, but with only a known group of candidate parents. Genetic evaluations depend upon good pedigree information to achieve their goals. Lack of pedigree information seriously impairs the success of breeding programs. The current work aimed at developing a methodology for pedigree reconstruction by using information of molecular markers, combining likelihood methods of parentage assignment and partitioning of individuals in a sample into full-sib and half-sib families. We analyzed a Girgentana goat flock composed of six groups (185 individuals with 10% males). Genomic DNA was isolated from blood with the GFX DNA extraction kit (Amersham). Fourteen microsatellite markers were used. The PCR product was visualized with an ABI3130 genetic analyzer (Applied Biosystems). Alleles were scored with GeneMapper software (Applied Biosystems). Results indicate that although pedigree reconstruction seems to be a possible task, missing individuals from the flock sample compromised confidence of this reconstruction. For successful reconstruction marker information from multiple years is required.

Session 25 Poster 47

First results about a Help Desk Service on electronic identification in sheep

W. Pinna[1], M.G. Cappai[1], G. Garau[1], A. Sfuncia[1], M. Picciau[1] and M.P.L. Bitti[2], [1]University of Sassari, Dipartimento di Biologia Animale Sezione di Produzioni Animali, via Vienna n. 2, 07100 Sassari, Italy, [2]Associazione Provinciale Allevatori, via Alghero n. 6, 08100 Nuoro, Italy

Since the 1st of January 2008 the electronic identification in small ruminants will be compulsory in the EU Countries(Reg. CE n. 21/2004). The experimental Help Desk Service (HDS) started on Sept. 2006 as practical support concerning animal wellbeing, veterinary aspects and losses/breakage of electronic identifiers for the EID of 110.763 sheep from 276 farms of Nuoro Province (7044 Km^2) in Sardinia (Italy) by 21 technicians. First results on 7527 (6.8%) sheep electronically identified by 2 technicians from 38 (13.7%) farms within Oct.-Dec. 2006, by ceramic boluses (70×21 mm 75 g) with a HDX transponder (32.5 x 3.8 mm) are reported. When a transponder was unread at monthly controls by the same 2 technicians, the HDS supplied a veterinary staff directly in field to differentiate failures from breakages, by a different hand-held reader and to detect the bolus presence in reticulum by a X-ray analysis. A total of 35 (0.46%) transponders in 3 (7.9%) of the 38 farms were unread: 6 (0.08%) showed failures of functioning at static reading; 27 (0.36%) boluses were detected by radiographies; 2 (0.03%) boluses lost. As the HDS appears a useful tool to collect and solve the main technical hitches, a short practical manual to support activities for the implementation of EID on a large scale in sheep is proposed

---

Session 25 Poster 48

Estimation of (co) variance components of some economic traits using different animal models in Karakul sheep

S. Hassani[1] and O. Bakhtiari Fayendari[2], [1]Gorgan University of Agricultural Sciences and Natural Resources, Department of Animal Sciences, Gorgan, 49138, Iran, [2]Mazandaran University, Department of Animal Sciences, Sari, 91839, Iran

(Co)variance components, direct (h_a^2) and maternal heritability (h_m^2) and c^2 of some economic traits of Karakul sheep were estimated using 1314 to 2472 records collected during 1994 to 2000 in Karakul sheep breeding station, Sarakhs,Iran. (Co) Variance components for estimation of genetic parameters were estimated by REML procedure using different animal models and the six different fitted models were compared.Direct heritability of birth weight, 3,6,9 and 12 month weights,fleece weight and pelt score were estimated as 0.24 ± 0.04, 0.19 ± 0.05, 0.19 ± 0.06, 0.29 ± 0.05, 0.31 ± 0.05, 0.20 ± 0.06 and 0.57 ± 0.05, respectively.Maternal heritability estimates for 6 month weight and pelt score were 0.02 ± 0.03 and 0.08 ± 0.02, respectively.c^2 for birth weight, 3 month weight and fleece weight were estimated as 0.16, 0.11 and 0.07, respectively.

Session 25 Poster 49

Effect of feeding steam treated pith baggase (SPB) on bodyweight change, milk constituents and blood parameters of dairy goats in Khouzestan provience Iran

M. Faseleh Jahromi, M. Eslami, J. Fayazi and R. Ebrahimi, Rumin University, Animal Science, Ahvaz-Mollasani, 74146, Iran

Forty dairy goat in 4th mount of pregnancy and multiparouse with 34±3.2 kg body areused in complete randomize design whit 4 treatment and 2 replacement for experiment of different levels of steam-treated sugarcane bagasse pith in partial replacement with wheat bran and barley grain on milk composition, dry matter intake, blood metabolites and body weight change in last pregnancy. Experimental diets include 51% forage (20% alfalfa and 31%wheat straw) and 49% concentration that balanced with NRC (2001). Concentrates were included amounts of 0, 5, 10and 15 percent steam treated bagasse pith which replaced with wheat bran and barley grain in diet of Goat. goats were fed for 120d, 14 days for adaptation and 105 days for Sampling. Statistical analysis of dry matter intake, milk composition percent, blood metabolites and body weight change were not differently means with SAS program. However, With increasing steam-treated sugarcane bagasse pith in diets, no significant differences between treatment were observed for milk composition percent and blood metabolites ($p < 0.05$). Steam-treated sugarcane bagasse pith in diets increased BUN means that fourth treatment was maximum. Statistical analysis of body weight change were not differently means ($p < 0.05$).

Session 25 Poster 50

Utilization of steam treated bagasse pith on performance of Khouzestanian lactating Lori ewe

R. Ebrahimi, M. Eslami, J. Fayazi and M. Faseleh Jahromi, Rumin University, Animal Science, Ahvaz-Mollasani, 74146, Iran

This experiment was done to evaluate the potential use of steam treated pith of sugar cane bagasse (SPB) in feeding of dairy sheep in stage of post Lambing. Forty Eight ewes with the average live weight 53.72±2.75Kg and similar condition from many flocks of autumn of Lori sheep were used in a completely randomized design with 4 treatments and 3 replications during 3 months.. Concentrates were included amounts of 0,10,20 and 30 percent of SPB. Ewes live weight were measured at lambing and 2 weeks interval after that until weaning lamb. Venous blood was sampled during study, and subjected to Haematological analysis. Milk sampling for determination of milk constituents in 5 stage post-lambing. Results indicted that were not differences in milk composition, ewes fed 10 % SPB had high milk fat percentage and ewes fed 30 % SPB had high protein percentage. The means of constituents for fat, protein, lactose and solid not fat (SNF) percentages of different experimental groups following in Table 2. ewes fed 30% SPB had higher plasma BUN. Statistical analysis was indicated that, there were not significant differences between diet for milk composition percent (fat. Protein, carbohydrate and solid not fat) and blood parameter (BUN, triglyceride, cholesterol and glucose) and body change weight ($p < 0.05$).

Effects of age and fattening period on growth performance of Zel and Dalagh lambs

M. Mohajer[1], R. Kamali[1], A. Toghdory[2] and A. Kavian[1], [1]Member of scientific board of agriculture and natural recourse research center, Gorgan, 4915677555, Iran, [2]Member of young researchers club, Islamic Azad university, Gorgan branch, Gorgan, 4914739975, Iran

An experiment was conducted to determine the best age and length for fattening lambs and comparison between Zel and Dalagh breed. Experimental design was completely randomized design in factorial arrangement with three factors of breed (Zed and Dalagh), age (3, 6 and 9 month) and fattening period (60, 90 and 120 days) by eight replicate. Lambs were fed diet containing corn silage, wheat straw, barley grain, wheat grain, wheat bran, cotton seed meal and sugar beet pulp that were offered *ad libitum*. During the fattening period the lambs were weighted from each 10-day. The results showed that there were no significant different between two breed in Average daily gain (ADG) ($P > 0.05$). The highest ADG were in Zel lambs (6 month age), when fattened 90 days and Dalagh lambs (9 month age), when fattened 120 days ($P < 0.05$). The lower rate of ADG were in Zel lambs (3 and 6 month age), when fattened 60 and 120 days ($P < 0.05$). Zel lambs with 3 month age had the best feed conversion ratio, that were different by Dalagh lambs in this age ($P < 0.05$). The highest rate of feed conversion ratio was in Dalagh lambs with 9 month age ($P < 0.05$).

Characterization of the maximum test day yield in the East Friesian ewes in Macedonia

N. Pacinovski[1], G. Dimov[2] and E. Eftimova[1], [1]Institute of Animal Science, Ile Ilievski str. 92A, 1000, Skopje, Macedonia, [2]AgroBioInstitute, 8 Dragan Tsankov Blvd, 1164, Sofia, Bulgaria

The milk test day (TD) milk yield was measured on 98 East Friesian ewes for 2 years period: 2005 and 2006. Totally 139 lactations were included. The daily yield was recorded 2xday, once monthly, after weaning of the lambs at about 75 days of age. The number TD varied from 5 to 11. Only the maximum of these daily yields (MTD) were analyzed for the effects of the year (Y), parity (P), number of lambs born (NL), duration of the suckling period (SP). The period from lambing to the maximum TD yield was presented in days (DMY) as well as number of TD. P were from 1 to 6, NL from 1 to 3. It was found that the MTD appeared on the $110^{th}\pm3.1$ day of the lactation, the main significant factors were the year ($P < 0.001$) and parity ($P < 0.001$), while the NL did not affect the MTD. For younger ewes MTD appeared on the $96\text{-}102^{th}$ day, while for parities 4 and 5 it was on the $118 - 125^{th}$ day with a tendency for a decrease at parity 6. The average MTD was 2.04 ± 0.10 l and it was affected significantly by the year ($P < 0.05$) and parity ($P < 0.001$). The MTD of 2.66 ± 0.36 l was found for the 4th parity and lower as 1.7 ± 0.2 l for 1st parity and very close figures for 6th one. In most of lactations the MTD appeared on the 1st and 2nd TD and in about 10% of lactations on the 4th and higher TD.

Session 25 Poster 53

Carcass slaughter characteristics of sheep depending on age and sex

Z. Antunovic[1], J. Grbavac[2], I. Bogut[2], D. Sencic[1], M. Šperanda[1] and Z. Steiner[1], [1]Faculty of Agriculture, Dep. of Animal Sci., Trg sv. trojstva 3, 31000 Osijek, Croatia (Hrvatska), [2]Faculty of Agriculture, Dep. of Animal Sci., K. Zvonimira 14, 88000 Mostar, Bosnia and Herzegowina

The influence of age on slaughter carcass quality has been researched on: 3 and 3.5 month old weaning lambs, 4.5 mth old post-weaning lambs and adult sheep (>18 mth). The sex influence on the carcass slaughter characteristics was not investigated in the weaning lambs up to 3 months old. The age and sex of sheep had a significant influence on the slaughter quality of their carcasses. With increasing body weights (and ages), the carcass dressing percentages increased too. The weaning and post-weaning male lambs had a statistically higher body weight at the age of 3.5 and 4.5 mths in comparison to the female lambs. The rams had a statistically higher body weight at the age over 18 m. in comparison to the ewes of the same age group. The weaning male lambs at the age of 3.5 mths and rams had a statistically higher dressing percentage than the weaning female lambs and ewes. Sex had a significant influence on the conformation of the sheep carcasses. The relative proportion of ribs was higher in the carcasses of ewes than in the carcasses of rams. The legs of rams were significantly heavier than the legs of ewes and they had a significantly higher absolute and relative share of the muscle tissue and a statistically significant lower share of fat tissue compared to the legs of ewes.

Session 25 Poster 54

Lactation curves in Valle del Belice sheep using random regression models

M. Tolone, D.O. Maizon, V. Riggio and B. Portolano, Università degli Studi di Palermo, S.En. Fi.Mi.Zo. Sect. Animal Production, Viale delle Scienze, 90128 Palermo, Italy

In recent years, several mathematical functions have been proposed to fit lactation curves. The goal of this study was to find the function that best fits the test-day records collected by the University of Palermo in Valle del Belice sheep, in order to implement a genetic evaluation. The lactation curve was modelled with different cubic spline (7, 9, 11, 13 knots) and Legendre polynomial (of degree 3, 4, 5, and 6) functions using ASReml. The dataset used comprised test-day records of milk production traits of Valle del Belice sheep recorded in accordance with the A4 method. A total of 30,428 test-day records from 4,424 lactations belonging to 2,136 ewes, offspring of 140 sires, were used. The average number of test-days per ewe per lactation was equal to nine. A sire model was used considering in the construction of the pedigree sires and maternal grandsires. In the model, the days in milk and the permanent environmental effect were modelled with cubic splines and the Legendre polynomials. Based on minimising the Akaike criterion, a Legendre polynomial model of degree four was the most adequate model. However, there was not a great deal of difference among the models compared.

Session 25 Poster 55

Seasonal changes in estrus activity in Turkish Karayaka sheep

M. Olfaz, E. Soydan, H. Onder and N. Ocak, Ondokuz Mayis University Agricultural Faculty, Animal Science, Ondokuz Mayis University Agricultural Faculty Animal Science, 55139, Samsun, Turkey

Karayaka sheep, widely raised in the middle of the Black See Region of Turkey, is a native breed. Slaughter lambs of this breed are in demand throughout the year because Karayaka is the second most important breed among the native breeds with high meat quality and also adaptation ability. However, there is incomplete knowledge of the changes of estrus activities of the breed throughout the year. The objective of the present study was to determine frequencies of estrus activity throughout the year and estrus interval, and changes in these traits according to season. The experiment was conducted at the experimental farm of the Ondokuz Mayis University, Agricultural Faculty, Samsun, Turkey (41.2°N). A total of 10 Karayaka ewes, approximately 3 years old, were included in the study. Estrus was detected with teaser rams. Frequencies of estrus activity increased in the autumn (September, October and November) and spring (March, April and May) seasons (36.8 % and 28.7%, respectively). Estrus interval and average number of estrus cycle were 17.48±0.84 and 3.7±0.33 in the autumn and 28.41±3.27 and 2.3±0.33 in the spring season, respectively. It was concluded that Karayaka sheep may be mated in the spring and especially in the autumn seasons, taking into account demands for slaughter lambs and/or feeding of lambs in pasture.

Session 25 Poster 56

Utilization of low and high-roughage diets by two breeds of sheep: effects on internal fat depots

M.J. Gomes, A.L.G. Lourenço, S. Silva, J. Azevedo and A.A. Dias-Da-Silva, CECAV-UTAD, Animal Science, PO Box 1013, 5001-801 Vila Real, Portugal

The deposition of internal fat (IFD) is an advantage in energy retrieval and is related to the purpose for which a breed was developed. The breed and diet are factors that could influence the amount and distribution of fat depots. A trial was carried out using 15 female lambs from each of the breeds, the Ile de France (IF; meat breed) and the native Portuguese breed Churra da Terra Quente (CTQ; milk breed). The animals, with an initial body weight (BW) *ca* 45% of the mature BW (DM45), were assigned to be slaughtered as an initial group or fed either a low or a high energy diet (9 and 11.5 MJ ME/ kg DM, respectively) until they reach a BW *ca* 65% of the mature BW (DM65), and then slaughtered. Internal fat depots (IFD; mesenteric, omental and kidney plus pelvic) were removed and weighed. There was a tendency ($P < 0.1$) for a higher relative weight of the IFD (g/kg empty BW) for IFD in CTQ (22 and 16%, more fat at DM45 and DM65, respectively). Animals fed high energy diet had a higher relative weight of the IFD ($P < 0.01$). Differences between breeds were also observed for mesenteric fat, the content in CTQ lambs having higher content ($P < 0.05$). Animals fed high energy diet presented more kidney plus pelvic ($P < 0.01$) and omental ($P < 0.01$) fat.

Session 25 Poster 57

Intramuscular fat levels in sheep muscle during growth

D.L. Hopkins[1], M.J. Mcphee[2] and D.W. Pethick[3], [1]NSW Department of Primary Industries, Centre for Sheep Meat Development, PO Box 129,, Cowra, N.S.W., 2794, Australia, [2]NSW Department of Primary Industries, Beef Industry Centre of Excellence, JSF Barker Building, Trevenna Road, UNE, Armidale NSW 2351, Australia, [3]Murdoch University, 3Department of Veterinary Biology and Biomedical Science, South Street,, Murdoch, WA 6150, Australia

The level of intramuscular fat (IMF) in muscle is important as it impacts on eating quality. Serial slaughter studies in cattle have shown that the rate of IMF accretion increases significantly in a linear manner after approximately 150-200 kg HCW. The pattern of accretion in sheep is not well characterised. A longitudinal experiment was conducted with 595 animals of mixed sex (ewe; wether) and of five genotypes by slaughtering at 4 ages, from weaning (4 months) to 22 months. After carcase composition (fat, ash and protein (lean)) was determined by dual energy X-ray absorptiometry, the entire loin muscle removed and weighed and IMF% determined using NIR. Analysis revealed a significant, but relatively small increase in the IMF% as the sheep aged. The ratio of IMF to total carcase fat declined as HCW increased suggesting that IMF deposition was an early maturing feature. Genotype ($P < 0.01$) and the interaction between HCW x sex ($P < 0.01$) impacted on the curvilinear relationship ($R^2 = 61.0$) and the decrease in accretion of IMF as HCW increased suggests that excessive finishing systems will not lead to significant gains in IMF levels. This is an important outcome for the sheep industry.

Session 25 Poster 58

Genetic variability among four Egyptian sheep breeds using random amplified polymorphic DNA (RAPD) and PCR-RFLP techniques

S.A. El-Fiky[1], S.M. Abdel-Rahman[2], S.A. Hemeda[1], A.F. El-Nahas[1] and S.M. Nasr[1], [1]Faculty of Veterinary Medicine, Alexandria University, Animal Husbandry and Wealth Development, Edfina, Rasheed, Beheira, Egypt, [2]Genetic Engineering Institute, Mubarak City for Scientific Research and Technology Applications, Nucleic Acid Research, Research Area, New Borg El-Arab, 21934, Alexandria, Egypt

Genetic similarity among four Egyptian sheep breeds (Barki, Rahmani, Ossimi and Romanov) were studied using both random amplified polymorphic DNA (RAPD) and restriction fragment length polymorphism (RFLP) techniques. Sixteen random primers were used to amplify DNA fragments in these four sheep breeds. RAPD patterns with a level of polymorphism were detected among breeds. The results showed that the genetic similarity among these four breeds was as follow: 90% (Barki x Ossimi), 87% (Barki x Rahmani), 87% (Ossimi x Rahmani), 80% (Ossimi x Romanov), 80% (Rahmani x Romanov) and 74% (Barki x Romanov). However, the closer proximity or the highest genetic similarity was observed between Barki and Ossimi (90%), while the lowest was observed between Barki and Romanov (74%). On the other hand, the results of the RFLP for 18S rRNA gene technique showed that, no genetic variation were found among the four sheep breeds under study. Finding in this study confirms the phenotypic classification of Egyptian sheep to three related breeds (Barki, Ossimi and Rahmani) and the other breed (Romanov) is faraway from them.

Session 25 Poster 59

Mutations of the MC1R gene in Sicilian goat breeds, relationships with coat colours and perspectives for their use in breed traceability systems of goat products

F. Beretti[1,2], R. Finocchiaro[2], B. Portolano[2], V. Russo[1], R. Davoli[1] and L. Fontanesi[1], [1]University of Bologna, DIPROVAL, Sezione di Allevamenti Zootecnici, Via F.lli Rosselli 107, 42100 Reggio Emilia, Italy, [2]University of Palermo, Dep. S.En.Fi.Mi.Zo., Sezione di Produzioni Animali, Viale delle Scienze, 90128 Palermo, Italy

Mutations in the melanocortin receptor 1 (*MC1R*) gene have been shown to affect coat colours in diveffent mammals, including several farm animals. Due to the fact that most farm animal breeds are fixed for distinctive coat colours and pattern, polymorphisms of this gene have been proposed to assess between breed variability and as tools for breed identification and animal products authentication. We sequenced fragments of the *MC1R* gene in 30 goats of three Sicilian breeds showing differences on coat colours: Rossa Mediterranea, solid red colour; Girgentana, cream/light grey with few small red spots; Maltese, white with black spotted head. Five SNPs were identified in the coding region: one silent, one nonsense and three missense mutations. The nonsense mutation was genotyped by PCR-RFLP in a larger sample of goats of these breeds. All Girgentana goats carried this mutation that was also present in few Rossa Mediterranea animals but not in any analysed Maltese goats, suggesting that it might affect, at least in part, red coat colour. Further studies are underway to confirm the effect of this mutation and, eventually, to use it for breed traceability of goat products.

Session 25 Poster 60

Sheep and goat grazing in relation to riparian and watershed management

S.J. Sadatinejad, Shahrekord University, Natural resources, Shahrekord University, Shahrekord, Iran, 115, Iran

Grazing is an integral and natural process on rangelands. A critical measure of grazing management success is the functional capacity of rangeland watersheds and riparian ecosystems. Long-term studies have concluded that grazing can be managed to manipulate the vegetative composition of rangelands. The increasing demand for water and the importance of riparian areas for recreation, and their contribution to biodiversity will place increasing emphasis on riparian and especially watershed management. It is clear that overgrazing and poorly timed grazing by any livestock species can be detrimental to rangeland resources. However, on the basis of our research low to moderate sheep and goat grazing levels at optimum times for the vegetation community are not detrimental. The obvious conclusion is that sheep and goat can be used as a tool for improvement of rangeland resources. Many, but not all, environment has been shown to be responsive to managed grazing as an improvement tool.

Session 25 Poster 61

Farm and breed effects on milk yield and composition of Hungarian goats

T. Németh[1], G. Baranyai[2] and S. Kukovics[1], [1]Research Institute for Animal Breeding and Nutrition, Gesztenyés u. 1., 2053 Herceghalom, Hungary, [2]Hungarian Goat Breeders Association, Gesztenyés u. 1., 2053 Herceghalom, Hungary

Based on four weekly test milkings between April and October 2006, the authors collected milk quantity data from 5 goat farms, including two small (30 and 70 does) farms and three bigger ones carrying 150 - 250 does. During the test milkings individual milk samples were collected (twice a day) from the does belonging to 4 goat breeds like Hungarian Milking White, -Milking Brown, -Milking Multicolour, Anglo-Nubian. The milk samples were analysed by an official raw milk laboratory (Livestock Performance Testing Ltd. in Gödöllő). The fat, protein and lactose content as well as the somatic cell count were measured by Milkoscan 600 and Fossomatic instruments. Data collected were processed by using SPSS for Windows 10.0 software. Average milk yield and milk composition and their changes during the day and over the lactation were studied followed by the correlations between daily milk yields and contains. Farm, breed and age effects on the lactation yield and/or composition were also examined. The breed and age effect were evaluated within and between farms. There were significant differences found among breeds and farms in milk yield and composition as a consequence of nutritional deviations. The labour quality and goat handling technology caused big differences among farms in somatic cell count values. No significant differences were found among Hungarian breeds in the summarised data.

Session 25 Poster 62

Determination of milk yield in sheep using RFID identification system

Ü Yavuzer, University of Harran, Faculty of Agriculture, Dept. of Animal Science, Şanlıurfa, 63040, Turkey

This study aims to find the most suitable milk yield prediction method that can be uses by comparing different prediction methodologies using sheep daily milk yield data and recorde by an electronic system during lactation. The electronic identification and milk yield measuremet system makes use of RFID (Radio Frequency Identification) tags to identify each animal, and sends milk weight measured by electronic scales to a computer. The data can then be use in analysis. Electronic sheep identification and milk yield measuremet system available systems. Animals will be automatically identified using RFID tags. Milk weight data will be sent to a computer along with sheep identification information. The computer will store individual sheep's milk yield records, and display it on demand.

Session 25 Poster 63

A comparative study of ewe's, goat's and camel's milk during lactation stages in Egypt; 2 - nitrogen distribution

K. Soryal[1], A. Hagrass[2], A. Metwally[2] and A. Ibrahim[1], [1]Desert Res. Center, Animal Breeding Dept., Matarya, 31111 Cairo, Egypt, [2]Fac. of Agric., Ain Shams Univ., Food Sci. Dept., Shubra El Khema, 31111 Cairo, Egypt

Camel, sheep and goat are considered the important animals in arid and semi-arid regions for supplying milk. Their ability to live on poor grazing land was considered and their milk played a key role in the nutrition of the inhabitants. In Egypt, there is a significant milk gap in spite of the low kg /year per capita consumption therefore; interest was directed to these animals as a source of milk. Milk samples of ewes, does and camels were collected during early, mid and late lactation stages from the flock of Desert Research Center. In sheep, total and protein nitrogen showed the greatest values during mid lactation while, casein nitrogen recorded the lowest values during late lactation ($P < 0.05$). Whey protein nitrogen recorded the highest value during mid lactation ($P < 0.05$). The non-protein nitrogen recorded the maximum values during late lactation ($P < 0.05$). Casein nitrogen of goats recorded the highest values during mid lactation ($P < 0.05$). Lowest values of proteose peptone and highest values of non-protein nitrogen were recorded at late lactation ($P < 0.05$). All the nitrogen fractions of camel milk showed maximum values during early lactation and remarkable decrease during late lactation ($P < 0.05$). Ewe milk showed the highest values of all nitrogen fractions ($P < 0.05$).

Session 26 Theatre 1

Genetic parameters for carcass traits, bone strength and osteochondrosis in Finnish Landrace and Finnish Large White pigs

M.-L. Sevón-Aimonen, A. Storskrubb, A. Mäki-Tanila, M. Honkavaara and M. Puonti,

The objective of this study was to estimate genetic parameters for carcass traits (Hennessy meat percentage, and proportions of valuable cuts, bones and fat in commercially dissected carcasses), bone strength (breaking force of *fibula*) and osteochondrosis (OD, at proximal end of *humerus* and distal end of *femur* using a visual analogue scale of 1 to 6, where 1 = no changes and 6 = severe changes). Data were records from 464 Finnish Landrace and 326 Finnish Large White station test pigs slaughtered at 100 kg live weight. (Co)variances were estimated using an animal model REML and DMU software. The statistical model contained breed, sex, carcass weight and rearing batch as fixed effects and litter, additive animal and error as random effects. The heritabilities for carcass traits were high (h^2 0.31 to 0.71) and for bone strength were moderate (h^2 0.26). The heritability for OD in *humerus* was almost zero but in *femur* it was moderate (h^2 0.26). The genetic correlations between breaking force and Hennessy meat percentage, and proportion of valuable cuts, bones and fat in carcass were -0.18, 0.04, 0.52 and -0.07, respectively. These results demonstrate that decreasing fat content by selection presumably does not impair bone strength.

Session 26 Theatre 2

Genetic parameters for litter size in pigs by joining purebred and crossbred data

Š. Malovrh[1], E. Groeneveld[2], N. Mielenz[3] and M. Kovač[1], [1]University of Ljubljana, Groblje 3, SI-1230 Domžale, Slovenia, [2]Federal Agricultural Research Center, Höltystrasse 11, D-31535 Neustadt, Germany, [3]Martin-Luther-University, Halle-Wittenberg, Adam-Kuckhoff-Strasse 35, D-06108 Halle, Germany

The aim of study was genetic evaluation of purebred animals for litter size using records of pure- and crossbred relatives from crossbreeding scheme. Data was obtained for two breeds Slovenian Landrace (line 11, SL) and Large White (LW), and their crosses SL (dam) x LW (sire) from three Slovenian pig farms. Heterogeneous genetic (co)variances among populations and the other covariance components were estimated by the REML method using the VCE-5 package. Litter size in purebred and crossbred populations was treated as different traits with different models. The repeatability model was utilised, the fixed effects differed for gilts and sows. Common litter and permanent environment were included as trivial random effects. Animal model with direct additive genetic effect was used for purebreds, while reduced animal model was applied for crossbreds. Common litter effect accounted for 0.2% to 1.7% of variation in litter size, while permanent environment explained between 4.3 and 7.9% of variation for purebred animals and between 10.7% and 12.1% for crossbreds. Heritabilities reached values between 10.2% and 11.9% in purebreds and between 8.8% and 13.5% in crossbreds. Genetic correlations between SL and SLxLW were from 0.89 to 0.99, while between LW and SLxLW, they were lower and ranged from 0.84 to 0.92.

Session 26 Theatre 3

Parameters of AI boars and predicted correlated responses of selection against boar taint

R. Bergsma[1], E.F. Knol[1] and H. Feitsma[2], [1]Institute for Pig Genetics, P.O. Box 43, 6640 AA Beuningen, Netherlands, [2]Dutch Association of Co-operative Pig AI Centres, Maasbreeseweg 65, 5988 PA Helden, Netherlands

From the IPG database all ejaculates of 1 sire line were selected. This resulted in a dataset of 746 boars with known parentage with, on average, 66.3 ejaculates per sire. Information on volume, sperm concentration, motility and decrease of motility in time were available, lifetime production was calculated. Heritability's were calculated using a repeatability model. Information on androstenon, skatol and indol levels in 700 pure line slaughter pigs of the same genetic line were available. Sperm concentration and sperm motility were repeatable ($r^2 > 0.4$), as was sperm volume ($r^2=0.35$). Decrease of motility in time was not. About half of the repeatability had a heritable background. Heritability's of androstenon, skatol and indol levels in slaughter pigs were high (0.75, 0.53 and 0.29 respectively). To prevent inefficient semen production on days of peak production, guaranteed high semen quality boars can be selected, since concentration and motility are rather repeatable traits. Suitability for storage does not depend on the individual boar. So evaluation will be necessary for every ejaculate. The impact of selection against boar taint on male fertility characteristics will not be very high, since genetic correlations were low (-0.19 - +0.25) but, in general, unfavourable.

Session 26 Theatre 4

Piglet preweaning survival and its relationship with the breeding goal of a Large White boar line

A. Cecchinato, V. Bonfatti, L. Gallo and P. Carnier, University of Padova, Department of Animal Science, Viale dell'Università 16, 35020 Legnaro Padova, Italy

Aim of this study was to investigate piglet preweaning survival and its relationship with a total merit index (TMI) used for selection of breeding candidates in the C21 Large White boar line. Data on 13,924 crossbred piglets (1,347 litters), originated by 189 Large White boars and 328 Large White-derived crossbred sow, were analyzed under a frailty proportional hazards model, assuming a piecewise Weibull distribution and including sire and litter as random effects. Estimated hazard ratios (HR) indicated that sex, cross-fostering, year-month of birth, parity of the nurse sow, size of the litter and class of TMI were significant effects for piglet survival. Female piglets (HR = 0.80) had less risk of dying than males as well as cross-fostered piglets (HR = 0.50). Survival increased when parity of the nurse sow increased up to the fourth parity (HR = 0.70), but further increases of parity did not enhance pre-weaning survival. Piglets of small (HR = 3.13) or very large (HR = 3.69) litters had less chances of surviving in comparison with litters of intermediate size. Class of TPI exhibited an unfavourable relationship with survival (HR = 1.23 for the TMI top class). Estimated heritability of survival was 0.025. These results suggest that piglet preweaning survival should be included in the breeding goal of the C21 boar line.

Session 26 Theatre 5

Claw lesions as a predictor of lameness in breeding sows

J. Deen, S.S. Anil and L. Anil, University of Minnesota, Veterinary Population Medicine, 1998 Fitch Ave, St Paul, MN 55108, USA

Claw lesions are very common in pigs and may cause lameness. Claws of 771 sows were individually examined for lesions on the side wall, heel, sole, junction between heel and sole, white line and toe and the association of lesion scores (< median *vs* ≥ median) with lameness was analyzed using multivariate logistic regression analysis (Proc logistic, SAS v 9.1). Claws were examined on day 110 of gestation. Lesions were scored on a scale of 0 (no lesions) to 4 (severe). The numbers of lesions were multiplied by their severities to obtain the final lesion score. The lesion scores on different claw areas of lame and non-lame sows were also compared using a Kruskal-Wallis test. Sows with less than median heel lesion scores had 34% lower ($P \leq 0.05$) likelihood of being lame. Sows with less than median white line lesion scores were 31% less ($P \leq 0.05$) likely to be lame. Likewise, less than median side wall lesions had a tendency to be 31% less likely to be lame ($P = 0.06$). However, other lesions were not associated with lameness. The comparison of lesion scores indicated differences ($P \leq 0.05$) between lame and non lame sows in terms of lesion scores on side wall and white line. This study indicated the association of claw lesions, especially on white line, heel and side wall with lameness. Measures to minimize incidence of claw lesions may reduce the adverse economic effects and compromised welfare due to lameness in swine breeding herds.

Session 26 Theatre 6

In vivo and post mortem electronic identification and DNA analysis in swines

W. Pinna[1], M.G. Cappai[1], G. Garau[1], A. Fraghi[2] and S. Miari[2], [1]University of Sassari, Dipartimento di Biologia Animale Sezione di Produzioni Animali, via Vienna n. 2, 07100 Sassari, Italy, [2]Istituto Zootecnico e Caseario della Sardegna, Strada Sassari-Fertilia Km 18,600, 07040 Bonassai (Olmedo), Italy

71 piglets (1d to 4d aged and at 2.9 to 3.7 kg weighed) after being identified by eartags were divided into two groups: 52 piglets (T group) were also electronically identified by injectable transponders HDX 32.5×3.8 mm 134.2 kHz and sampled for DNA analysis on auricle tissue; 19 piglets (C group) as control. Averaged time for transponder's application and auricle sampling took 2'±9"/pig. On 4 T group animals, x-ray analysis was carried out at 0d, 28d and 70d after EID to detect its position in abdomen. Live weights: 122.1±4.0 kg *vs* 123.6±3.6 kg; mean daily gain: 624±0.9 gr *vs* 632±1.7 gr; carcass yield: 78.2% *vs* 77.6% of T and C group pigs, respectively slaughtered at 203d, showed no statistic significance. Eartag loss on pigs from both groups was 11.2% at 28d, 14.1% 70d and 22.5% 203d. *In vivo* and *post mortem* transponders readability was 100%. Transponders recovery in the abattoir was 100%: 75% found free in abdomen, 17.3% embedded in omentum fat and 7.7% extraperitoneal. Genetic profiles by 6 microsatellites among FAO-ISAG panel confirmed the correspondence between *in vivo* and post mortem samples. DNA analysis on muscles randomly sampled from 11 T group carcasses matched with genetic profiles from auricles, while muscles' genetic profiles from 7 C group carcasses showed genetic non identity

---

Session 26 Theatre 7

Performance and carcass quality of castrates and boars fed a standard or a potato starch enriched diet

C. Pauly[1], P. Spring[1], J. O'Doherty[2] and G. Bee[3], [1]SCA, Länggasse 85, 3052 Zollikofen, Switzerland, [2]UCD, Belfield, Dublin 4, Ireland, [3]ALP, Rte de la Tioleyre 4, 1725 Posieux, Switzerland

The objectives of this trial were to evaluate the growth performance, carcass quality and meat quality traits of group-penned boars and to determine the effects of raw potato starch (RPS) on skatole levels in the back fat. At weaning, 36 pigs were blocked by BW into 12 blocks (3 littermates/block) and assigned to 3 treatments: T1: surgical castrates; T2: boars; T3: boars offered RPS (30% RPS the last wk before slaughter). Throughout the experiment, the pigs had *ad libitum* access to the diet. Individual feed intake was recorded and BW was determined weekly. Results were analysed by ANOVA and Mann Whitney U test ($P < 0.05$). From 20 to 65 kg and 20 to 105 kg BW, boars grew slower (766 *vs* 828 g/d and 774 *vs* 830 g/d, respectively; $P < 0.01$), consumed less feed (1.88 *vs* 2.23 kg/d;$P < 0.01$) and were more efficient (2.43 *vs* 2.69; $P < 0.01$) than castrates. Dressing percentage was lower (79.4 *vs* 81.6%; $P < 0.01$) and the percentage valuable cuts were higher (56.9 *vs* 52.6%; $P < 0.01$) in carcasses of boars than castrates. Initial and ultimate pH, meat colour, and drip loss percentage after 48 h did not ($P > 0.05$) differ among treatments. The back fat of T3-boars fed RPS had lower (P=0.05) skatole concentrations (0.22 *vs* 0.85 ppm in 'water-free' fat) than T2-boars. Future research is needed to determine if the low feed intake of the boars was due to more active behaviour or reduced intake capacity.

Session 26 Theatre 8

Growth and product quality of different pig breeds fattened according to the regulations of organic farming

S. Küster[1,2], U. Baulain[1], M. Henning[1] and H. Brandt[2], [1]Institute for Animal Breeding, Höltystraße 10, 31535 Neustadt, Germany, [2]Institute of Animal Breeding and Genetics, Ludwigstraße 21b, 35390 Gießen, Germany

In order to investigate which pig breeds are suitable for organic pig farming a total of 90 pigs were housed and fed under the regulations of organic farming. Breeds were: Swabian Hall (SH) as a rare breed, Piétrain*Swabian Hall (Pi*SH), Duroc*German Landrace (Du*DL) and pigs of the German Federal Hybrid Pig Breeding Programme (BHZP) as common breeds. During the fattening period all animals were weighed once a week and scanned for tissue composition with a Magnetic Resonance Tomograph (MRT). Pigs were slaughtered at a live weight of about 120 kilograms. One carcass side was scanned again by means of MRT, carcass quality was determined and samples taken for investigation of meat quality characteristics. Evaluation of growth, carcass - and meat quality traits showed that SH pigs had the highest growth of fat volume and lowest in muscle volume right from the beginning of the fattening period. Du*DL had the significantly highest average daily gain while highest muscle growth was found in Pi*SH and BHZP. This leads to a superior carcass quality, while Swabian Hall had the significantly lowest lean content. In meat quality Pi*SH had a significantly lower pH_1 in the loin muscle 45 minutes post mortem and highest drip loss 24 hours post mortem. Although differences between the breeds in meat quality were found, no general quality deficiencies were noticed.

Session 26 Theatre 9

Effects of the water-feed ratio and of a rheological sepiolite on some physical parameters of liquid feed and performance of pigs

E. Royer[1], V. Ernandorena[1] and F. Escribano[2], [1]Ifip, 34 bd de la Gare, 31500 Toulouse, France, [2]Tolsa SA, Box 38017, 28080 Madrid, Spain

This study evaluated the effects of the water-feed ratio (WFR) and of a rheological sepiolite (Splf, Tolsa SA) on physical parameters of the liquid diet, reliability of liquid feeding systems and performances in growing-finishing units. In exp.1 (23 to 115 kg), 48 pigs per group received *ad lib* a diet mixed with water in a ratio of 2.8:1 l/kg, or the same diet with 1 % Splf at 2.8:1 or 2.2:1 l/kg. In exp.2 (27 to 69 kg), 36 pigs per group were fed *ad lib* in a 2x2 factorial study: inclusion or not of 1% Splf and WFR (2.7:1 or 2.35:1 l/kg). From d0 to d21 in exp.1, pigs receiving Splf at a 2.8 or 2.2:1 ratio had higher intakes (FI: respectively 1.64 and 1.64 *vs* 1.59 kg/d; $p < 0.05$) and daily gains (ADG: 796 and 798 *vs* 772 g/d; $p < 0.05$) as feed conversion ratio was not affected. From d0 to d47, FI (2.08 and 2.11 *vs* 2.04 kg/d; $p < 0.01$) and ADG (897 and 905 *vs* 886 g/d; not significant) stay higher. From d0 to d15 in exp.2, FI were equal for all diets and ADG of Splf treatments (783 *vs* 759 g/d) did not differ statistically. From d0 to d43, ADG and FCR were unchanged by WFR and were slightly but not significantly improved by Splf (ADG: 992 *vs* 974 g/d). It was concluded that rheological properties of Splf increase water retention by diet, setting of particles in suspension, fluidity of the liquid mixture, facilitate accuracy of the distribution and a lower WFR, and improve performances of growing pigs.

Session 26 Theatre 10

Evaluation of different farrowing systems regarding productivity and animal welfare

J. Baumgartner, E. Ofner, E. Quendler and C. Winckler, University of Veterinary Medicine Vienna, Veterinaerplatz 1, A-1210 Vienna, Austria

In order to save costs most sows are kept in crates during farrowing and lactating. This leads to a number of welfare problems. In this study the effect of different farrowing pens on animal health and behaviour, on productivity and working time requirement of the farm was investigated. Data collection took place in a commercial farm with 109 farrowing places in 9 different types of farrowing pens: 5 systems with a crate and 4 free farrowing systems. The herd consisted of 600 sows which were batch farrowed in 5 groups and a 4 weeks cycle. Productivity data of 1500 sows were analysed. The working time requirement was measured on the level of work elements. Clinical examination of 455 sows and litters were carried out before birth, in the first and third week of lactation. In each type of farrowing pen the behaviour of at least 20 sows was video taped. From 9.1 to 9.9 piglets/litter were weaned in the farrowing pens investigated. Piglet loss from crushing ranged from 4.5 to 13.2 % with significant effects of farrowing system, management procedure and time of farrowing. Differences of 20 to 40 % in working time requirement could be found between systems. The prevalence of skin lesions of the animals differed considerably between the different farrowing pens. It is concluded, that both farrowing crates and free farrowing pens have positive and negative effects on economy and animal welfare. Thus a final evaluation of farrowing systems has to be based on ethical judgement.

Session 26 Theatre 11

Performance and carcass quality of genetically different pigs under conventional and organic conditions

D. Werner[1], W. Brade[2], F. Weismann[3] and H. Brandt[1], [1]University Giessen, Institut für Tierzucht und Haustiergenetik, Ludwigstr. 21 B, 35390 Giessen, Germany, [2]Landwirtschaftskammer Niedersachsen, Johannssenstr. 10, 30159 Hannover, Germany, [3]Bundesforschungsanstalt für Landwirtschaft, Institut für ökologischen Landbau, Trenthorst, 23847 Westerau, Germany

A study was conducted to examine the performance and carcass characteristics of genetically different pigs under conventional and organic conditions. The pig genotypes were Bundeshybridzuchtprogramm, Schwäbisch Hällisches Schwein (SH), Angler Sattelschwein (AS), Piétrain x SH (PIxSH), Piétrain x AS (PIxAS), Piétrain x Deutsches Edelschwein and Duroc x Deutsche Landrasse. A total number of 682 pigs were kept at two test stations under conventional and organic housing and feeding. The fattening performance (daily gain and feed conversion) and carcass characteristics (meat percentage, fat measurements and meat quality) of all pigs were analysed using a linear model including the fixed effects of genotype, sex, environment and the interaction of genotype and environment. Although significant interactions between genotype and environment were found for most of the traits no differences in ranking between breeds within the two environments are observed. So in conclusion the best breeds under conventional housing and feeding will also perform well under organic housing and feeding conditions and no special breeding program for organic pig production would be necessary.

Motivation for additional water intake in dry and liquid fed finishing pigs

H.M. Vermeer and N. Kuijken, Animal Sciences Group - Wageningen UR, Animal Husbandry, PO Box 65, 8200 AB, Netherlands

The EU minimum standards for the protection of pigs (2003) state that pigs must have permanent access to a sufficient quantity of fresh water. An experiment with additional water supply to finishing pigs in 2 batches with 48 pens (566 pigs) was carried out. The aim was to measure motivation for additional water intake by reducing the flow rate in: A–Long trough: Liquid feed, 3 times per day, 12 feeding places; B–Sensorfeeding: Liquid feed, from 3 to 10 meals/day, 4 feeding places; C–Variomix: Liquid feed, 1 feeding place; D–Dry feed: Dry pelleted feed + drinker in trough, 1 feeding place. Each pen had an additional drinker with a weekly changing flow rate of 134, 356, 733 or 1041 ml/min. From the additional drinker pigs in D drank 3.39 l and A, B and C 0.76, 0.58 and 0.44 l per pig per day. Pigs in D also drank 1.09 l from the nipple drinker in the feeder and the liquid fed pigs received 6.5 l in the feed. The highest flow rate resulted in 4.37 l for D and 1.05, 0.88 and 0.57 for A, B and C. The lowest flow rate gave a 3 times higher drinking time and a significant reduction in water intake: 1.95 for D and 0.40, 0.20 and 0.28 l/pig*day for A, B and C. The conclusion is that the additional water intake was lower for the liquid than for the dry fed pigs. At lower flow rates the water intake was lower, but drinking time (motivation) was higher than at high flow rates in all treatments. However, the water intake of liquid fed pigs is so low that it does not justify an extra drinker.

Relationship between behaviour of sows and piglet losses

D. Wischner[1], B. Hellbrügge[1], K.-H. Tölle[2], U. Presuhn[3] and J. Krieter[1], [1]Christian-Albrechts-Universität Kiel, Institute of Animal Breeding and Husbandry, Olshausenstrasse 40, 24098 Kiel, Germany, [2]Chamber of Agriculture Schleswig-Holstein, Blekendorf, 24327 Blekendorf, Germany, [3]farm concepts, Wahlstedt, 23812 Wahlstedt, Germany

The behaviour of 386 German Landrace sows with 438 pure-bred litters were videotaped continuously starting 12 hours ante partum until 48 hours post partum in a nucleus herd for one year. From the population sows were randomly sampled in a block data design considering different matching criteria (number of piglets born alive, parities and farrowing date). Two blocks (groups) of sows were compared: one group of sows that crushed more than one piglet (C) the other group crushed none (NC). In total 24.000 observations were analysed of sow's positions and maternal behaviour. NC sows performed a longer duration of nest-building behaviour (39 sec. *vs* 35 sec.) as well as increased frequencies of nest-building behaviour ante partum (89 times *vs* 58 times). Additionally, these sows showed significant fewer movements like 'standing' or 'rolling' post partum. Further NC sows showed a higher attentiveness towards their piglets by e.g. looking more often to the piglet nest and taking higher nose-to-nose contacts. In conclusion, NC sows acted more carefully and showed less risky body movements (e.g. rolling). In spite of different patterns of behaviour a general measurement of activity (e.g. lying behaviour) is useful to characterize the mothering ability of sows to minimise piglet losses.

Changing from chemical to near infrared transmittance analysis for intramuscular fat content

E.J. Gjerlaug-Enger[1], B. Holm[1] and O. Vangen[2], [1]NORSVIN, P.Box. 504, No-2304 Hamar, Norway, [2]University of Life Sciences, Dep. of Animal and Aquacultural Sciences, P.Box. 5003, No-1432 Aas, Norway

As a result of a larger research project on meat- and fat quality, Norsvin has changed from analysing intramuscular fat (IMF) content by chemical analysis to using near infrared transmittance (NIT), Foodscan. The Norsvin test system includes a complete carcass dissection of 2.800 purebred animals annually. IMF content has been chemically determined on these animals since 1998, but for cost and time reasons, Norsvin wanted to evaluate a faster and cheaper method. The phenotypic correlation between the chemical- and NIT IMF determined content were 0.87 and 0.95 for Norsvin Landrace and Norsvin Duroc, respectively. Single trait analysis resulted in a substantially higher heritability for NIT IMF. Multivariate evaluations revealed that genetic correlations were close to 1. Norsvin has therefore included NIT IMF in the genetic evaluation models for both Norsvin Landrace and Norsvin Duroc. The heritability for NIT IMF in the Norsvin Duroc and Norsvin Landrace are 0.53 and 0.44, respectively. Genetic correlations to other carcass-, production- and efficiency traits revealed that IMF content is highly unfavourably linked to several traits, and that the correlations are more unfavourable the leaner the breed is. In the same project, Norsvin currently work on further research to implement effective selection for drip loss, fat quality and meat colour in general in the selection programme.

Performance and carcass quality of entire male pigs fattened on a commercial farm in Switzerland

C. Pauly[1], J. O'Doherty[2] and P. Spring[1], [1]SCA, Länggasse 85, 3052 Zollikofen, Switzerland, [2]UCD, Belfield, Dublin 4, Ireland

The aim of this trial was to determine the level and seasonal variability of boar taint under Swiss commercial conditions. Five hundred and eight boars were fattened in a commercial barn with 10 pens of 20 animals each. The diet consisted of corn cob mix, protein concentrate (soybean cake, wheat starch, rapeseed cake, broken rice, sugar beet pulp) and water (15.5 MJ DE, 188 g CP and 10 g dLys per kg DM). Androstenone, indole and skatole concentrations were determined by HPLC from 180 animals randomly selected over time. Multiple regressions were used to estimate the effect on androstenone level and ANCOVA was conducted to evaluate seasonal effects on androstenone concentrations. Average daily gain from birth to slaughter (average age 164 d and 72 kg hot carcass weight) was 543 g/d. Lean meat percentage was 56.5 %. Skatole levels ranged from 0.03 to 0.48 ppm while androstenone levels ranged from 0.3 - 5.2 ppm. Out of 180 boars, only two animals exceeded the acceptable skatole limit of 0.17 ppm and 7 animals exceeded the androstenone limit of 1.7 ppm (limits are set for 'water-free' fat). Carcass weight, slaughter age, and birth period influenced androstenone levels in the backfat. Boars slaughtered in March and April had higher ($P < 0.05$) androstenone levels than those slaughtered in May and October (0.98 and 0.68 *vs* 0.54 and 0.52 ppm). Slaughtering the boars at a maximum of 170 d of age allows keeping boar taint reasonably low; however, carcass weights have to be improved.

Session 26 Poster 16

A comparison of vitality and growth performance before weaning of crossbred piglets obtained from Piétrain or crossbred Large White x Piétrain boars and Large White x Landrace sows

N. Quiniou, I. Mérour and S. Boulot, IFIP Institut du Porc, BP35104, 35561 Le Rheu cedex, France

A trial was carried out to compare gestation length (without farrowing induction), farrowing and lactation progress when sows were artificially inseminated with semen from either Yorkshire x Piétrain crossbred sires (treatment LWxPP) or purebred Piétrain ones (treatment PP) homozygote halothane reacting pigs. Sows were allocated to one of both types of sires depending on their parity and body condition. Gestation length averaged 114.3 days and farrowing 5.0 hours, whichever the sire concerned. Stillbirth rate did not differ among type of boars (7 and 9% in treatments LWxPP and PP, respectively). Losses over the first 24 hours of life were higher in piglets from PP sire (8 *vs* 5% in LWxPP, $P < 0.05$) but the difference was not high enough to induce a significant difference on litter size at weaning (10.9 piglets on average). Growth rate averaged 270 g/d/piglet for both treatments. Proportion of splayleg piglets was not higher with PP sire but piglets seemed to be more sensitive to infectious lameness (14 *vs* 8% of born alive and uncrushed piglets). Estimation for genetic sensitivity to these pathologies needs to be more specifically studied. According to present results, changing the sire's breed from LWxPP to PP does not imply any change in farrowing induction schedule. Induction would even be more interesting with PP semen as a more intensive supervision is required to prevent from peri-partum losses.

Session 26 Poster 17

Effect of providing turkey drinkers as a source of supplementary water to newly weaned pigs

P.G. Lawlor[1], P.B. Lynch[1] and A. Guilhot[1,2], [1]Teagasc, Pig Production Development Unit, Moorepark Research Centre, Fermoy, Co. Cork, Ireland, [2]ENITA de Bordeaux, 1 Cours du Général de Gaulle, CS 40201, 33175 Gradignan Cedex, France

The objective of the study was to assess the effect on pig growth performance of providing a supplementary turkey drinker (Rotecna, Lleida, Spain) to weaned pigs for 21 days post-weaning. Twenty four groups of 14 pigs were formed from pigs weaned at 28 days of age. The experiment was designed as a 3 (weaning weight categories; light, medium and heavy) x 2 (supplementary water; with and without) factorial. Each pen, regardless of treatment, was serviced with one drinking bowl (BALP, La Buvette, Charleville Nord, France). All pigs were provided with the same diets for the duration of the experiment. The mean water usage from the Rotecna drinkers was 677 s.d. 51.1 ml/pig/day. However, providing supplementary water did not effect daily feed intake ($P > 0.05$), average daily gain ($P > 0.05$) or feed conversion efficiency ($P > 0.05$). Pig weight was 6.3, 7.7 and 9.0 Kg (s.e.d. 0.24; $P < 0.001$) at weaning and 12.2, 13.7 and 15.3 Kg (s.e.d. 0.37; $P < 0.001$) at 21 days post-weaning for light, medium and heavy pigs, respectively. Daily feed intake was higher ($P < 0.001$) for the medium and heavy pigs than the light pigs. In conclusion, water availability from the fixed BALP bowl drinkers was sufficient to meet piglet requirements. The weight differential between groups of pigs graded on weight at weaning persisted to 21 days post-weaning.

Session 26 — Poster 18

Pig hybridization in Lithuania

A. Klimienė and R. Klimas, Šiauliai University, Biological Research Centre, P. Višinskio 19, LT-77159, Šiauliai, Lithuania

By the end of the year 2006, in the Lithuania breeding centres about 25 % of all purebred pigs consisted of Lithuanian White (LW), 31 % - Large White –Yorkshire (Y), 42 % - Landrace (L) and 2 % - Duroc (D) and Pietrain (P). Besides that, there are 7 breeding enterprises where sires of the mentioned breeds are housed for semen collection. The goal of this work was to determine the most effective combinations of pig hybridization. The work has been carried out in the years 2004-2005. Litter size of sows, crossing with the boars of other breeds, ranged from 10.3 (YxP) to 11.2 (Yx L). The difference is statistically reliable ($P < 0.05$). Better growth of crossbreds was determined in these combinations, where boars of Duroc breed were used. Milk yield of sows, crossing with Duroc, ranged from 64.8 to 66.0 kg. Crossbreds of Y, L and D (three breeds) distinguished by the speediest growth and highest daily gain (994 g), and crossbreeds of Y and P - by the least consumption of feed per kg gain (2.5kg), ($P < 0.01$-0.001). Crossbreeds of Y x P and L x P distinguished by the biggest muscularity (57.4 and 57.2 %) among the crossbreds. The least muscularity (54.0 %) was noticed namely among hybrids of those combinations (YxLxD), which had the highest daily gain ($P < 0.05$-0.001). Thus, solving relevant questions of meat quality in the future, we will maybe have to refuse P breed in the hybridization combinations. D and L x D crossbreds are turning to be the most promising paternal breeds.

Session 26 — Poster 19

Physical changes in gestating sow body size

M.K. O'Connell[1], P.B. Lynch[1], S. Bertholet[2], F. Verlait[2] and P.G. Lawlor[1], [1]Teagasc, Pig Production Development Unit, Moorepark Research Centre, Fermoy, Co. Cork, Ireland, [2]ENITA de Bordeaux, Cours du Général de Gaulle, C540201, 33175 Gradignan Cedex, France

The objectives were to model changes in gestating sow size and document the morphometric measurements of modern sows for use in sow housing design. Sets of measurements were taken on days 0 (at service; n=74), 25 (n=78), 50 (n=80), 80 (n=140) and 110 (n=157) of gestation. Parity was 1 to 8. The 12 measurements were: BW, P2 backfat depth (over last rib, 65 mm from dorsal midline), heart girth, height (floor to: dorsal surface of last rib, ventral surface of sow, dorsal surface of hip), depth at last rib, length (from snout and shoulder to posterior) and body width (ham, last rib, shoulder). All measurements were taken while sows stood level. The effect of day of gestation (DAY) on measurements of sow size were analysed by the PROC GLM procedure of SAS 9.1 (SAS Inc., Cary, NC), with sow as the experimental unit. Mean values for DAY were used in PROC REG. Mean sow body depth increased 1.03 (±0.197) mm per DAY (R^2 0.90, $P < 0.05$, RSD 17.2 mm). Sow length from shoulder to posterior increased 0.76 (±0.095) mm per DAY (R^2 0.95, $P < 0.01$, RSD 8.3 mm). Height from floor to ventral surface of the sow decreased 0.96 (±0.177) mm per DAY (R^2 0.91, $P < 0.05$, RSD 15.4 mm). Width at last rib increased 1.09 (±0.159) mm per DAY (R^2 0.94, $P < 0.01$, RSD 13.9 mm), and sow girth increased 1.53 (±0.104) mm per DAY (R^2 0.99, $P < 0.001$, RSD 9.1 mm).

Predicting body weight of gestating sows from morphometric measurements

M.K. O'Connell[1], P.B. Lynch[1], S. Berthelot[2], F. Verlait[2] and P.G. Lawlor[1], [1]Teagasc, Pig Production Development Unit, Moorepark Research Centre, Fermoy, Co.Cork, Ireland, [2]ENITA de Bordeaux, 1 Cours du Général de Gaulle, C540201, 33175 Gradignan Cedex, France

This study examined a number of morphometric measurements of gestating sows that could be used to estimate sow body weight (BW). Measurements were taken on day 0 (at service; n=74), 25 (n=78), 50 (n=80), 80 (n=140) and 110 (n=157) of gestation. Parity was 1 to 8. The 12 measurements were: BW, P2 backfat depth (over last rib, 65 mm from dorsal midline), heart girth, height (floor to: dorsal surface of last rib, ventral surface of sow, dorsal surface of hip), depth at last rib, length (snout and shoulder to posterior) and body width (ham, last rib, shoulder). Sows were measured with specially designed digital calipers, a ruler, a fibre glass tape measure, a scales and a Renco Lean Meater®. All measurements were taken as sows stood level. PROC REG (SAS v9.1, SAS Inc., Cary, NC) was used to predict BW from each measurement. Stepwise regression was used to develop a model to accurately predict BW using a number of morphometric measures. Heart girth was the best individual predictor of BW: BW=-254+0.35*girth; R^2 0.81, RSD 16.5 kg, $P < 0.001$. Including a second measurement, with parity and day of gestation, increased adjusted R^2 to 0.89 and reduced RSD to 12.4 kg: BW=-133+3.77*parity+0.32*day+1.17*backfat+0.23*girth, $P < 0.001$.

Monitoring physical characteristics and productivity of multiparous sows

M.K. O'Connell, P.B. Lynch and P.G. Lawlor, Teagasc, Pig Production Development Unit, Moorepark Research Centre, Fermoy, Co. Cork, Ireland

The objective was to monitor the relationship between sow physical characteristics, productivity and parity. Four hundred and twenty four sets of sow data were included. Parities ≥6 were combined. Sows were weighed (BW), had P2 backfat depth (over last rib, 65 mm from dorsal midline) and flank size (in front of hind legs, from bottom of flank on one side to other, over hip) taken at pre-farrowing and weaning. Measurements were taken as sows stood level. Productivity data were retrieved from the PigCHAMP herd recording system (PigCHAMP, IA). The sow was the experimental unit. PROC REG (SAS v9.1, SAS Inc., NC) was used to model relationships between parity, physical characteristics and productivity. Pre-farrowing BW (Adj. R^2 0.98; $P < 0.001$, RSD 2.5 kg), weaning BW (Adj. R^2 0.99; $P < 0.001$, RSD 2.8 kg) and weaning flank size (Adj. R^2 0.93; $P < 0.001$, RSD 0.74 cm) increased and pre-farrowing backfat depth (Adj. R^2 0.93; $P < 0.01$, RSD 0.15 mm) decreased quadratically with parity. Pre-farrowing flank size increased linearly with parity (Adj. R^2 0.94; $P < 0.001$, RSD 0.73 cm). Lactation BW loss (Adj. R^2 0.77; $P < 0.05$, RSD 2.5 kg) and lactation backfat loss (Adj. R^2 0.85; $P < 0.01$, RSD 0.25 mm) declined linearly with parity. There were no relationships between parity, BW, backfat or flank size and productivity ($P > 0.05$).

Session 26 Poster 22

Evaluation of selected factors influenced on the reproduction of sows in pork production farm in CR

A. Jezkova[1] and J. Vitasek[2], [1]Czech University of Agriculture Prague, Kamycka 129, Prague 6–Suchdol, 165 21, Czech Republic, [2]Czech Animal Breeding Inspection, Přátelství 815, Praha 10- Uhříněves, 104 00, Czech Republic

The reproductive results (percentage of repeated heats, number of live and dead born piglets/litter, and the length of the interval since birth to the conception, intervals of repeated heats, and the losses of piglets till weaning) in 2002-2005 years were used for evaluation. The results obtained in 2005 were evaluated with the special attention (the age structure of the herd, effect of sequence of the litter, interval of repeated heats, effect of the season, influence of the technicians, influence of the artificial collection centre) to detect some key problems in farm. The reproductive indicators in the total are: number of weaning piglets 17.85 pc/sow/year, number of birth piglets in average 11.8 pc/all born/litter and 10.7 pc/live born/litter, the losses before weaning 21.9 % and the turn over 2.1, interval of repeated heats 44.1 days, interval since weaning till conception 7.5 days, the gravidity after the first artificial insemination 83.6%.The differences of conception rate after the first artificial insemination and number of new born piglets were registered as an effect of insemination technician work. It is necessary to improve the work of artificial insemination technicians during the reproductive period, to improve heat detection, to ensure uninterrupted assistance to farrowing sows to ensure the lowering of losses of piglets.

Session 26 Poster 23

Effect of slaughter weight and gender on the concentration of skatole and androstenone in the back-fat of pigs

J. Mullane[1,2,3], P.G. Lawlor[1], P.B. Lynch[1], J.P. Kerry[2] and P. Allen[3], [1]Teagasc, Pig Production Development Unit, Moorepark Research Centre, Fermoy, Co. Cork, Ireland, [2]University College Cork, Department of Food Technology, Cork, Ireland, [3]Teagasc, Ashtown Food Research Centre, Ashtown, Dublin 15, Ireland

The objective of this study was to determine the effect of live-weight at slaughter on skatole and androstenone concentrations in the backfat of entire male, castrate and gilt pigs. Seventy two pigs (Meatline Landrace sire on Landrace x Large White sows) were used in a 3 gender (boar, castrate and gilt) x 3 slaughter weight (80, 100 and 120 kg live-weight) factorial design. All pigs were fed the same sequence of diets from weaning until they reached their prescribed slaughter weight. Samples of back-fat were taken from each split carcass at the level of the 3rd and 4th last rib and analysed for skatole and androstenone using high-performance liquid chromatography. The concentrations of skatole and androstenone in castrates and gilts of each slaughter weight were below the concentrations detectable by a trained sensory panel. Androstenone and skatole concentrations for entire males were at or above the concentrations detectible by a trained sensory panel at all the weight categories examined. With entire males, skatole concentration was 0.20, 0.37 and 0.26 (s.e. 0.123 mg/kg; $P > 0.05$). Androstenone concentration in the backfat of entire males was 0.79, 1.16 and 1.62 (s.e. 0.383 mg/kg; $P > 0.05$).

Session 26 Poster 24

Evaluation of different ultrasound intensity level to predict intramuscular fat in live pigs

I. Bahelka, P. Demo, E. Krupa and L. Hetényi, Slovak Agricultural Research Centre-Research Institute for Animal Production, Department of Animal Breeding, Hlohovska 2, Nitra, 94992 Nitra, Slovakia (Slovak Republic)

In the present study, 189 hybrid pigs were used to predict the intramuscular fat content in live animals. One day before slaughter, pigs were scanned above right last rib area using ultrasound device ALOKA SSD-500 with probe 3.5 MHz/172 mm. Each pig was measured at the same frequency (3.5 MHz) but at three different ultrasound intensity (70, 80 and/or 90 % of total intensity). After slaughter, the cross-sectional slices of the longissimus dorsi muscle at the same place as *in vivo* measurements were taken and chemical extraction to determine real intramuscular fat content was done as well. The computer image analysis was used to estimate the percentage of intramuscular fat on sonographic pictures. The mean absolute differences between real and predicted intramuscular fat at three intensities were 0.88, 0.06 and 1.11, respectively. Correlations between chemically analysed intramuscular fat content and that estimated *in vivo* were 0.45, 0.51 and/or 0.15. The first two correlations were statistically significant. Results suggest that two ultrasound intensities (70 and/or 80 % of total intensity) may be used to predict intramuscular fat content in live pigs.

Session 26 Poster 25

Influence of five gestation feeding regimes for sows on lactation feed intake, lactation weight loss and subsequent farrowing rate and litter size

L. Mc Namara[1,2], P.G. Lawlor[1], P.B. Lynch[1], M.K. O'Connell[1] and N.C. Stickland[2], [1]Teagasc, Pig Production Development Unit, Moorepark Research Centre, Fermoy, Co.Cork, Ireland, [2]The Royal Veterinary College, Royal College Street, London NW1 0TU, United Kingdom

Excessive feed intake during gestation may reduce sow feed intake during lactation thus increasing lactation weight loss which has been shown to reduce subsequent reproductive performance. The objective was to determine the influence of five gestation feeding regimes on lactation feed intake, lactation weight change and the reproductive performance of sows. Sows (n=238) were assigned to treatments: (1) 30 MJ digestible energy (DE)/day throughout gestation, (2) 60 MJ DE/day from day 25-50, (3) 60 MJ DE/day from day 50-80, (4) 60 MJ DE/day from day 25-80 and (5) 45 MJ DE/day from day 80-110. Lactation energy intake was less for Treatment 4 than any of the other treatments ($P < 0.05$). Lactation weight loss of the sows was not affected ($P > 0.05$) by the gestation treatment. There was a tendency towards a treatment effect for subsequent farrowing rate ($P=0.10$) where treatment 1 was lowest and treatment 5 highest. Mean daily lactation energy intake was higher for parities 2-4 than parity 1 ($P < 0.05$), however, lactation weight loss was not affected by parity grouping ($P > 0.05$). In conclusion, increasing the energy allowance of sows from day 25-80 of gestation reduced appetite during lactation; however, this did not increase lactation weight loss or reduce subsequent performance.

Session 26 Poster 26

Behavioural and physiological changes in pigs subjected to different fasting or lairage times

J. Cros, J. Tibau, J. Soler, X. Puigvert, M. Gispert, A. Velarde and E. Fabrega, IRTA, Veïnat Sies, 17121 Monells, Spain

Pre-slaughter conditions affect both welfare and meat and carcass quality. The aim of this study was to evaluate the effect on behaviour and physiology of different fasting and lairage times. Seventy-five male pigs were housed in 6 groups and fasted for either 12 or 2 hours (12F and 2F). Pigs were videotaped before and during fasting to record by scan sampling behavioural data. Pigs were also observed by focal sampling during fasting to determine total number of aggressions. Blood samples were obtained before transportation and at exsanguination to determine cortisol, creatine kinase (CK) and lactate dehidrogenase (LDH). Animals were transported in 3 groups without mixing. Three lairage times were applied 0, 5 or 10 hours (0L, 5L, 10L) to each of the fasting times. No differences in general behaviour between 12F and 2F groups were observed, but more aggressive interactions were recorded in the 12F group. This finding was consistent with a tendency for higher skin lesions in the 12F group. Moreover, a higher increase ($p < 0.001$) in CK and LDH was observed for the 12F group, suggesting a major physical stress for those pigs. Cortisol increase ($p < 0.001$) was higher for the 0L group compared to 5L or 10L, suggesting that not allowing lairage may imply more psychological stress. These findings suggest that fasting and lairage times had an effect on pig welfare under this minimal stress conditions.

Session 26 Poster 27

Studies of meat and fat quality of different crossbred pigs

D. Ribikauskiene and B. Zapasnikiene, Institute of Animal Science of LVA, Animal Breeding and Genetics, R. Zebenkos 12, Baisogala, Radviliskis distr., LT-82317, Lithuania

The purpose of the present study was to investigate the effect of boars of various imported breeds on the meat and fat quality of crossbred pigs. In 2005, two experimental groups of pigs were formed: group 1 - hybrids from Lithuanian White and English Large White (LWxELW), group 2 - hybrids from English Large White and Norwegian Landrace (LWxNL). LWxELW crossbred pigs were distinguished by higher contents of dry matter (by 0.52 %, $P < 0.05$), intramuscular fat (by 1.29 %, $P < 0.001$) compared with ELWxNL crossbred pigs. The lower protein content (21.9 %) was determined in the meat of LWxELW group of pigs ($P < 0.025$). The analysis of the biological value of meat protein indicated that according to this indicator LWxELW crossbreds surpassed ELWxNL crossbred pigs. The tryptophan – oxyprolin ratio in the meat of LWxELW crossbreds was by 0.1 higher compared with that of ELWxNL crossbred pigs. The higher water binding capacity of meat (11.9 %, $P < 0.01$) and the lower cooking losses of meat (1.8%, $P < 0.005$) were determined for LWxELW crossbred pigs. The higher melting temperature of subcutaneous fat and saponification number, respectively, 40.5 oC ($P < 0.005$) and 17.9 ($P < 0.001$) were determined for ELWxNL crossbred pigs. The higher content of polyunsaturated fatty acids (6.9 %) was found in the lipids of *M. longissimus dorsi* of ELWxNL crossbred pigs.

Semi-parametric survival analysis of sow longevity on the basis of the accumulated number of piglets born

J. Casellas, N. Ibáñez-escriche and J.L. Noguera, IRTA-Lleida, Genètica i Millora Animal, Alcalde Rovira Roure 191, 25198 Lleida, Spain

Although sow longevity has been commonly defined as a temporal interval, it also could be stated as the accumulated numbers of piglets born (ANPB) during sow's productive live. This alternative definition can be regarded as a survival-like measurement of sow ability to by-pass the biological damage caused by successive pregnancies. The stayability measured as ACNB of 831 hyperprolific Large White sows (average litter size = 13.3 piglets) was analyzed under a Cox survival sire model. Significant systematic sources of variation were: number of farrowings of the sow ($p < 0.001$), year of birth of the sow ($p < 0.05$), percentage of stillbirths along its productive life ($p < 0.001$), and average lactation length ($p < 0.01$). Survivability increased with farrowing number, incidence of stillbirths and average lactation length, whereas a clear tendency through the years of birth was not found. These results allow calculating easily the maximum average litter size at each parturition in order to assure a probability of survival greater than 0.5 (11.0 piglets at 1st farrowing, 12.5 at 2nd, 12.7 at 3rd, 13.0 at 4th, 13.4 at 5th, 13.7 at 6th, 13.6 at 7th, 13.8 at 8th and 13.3 at 9th). The modal estimate of the genetic variance between sires was 0.054, with an approximate heritability in logarithmic scale of 0.127. This moderate genetic determinism suggests that selective breeding on this trait could be feasible.

Effect of breed, sex and slaughter weight on performances of Canadian purebred pigs

L. Maignel[1], P. Mathur[1], F. Fortin[2] and B. Sullivan[1], [1]Canadian Centre for Swine Improvement Inc, Central Experimental Farm, Bldg#54 Maple Drive, Ottawa, Ontario K1A0C6, Canada, [2]Centre de Développement du Porc du Québec, Génétique, 2795, boul. Laurier, bureau 340, Sainte-Foy (Québec) G1V 4M7, Canada

A study was conducted to evaluate the effect of increasing slaughter weight on various performance traits such as growth, carcass characteristics and meat quality. About five hundred purebred pigs from Duroc, Yorkshire and Landrace breeds from breeders across Canada were tested at the Deschambault test station of CDPQ, Quebec. The pigs were slaughtered at either the current slaughter weight (107 kg) or at a heavier weight (125kg). Increase in slaughter weight was associated with higher total feed intake, feed conversion, backfat, loin eye area and dressing percentage but a lower lean yield and very small changes on meat quality. Very little or no difference was found between breeds on growth and carcass quality traits, whereas large differences were found between sexes, in agreement with previous research. Significant interactions were found, especially between breed and slaughter weight, on several growth and carcass traits. Based on this study, modern breeds seem to be able to be raised to heavier weights, with some substantial changes in efficiency and carcass characteristics. Depending on the payment grid in use, a cost-efficiency analysis is required in order to find the optimal slaughter weight and ways to reach it using management practices as well as breed and sex differentiation.

Session 26 Poster 30

Evaluation of lying behaviour of sows housed in farrowing pens with different design

L. Botto, P. Kisac, M. Knizatova, L. Macuhova, V. Brestensky and S. Mihina, Slovak Agriculture Research Center, Hlohovska 2, 94992 Nitra, Slovakia (Slovak Republic)

Lying behaviour of sows housed in 3 types of straw-bedded farrowing pens with free movement and with different disposition design was evaluated. The space for sow in the K1, K2 and K3-pens had 4.7 m2, 4.1 m2 and 4 m2. The sows were observed once before and 5-times after farrowing for 6 hours. For statistical evaluation of the data we used the Anova. The longest time of sows lying was in the K2-pens and the shortest one was in the K1-pens (97.19 % *vs* 66.68 % of time, $P < 0.01$). There were no significant differences in the lying of sows on their side among individual type of pens. Sows lay on their side the shortest time in the K1-pens ($P < 0.05$). In these pens sows lay on their right side the shortest time ($P < 0.05$). In the K1-pens sows lay 2-times longer on their side than on their belly (46.25 *vs* 20.43 % of time, $P < 0.01$). In the K2-pens sows lay on their side 4-times longer than on their belly (69.87 *vs* 17.33 % of time, $P < 0.001$). The time of sows lying on their right side was longer than lying on their left one (66.03 *vs* 25.33 % of time). The sows in the K3-pens lay on their side 4-times longer than on their belly (66.03 *vs* 16.05 of time). Lying on the right and left side was equal (55 % of time). The longest lying was noticed in the K2-pens with area for sow in the form of "L" letter, which is not the most optimal design for movement and activities for sows. Lying of sows on their side was longer than lying on their belly.

Session 26 Poster 31

Inhibition of androstenone production as an alternative to castration of piglets

B. Bucher, H. Joerg and C. Wenk, Institute of Animal Sciences/ETH Zurich, DAGRL, Universitaetstrasse 2, 8092 Zurich, Switzerland

Androstenone is member of the 16-ene steroids and is supposed to contribute to boar taint, which can occur in carcasses of intact boars. As androstenone is synthesized in the testis, in Switzerland, male pigs are being castrated to prevent boar taint. This castration is being criticized by ethical and animal welfare arguments. The objective of this study was to clone and express the CYP17 and 5aR genes, which are involved in the synthesis of pregnenolone to androstenone, into *in vitro* cell cultures. Different substrates were used in *in vitro* incubation trials to calculate the conversion efficiency. The cell cultures were treated with putative inhibitors to detect an inhibitory effect to the activity of CYP17 and 5aR. The *in vitro* system with CYP17 converted neither pregnenolone nor pregnenediol in a required amount to androstadienol and wasn't used for inhibition trials. 5aR converted the substrate testosterone to the expected product dihydrotestosterone and was used for inhibition trials, in which Finasteride, developed and used in the human medicine, and andi23, a not commercial available potentially 5aR inhibitor, were the best inhibitors. Andi24, F8 and F9 inhibited the enzyme too. The results of the *in vitro* inhibition trials and experiences in the human medicine demonstrate the successful application of inhibitors. With this established *in vitro* system and its possible extension, more enzymes and inhibitors can be tested.

Session 26 Poster 32

Analysis of sex influence on chemical composition of pig meat

M. Okrouhlá, R. Stupka, J. Čítek, M. Sprysl, E. Kluzáková and M. Trnka, Czech University of Life Sciences Prague, Department of Animal Husbandry, Kamýcká 975, 16000 Praha, 6, Suchdol, Czech Republic

The objective of the work was to quantity the sex influence on the chemical composition of the main meat parts (MMP) of the pigs carcass. The total of 120 hybrid pigs are commonly used in the Czech Republic. After finishing the test all right half-carcasses were dissected onto individual parts. The representative samples (neck, roast, shoulder and leg) were homogenized and subjected to chemical analyses to determine water content, total musculus fat (TMF), crude protein (CP), ashes and selected amino-acids (AA). Results show that the barrows achieved lower water content values in the neck, leg and shoulder than the gilts. Different values were assessed in the TMF content of MMP, where the highest content was find out for neck (barrows, 9.25 %), contrary the lowest one for roast (gilts, 1.65%). In the content of CP there were no statistically significant differences in terms of the evaluation of the intersexuality differences. The values varied among barrows, resp gilts in the interval 19.71 – 23.30 %, resp 19.81 – 23.21 %. The highest/lowest value of the ash matters (1.53/1.12 %) was monitored among the barrows for leg/shoulder. Concerned of amino-acid contents in frame of the evaluation of the MMP intersexuality can be stated that all differences were not statistically significant.

Session 27 Theatre 1

Evolution of body weigh and withers height of Lusitano horses from birth to two years of age

A.S. Santos[1,2] and V.C. Alves[1], [1]CECAV, Department of Animal Science, Univ. de Trás-os-Montes e Alto Douro, Apartado 1013, 5001-801 Vila Real, Portugal, [2] Esc. Univ. Vasco da Gama, Department of Veterinary, Quinta de S. Jorge, Estrada da Conraria, 3040-714 Castelo viegas, Coimbra, Portugal

The objective of this study was to evaluate the evolution of growth parameters in Lusitano horses from birth till 2 years of age. For this purpose, body weight (BW) and withers height (WH) changes were measured on 13 Lusitano foals of four different studs during three consecutive years. Management conditions were similar in all four studs. Mares and foals were kept on pasture. Foals were weighted every 15 days from foaling till 6 months of age. Measurements for WH were made every 15 days since birth till 6 months of age and at 1 and 2 years of age. No significant differences were found ($P > 0.05$) between the two genders for BW and for WH. Average BW at foaling and at 6 months of age was 53±6.4kg and 178±20.7kg, representing 0.105 and 0.356 of mature BW. Average WH at foaling, at 6 months and at 2 years of age was 101±3.4cm, 133±2.5cm and 142±2.5cm (0.628, 0.833 and 0.90 of adult WH respectively). These results may contribute to a better understanding of the growth patterns for the Lusitano breed.

Session 27 Theatre 2

Growth rates in Thoroughbred and Trotter horses raised in Italy

F. Martuzzi, F. Vaccari Simonini, A. Sabbioni and A.L. Catalano, Università di Parma, Dipartimento di Produzioni Animali, via del Taglio 8, 43100 Parma, Italy

Growth rates of Thoroughbred horses are considered in several studies, while available data regarding Trotters are few. A survey was carried out in stud farms of Thoroughbreds (TB) and Italian Trotters (IT) in Northern Italy, to assess growth and development of young horses over a prolonged period. The following body measures were recorded from birth to 18 (IT) and 20 (TB) months of age: wither height; hip height; body length; hearth girth; cannon bone girth; body weight. 45 TB (24 colts and 21 fillies) and 77 IT (41 colts and 36 fillies) were measured. Herd management conditions were investigated: vaccinations, anthelmintic treatments, feeding, housing and pasture management. With a linear model the significance of the following factors was tested: stud farm nested within breed; breed; year of birth; month of birth; sex; age class. Until 18 months of age, no significant difference was found between colts and fillies of the same breed regarding withers height and hearth girth. In comparison with TB foals raised in USA, body weight (54.38±8.18 kg)and withers height (102.69±4.55 cm) of the Italian TB in the first month of life are lower than those of foals bred in Kentucky, similar to those of TB bred in Florida. Compared to TB foals, in the first month of life IT are taller, heavier, with a wider hearth girth. Season of mare's parturition (winter *versus* spring) doesn't seem to affect foal birth weight.

Session 27 Theatre 3

Correlations of osteochondrosis between joints and body measurements in Dutch Warmblood horses (KWPN)

E.M. van Grevenhof[1], B. Ducro[1], P. Bijma[1] and J.M.F.M. van Tartwijk[2], [1]Wageningen University, Animal Breeding and Genomics Centre, P.O. 338, 6700 AH Wageningen, Netherlands, [2]Royal Dutch Warmblood Studbook, P.O. 156, 3840 AD Harderwijk, Netherlands

Osteochondrosis (OC) is the most important orthopedic developmental disease, with an incidence of about 30% in warmblood populations. OC is thought to be related to withers height, and body condition score during rearing and later in life. About 15,000 Dutch warmblood horses are born each year. Objectives were to establish frequencies, identify relations between OC and body measurements, and between OC in stifle, hock, and fetlock. Data were collected on 811 randomly selected KWPN yearlings, with a mean age of 12 months. Yearlings descended from 32 stallions, with an average of 25 yearlings per stallion. OC was measured on a scale from A to E; where A is negative for OC, B smooth lesion, C irregular lesion, D small osseous fragment and E large osseous fragment. Each horse was measured at 28 locations, using 14 radiographs. Information on withers height and body condition score were also available. Cluster analyses showed that left and right joints were highly similar. OC was affected by age, body condition score, gender, withers height and the veterinary horse clinic. The effect of body measurements differed between joints. The prevalence of yearlings with only A scores for hocks was higher than for stifles. Prevalence of OC differed between offspring of stallions.

Session 27 Theatre 4

Effect of exercise and age on bone markers in Lusitano stallions

N. Bernardes[1], F. Figueiredo[2], J. Robalo[1], A. Ferreira[1] and G. Ferreira-Dias[1], [1]CIISA, FMV, Lisbon, 1300 Lisboa, Portugal, [2]EPAE, Queluz, 2745 Queluz, Portugal

Since exercise and age have been related to bone remodelling, we evaluated the effect of exercise and age on serum bone markers in Lusitano stallions. Fifty stallions, aging from 5 to 15 years old, were submitted to different exercise programmes: (i)GI-30' of free exercise; and (ii)GII-30' of free exercise and 15' of riding, for one year. Venous blood was collected at the end of the trial for determination of osteocalcin (OC) and bone specific alkaline phosphatase (BAP) by EIA; and C-terminal telopeptide of type I collagen (ICTP) by RIA. Stallions were grouped according to their age: A-<10 years old; and B-> 10 years old. Data were analysed by one-way ANOVA. Significance was defined as values of $P < 0.05$. Whenever a significant difference was detected, a post-hoc comparison test (Scheffé F test) was performed. ICTP decreased with age (A:5.1±0.9mg/L; B:4.4±1.0mg/L) ($p < 0.01$), osteocalcin had a tendency to decrease (A:15.3±4.7ng/mL; B:13,1± 3.5ng/ml)(p=0.08), while BAP did not change. When exercise was considered, GI presented a lower BAP concentration (30.7±6.9 U/L), when compared to GII (36.2±10.2 U/L)($p < 0.05$). No changes in other markers were observed. These data show that even though reabsorption was decreased in older Lusitano horses, bone formation was also impaired suggesting an equilibrium in bone cell metabolism. Besides, exercise was responsible for an increase in bone formation.

Session 27 Theatre 5

Evaluation of osteochondrosis and bone markers in Lusitano stallions

N. Bernardes[1], F. Figueiredo[2], J. Robalo[1], A. Ferreira[1] and G. Ferreira-Dias[1], [1]Faculty of Veterinary Medicine-Lisbon, Avenida da Universidade Técnica - Polo da Ajuda, 1300-477 Lisboa, Portugal, [2]Escola Portuguesa de Arte Equestre, Palácio Nacional de Queluz, 2745-191 Queluz, Portugal

Bone markers are very valuable indicators of bone turnover, at very early stages. Osteochondrosis is common in young horses and can lead to secondary osteoarthritis when not diagnosed and left untreated. The aim of this study was to assess the value of blood bone markers on the early diagnosis of osteochondrosis (OC) in the Lusitano horse.Fifty Lusitano stallions, aging from 5 to 15 years old, were submitted to radiologic examinations of all four fetlocks and both tarsus. Simultaneously, venous blood was collected for determination of osteocalcin and bone specific alkaline phosphatase (BAP) by EIA; and C-terminal telopeptide of type I collagen (ICTP) by RIA. Osteochondrosis was present in 16% of all animals. Among the 200 fetlock joints evaluated, 5% had OC lesions. Among the 100 tarsus studied, 4% of OC lesion were found. Data were statisitcally analysed by one-way ANOVA. Significance was defined as values of $P < 0.05$. Whenever a significant difference was detected, a post-hoc comparison test (Scheffé F test) was performed(Statistic for Windows, U.S.A.). Blood biochemical bone markers did not change between horses affected with OC and not affected ones ($p > 0.05$). Despite the reported usefulness of bone biochemical markers, as early indicators of OC, this study did not confirm its value in the Lusitano stallion.

Session 27 Theatre 6

Apparent digestibility of a dietetic feed claimed for horse clinical nutrition

N. Miraglia[1], M. Costantini[1], M. Polidori[1], D. Bergero[2], G. Meineri[2] and P.G. Peiretti[3], [1]Molise University, SAVA, Via De Sanctis, 86100 Campobasso, Italy, [2]Torino University, DIPAEE, Via L. Da Vinci 44, 10095 Grugliasco(TO), Italy, [3]CNR, ISPA, Via L. Da Vinci 44, 10095 Grugliasco(TO), Italy

The management of horses in the post-surgical phase involves many aspects concerning stress reactions and the recovery of the intestinal functions in the case of colics, enteritis and dismicrobial problems. The industries producing dietetic feeds are more and more involved in the research of formulae concerning mixed feeds for the recovery of horses in the post-surgical phase. Rough materials characterized by high digestibility together with probiotics are often employed to satisfy these exigence. The apparent digestibility of meadow hay or meadow hay plus dietetic feed (75:25 and 62:38 on DM basis, respectively) rich in processed cereals, probiotics, inulin, minerals and vitamins was determined by means of three *in vivo* trials, each performed on 6 adult healthy horses over a 6 day faeces total collection period with a previous 14 day adaptation period. The dietetic feed was tested at two different amounts: 0.5 and 0.8 kg/q live weight as fed basis. Faeces and feed dried samples were analysed to determine dry matter, organic matter, crude protein, crude fibre, fibrous fractions and gross energy. The results showed that the addition of the dietetic feed to the hay improves the digestibility of the 75:25 ration, while the 62:38 ration digestibility was lower than those of hay alone and 75:25 ration.

Session 27 Theatre 7

First pressure linseed oil for horse feeding: effects on fatty acids in plasma glycerides

J. Fayt[1], O. Dotreppe[2], J.L. Hornick[2] and L. Istasse[2], [1]S.A. Fayt Carlier, Rue Déportés, 6120 Jamioulx, Belgium, [2]Liege University, Animal Production - Nutrition Unit, Colonster, B43, 4000 Liege, Belgium

In heavily exercised horses, the amount of offered cereals is high in order to meet the energy requirements. Fat has been suggested as an alternate compound to provide energy. Corn oil, soja oil and coconuts oils are often used while only a limited number of studies were carried out in horses with linseed oil, a source of ω3 fatty acids. Four adult horses were fed either a control diet or a diet in which linseed oil was incorporated at a rate of 8% in the compound feedstuff. The diet was made of 50% grass hay and 50% compound feedstuff. The feeding of linseed oil in the diet significantly affected the concentrations of the different fatty acids except C16:0. The supplementation with linseed oil increased the concentrations in C18:3n-3 (3.6 *vs* 2.1 mg/100 ml; $P < 0.001$) and in C18:2 n-6 (83.98 *vs* 63.26 mg/100ml; $P < 0.001$) in plasma glycerides. Linseed oil induced also significant ($P < 0.001$) increases in the concentration of total fatty acids, of PUFA and of SFA. By contrast, linseed oil decreased the concentrations of C18:1n-7 + n-9 (11.4 *vs* 12.9 mg/100ml; $P < 0.05$), of C20:4 n-6 (0.9 *vs* 1.1 mg/100ml; $P < 0.001$) and of MUFA (12.2 *vs* 14.3 mg/100ml; $P < 0.01$). It is concluded that first pressure linseed oil is a compound of interest to induce changes in the fatty acids profiles of horses plasma and therefore to contribute to reduction in disorders such as allergy or inflammatory processes.

Dealing with stereotypic behaviour in the horse: What can we do?

J. Murphy, School of Agriculture, Food Science and Veterinary Medicine, UCD, D4, Ireland

Although unobserved in populations of wild equids, some horses exhibit stereotypic or Abnormal Repetitive Behaviours (ARBs). These behaviours are invariant, repetitive and seemingly functionless in character. Equine ARBs include box-walking, weaving, wind-sucking and crib-biting, which were originally labelled 'stable vices' and are classified as clinical 'unsoundness' in horses. In general, they tend to be associated with higher endorphin levels, appear in horses exposed to less than optimal conditions and/or may represent some form of coping mechanism. ARBs may have an energy cost, lead to un-thriftiness and have implications for animal welfare. Dealing with ARBs in the horse might include: 1. Punishment – probably the most common response to any undesired behaviour; 2. Pharmacological intervention – used in 'companion animals' but long-term pharmacological therapy is an unlikely solution for equine ARBs; 3. Environmental enrichment – provides the horse with opportunities to perform alternative activities and may address causative factors; 4. Reinforcement of alternative behaviour – successful in other species and could enhance welfare by offering horses new behavioural opportunities; 5. Genetic selection – an approach that is in its infancy in applied ethology – but should it be encouraged as a future strategy in the horse? The selective breeding of horses that are less frustrated by the stress inherent in some traditional rearing systems may be one of the more appropriate solutions.

Session 27 Poster 9

Body condition and leptin in Lusitano mares during late pregnangy and lactation

M.J. Fradinho[1], L. Mateus[1], R. Agrícola[2], M.J. Correia[3], M.J.C. Vila-Viçosa[4], R.M. Caldeira[1] and G. Ferreira-dias[1], [1]Faculdade de Medicina Veterinária, TULisbon, CIISA, Av. Universidade Técnica, 1300-477 Lisboa, Portugal, [2]Coudelaria Nacional, Fonte Boa, Santarém, Portugal, [3]Coudelaria de Alter, C Arneiro, Alter-do-Chão, Portugal, [4]U Évora, Herd. Mitra, Évora, Portugal

The main objective of this study was to evaluate the relationship between leptin and BCS in Lusitano mares during late pregnancy and lactation. BCS was monthly assessed in 18 Lusitano mares from the 9th month of gestation to the 5th month of lactation. At the same time, blood samples were collected for measurement of plasma leptin concentrations. According to BCS, mares were split into two categories: lower BCS (LBCS) and higher BCS (HBCS). Data were statistically analysed by one-way ANOVA (Statistica). HBCS mares presented higher leptin concentrations than LBCS mares, after the 9th month of gestation ($P < 0.05$). Leptin concentrations of LBCS mares were lower during the 9th and 10th month of gestation and increased before foaling ($P < 0.05$), decreasing afterwards to similar levels to the 9th and 10th months. In HBCS mares, leptin increased from the 9th to 11th month of gestation ($P < 0.001$). This was followed by a decrease until the 5th month of lactation ($P < 0.05$). There was a positive correlation between BCS and plasma leptin concentrations ($R2 = 0.48$, $P < 0.0001$). These results in Lusitano mares suggest a similar pattern in leptin plasma concentrations during this period, when compared with other light breeds.

Milk yield in Martina Franca asses: quantitative characteristics of lactation

A.G. D'Alessandro, G. Martemucci and G. Di Lena, University of Bari, Dept. PRO.GE.S.A., Campus, Via Amendola 165/A, 70126 Bari, Italy

The study was carried out to evaluate milk yield during a whole lactation (8 months) in Martina Franca asses. Twelve adult animals were milked once a day by machine milking (at 10: a.m.), with foals previously separated from dams for 3h. Individual milk secretion from the two half-udders was recorded daily, starting the first month from foaling. The Wood model was applied to fit milk yield data of individual lactations and the pooled total milk yield. Milk yield per milking varied significantly ($P < 0.01$) during lactation. The average of the pooled total milk yield per milking was higher in the first four months (403-470 mL/milking) compared to the other months of lactation ($0.05 > P < 0.01$), and a high persistence during the whole lactation was shown. The peak yield was reached at day 48 of lactation. There was a large individual variation ($P < 0.01$) in milk yield per milking, as expressed also by different shapes of lactation curves of the asses. Average milk yield per half-udder was not significantly different ($P > 0.05$).

Chemical composition of liquid and solid associated bacteria in the cecum and colon of horses

A.S. Santos[1,2], E. Jerónimo[3], L.M. Ferreira[1], M.A.M. Rodrigues[1] and R.J.B. Bessa[3], [1]CECAV, Department of Animal Science,, Univ. de Trás-os-Montes e Alto Douro Apartado 1013, 5001- 801 Vila Real, Portugal, [2]EUVG, Department of Veterinary, Estrada da Conraria, 3040-714 Castelo Viegas, Portugal, [3]Estação Zootécnica Nacional – INIAP, Vale de Santarém, 2005-048 Vale de Santarém, Portugal

This study intended to characterize fatty acids of the solid (SAB) and liquid (LAB) associated bacteria from caecum and colon of horses. Contents were collected from two adult horses after slaughter. LAB were obtained by differential centrifugation (first at 500*g*, 5 min, 4 °C, and then centrifugation of supernatant at 20000*g*, 20 min, 4 °C). SAB were isolated from solid phase formerly suspended 24h with saline solution (0,85%) at 8°C and submitted to thermal shock. The suspensions were filtered and the resulting liquid submitted to differential centrifugation as described for LAB. Bacterial pellets were freeze-dried, grinded and analysed for fatty acids. Fatty acid concentration in LAB and SAB was 161 and 185 mg/g of bacterial DM respectively. Caecum bacteria tended to had lower concentration of odd and branched chain fatty acids and higher rumenic and trans-vaccenic acid than colon. Caecum bacteria fatty acid pattern seems more closely related to rumen SAB, and colon patterns with rumen LAB. These results may help to clarify determinant factors of the efficiency of horse hindgut microbial ecosystem.

Growth and body progress of young stallions the Czech warmblood horses

J. Navratil[1], J. Maresova[1], M. Krejci[1] and F. Louda[2], [1]Czech University of Agriculture Prague, Kamycka 129, Prague 6–Suchdol, 165 21, Czech Republic, [2]Research Institute for Cattle Breeding, Vyzkumniku 267, 788 13 Vikyrovice, 788 13, Czech Republic

There are 19 breed of horses at present in Czech republic, the biggest part of this population of horses is Czech warmblood. Grow of selected group of young stallions are tested in our country from year 2001. There are tested in nine official rearing houses in standard conditions, is find growth intensity and qualities index until three years age (n = 418). Results of tests of foals rearing houses give in the first information about quality of breeders used in breeding especially in objective breeding in the acceleration program. It was analysed so far 1467 note mesurement height at withers by stick after individual stallions. This results was confronted with grow curve publicated by Dušek (1970). It was finded moving to plus values (zone +1) to higher growth area. This curve will be modify after recapitulated monitoring of other year – class of young stallions.

Potential for meat production of Spanish horse breeds: Preliminary results

M. Valera[1], M.D. Gómez[2], F. Romero[1], M.J. Alcalde[1], M. Juárez[1] and M. Villanueva[3], [1]University Seville, Ctra Utrera km1, 41013 Sevilla, Spain, [2]University Cordoba, C.U.Rabanales, 14071 Cordoba, Spain, [3]I.T.G.Ganadero, Av.Serrapio Huici 22, 31610 Villava, Spain

The consumption of horse meat is located in a restricted area in North/NorthEast of Spain. The traditional production is based on autochthonous horse breeds, with free ranching until weaning (6-7months) and intensively fattened until an optimum weight (12-24). Since the approval of a measure for the Organization and Promotion of Equine Sector in Spain by the Ministry of Agriculture, Fisheries and Food, horse breeding is being more selective in sport and meat producer breeds. In order to contribute to genetic improvement of meat producer horses, 50 equine carcasses are analysed. Data collected in the abattoir include: 6 zoometric measures (hindleg length, hindleg width, back and loin length, thorax depth and carcass length) and the carcass weight. Another 23 different measures are also analysed in some live animals, in order to correlate them with the carcass characteristics. Analysed traits have medium level of variation. There are no differences by sex. Carcass weight, length and compactness index are positively correlated with all traits. There are important correlations between carcass and live animal measures. Basic statistics and phenotypic correlations between variables are used to preselect the traits more related to meat production, to design a new evaluation system to evaluate the meat production potential in live animals.

Session 27 Poster 14

Clitoral isolated bacteria from problem and pregnant mares in Iran

M. Mohammadsadegh and S. Esmaeily, Islamic Azad University, Garmsar Branch, Veterinary Medicine, Engelab Street Garmsar, Semnan Province, Iran

In order to determine the kind of clitoral and uterine bacteria, 41 pure or crossbred Iranian mares were selected and 20 pregnant mares were encountered as control and 21 barren and/or repeated breeder mares as a test group. Clitoral bacterial samples were collected from pregnant and problem mares and uterine swabs and cytology smear samples were collected only from problem mares to determine the existing bacteria. The kind and numbers of clitoral bacteria were compared in control and test group with Chi-square and Fisher exact test. Findings showed that *E. coli* were the most frequent isolated bacteria in 80.9% of clitoral samples of barren, 68% of clitoral samples of pregnant mare and 61.9% of uterine samples of barren mares. However, they would be a secondary contamination. The most important isolated bacteria were β-hemolytic Streptococci, which were isolated from uterine and clitoral samples of problem mares and were not isolated from pregnant mares. Furthermore, examination of clitoral bacteria prior to breeding could not be an accurate substituting for uterine culture and cytology.

Session 28 Theatre 1

Maximising forage and pasture use to produce milk

J.L. Peyraud and L. Delaby, INRA, UMR 1080 Production du lait, Agrocampus, 35590 St Gilles, France

The aim of this paper is to discuss conditions for efficient dairying systems based on grazing and to direct attention to recent advances towards increasing the proportion of grass that will meet the animals requirements for milk production. On a short term basis, the possibilities to increase herbage intake by increasing pasture allocation are limited because this will impair the quality of grass in subsequent regrowths. Decreasing the supply of concentrate will increase grass intake. On a medium term basis, there is interest in extending the grazing season as much as possible. Several experiments have shown that access to grass for some hours per day between mid February to mid April or during November and December increase milk yield while decreasing intake of conserved forages. Moreover, well managed early spring grazing can have beneficial effects on sward quality as long as it does not affect DM production, in the period up to mid June. However, turnout date is critical for good grass utilisation in subsequent rotations and should be decided according to the area available per cow during the entire grazing season. At the system level, several experiments have shown that relatively high milk production (7000 kg/lactation) is achievable with less than 500 kg of concentrate when good quality forages (fresh and conserved) are available. In conclusion, there is considerable scope to increase forage intake to produce milk given recent developments in our understanding of herd and grazing management factors. influencing grass intake.

Session 28 Theatre 2

Effect of daily herbage allowance and concentrate supplementation offered to spring calving dairy cows in early lactation

E. Kennedy[1,2], M. O'Donovan[1], F. O'Mara[2] and L. Delaby[3], [1]Teagasc, Dairy Production Research Centre, Moorepark, Fermoy, Co. Cork, Ireland, [2]School of Agriculture, Food Science and Veterinary Medicine, UCD, Belfield, Dublin 4, Ireland, [3]INRA, UMR Production du Lait, 35590 St. Gilles, France

This study established the influence of daily herbage allowance (DHA) and concentrate level on the milk production and dry matter intake (DMI) of spring calving dairy cows in early lactation. Sixty-six cows were randomly assigned to a 6 treatment grazing study (3 DHAs-13, 16 and 19kg DM/cow/day >4cm) and 2 concentrate levels (0 and 4kg DM/day)). Treatments were imposed from 21 February to 8 May. Following this all animals were offered 20kg herbageDM/cow/day and no concentrate for a further 4-week period. Milk yield was recorded daily; milk composition was determined weekly. Intake was measured twice. The experiment was a randomised block design and was analysed using covariate analysis. There was no effect of 1st grazing rotation DHA on any measured variable. However, a low DHA during the 2nd grazing rotation decreased ($P < 0.05$) animal performance. Offering concentrate significantly increased total DMI, milk, SCM, fat, protein and lactose yield. These results suggest that dairy cows in early lactation should be offered a low DHA throughout the first grazing rotation. DHA should then be increased for subsequent grazing rotations. Animals should also be supplemented with concentrate as it increases milk production, the effect of which continues into mid-lactation.

Session 28 Theatre 3

Effect of daily herbage allowance and concentrate level on dry matter intake and milk performance of spring calving dairy cows in early lactation

M. Mc Evoy[1,2], M. O'Donovan[1], T. Boland[2] and L. Delaby[3], [1]Teagasc, Dairy Production Research Centre, Moorepark, Fermoy, Co. Cork, Ireland, [2]University College Dublin, School of Agriculture, Food Science and Veterinary Medicine,, Belfield, Dublin 4, Ireland, [3]INRA, UMR Production du Lait, St Gilles, 35590, France

The objective of this experiment was to establish the effect of concentrate level and daily herbage allowance (DHA) on dry matter intake (DMI) and milk performance of spring-calving cows in early lactation. Seventy-two spring calving Holstein Friesian dairy cows were randomised across 6 treatments (n=12) from 20 Feb. to 7 May. Animals were balanced on calving date [2 February (s.d. 9.4 days)], parity [2.5 (s.d.1.65)], first ten days milk yield [25.7kg (s.d.0.49)], bodyweight [541kg (s.d.77.5)] and body-condition score (BCS) [2.9 (s.d.0.49)]. The treatments were: 13kg (L) or 17 kg (H) DM/cow/day DHA and 0, 3kg or 6kg DM/cow/day concentrate. Milk yields were recorded daily. Milk fat, protein and lactose concentrations, bodyweight and BCS were recorded weekly. DMI was measured at 35, 55 and 85 days in milk. A linear response in milk yield to concentrate supplementation did occur. DHA and concentrate had a significant ($P < 0.05$) effect on DMI. DHA and concentrate had a significant ($P < 0.01$) effect on milk protein and lactose yield, SCM and bodyweight. Results indicate that offering spring calving dairy cows 17kg DM DHA and 3kg DM concentrate in early lactation will achieve high cow performance and high grass utilisation.

Session 28 Theatre 4

Effect of perennial ryegrass cultivar maturity and defoliation pattern on sward nutritive composition

G. Hurley[1,2], M. O'Donovan[1] and T.J. Gilliland[2,3], [1]Teagasc, Dairy Production Research Centre, Moorepark, Fermoy, Co. Cork, Ireland, [2]QUB, Faculty of Science and Agriculture, Newforge Lane, Belfast, Ireland, [3]Agri-Food & Biosciences Institute, Crossnacreevy, Belfast, Ireland

The objective was to investigate the changes induced by four cultivars and three spring defoliation dates on the digestibility and nutritive content of plant parts within the vertical profile of mid season swards. Cultivars investigated included Fennema, Corbet, Foxtrot and Melle. Three different spring 'start-date' managements were imposed; 15 February (early), 1 March (medium) and 29 March (late). The sward was cut into three horizons 0-8cm (lower), 8-15cm (middle) and 15+cm (upper) during four mid-season rotations. Each horizon was separated into leaf, stem and dead and the nutritive composition of each plant part was analysed. Cultivar had no effect on the nutritive composition while management had a significant effect. In the upper horizon, swards initially defoliated late had a lower leaf digestibility than swards defoliated earlier. In the middle horizon, late initial spring defoliation resulted in leaf having a higher ash and neutral detergent fibre (NDF) concentration with a lower crude protein (CP) concentration than swards defoliated early. Leaf and stem CP in the lower horizon was higher under the early management with a lower NDF concentration. An early initial spring defoliation results in a higher quality leaf mid-season with greater digestibilities and CP content.

Session 28 Theatre 5

A strategic model to maximise pasture use in the diet of dairy cows

A. van den Pol-van Dasselaar, G. Holshof, A.P. Philipsen and R.L.G. Zom, Animal Sciences Group, Division Animal Production, Wageningen UR, P.O. Box 65, 8200 AB Lelystad, Netherlands

The decline in the number of grazing dairy cows in the Netherlands is a matter of concern to many inhabitants. Communication and research on grazing is therefore supported by various partners from government, society and industry, for example in the project Koe & Wij ("Cow and Us"). "Cow and Us" aims to facilitate dairy farmers in their choice between grazing and zero-grazing. Various tools are developed to support the dairy farmer. We present a mechanistic model which enables farmers to see the results of changes in grazing management and supplementary feeding on the utilisation of grazed herbage by dairy cows. The model requires simple inputs regarding specific farm conditions (soil type, drainage), herd characteristics (size, breed, age and calving pattern), grazing management (i.e. pre-grazing herbage mass and height, paddock size, access time to pasture (h/day), age of regrowth, topping and previous usage) and supplementary feeding. Using these inputs, the model calculates herbage intake (kg DM/cow), post-grazing herbage mass, herbage loss due to fouling with dung and urine, poaching and trampling and finally herbage utilisation defined as the amount of herbage consumed as proportion amount of herbage grown. This model provides a tool for farmers to maximise the pasture use in the diet of grazing dairy cows. The model will be available on the internet from summer 2007 onwards (www.koeenwij.nl).

Feed intake and milk yield of dairy goats and sheep depending on quality of forage and level of supplement

F. Ringdorfer, L. Gruber, G. Maierhofer and E. Pöckl, HBLFA Raumberg-Gumpenstein, Raumberg 38, 8952 Irdning, Austria

Profitability of goat and sheep husbandry depends on revenues for milk and expenses for feeding, which normally consists of preserved forage supplemented with concentrate. Forage quality determines feed intake and hence also milk production. Apart from Saanen goats (DG) and East Friesian Milk Sheep (EFMS), Austrian Mountain Sheep (AMS) were included in the study. A full lactation lasted 240 d for EFMS and DG whereas it was 150 d for AMS. Diets consisted of hay of two different qualities resulting from either two or three cuttings of a homogenous grassland area. Additionally, three levels of concentrate (C) were administered to the animals (5 %, 25 % or 50 % of daily feed intake). Intake of forage per kg LW during lactation was significantly higher in DG than in EFMS, which, in turn, consumed significantly more hay than AMS. Forage quality did not influence average daily hay intake in AMS and DG, but in EFMS intake was significantly higher with the 3-cuts hay. As concentrate supplementation rose, hay ingestion was significantly reduced in all breeds (C5: 28.2 g/kg LW, C25: 24.3 g/kg LW C50: 18.3 g/kg LW). Substitution rates during lactation were highest in EFMS (0.423) while in the same range in AMS and DG (0.381 and 0.363). Forage quality significantly influenced substitution rate, being more pronounced with the 3-cuts hay (0.437) than with the 2-cuts hay (0.325). Higher forage quality promoted milk yield in EFMS (+27.6 %) and DG (+ 11.6 %), but not in AMS.

Session 28 Theatre 7

How does sward accessibility affect intake and feeding choices in horses?

N. Edouard[1,2], G. Fleurance[1,3], P. Duncan[2], B. Dumont[1] and R. Baumont[1], [1]INRA, centre Clermont/Theix, 63122 St-Genès-Champanelle, France, [2]CEBC, UPR1934, 79360 Beauvoir sur Niort, France, [3]Les Haras Nationaux, Direction des connaissances, 19230 Arnac-Pompadour, France

Grass represents a large part of the diet of horses. It can match their nutritional requirements while promoting the appropriate management of grassland. However, very little is known on the factors affecting their daily intake and diet selection at pasture, including the effect of classical vegetation characteristics such as sward height. Three groups of three 2-yr-old saddle horses were thus grazed on a semi-natural pasture that was exploited at three contrasting sward heights, i.e. 6, 11 and 17 cm, in a Latin-square design. The different sward heights were either offered alone to the animals or in binary preference tests. Diet preferences were established at the daytime scale based on the time animals spent grazing each sward type. Daily food intake was measured individually using the total collection of faeces method, for swards being offered in choice and alone. Results were analysed using mixed models. In preference tests, animals spent more time grazing on the taller sward whatever the sward height contrast. Voluntary intake was not affected by sward height and averaged 21gDM/kgLW/day whether the swards were offered alone or in binary choice tests. These results obtained at the daily scale are discussed in relation to ingestive behaviour parameters, i.e. bite mass, biting rate and daily grazing time.

Effect of maturity on cell-wall digestibility of guineagrass

S.S. Stabile[1], D.R. Salazar[1], L. Jank[2] and L.F.P. Silva[1], [1]University of São Paulo, Pirassununga, 13635-900, Brazil, [2]EMBRAPA, Campo Grande, 79002-970, Brazil

Our objective was to evaluate the effect of maturity on *in vitro* neutral detergent fiber ruminal digestibility (NDFD) of Guineagrass (*Panicum maximum* Jacq). The field study was established in 2006 at EMBRAPA-Brazil, as a completely randomized block design with three repetitions of 12 accessions. After a leveling cut, two lines from each plot (4 x 2,5 m) were harvested at 30, 60 and 90 days of regrowth. Samples were divided in three fractions: leaf blades, stems and senescent material. After drying and grinding, the fractions were incubated for 30h at 39°C in ruminal fluid and MacDougall solution. There was a linear effect of age on NDFD of leaves ($P < 0.05$), reducing from 38.6% to 31.4% when harvested after 30 and 90 days regrowth, respectively. There was no variation on NDFD of leaves, and the interaction of regrowth age with accession was not significant. There was also a linear effect of age on NDFD of stems ($P < 0.05$), reducing from 45% to 32.1% with advanced maturity. For the stems, there was a significant effect of accession for NDFD, but only when harvested after 90 days regrowth ($P < 0.05$), indicating the existence of genetic diversity for this trait. There was no difference among accessions harvested after 30 or 60 days regrowth. These results demonstrate that there is little genetic variation in the effect of maturity in cell-wall digestibility of leaf blades or young stems, but that breeding programs could use NDFD of stems with more than 90 days of regrowth as a selection parameter.

Effects of tanniniferous oak (*Quercus hartwissiana*) leaves on leptin, insulin-like growth factor-I and pulsatile luteinizing hormone secretion in lambs

S. Yildiz[1], M. Cenesiz[1], M. Kaya[1], F. Onder[1], O. Ucar[2], M. Uzun[1], D. Blache[3], M. Blackberry[3], I. Kaya[4], Y. Unal[4] and G.B. Martin[3], [1]Kafkas University, Faculty of Veterinary Medicine, Department of Physiology, Pasacayiri Kampusu, 36040 Kars, Turkey, [2]Kafkas University, Faculty of Veterinary Medicine, Department of Reproduction and Artificial Insemination, Pasacayiri Kampusu, 36040 Kars, Turkey, [3]University of Western Australia, Animal Biology, Crawley, Perth, Australia, [4]Kafkas University, Faculty of Veterinary Medicine, Department of Animal Nutrition, Pasacayiri Kampusu, 36040, Turkey

Effects of replacing hay diet with tanniniferous oak leaves (*Quercuss hartwissiana*) and polyethylene glycol, a tannin binding substance, on plasma levels of leptin and IGF-I and on pulsatility of LH secretion were investigated in fat-tailed Tuj ewe-lambs. Lambs were kept in individual metabolism cages for 60 days and a total of 7 equal groups (n = 6) were formed (Group I, 645 g hay; Group II, 185 g leaf; Group III, 370 g leaf; Group IV, 185 g leaf plus 10 g PEG; Group V, 185 g leaf plus 20 g PEG; Group VI, 370 g leaf plus 20 g PEG; Group VII as 370 g leaf plus 40 g PEG). IGF-I levels, leptin concentration and LH pulsatility did not differ between the groups. However, body condition score positively influenced leptin levels, which in turn positively affected LH pulse frequency. IGF-I levels were also positively corraleted with LH pulse frequency. The results suggest that oak leaves can replace hay during the scarcity of feedstuffs.

Faecal NIRS: a practical method for monitoring goat nutrition under free-ranging conditions

T.A. Glasser, S. Landau, E.D. Ungar, A. Perevolotsky, L. Dvash, H. Muklada and D. Kababya, Agricultural Research Organization - the Volcani Center, Department of Agronomy and Natural Resources, Institute of Plant Science, P.O.B 6, 50250, Israel

Faecal NIRS was examined as a practical method to determine diet quality under free-ranging conditions. The need to understand behavior of browsers such as goats brought us to look for a practical method to monitor individual animals. The aim of this research was to develop faecal NIRS equations to predict chemical and botanical composition of diets, and to apply them to investigate differences between goat breeds. Reference values that served to calibrate faecal NIRS were based on 43 observations of goats of three breeds: Damascus, Mamber & Boer. Calibrations were based on diet composition (botanical & chemical) data and faecal NIRS scans. The resulting R^2 and SECV values for the percent of CP, NDF, ADF, IVDMD & Peg-binding tannins in diet were between 0.74-0.93 & 0.87-4.27, and between 0.77-0.89 & 5.6-7.8 for the main botanical components (*Pistacia lentiscus* L. *Phillyrea.latifolia* L. and herbaceous plants). The faecal NIRS method enabled us to greatly extend the database for analysis of breed differences by obtaining faecal samples from all the non-observed goats and applying the calibration equations to spectra of these samples. Comparison of main dietary components between the three breeds was possible and differences were found suggesting some breeds are much more adapted to eat plants with high tannin content than others.

Chelated trace elements improve hill lamb performance

L.L. Masson[1,2], J.C. Alliston[1], J. White[3] and G.P.F. Lane[1], [1]Royal Agriculture College, Cirencester, Gloucestershire, GL7 6JS, United Kingdom, [2]Agri-Lloyd International Ltd, Glendower Road, Leominster, Herefordshire, HR6 0RL, United Kingdom, [3]Northumberland College, Kirkley Hall, Ponteland, Northumberland, NE20 0AQ, United Kingdom

Trace element deficiencies can have a significant impact on growth performance of lambs. Chelated trace element and vitamin drenches were used in a hill flock of 438 Swaledale x Blackface ewes and their lambs, to determine their effect on lamb growth. Half of the single-bearing and twin-bearing ewes were drenched with 20ml Liquithrive Sheep four weeks before lambing. The rest of the ewes were not drenched (control groups). All lambs were weighed at eight weeks of age. Lambs born to drenched ewes were treated with Liquithrive Lamb at eight weeks (7.5ml) and at weaning (15ml). Lambs from undrenched ewes were not treated (control group). All lambs were weighed at weaning and at four and eight weeks post-weaning. Single lambs born to treated ewes were significantly heavier at eight weeks (1.15kg, $P < 0.01$). At weaning, single lambs averaged 2.66kg heavier ($P < 0.001$) and twin lambs 1.22kg heavier ($P > 0.05$) compared to control lambs. At eight weeks post- weaning, drenched single lambs were 3.03kg heavier ($P < 0.001$) and twins were 2.53kg heavier ($P < 0.05$) compared to control lambs. Treating lambs with chelated trace elements and vitamins helped to overcome the natural deficiencies in the forage to benefit growth performance, resulting in a significant financial benefit to the hill sheep farmer.

Session 28 Theatre 12

Effects of chopping and stage of maturity of whole-crop barley silage on feed intake and eating rate in dairy steers

B.-O. Rustas[1], A. Sahlin[1], E. Nadeau[1] and P. Nørgaard[2], [1]Swedish University of Agricultural Sciences, Dept of Animal Environment and Health, P.O. Box 234, SE-532 23 Skara, Sweden, [2]University of Copenhagen, Dept of Basic Animal and Veterinary Sciences, Grønnegårdsvej 7, 1870 Frederiksberg C, Denmark

The objective of this study was to evaluate how chopping of whole-crop barley (WCB) silage, harvested at different maturity stages, affects feed intake and eating rate in dairy steers. Eight steers (Swedish red breed, LW 300 kg) were offered long (L) or chopped (C) round baled WCB silages harvested at heading (H) and at dough stage (D) of maturity, resulting in four treatments: HL, HC, DL and DC. Animals were randomly assigned to a 2 x 2 factorial arrangement of treatments in a 4 x 4 Latin square design with four 3-wk periods. Silage *ad libitum* intake was registered from day 8 through 17. After three days of restrictive feeding, eating rate was measured on day 21 by offering animals 25% of the daily silage allowance four times throughout the day. At each feeding, amounts consumed during 20 minutes were registered. Results were analysed by ANOVA using GLM. The DM intake in kg/day tended to differ ($p < 0.07$) between HL and HC (7.97 *vs* 7.44) and between DC and DL (7.84 *vs* 7.32). The NDF intake was higher for H than for D (4.03 *vs* 3.68 kg/d, $p < 0.001$). Eating rate was higher for C than for L (49 *vs* 35 g DM/min; $p < 0.001$). Chopping of WCB silage increased eating rate but did not consistently increase intake, suggesting that time spent eating did not limit feed intake.

Session 28 Theatre 13

Use of hay-rich diet for Piemontese beef production

C. Lazzaroni and D. Biagini, University of Torino, Dept. Animal Science, Via L. da Vinci 44, 10095 Grugliasco, Italy

Diet more rich in forages than in concentrate are now used also in specialised beef production, so the possibility to apply such diet in fattening Piemontese young bulls was verified. Two groups of 10 calves (8 months of age, 200 kg l.w.) were fed hay and concentrate, following a diet rich in forage (4-8 kg/d, 60% DM intake) compared to a traditional one (2 kg/d) more rich in concentrate (2-3 *vs* 3-8 kg/d), then slaughtered (about 550 kg l.w.). Monthly individual weights, average daily weight gain, daily feed consumption, and feed conversion rate were recorded. During the first part of the trial, the performances of both groups were comparable, but at the 15th month of age the forage-fed group start to show meagre weight gains (0.6 *vs* 1.3 kg/d; $P < 0.001$) and reduced hay consumption (-1 kg/d DM), so they didn't reach the proper muscular development and fattening degree at the same age of the traditional-fed group (about 16 month of age), even if the animals reached the same weight. Only after a period of about 4-5 months following a traditional fattening diet they were ready for the market, showing higher live weight (580 *vs* 530 kg, $P < 0.05$), similar dressing percentage (68.8 *vs* 66.6), but higher feed consumption, both as hay (1657 *vs* 595 kg DM) and concentrate (1527 *vs* 1488 kg DM). A diet so rich in forage didn't suit to fattening Piemontese young bulls, as it didn't allow obtaining a product appraised on the market, but it could be interesting for the first part of the fattening period and in a not so high percentage.

Session 28 Poster 14

Evaluation of different kinds of silage supplemented with limestone

N.M. Eweedah, M.S. Saleh, H.M. Gaafar, E.M. Abdel-Raouf and W.E. Hagag, Faculty of Agriculture, Kafre El-Sheikh University, Animal Production, Kafre El-Sheikh, 33516, Kafre El-Sheikh, Egypt

Four fresh cereal crops, whole plant corn, corn stover, fodder corn and sorghum were used to evaluate the effect of limestone supplementation on chemical composition, silage quality and *in situ* disappearance of DM, CP and CF. Forage crops were chopped to 1.5-2.0 cm length. Ground limestone was added to the different crops at levels of 0.0, 0.5, 1.0, 1.5 and 2.0 % of wet weight, then mixed and ensiled. After two months, the silos were opened and silage quality was examined. The rate of ruminal degradation of DM, CP and CF was determined. The concentration of lactic acid decreased with increasing DM content of silage and with increasing limestone level up to 1%. The concentration of NH_3-N decreased with increasing DM content, but it increased with increasing CP content. DM disappearance was significantly increased with increasing limestone level up to 1%, but decreased afterwards. Meanwhile, the CP disappearance was significantly decreased with increasing limestone level. As the undegradable protein fractions increased and effective degradability at different outflow rates decreased, there was a resultant increase in the amount of protein escaping ruminal degradation. The CF disappearance increased with increasing CF content. However, it decreased with increasing NFE content. Results indicated that the addition of 1.0% limestone at silage making improved silage quality, decreased ensiling losses and enhanced DM, CP and CF disappearance.

Session 28 Poster 15

Isolation of genes related to lignin biosynthesis in tropical forages

P. Lazarini[1], L. Jank[2] and L.F.P. Silva[1], [1]University of São Paulo, Pirassununga, 13635-900, Brazil, [2]EMBRAPA, Campo Grande, 79002-970, Brazil

Our objective was to isolate and sequence mRNAs coding the main enzymes from the lignin biosynthesis pathway, in order to determine the similarity between guineagrass and other grass species with more advanced genomic resources. Total RNA was isolated from guineagrass (*Panicum maximum* Jacq. cv. Tanzânia) and treated with DNaseI. Degenerated oligonucleotide primers for glyceraldehyde-3-phosphate dehydrogenase (GAPDH), phenylalanine ammonia-lyase (PAL), cafeic acid O-methyltransferase (COMT), and cinnamyl alcohol dehydrogenase (CAD) were designed based on conserved regions of maize and rice. cDNAs were amplified by PCR, and the products obtained were purified, precipitated and sequenced in both directions. Quality and alignment of sequences were obtained using Phred and Cap3 software's. Guineagrass sequences were compared with other grass species using the BLAST program. There was a high degree of similarity between guineagrass sequences and those from maize and rice. Similarity of guineagrass GAPDH was 87% and 89% with rice and maize, PAL was 88% and 87% similar, CAD was 84% and 90% similar, and COMT was 90% and 95% similar with rice and maize, respectively. On average, guineagrass sequences were 87.2% similar with rice and 90.2% similar with maize. The high similarity observed suggest that genetic resources developed for maize and rice, such as microarrays, primers and genes databases, could be used for genetic studies with guineagrass.

Session 28 Poster 16

Effect of hours at pasture on production, milk composition and behaviour of dairy cows

T. Kristensen[1], F. Oudshoorn[2] and J. Sahana[1], [1]University of Aarhus, Research Centre Foulum, PO Box 50, DK 8830 Tjele, Denmark, [2]University of Aarhus, Research Centre Bygholm, DK 8730 Horsens, Djibouti

During a six-week period from the start of the 2005 grazing season an experiment was conducted to investigate the effect of restrictive indoor feeding combined with limiting the time at pasture in a system based on continuous stocking on productivity and behavior of high yielding dairy cows (31.0 ± 5.4 kg ECM). The herd was split into three groups allocated to three treatments, 4, 6.5 and 9 hours at pasture respectively. Only cows with the longest period at pasture (9 hours) had access to *ad libitum* indoor feeding. The allowance of herbage was on average 1660 kg DM per ha and the intake of supplemental feed was in average 9.1 kg DM per cow daily. It was concluded that the limitation of the time at pasture in combination with restrictive feeding indoor of high yielding dairy cows reduced the daily milk production significantly and tended to reduce the live weight gain. The content of individual fatty acid was significantly influenced by treatment. The proportion of time during which the cows were grazing at pasture increased with reduced time at pasture as well as the cows walked faster, and a longer distance per hour. The observed grazing intensity, bites per minutes, of grazing were unaffected by hours at pasture.

Session 28 Poster 17

Chemical composition, tannin content, and in sacco, *in vitro* and short-term *in vivo* digestibility of oak (*Quercuss hartwisiana*) leaves

S. Yildiz[1], I. Kaya[2], Y. Unal[2], C. Arslan[2] and A. Oncuer[2], [1]Kafkas University, Faculty of Veterinary Medicine, Department of Physiology, Pasacayiri Kampusu, KARS, Turkey, [2]Kafkas University, Faculty of Veterinary Medicine, Department of Animal Nutrition, Pasacayiri Kampusu, KARS, Turkey

Oak leaves exist as an alternative feedstuff in North-East of Turkey but they have been reported to contain high amounts of tannins which might offset their beneficial effects. Therefore, we analysed these leaves in terms of chemical composition and tannin type and content by using a battery of assays. Additionally, in sacco degradability and *in vitro* gas production were assessed. Afterwards, these leaves were offered to rams in short-term (15 days) sequential feeding trials by increasing the proportion of oak leaves each time. On a dry matter basis oak leaves on average contained 13.4% crude protein, 9.6 % total phenolics, 6.9 % total tannins, 1.3% condansed tannins and 0.2% gallic acid. The increase in gas production with the addition of PEG was 73%. In *in vivo* experiments oak leaves substantially reduced protein digestibility. The results suggest that negative effects of tanniniferous oak leaves need to be overcome when fed to sheep.

Session 28 Poster 18

Nutritive value assessment of some trees and shrubs grown in Saudi Arabia as alternative feed resources for camels and sheep

A.A. Al-Soqeer[1], S.N. Al-Dobaib[2] and H.E.M. Kamel[2], [1]Qaseem Univ. Agric. & Vet. Faculty, Plant Production & Protection, P.O. Box 1482, Buriedah 51431, Saudi Arabia, [2]Qaseem Univ. Agric. & Vet. Faculty, Animal Production and Breeding, P.O. Box 1482, Buriedah 51431, Saudi Arabia

The *in vitro* gas production and nutritive value for some trees and shrubs grown under the Saudi Arabian environment were evaluated as alternative feed resources for both camels and sheep. *Acacia saligna, Atriplex halimus, Conocarpus erectus, Leucaena leucocephala, Prosopis juliflora* (pods) and *Prosopis juliflora* (shoots) were tested. Values of protein content ranged from 7.4 to 22.8%, with L. leucocephala having the highest value, while the lowest amount was found in *C. erectus* On the other hand, *L. leucocephala* had the lowest neutral detergent fiber, with the highest value for *C. erectus*. Condensed tannins content (mg/g DM) was ranged from 0.12 in P. juliflora (pods) to 40.7 in A. saligna. The extent of ruminal degradation for *A. saligna, A. halimus, C. erectus, L. leucocephala, P. juliflora* (pods) and *P. juliflora* (shoots) as evidenced by gas production (ml/200mg) were 70.9, 44.2, 38.8, 58.9, 90.5 and 22.7 for camels, and 56.5, 41.6, 30.0, 52.6, 92.8 and 19.1 for sheep, respectively. Estimated metabolizable energy was significantly ($P < 0.05$) higher when tested feed was incubated with rumen liquor obtained from camels rather than sheep.

Session 28 Poster 19

Management strategies in hill pastures of Central Italy grazed by rotational-stocked cattle

P. D'ottavio[1], M.F. Trombetta[2] and R. Santilocchi[1], [1]Università Politecnica delle Marche, Dipartimento di Scienze ambientali e delle Produzioni vegetali, Via Brecce Bianche, 60131 Ancona, Italy, [2]Università Politecnica delle Marche, Dipartimento di Scienze degli Alimenti, Via Brecce Bianche, 60131 Ancona, Italy

The aim of the paper is to discuss various management aspects for beef cattle breeding in a multi-paddock rotational stocking unit. The experiment was performed on clay soils characterised by different morphology and slope located at a mean altitude of about 500 m a.s.l.. The climate of the study area is characterised by a mean annual temperature of 12.6 °C and a mean annual precipitation of 945 mm. The pasture surface was divided into paddocks by electric fences, with 2 mobile water-points. It was rotationally grazed by a herd of about 30 cows and 10 calves of the local Marchigiana breed. In each paddock botanical composition, DM yield and forage quality were assessed before, during and after utilisation throughout the grazing periods of 2005 and 2006. The main results deal with the following aspects: grazing management, forage yield and quality, mean daily intake and forage balance.

Session 28 Poster 20

Effect of preference for white clover or red clover silage over ryegrass silage on nutrient supply, milk yield and composition in dairy cows

H.A. Van Dorland, H.-R. Wettstein, H. Leuenberger and M. Kreuzer, ETH Zurich, Institute of Animal Science, Universitätstrasse 2, 8092 Zurich, Switzerland

An experiment was conducted to determine the effect of giving cows the choice between ryegrass silage, and either white clover-, or red clover silage on nutrient supply and milk composition in dairy cows. The experiment consisted of four treatments, two choice-, and two mixed diets (40% clover on DM basis) with ryegrass silage and either white or red clover silage. Intake of forages and total DMI and were not affected by treatment. With choice diets, the cows preferred white clover and red clover (73.0 and 69.2%, respectively) over ryegrass. The choice diet with white clover enhanced dietary contents of crude protein (+16.7%; $P < 0.001$) and NE_L (+8.0%; $P < 0.05$), and decreased contents of DM (-12.0%; $P < 0.05$) and NDF (-17.1%; $P < 0.001$) compared to the mixed diet with white clover. Similar changes were observed for the red clover choice diet, but to a lesser extent. Milk yield and gross composition were not affected by treatment, except for the protein content, which was higher (+12.6%) in case of the choice diet compared to the mixed diet with white clover ($P < 0.05$). Both choice diets enhanced the content of *n*–3 fatty acids in milk fat compared to the mixed diets. This study suggests that clover silage intake (either white or red clover) as preferred by dairy cows, although not necessarily benefiting their performance, could increase *n*–3 fatty acids proportions of milk fat.

Session 29 Theatre 1

Horse genome sequence: watershed for horse breeding

E. Bailey, University of Kentucky, Gluck Equine Res. Center, 40546-0099 Lexington, KY, USA

The post genomics age is upon us. We now study biology in light of whole genome sequences for several species. Consequently, the recent assembly of the horse genome sequence has a special significance for those of us studying horses. During the preceding decade, the molecular causes for many simple, Mendelian diseases and color traits were uncovered for horses. However, the applications have not addressed the major problems encountered by horse breeders, specifically, diseases of the respiratory and muscle systems, skeletal development, allergy, infectious diseases and performance. The whole genome sequence for the horse will accelerate discoveries benefiting the health and welfare of horses. The information is freely available in public databases and allows us to identify the chromosome position, DNA sequences and proximal DNA sequences for any gene of interest. If we have better funding we can do complex studies of gene expression or genetic variation using arrays of oligo-nucleotides. We are also in an excellent position to apply the information developing from human and mouse studies on the conserved-non-coding DNA sequences that control timing and tissue localization of gene expression. Within the next 5 years we can develop applications for the complex management-hereditary problems confronting horse owners if we learn to use these tools. Of course, a watershed also needs rain.

Development of a BAC-based physical map of the horse genome

O. Distl[1], T. Leeb[1], M. Scharfe[2], M. Jarek[2], G. Nordsiek[2], F. Schrader[2], P.J. de Jong[3], B.P. Chowdhary[4], C. Vogel[5], B. Zhou[3], A. Wöhlke[1] and H. Blöcker[2], [1]Institute of Animal Breeding and Genetics, Buenteweg 17p, 30559 Hannover, Germany, [2]Helmholtz Centre for Infection Research, Inhoffenstraße 7, 38124 Braunschweig, Germany, [3]Childrens Hospital Oakland, 747 52nd Street, Oakland, CA 94609, USA, [4]Department of Veterinary Integrative Biosciences, Texas A&M University, College Station, TX 77843, USA, [5]Institute of Animal Breeding and Genetics, Veterinaerplatz 1, 1210 Wien, Austria

Recently, for the horse a 7x whole genome shotgun sequence became available. High-resolution BAC-based physical maps have the potential to complement the assembly of whole genome-shotgun sequences, which will lead to an improved long-range contiguity of the genome sequence. In a Lower Saxonian effort a physical map of the horse genome will be created based upon a combination of BAC fluorescent fingerprinting and BAC end sequencing of the CHORI-241 library. BAC end sequences (BESs) and fluorescent fingerprints and of 150,000 BAC clones (~10x genome coverage) will be obtained by using the 4-restriction enzyme 4-color technique and separating the resulting fragments on capillary sequencers. The BESs will enable the anchoring of the emerging BAC contigs to the equine RH map as well as the comparative analysis with respect to the human genome. So far, ~170,000 BESs have been submitted to the public databases and can be accessed through the website of the Institute for Animal Breeding and Genetics at Hannover.

A whole genome scan to identify quantitative trait loci for guttural pouch tympany in German warmblood and Arabian horses

A. Zeitz[1], A. Spötter[1], B. Ohnesorge[2], H. Hamann[1] and O. Distl[1], [1]Institute for Animal Breeding and Genetics, Buenteweg 17p, 30559 Hannover, Germany, [2]Clinic for horses, Bischofsholer Damm 15, 30173 Hannover, Germany

We performed a whole genome scan to identify quantitative trait loci for guttural pouch tympany using 69 German warmblood and 77 Arabian horses. The whole genome scan included 185 microsatellites equidistantly distributed over the horse genome. The number of affected foals was 59 distributed over eleven pedigrees. Guttural pouch tympany (GPT) is an inherited disease which is characterized by a nonpainful inflation of one or both guttural pouches occurring in the first weeks of life in foals. GPT can be acute life-threatening in affected foals without surgical treatment. Samples of affected foals analyzed were from hospitalized foals at the Clinic for Horses, Hannover. Pedigree analyses based on complex segregation analyses could show a major gene besides a polygenic component and in addition, a significant influence of Arabian blood proportion in German warmblood horses. Chromosome-wide significant QTL were located on horse chromosomes (ECA) 2, 15 and 22. Furthermore, a significant QTL on ECA26 could be detected for German warmblood horses. The significant QTL will be refined using additional microsatellites and SNP markers. The recently released sequences from the horse genome project will greatly enhance the positional cloning of the QTL and unravelling the responsible genes.

Session 29 Theatre 4

SNP markers for osteochondrosis in horses

V. Lampe, C. Dierks, C. Wittwer and O. Distl, Institute for Animal Breeding and Genetics, Buenteweg 17p, 30559 Hannover, Germany

The objectives of this work are to develop intragenic single nucleotide polymorphism (SNP) markers for osteochondrosis in horses. We used positional candidate genes located in quantitative trait loci (QTL) identified through a whole genome scan in Hanoverian warmblood and South German coldblood horses. The whole genome scan included 24 paternal half-sib families and more than 350 horses. The marker set consisted of more than 200 highly polymorphic microsatellites. Traits regarded were osteochondrosis (OC) in fetlock and hock joints. SNP markers were developed using equine BAC end, EST and whole genome shotgun sequences. QTL chosen were located on equine chromosomes 2, 4, 5, 16 and 18. Equine maps had to be refined for the QTL regions to be able to correctly locate the candidate genes. We were able to identify significantly associated SNP markers on ECA4 and 18 in the same candidate genes in both horse breeds analysed. Association analyses were performed using an animal model and the genotypes of the SNPs for both breeds separately. Here, we used horses which were not closely related with each other. The additive genetic effects of the SNP markers were between 0.20 and 0.43 for occurrence of fetlock or hock OC as 0/1 traits. Further SNPs will be developed and tested for association with OC. This work is an important step towards an equine SNP marker set to be employed in horse breeding.

Session 29 Theatre 5

Molecular and association analysis in an endangered breed: summer dermatitis in Old Kladruber horses

P. Horin[1], L. Vychodilova[1], L. Putnova[2], I. Vrtkova[2], M. Vyskocil[1], M. Sedlinska[1], J. Osickova[1] and J. Hanak[1], [1]Faculty of Veterinary Medicine, Palackeho, 61242 Brno, Czech Republic, [2]Mendel University, Zemedelska, 61300 Brno, Czech Republic

Various types of diseases represent a serious threat for the Old Kladruber breed, due to it´s isolation and small size. Summer dermatitis (SD) occurring also in the Old Kladruber grey population is an allergic reaction to insect bites, with possible involvement of specific susceptible genotypes. Based on clinical signs of SD recorded over last 8 years, 73 breeding mares were classified as susceptible and/or resistant. 50 microsatellites and 9 single nucleotide polymorphisms (SNP) in 5 candidate genes, the lipopolysaccharide receptor gene *CD14*, the toll-like receptor 4 gene *TLR4*, the gene *Cε* encoding the IgE heavy chain molecule and the gene *FcεR1 alpha* coding for the alpha subunit of the IgE receptor molecule, were used for association analysis. Standard Fisher test with appropriate corrections for multiple comparisons was used for this purpose. Significant associations were found between the microsatellite *AHT04* ($p_{corr}<0.001$) and for the *T546C* SNP within the *FcεR1 alpha* gene ($p_{corr}<0.03$). The results suggest involvement of IgE controlling loci in susceptibility to horse SD. Currently, associations with total IgE levels are investigated. Conservation issues for the breeding programme of this endangered population are discussed. Supported by projects GACR 523/06/1402 and Ministry of Agriculture of the CR (1G58073).

Session 29 — Theatre 6

Candidate gene markers for stallion fertility

K. Giesecke[1], H. Hamann[1], H. Sieme[2] and O. Distl[1], [1]Institute for Animal Breeding and Genetics, Buenteweg 17p, 30559 Hannover, Germany, [2]Institute for Reproduction Medicine/Clinic for horses, Buenteweg 15, 30559 Hannover, Germany

Implementation of artificial insemination (AI) in horse industry increased the impact of individual reproduction performance and requires consequent evaluation of semen quality for AI. Fertilization of eggs by spermatozoa is a complex process influenced by sperm maturation, capacitation and seminal plasma proteins. The objective of this work is to investigate equine candidate genes for their possible association with pregnancy rate per oestrus and per breeding season. Intragenic polymorphisms will be identified in genes encoding the major and minor seminal plasma proteins including Fn-2 type proteins and spermadhesins. In addition, genes encoding proteins involved in recognition and adhesion between spermatozoa and the zona pellucida will be regarded. We have selected 15 candidate genes, localized them on the equine genome and used equine whole genome sequences, BAC end sequences and ESTs to annotate their structure. The association study will be based on AI records collected between 1997 and 2005 from 19,897 mares and 246 stallions belonging to the Hanoverian warmblood. Heritability estimates for the pregnancy rate per oestrus were at 1% for the paternal and maternal component and at 4% for the direct component. The study should provide evidence for the effects of genes encoding important proteins for stallion fertility and support predicting stallion fertility.

Session 29 — Theatre 7

Corticotropin-releasing hormone (CRH) family of peptides: sequence variation in cattle and horse revealed by comparative genomic approach

T. Kunej, Z. Jiang and P. Dovc, Biotechnical Faculty, Deparmtment of Animal Science, Groblje 3, 1230 Domzale, Slovenia

The corticotropin-releasing hormone (CRH) family of peptides plays roles in physiological, developmental and behavioral events, such as activation of the hypothalamic-pituitary-adrenal axis, modulation of gastrointestinal functions, stimulation of anxiety-related behavior, control of locomotor activity and regulation of food intake and energy balance. Characterization of the CRH family of peptides is of interest in animal species where fat deposition, meat quality and physical performance plays an important role. We determined the genomic organization and genetic variability of *UCN3* (urocortin 3) and *CRHR2* (corticotropin-releasing hormone receptor 2) genes in cattle and horse. Marker genotyping in cattle revealed significant association between urocortin 3 (*UCN3)* gene and marbling score as well as subcutaneous fat depth. In bovine *UCN3* and *CRHR2* genes we also discovered a novel type of sequence variation, i.e., multiple nucleotide length polymorphism (MNLP). Both MNLPs involved multiple nucleotides and length polymorphisms between two alleles (5 bp/10 bp allele in *UCN3*, and 12 bp/18 bp allele in *CRHR2*,) and changes in promoter activity between two alleles. Polymorphisms in the regulatory regions of the CRH family of genes in horse may affect physical performance of animals. This novel genetic complexity would contribute significantly to the evolutionary, functional and phenotypic complexity of genomes within and among species.

Session 29 Theatre 8

Characterisation of the Kerry Bog Pony

A. Heffernan, Weatherbys DNA Laboratory, c/o Irish Equine Centre, Johnstown, Naas, Co. Kildare, Ireland

In this study nuclear DNA (nDNA) microsatellite markers and mitochondrial DNA (mtDNA) sequence analysis were implemented in order to carry out an extensive genetic characterisation of the Kerry Bog Pony (KBP) population in Ireland. The nDNA of 172 KBPs was genotyped using a panel of 17 microsatellite markers. The allele frequency data was analysed to reveal the levels of genetic diversity, differentiation and inbreeding. Breed assignment and bottleneck tests were performed and the genetic relationship between the KBP and three other Northern European pony breeds was established. 39 unrelated foundation stock KBPs were selected to determine the mtDNA sequence variation in the breed. A 511bp region of the mitochondrial D-loop was amplified, sequenced and analysed. The haplogroup diversity was compared to the diversity in 68 other globally widespread horse and pony breeds. The results from the nDNA analysis revealed high levels of genetic diversity and differentiation and low levels of inbreeding in the KBP population. The phylogenetic analysis revealed that the Welsh Pony is genetically the closest of the studied breeds to the KBP. The majority of the KBP sequences (30%) were assigned to haplogroup E. This is a rare equine haplogroup with only 3% of horses falling into this category in a global context. The information gained during this genetic characterisation of the KBP when used in conjunction with the breeding strategy of KBP Co-operative Society will ensure a safe and successful increase in the population of KBPs in Ireland.

Session 29 Theatre 9

Genetic diversity and admixture analysis of Sanfratellano horse population assessed by microsatellite markers

D. Marletta[1], A. Zuccaro[1], S. Bordonaro[1], A. Criscione[1], A.M. Guastella[1], G. Perrotta[2], M. Blasi[2] and G. D'urso[1], [1]DACPA: sezione di Scienze delle Produzioni Animali. Università di Catania, via Valdisavoia,5, 95123 Catania, Italy, [2]LGS - Laboratorio di Genetica e Servizi, via Bergamo, 292, 26100 Cremona, Italy

Sanfratellano is a native Sicilian horse population, mainly bred in the North East of the island, developed in the 19th century from local mares and sires with a restricted introgression of Oriental, African and, more recently, Maremmano stallions. In this study the genetic relationships and admixture among Sanfratellano, the other two Sicilian autochthonous breeds and Maremmano were assessed using 11 microsatellites. The entire sample included 384 horses (238 Sanfratellano, 50 Sicilian Oriental Purebred, 30 Sicilian Indigenous, 66 Maremmano) chosen avoiding closed related animals.A total of 111 alleles have been detected. The average number of alleles was the lowest in Oriental Purebred (6.7), the highest in Sanfratellano (8.2). All the breeds showed high levels of heterozygosity, ranging from 0.71 in Oriental Purebred to 0.81 in Sicilian Indigenous. The gene differentiation coefficient was low, only 5% of the diversity being among breeds. A NJ tree based on individual Dps' distances was built. Bayesian approach, used for admixture analysis, revealed within Sanfratellano population individuals with different proportions of Maremmano, Oriental and Sicilian Indigenous blood.

Session 29 — Theatre 10

Molecular genetic characterisation of Mezőhegyes horse breeds

S. Mihók[1], B. Bán[2], I. Bodó[1], C.S. Józsa[2], I. Péntek[2] and I. Komlósi[1], [1]Debrecen University, Animal Breeding, Böszörményi út 138, 4032 Debrecen, Hungary, [2]Agricultural Administration Office, Genetic Lab, Keleti Károly u. 4, 1024 Budapest, Hungary

At the Mezőhegyes National Stud (1785) three horse breeds were founded: the Nonius, Gidran and Furioso-North Star based upon Spanish mares and different founder stallions (Anglo-Norman, Arabian, Thoroughbred) selected for different breeding goals (working, riding, and dual purpose). They were improved later by Thoroughbred. These breeds were compared to the English Thoroughbred and to the Hutzul using the D blood group, polymorphic systems (transferrin, albumin, etc and 12 DNA microsatellites. Blood samples of 95 Gidrans, 136 Nonius, and 58 Furioso horses were analysed. The data were analysed by Popgen32. On the basis of the D blood group, transferrin, albumin, and esterase the Nonius shows a great difference from the other two local breeds. The frequency of the GC and A1B alleles were similar in the three breeds. The frequency of Gidran alleles showed a more homogenous picture than that of Nonius or Furioso. In some cases the common maternal origin was expressed in the similar allele frequencies. The allele frequency observed in the Gidran was closest to Thoroughbred. The microsatellite profile of the Hutzul was the most distant. The genetic distance between breeds, lines and families is useful for breeding strategies, mating plans and long term decisions. The maintenance of frequency of rare alleles can be used to reduce the impact of genetic drift.

Session 29 — Theatre 11

How to estimate kinship and inbreeding with SNPs

B. Langlois, INRA, GA-SGQA, Domaine de Vilvert, 78 350 Jouy en Josas, France

I recently developed (8th wgalp) how to choose genetic markers to infer kinship or inbreeding coefficients. The first draft of the horse genome sequence has now been deposited in public databases and is freely available for use. In addition to sequencing the horse genome, a map was produced which should comprise one million of SNPs. We will show in this paper how it could be used to estimate kinship and inbreeding coefficients.

Session 29 — Theatre 12

Equine cDNA microarrays and the transcriptomic response of thoroughbred skeletal muscle to exercise

B. Mcgivney and E. Hill, University College Dublin, Agriculture and veterinary, Belfield Dublin 4, 0000, Ireland

With the recent advances in the sequencing of the equine genome it is important to utilize this information in the laboratory in conjunction with *in silico* studies. We are using equine cDNA microarrays to identify variations in gene expression in equine skeletal muscle in response to exercise. This will lead to a greater understanding of the molecular networks that control cellular function relating to muscle physiology in the horse. Eight untrained four year old thoroughbred geldings were exercised to maximal heart rate or fatigue on an equine high-speed treadmill. Skeletal muscle biopsies were taken from the middle gluteal muscle before, immediately after and four hours after exercise. Individual comparisons between timepoints will enable an understanding of each individual's response to exercise. This reduces the statistical importance of inter-individual variability in mRNA level which is high in untrained human subjects. mRNA abundance will be analysed using microarray data analysis software such as BRB array tools to determine statistically significant expression changes and to identify gene pathways relevant to exercise. Results will be confirmed using quantitative realtime RT-PCR. This study will be the first to characterize global mRNA expression profiles in equine skeletal muscle using an equine-specific microarray platform and will provide valuable information regarding the response to intense exercise and mRNA expression during recovery from exercise.

Session 29 — Poster 13

Molecular sex determination using sexual dimorphism of the ZFX and ZFY genes in horses

S.H. Han, I.C. Cho, S.S. Lee and M.S. Ko, National Institute of Subtropical Agriculture, Rural Development Administration, Livestock Division, San 175-6 Odeung-dong, Jeju, 690-150, Korea, South

The sex chromosome-encoding *zinc finger* genes (*ZFX* and *ZFY*) are analyzed and tested molecular sexing using the amplification patterns of intron 9 of *ZFX* and*ZFY* in the horse (*Equus caballus*). The amplification of the *ZFX* and *ZFY* genes produced two distinct patterns, reflecting chromosome-linked sexual dimorphism based on a length difference between the X and Y chromosomes. The amplification products from foals showed two distinct bands: one was common to all foals and mares, indicating that this band was amplified from *ZFX* on the X chromosome, while the other was specific to foals, indicating that it was from *ZFY* on the Y chromosome. There was no complete deletion of transposable element in the intron 9 sequences of the *ZFX* and *ZFY* genes, as previously reported for other mammals. However, multiple deletion mutations were found in the equine intron 9 sequences of both genes comparing to that of bovine *ZFY* gene. In addition, the PCR sex typing showed identical results to those of amplification patterns of the Y chromosome-specific *SRY* gene and those of investigations of the phenotypic gender in two different horse populations (thoroughbreds and Jeju native horses). We suggest that this PCR strategy for determining sexes by comparing the amplification patterns of the dimorphic *ZFX* and *ZFY* genes on the X and Y chromosomes is a convenient and precise method for discriminating sexes in horses.

Session 29 Poster 14

Coat colour genetics: application to Maremmano horse breeding program

A. Verini Supplizi[1], K. Cappelli[1], S. Capomaccio[2], C. Pieramati[1] and M. Silvestrelli[1], [1]University of Perugia, Sport Horse Research Centre, San Costanzo 4, Perugia, 06126, Italy, [2]University of Perugia, DBVBAZ, XX Giugno 74, Perugia, 06121, Italy

The Maremmano horse has a Stud Book since 1980. This breed is used for Saddle horse production, as purebred or crossbred (Italian Saddlebred). The coat colours are: black, bay and, only for female, chestnut. The Stud Book consider the coat colour important as breeding goal and today molecular analysis is used to help selective breeding for black and against chestnut. The extension (E) and agouti (A) loci determine the relative amount of red/yellow pigment (phaeomelanin) and black pigment (eumelanin) in mammals. In horse, mutation at locus E leading to the chestnut is a single base substitution in melanocyte-stimulating hormone receptor (MC1R) whereas an 11bp deletion in agouti-signaling-protein (ASIP) exon 2 lead to a black coat (Rieger *et al.*, 2001). In order to evaluate the allelic state of this population at locus E and locus A, we used a protoçol, based on a simple PCR step (locus A and E) and a digestion with a specific enzyme (locus E). The E locus assay is based on the amplification of a target sequence that if mutated creates a restriction site for *Taq*I. In a population of Maremmano horses (10% breeding stock) 53 % were E/E, 45% were E/e and only 2% were e/e (chestnut). The A locus screening is based on the detection of a 11 base pair deletion in agouti exon 2; in our population 54% of individuals were A/a, 28% were AA, and 18% resulted a/a.

Session 29 Poster 15

Structure, expression and polymorphism of a gene of the glycogenolysis pathway (AGL) in the horse

B. Herszberg, X. Mata, B. Langlois, S. Chaffaux and G. Guérin, INRA, Centre de Recherches de Jouy, Génétique animale, Laboratoire de Génétique biochimique et de Cytogénétique, 78350, Jouy-en-Josas, France

Mutations in enzymes of the glycogen metabolic pathway may introduce hereditary affections among which is the polysaccharide storage myopathy (PSSM). While a mutation of the glycogen branching enzyme produces a fatal neonatal disease in Quarter Horses, the etiology of an adult form of glycogenosis inducing myopathy in draft horses is still unknown. PSSM affected draft horses accumulate a glycogen-like polysaccharide in striated muscles and present myositis symptoms under specific conditions of exercising and nutrition. We determined the structure, expression, transcriptional regulation and polymorphism of a candidate gene for the adult form of the disease: the amylo-alpha-1,6-glucosidase (AGL). The equine AGL mRNA undergoes alternative splicing in its 5'extremity. We could identify 4 of the 6 variants described in human and a new horse variant was detected in the striated muscle and in the heart. AGL has an ubiquist expression in horse but specific variants are preferentially expressed in liver or muscle. The 5 equine splicing variants encode two potential proteic isoforms of 1533 and 1377 amino acid. We identified 4 single nucleotide polymorphims (SNP) in the cDNA of the equine gene but none of them presents a strict association with the affected phenotype. The role of AGL in the etiology of PSSM horses needs further investigations.

Session 29 Poster 16

Genetic characterization of endangered Spanish horse breeds for meat production

P.J. Azor[1], M.D. Gómez[1], M.E. Alonso[2], J. Jordana[3], A. Pérez De Muniaín[4] and M. Valera[5], [1]University Cordoba, C.U.Rabanales, 14071 Cordoba, Spain, [2]UNIVERSITY LEON, C.U.VEGAZANA, 24071 Leon, Spain, [3]University Barcelona, Edifici A, 08193 Bellaterra, Spain, [4]ITG, Av.Serapio Huici 22, 31610 Villava, Spain, [5]University Seville, Ctra Utrera km1, 41013 Seville, Spain

The horse breeds raised for meat production in Spain are located in the north of the country and all of them are endangered breeds. The aims of this study is to know the genetic variability of these horse breeds and their genetic relationship in order to assist in their breeding programs for meat production. We amplified 15 polymorphic microsatellite markers in 120 unrelated horses belonging to four breeds, 30 horses per breed (Jaca Navarra (JAC), Burguete (BUR), Hispano Bretón (HB) and Agrupación Hipermétrica del Pirineo (AHP)). The observed and expected heterozygosity ranged from 0.645 and 0.719 in AHP to 0.807 and 0.788 in JAC respectively. Average number of alleles per locus was 7.00 in AHP, 8.13 in BUR, 8.4 in HB and 8.06 in JAC. Only 7.18% of the total genetic variability could be attributed to differences between breeds (F_{ST} = 0.0718). The highest effective number of migrants per generation (*Nm* = 16.11) was between the BUR-HB pair.

Session 30 Theatre 1

Crossbreeding in New Zealand dairy cattle

N. Lopez-Villalobos[1], D.J. Garrick[1,2] and C.W. Holmes[1], [1]Institute of Veterinary, Animal and Biomedical Sciences, Massey University, Private Bag 11222, 4442 Palmerston North, New Zealand, [2]Department of Animal Sciences, Colorado State University, 80523-1171 Fort Collins, Colorado, USA

Heterosis effects for traits of economic importance in New Zealand (NZ) dairy cattle have been reported to be (in phenotypic sd): 0.27, 0.38 and 0.38 for lactation yields of milk, fat and protein, 0.20 for cow live weight, 0.09 for cow fertility, -0.05 for somatic cell score and 0.25 for longevity. Simulation shows that a two-breed rotation with straightbred Holstein-Friesian and Jersey bulls had highest net income per hectare (NZ$1068) followed by the synthetic Holstein-Friesian×Jersey (NZ$1047). The Holstein-Friesian herd had lowest net income per hectare (NZ$993). These results agree well with Production Worth or overall merit expressed in profit per 4.5 t dry matter produced by the national genetic evaluation. The ranking of alternate crossbreeding systems is however sensitive to relative values of milk volume, fat and protein. Crossbreeding in NZ has increased the proportion of Holstein-Friesian×Jersey crossbred cows to 30% of the national herd in 2006. Three major changes in the industry have been required or have resulted from adoption of crossbreeding: implementation of an across-breed database and across-breed genetic and economic evaluation systems, modification of the selection scheme to progeny test crossbred as well as straightbred bulls and significant shifts in industry production of milk components and mix of dairy products.

Milk production, udder health, body condition score at breeding and fertility of first lactation Holstein-Friesian, Norwegian Red and Holstein-Friesian×Norwegian Red cows on Irish dairy farms

N. Begley[1,2], M. Rath[2] and F. Buckley[1], [1]Teagasc, Moorepark, Dairy Production Research Centre, Fermoy, Co. Cork, Ireland, [2]College of Life Sciences, University College Dublin, Belfield, Dublin 4, Ireland

The objective of this study was to compare milk production, udder health, body condition score at breeding and fertility of first lactation Holstein-Friesian (HF) (n=710), Norwegian Red (NRF) (n=325) and Holstein-Friesian×Norwegian Red (F_1) (n=292) cows across 46 dairy herds. Predicted 305 d yields were obtained from ICBF. The 305 d milk yields of the HF and F_1 were similar at 5,353 kg and 5345 kg, respectively. That of the NRF was lower at 5149 kg ($P < 0.001$). Fat content was higher for the HF at 4.00%, compared to the NRF at 3.94% ($P < 0.05$). That of the F1 was intermediate. Protein content was similar for all breeds at 3.46%, 3.45% and 3.45% for HF, NRF and F_1, respectively. Compared to the HF, the F_1 and NRF cows had lower somatic cell score ($P < 0.01$) and lower incidence of mastitis ($P \leq 0.062$). BCS at breeding was 2.85, 3.03 and 2.98 for the HF, NRF and F_1, respectively, significantly higher for the NRF and F_1 ($P < 0.001$) compared to the HF. Fertility data was available for 43 of the 46 herds. In calf rate was 91%, 95% and 93% for the HF, NRF and F1 respectively, significantly higher for the NRF compared to the HF (P=0.051). In summary the F_1 cows had similar levels of milk production to HF. Both the NRF and F_1 had better udder health and BCS at breeding. In calf rate was highest for the NRF.

First lactation performance of Holstein-Friesian, Jersey and Jersey×Holstein-Friesian cows under grass-based systems

R. Prendiville[1,2], M. Rath[2], N. Byrne[1] and F. Buckley[1], [1]Teagasc, Dairy Production Research Centre, Moorepark, Fermoy, Co. Cork, Ireland, [2]College of Life Sciences, UCD, Belfield, Dublin 4, Ireland

The study comprised primiparous cows; 29 Holstein-Friesian (HF), 27 Jersey (J), and 28 J×HF (F_1), across two seasonal grass-based systems, in 2006. Mean calving date was February 20. Large differences ($P < 0.001$) in milk production characteristics were observed. Milk yield ranged from 4700 kg for the HF to 3836 kg for the J. The F_1 produced 4294 kg. Fat and protein content were highest for the J at 5.24% and 3.97%, 4.67% and 3.71% for the F_1, and 3.95% and 3.42% for the HF. Solids corrected milk yield (SCM) was similar for all groups. Averaged over lactation, live weight (LW) was 465 kg for the HF, 350 kg for the J and 417 kg for the F_1. The J produced 12.4 times their LW in SCM. Comparable values for the HF and F_1 were 9.5 and 10.8, respectively. Body condition score during lactation was highest ($P < 0.05$) for the F_1 (3.09), compared to the HF (2.92) and J (2.99). Individual intakes were measured once at pasture in late lactation (October) using the n-alkane technique. Total DM intake was 15.4 kg, 13.8 kg and 14.7 kg for the HF, J and F_1, respectively. No difference in residual feed intake was observed. Breeding started in late April and lasted 13 weeks. The HF had the lowest fertility performance. Pregnancy rate to first service (PRFS) was 43% and in-calf rate was 83%. Comparable values for the J and F_1 were, PRFS 64% and 65%, and in-calf rate 93% and 97%, respectively.

Session 30 Theatre 4

Breed and heterosis effects on different production levels in dairy cattle

M. Penasa[1], R. Dal Zotto[1], G. De Jong[2], M. Cassandro[1] and G. Bittante[1], [1]University of Padova, Department of Animal Science, Viale Università 16, 35020, Agripolis, Legnaro (PD), Italy, [2]Nederlands Rundvee Syndicate (NRS), P.O. Box 454, 6800, AL Arnhem, Netherlands

Genotype by environment interaction for milk yield traits was investigated in primiparous Dutch dairy cows to quantify the influence of the environmental production level (EPL) on the expression of breed and heterosis effects. Three purebreds and two F1 crossbreds were investigated: Holstein-Friesian (HF), Dutch Friesian (DF), Mosa-Reno-Yssel (MRY), HFxDF, and MRYxHF. Data included 305-d milk, fat, and protein lactation yields of 37,542 heifers obtained from a random sample of 2000 herds. All cows calved between 1990 and 2000. Herd-year-season (HYS) groups (5149) with at least 5 animals were formed. Two seasons were considered for each herd-year: March to August and September to February. Three different environmental production levels were formed: high, medium and low. A linear model including fixed effects of HYS, age at calving and the genotype by EPL interaction was applied. The analysis evidenced that HF breed performed better in the high EPL than in the low one, while DF and MRY breeds expressed better their potential in the low EPL. Heterosis estimates seemed larger in the low EPL than in the high one.

Session 30 Theatre 5

Additive and heterotic effects on production and reproduction in Friesian × Sahiwal crossbreds and optimum Friesian contribution

S. Hassani[1] and M.G. Govindaiah[2], [1]GUASNR, Dept. Ani. Sci., Gorgan, 49138, Iran, [2]UAS, Ani. Breeding, Genetics and Biost., Bangalore, 560024, India

Data on 1912 Friesian × Sahiwal crossbred cows having different Friesian inheritance from 3/8 to 7/8, calved between November 1983 and December 1998 (both inclusive) and reared at Bangalore, Secunderabad and Pimpri military dairy farms located in South India were utilized for the analysis. Mid-parent value, overall additive effect, overall dominant effect and heterosis of productive and reproductive traits were estimated using additive-dominance model. Heterosis estimated for first lactation milk yield, first standard lactation milk yield, first lactation length, first dry period, milk yield per day of first calving interval, age at first calving, first calving interval, first service period and breeding efficiency were 88.33, 72.49, 6.75, -16.2, 67.78,-13.79, 3.52, 4.76 and 2.32 percent, respectively. Additive-dominance model was found to be adequate for all productive and reproductive traits. To determine the optimum Friesian contribution a selection index was constructed including first lactation milk yield, age at first calving, milk yield per day of first calving interval, breeding efficiency, first calving interval and first service period. The quadratic regression fitted revealed the maximum index value at about 71 percent Friesian inheritance, which appears to be the optimum Friesian inheritance.

Session 30 Theatre 6

Crossbreeding in the sheep sector

A.F. Carson and L.E.R. Dawson, Agri-Food and Biosciences Institute, Agriculture Branch, Hillsborough, Co. Down, BT26 6DR, United Kingdom

The stratified breeding structure within the sheep sector in the UK and, to a lesser extent, Ireland has embedded crossbreeding into the industry. Recently this structure has become somewhat less well defined with growing interest in separate breeding structures for the hill and lowland sectors. Using crossbred females in the hill sector enables the benefits of maternal heterosis to be exploited in harsh conditions where their benefits are likely to be largest. Moving from purebred Scottish Blackface to F1 crossbred ewes has improved individual ewe productivity in terms of weaned lamb output by up to 10% with additional improvements in longevity (Swaledale X) and carcass quality (Lleyn X and Texel X). Options for breeding structures to gain the benefits of crossbreeds include stratification within hill farms or the use of 2 or more breeds in rotation. In the lowland sector, 3-breed crosses including a proportion of Texel genes have been found to improve carcass characteristics whilst maintaining high levels of prolificacy. In rotational lowland breeding programmes particular emphasis is now being placed on the inclusion of breeds with improved maternal traits. In this regard, Lleyn and Belclare sires have been shown to promote high levels of prolificacy in their crossbred female progeny. Currently the potential to include breeds with easier-care traits is under investigation.

Session 30 Theatre 7

Production and functional traits of improved Valachian ewes and crossbreds with different genetic portion of Lacaune and East Friesian breed

M. Margetín[1], A. Čapistrák[1], D. Apolen[1], M. Milerski[2] and M. Oravcová[1], [1]Slovak Agricultural Research Centre, Teplická 103, 91401 Trenčianska Teplá, Slovakia (Slovak Republic), [2]Institute of Animal Science, Pratelstvi 815, 10400 Praque, Czech Republic

The paper was aimed at study of milk production and functional traits of crossbreds of Improved Valachian (IV) with Lacaune (LC) and East Friesian (EF). General linear model was applied to study milk yield of 12,210 lactations of ewes of 23 genotypes which belonged to seven flocks. Proportion of crossbreds was as follows: IVxLC 22.8 %, IVxEF 13.9 %, IVxLCxEF 2.8 %. Milkability and morphological traits of the udder were studied in one experimental flock. Increasing portion of LC and EF caused increase of milk yield. The increase, however, was non-linear. Statistically highly significant difference in milk yield ($P < 0,001$; increase by 18.2 to 49 %) was found between crossbreds IVxLC (portion of LC between 25 and 87.5 %) and purebred IV ewes which gave milk yield 118.41±1.653 l. Milk yield of crossbreds IVxLC was higher than milk yield of crossbreds IVxEF. The highest milk yield was found for crossbreds built on base of all three breeds (IV, LC, EF) and depended on genotype composition to great extent. The decline of fat and protein content in crossbreds was not significant. Morphological and functional traits depended highly significantly on genotype ($P < 0.001$). Udders of crossbreds fit better to machine milking in comparison with purebred IV and LC ewes.

Session 30 Theatre 8

Crossbreeding trials aiming to improve meat quality of sheep in Hungary

S. Kukovics[1], S. Nagy[2], A. Javor[3], A. Lengyel[4], T. Németh[1] and A. Molnár[1], [1]Research Institute for Animal Breeding and Nutrition, Seep and Goat Breeding, Gesztenyés u. 1., 2053 Herceghalom, Hungary, [2]Bakonszegi Awassi Corporation, Hunyadi u. 83., 4164 Bakonszeg, Hungary, [3]University of Debrecen, Centre of Agricultural Sciences, Böszörményi út 138., 4032 Debrecen, Hungary, [4]University of Kaposvár, Faculty of Animal Sciences, Guba Sándor u. 40., 7400 Kaposvár, Hungary

In order to improve the meat quality and quantity of the extensive Gyimesi Racka (GyR) breed without changing the sheep keeping technology a crossbreeding program was carried out using the rams of following breeds: GyR (control), Beltex (BX), British Milksheep (BMS), Charollais (CH), Dorper (DR), German Blackhead Mutton Sheep(GBM), Ile de France (IDF), Suffolk (SK) and Texel (TX). In intensive fattening, the GBM F_1 lambs reached the best (177%) and IDF F_1 ones (129%) the weakest result. In extensive fattening BX F_1 lambs had the highest (142%) and the IDF F_1 lambs (79%) the worst result. According to CT results the largest carcass volume, the boneless meat content and the loin value were belong to TX F_1, and SK F_1 lambs gave the smallest data. In the leg volume data CH F_1 lambs gave the best, and SK F_1 lambs the weakest results. The hot carcass weight and the dressing % of the crossbred lambs were significantly higher in both fattening trials. In S/EUROP carcass classification the crossbred lambs had one class advantage above control. Meat content of the carcass was increased, but in some cases the bone weight was also elevated.

Session 30 Poster 9

Effect of breed fraction on dairy traits

B. Logar, Agricultural Institute of Slovenia, Animal Science Department, Hacquetova 17, 1001 Ljubljana, Slovenia

In order to improve milk traits and some dairy functional traits, upgrading Simmental and Holstein (red) bulls have often been mated to Simmental dual purpose cows in Slovenia since 1976. The objective of the study was to asses the contribution of breed fraction and to estimate genetic and environmental parameters for milk yield traits using animal models methodology. Nearly half a million lactation records (201 to 305-day) with known gene proportions of Simmental and Holstein breeds included in milk recording scheme in the period 1990-2006 were selected from the national milk recording data base. Pure breed Simmental and Holstein animal was defined if more than 87 % of genes come from the same breed. For the evaluation of breed fraction the linear regression approach was applied. The yield was increasing for 25.6 kg milk, 0.9 kg fat and 0.7 kg protein with higher percentage of Holstein genes. The results obtained indicated that there is a significant distance in the production of Simmental and upgraded Simmental cows. Further results will be presented in a paper.

Session 30 Poster 10

Economic values for production and functional traits for crossbred (Holstein x Gir), Holstein and Gir cattle in Southeast Brazil

V.L. Cardoso[1], M.L. Pereira Lima[1], L. El Faro[1], A.E. Vercesi Filho[1], J.R. Nogueira[1], P.F. Machado[2] and J.A.M. van Arendonk[3], [1]Instituto de Zootecnia, Av Bandeirantes 2419, 14030 670 Ribeirao Preto SP, Brazil, [2]ESALQ USP, DZ, Piracicaba SP, Brazil, [3]Wageningen University, PO Box 338, 6700 AH Wageningen, Netherlands

Commercial herds of crossbred cattle compose the main group of milk producers in Brazil. Production systems are usually pasture based due to the high production potential of tropical grasses and adaptative capacity of animals. Gir (G) and Holstein (H) breeds are used to produce crossbred dairy cows (HG). To evaluate the relative economic importance of different production and functional traits for the three main genetic groups involved in the milk production in the Southeast Brazil (HG, H and G), economic values (EV) for milk (M), fat (F), protein (P), adult body weight (ABW), conception rate (CR) and herd life (HL) were calculated using a bio-economic model that describes performance, revenues and costs for pasture based milk production systems. The differences between genetic groups for production traits were small. EV (US$) for M, F and P (kg/cow/year) were, 0.18, -0.29 and –0.17 for H; 0.19, -0.28 and -0.17 for HG and 0.19, -0.24and -0.15 for G, respectively. Larger differences were found for ABW and CR. EV (US$) for ABW (kg/cow/year), CR (%) and HL (cow/day HL) were, –0.59, 1.16 and 1.29 for H; -0.24, 1.06 and 1.32 for HG and –0.41, 2.60 and 1.24 for G, respectively. Results provide solid basis for supporting breeding programmes.

Session 30 Poster 11

Economic values for milk production traits for crossbred (Holstein x Gir), Holstein and Gir cattle in Southeast Brazil under different milk payment policies

V.L. Cardoso[1], L. El Faro[1], J.R. Nogueira[1], P.F. Machado[2] and J.A.M. Van Arendonk[3], [1]Instituto de Zootecnia, Av Bandeirantes 2419, 14030670 Ribeirao Preto SP, Brazil, [2]ESALQ USP, DZ, Piracicaba SP, Brazil, [3]Wageningen University, PO Box 338, 6700 AH wageningen, Netherlands

Until recently milk payment policies in Brazil did not include payment for components. Due to changes in regulation on milk quality requirements by the Government, industries have introduced quality payment policies. To evaluate the relative economic importance of milk and components under different milk payment policies (PP), in the three main genetic groups involved in the milk production in the Southeast Brazil, i.e., Holstein x Gir crosses (HG), Holstein (H) and Gir (G), economic values (EV) for milk (M), fat (F), protein (P) were calculated using a bio-economic model. PP in the basic situation (BS) was based exclusively on the volume (V). Alternative PPs were based on V plus different proportions of F: P values (1: 1, 1: 2, and 2: 4). Relative EVs for M, F and P were, 1.00, -1.59, -0.92 (BS), 1.00, 0.79, 1.45 (1:1), 1.00, 0.79, 3.83 (1:2) and 1.00, 3.09, 8.40 (2:4), respectively for H; 1.00, -1.47, -0.89 (BS); 1.00, 0.74, 1.30 (1:1), 1.00, 0.74, and 3.48 (1:2) and 1.00, 2.85, 7.66 (2:4) respectively for HG and 1.00, -1.27, -0.8 (BS), 1.00, 0.91, 1.49 (1:1), 1.00, 0.91, 3.49 (1:2) and 1.00, 3.33 and 7.43 (2:4) respectively for G. Differences between genetic groups were observed for all PPs and should be accounted when designing breeding programs.

Session 30 Poster 12

The competitiveness of F1 Brown-Swiss x Holstein crosses in the intensive environment of a high-yielding dairy herd

S. Bloettner[1], B. Fischer[2], T. Engelhard[2] and H.H. Swalve[1], [1]Institute of Agricultural and Nutritional Sciences, Animal Breeding, Adam-Kuckhoff-Str. 35, 06108 Halle, Germany, [2]Institute for Agriculture, Forestry and Gardening, Centre of Livestock and Housing Technology (LLFG-ZTT), Lindenstr. 18, 39606 Iden, Germany

An experiment to evaluate the competitiveness of F1-BS x HOL crosses in the intensive environment of a high-yielding dairy herd was designed in cooperation between Halle University and LLFG-ZTT. In the herd of the experimental station Iden (herd average > 11,000 kg) for the control group, 10 Holstein bulls were mated to Holstein cows while the 'treatment' group was mated to 10 Brown Swiss bulls. The experiment started with the matings in November, 2002. Calving of females began in July, 2005. Currently, 50 F1-crosses (F1-BS) and 45 Holstein controls (HOL) are still milking First lactations have been completed by all animals. Growth parameters during the rearing period did not reveal any significant differences. Dairy production traits evaluated using test day models based on weekly milk recordings showed nearly equal fat and protein yields with slight differences in milk yield (HOL > F1-BS) and significant differences in fat and protein content (F1-BS > HOL). F1-BS had lower milking speed than HOL. Metabolic parameters in general were within tolerance levels for both groups although significant differences existed with respect to blood and milk urea content as well as the acid base quotient, being higher in F1-BS as compared to HOL.

Session 30 Poster 13

Crossing with colored breeds in Hungary

Z. Szendrei, S. Harangi, A. Radácsi, B. Béri and I. Bodó, University of Debrecen, Institute of Animal Science, Böszörményi út 138., 4032 Debrecen, Hungary

Milk production is done with the Holstein-Friesian (HF) in 99% in Hungary. Index-selection for high milk yield in dairy cattle resulted in increasing inbreeding and declining fertility and -in most cases- decreasing milk fat and protein percentages. One way for farmers to improve these traits is crossbreeding, utilizing heterosis in fertility and fitness traits. Crossbred cattle outperform purebred ones in several traits. Value of these traits is always changing but recently increases. In Hungary of the dairy breeds Brown Swiss (BS) and Jersey (JE) are considered for crossbreeding partners. Our aim was to determine which breed and breeding scheme is the most suitable in present and expected conditions. Initial steps has been taken on some farms towards upgrading HF to JE or to BS. The effects of crossbreeding on milk production (milk yield, fat and protein yield, somatic cell count) and fertility (age at first calving, days open, conception rate) were examined in this study. Type traits of the crossbred cows were also evaluated. Primiparous crossbred cows were compared to HF ones kept on the same farm. Data of JE×HF and BS×HF crosses were analyzed. Data were provided by the National Institute for Agricultural Quality Control and from the herd management software.

Session 30 — Poster 14

Production traits in German Angus, German Simmental, and their crosses

A. Müllenhoff[1], H. Brandt[1], G. Erhardt[1] and M. Gauly[2], [1]Justus-Liebig-University, Department of Animal Breeding and Genetics, Ludwigstraße 21 B, 35390 Giessen, Germany, [2]Georg-August-University, Institute for Animal Breeding and Genetics, Albrecht-Thaer-Weg 3, 37075 Göttingen, Germany

For a total of 795 German Angus (GA) and 730 German Simmental (GS) pure-bred calves born between 1998 and 2004 at the Research Station Rudlos and for reciprocal crosses born in 2003 and 2004 (GAxGS n=124 and GSxGA n=188) birth weight, weight at weaning and daily weight gain were analysed and crossbred parameters estimated. The reproduction parameters (calving rate) of the cows were significantly different between the breeds (GA - 96.2%; GS - 92.6%) while there was no significant difference in the calving. Stillbirth was 5.4% in GS and 2.6% in GA while the rate of twinning was 1.17% in GA and 4.60% in GS. Sire, breed, sex, and birth type had a significant influence on birth weight, weaning weight and daily gain until weaning. Individual heterosis, maternal and paternal positional effects were estimated for all traits. For birth weight (3.7%) and weaning weight (-0.7%) non significant individual heterosis effects were estimated. However the GAxGS crossbred calves showed significant higher weaning weights (8.4%) than the GSxGA calves which can be explained by maternal effects.

Session 31 — Theatre 1

Adding value to Irelands rural environment protection scheme

W. Dunne, Teagasc, Rural Economy Research Centre, Malahide Road, Dublin 17, Ireland

Ireland implemented its Rural Environment Protection Scheme (REPS) in the 1990's. Unlike Agri-Environment Schemes (AES) in most EU countries, REPS uses a whole farm approach, all land farmed must enter AES management, most measures are compulsory, additional optional measures are available, the contract is for five years and farmer participation is voluntary. This paper outlines new measures on (a) land use and crop mix (b) livestock enterprise mix (c) built environment, and (d) linkages between REPS farmers, with the underdeveloped food and product markets plus public goods "markets". Measures (a) and (b) would arrest the drift towards monoculture land use, single enterprise livestock farms, preserve eco-systems, maintain a mix of: food, fodder and energy crops, support the integration of crop and livestock farming, exploit complementarities of: mixed grazing, organic nutrients recycling, and conserve traditional farm practices and equipment. Measures (a), (b) and (c) would appeal to contiguous farmers with similar resource conditions, resulting in a scale effect for land use, livestock enterprises, plus related structures and ecosystems. This could be further exploited through (d) to achieve a scope or "REPS brand" effect. Measure (d) would also strengthen: the links between farmers, with underdeveloped local food markets, community and community based recreational facilities, the wider food and non-food product markets, and public good "markets".

Session 31 Theatre 2

Transition of multifunctional agriculture in The Netherlands through an innovation network of rural entrepreneurs

D. de Jong, G. Michgels and A. Visser, Wageningen UR, Animal Sciences Group & PPO, P.O. Box 65, 8200 AB Lelystad, Netherlands

During the second half of the 20th century the knowledge transfer towards farmers was driven from government and researchers towards farmers. This was a top down process where new technology and policy was implemented in the agricultural sector. Nowadays, a growing part of the research and policy agenda is set by networks of rural entrepreneurs. In the research programme on system innovations in multifunctional agriculture, the innovative network Waardewerken is a grindstone for research and policy. The input of the network is appreciated by researchers as well as by policy makers. The participants of Waardewerken are facing struggles in the process of professionalizing multifunctional agriculture. As a network, they are identifying transition points for the development of the multifunctional sector. The network is source of learning for policy makers. An example is the necessity to adjust legislation which is presently designed for traditional agricultural activities. Researchers are learning to develop knowledge in participative projects with members from the network. This results in a high level of ownership of the development and transfer of knowledge by the network toward the end-users. The experiences, results and impact of this network will be discussed in the presentation.

Session 31 Theatre 3

Landscape as indicator for multifunctionality related to pig farming in Midi Pyrenees, France

M.A.M. Commandeur, F. Casabianca and A.G.T.M. Bruins, INRA LRDE, SAD, Quartier Grossetti, 20250, France

The Midi Pyrenees are a particular scenic part of France, so obviously, landscape production is a desired externality of agricultural activities. Pig farming is usually not associated with the construction of landscape, because of the non requirement of land. The indirect contribution of the pig farming activity on the landscape in Midi Pyrenees is however strong, because it is embedded the poly active farming system. During the 1980s pig farming fulfilled the need for an additional economic base on moderate sized farms in addition to crop farming or dairy farming. Based on semi-structured interviews with pig farmers we explore three themes: *i* the significance of pig farming in the landscape, *ii* the conflicts of interests with the neighbours and *iii* the role of the resources for commercial activities. Nowadays pig farming is in rapid decline, and pig farmers can no longer compete with the production in other regions, due to the higher production costs. So there is a contradiction between desired contribution to landscape and poor competitiveness of such farms. The paradoxically is that dramatic effects on the landscape can occur from decline of pig production as an indirect consequence. In conclusion, we show that pig farmers do not feel recognised as positive contributors to the regional landscape and excluded from the perspective of the region. There is interest for exploring their options for improvement of competitiveness.

Session 31 — Theatre 4

Breeding decisions of organic farmers

W.J. Nauta[1] and D. Roep[2], [1]Louis Bolk Institute, Animal Production, Hoofdstraat 24, 3972LA Driebergen, Netherlands, [2]Wageningen University, Rural Sociology Group, Hollandseweg 1, 6706 KN Wageningen, Netherlands

Organic farming is multifunctional by nature. In the Netherlands 80% of farms have one or more economic activities next to milk production. There are also large differences between farms in concentrated feed inputs. To ensure adequate availability of breeding bulls, organic dairy farmers were asked about their farm characteristics and activities, and their demand for breeding. Based on this information, farms were allocated to one of two groups for farming diversity (Specialized dairy *vs* Multifunctional) and again for farming intensity (High input *vs* Low input), two aspects which were expected to influence breeding decisions. The High input group purchased at least 1200 kg concentrates/cow/year; Low input farms less than 800 kg. Demands for breeding were compared. Although significant differences were found for several farm characteristics, few differences were found for breeding goal and production, conformation and functionality traits. Intensive and specialized dairy farms did prefer Holstein cows, while Multifunctional and Low input farms were more interested in using robust Dutch breeds over foreign breeds. Even if farmers shared similar farming goals, they nevertheless had different preferences for breeds and crossbreeds, indicating that organic farmers are trying out different breeds. They would be best served with information on the qualities of different breeds for organic conditions.

Session 31 — Theatre 5

Values of green care farms for demented elderly

S.R. de Bruin and S.J. Oosting, Wageningen University, Animal Sciences, P.O. Box 338, 6700 AH Wageningen, Netherlands

The number of demented elderly in the Netherlands increases. These elderly often loose part of their independence and therefore need e.g. day care in a residential home (so-called 'regular day care'). Since a few years a new type of day care has supplemented the regular day care: so-called "green care farms", farms that provide day care to demented elderly. It is claimed that farms offer meaningful values to this target group and that such values contribute to their quality of life (QoL). The objectives of the present study were to assess farm values that contribute to QoL of demented elderly and to compare these values with values of regular day care facilities. The present work is part of a study on the effects of green care farms on QoL of demented elderly. Values were gathered during interviews with demented elderly and their caregivers and by observations at farms and at regular day care facilities. Results show that although day structure at farms and at regular day care facilities is the same, the day program differed between both day care facilities. Compared with regular day care facilities, farms offer more opportunities to move freely and to go outside and offer more diverse and meaningful activities for demented elderly. Animals are part of the diversity of activities at green care farms but time spent on human-animal interactions was limited. Since we conclude the values of day care at green care farms differ from values of regular day care facilities, we expect different effects on QoL as well.

Session 31 Theatre 6

Conceptual approaches to the multifunctionality of livestock farming systems

H. Renting, Wageningen University, Rural Sociology Group, Hollandseweg 1, 6706 KN Wageningen, Netherlands

In the last decade the multifunctionality concept has emerged as key notion in debates on agricultural development. While at international level multifunctionality has mainly drawn attention due to its role within trade negotiations, at the European level it reflects the growing importance of social and environmental functions in rural policies. This contribution analyses livestock farming systems within the overall framework of a shift of European agriculture from productivist to multifunctional farming practices. For this it builds on insights and conceptual approaches from the EU research projects IMPACT and MULTAGRI. From a multifunctionality perspective farming systems are to be analysed within the framework of changing relations between agriculture and wider society. Evolving societal and consumer demands form the basis for new farm household strategies, who respond to these by "broadening" their activity base with new on-farm activities (e.g. care services, nature management) or "deepening" their relationships with food supply chains through the provisioning of foods with distinctive quality attributes (e.g. organic and quality foods). This contribution will highlight some important implications for the study of livestock farming systems, including the changing role of territorial resources, the growing importance of relations with non-agricultural stakeholders, required entrepreneurial skills and the need for new institutional arrangement to facilitate multifunctional farming systems.

Session 31 Poster 7

Be good and tell it? Re-establishment of the connection between livestock industry and society

D. de Jong, I. Enting, M. Mul and O. van Eijk, Wageningen UR, Animal Sciences Group, P.O. Box 65, 8200 AB Lelystad, Netherlands

The livestock production chains have been of major significance for the rural areas in Western Europe. They provided food in abundance, employment and economic activities. However, the last decennium there has grown a gap between animal production methods and perceptions of society. The project BGood is looking for new ways to re-establish the connection between the livestock industry and society. The ways concentrate on communication strategies. Repairing a disturbed relationship is a combined action between image and identity. Therefore, the communication strategies focus on changing the view of society on animal production and changing the attitude of the livestock industry towards society. Twenty-five interviews have been conducted to explore how people working in other, non-agricultural, industries have acted in situations to re-establish contact. The project is now in a stage that the plurality of ideas from the interviews will be used as inspirational examples in a in a conference with people from the agricultural industries and people from other sectors who are engaged with the problem definition. During the conference working groups on different themes will be formed which can take form their own ideas and take them into practise. One of the conclusions of the project so far is that the cooperation between people from agriculture and non-agriculture gave multiple eye-openers.

Session 31 Poster 8

The important role of organic farming on sustainable animal production in developing countries

G. Bengisu[1] and Ü Yavuzer[2], [1]Harran University, Faculty of Agriculture, Crop Science, Şanlıurfa, 63040, Turkey, [2]Harran University, Faculty of Agriculture, Animal Science, Şanlıurfa, 63040, Turkey

When compared with the traditional livestock breeding, organic livestock breeding has some limiting factors in obtaining sustainable competition, like the number of animals per hectare, the transition time of the livestock feed production to organic livestock, and the production of livestock feed via organic farming. Due to extensive livestock breeding, improving the organic livestock breeding is possible in developing countries. While there is a big potential in organic livestock, factors like the requirement that the breed and race should be from organic product manufacturers, fewness of the certified organic animal breeding firms are other limiting factors. In developing countries, where livestock breeding is wide-spread, there are many companies which perform breeding equivalent to organic livestock breeding, though they do not hold a certificate yet. In developing countries, the use of chemical fertilizers and pesticides are not widespread in the manufacturing of the animal products, especially of those based on pastures. Organic animal products should not be hard to reach or hard to purchase due to expensive production and due to other limiting factors. The land and vegetation analyses of the pastures where organic livestock breeding is considered must be performed, and the transition time to organic livestock breeding must be shortened.

Session 32 Theatre 1

Effect of environmental temperature on lactating sows voluntary feed intake and performance under commercial conditions

M. Anguita, A. Cerisuelo, E.G. Manzanilla and J. Gasa, Universitat Autònoma de Barcelona, Animal Nutrition, Management and Welfare Research Group, Facultat de Veterinària, 08193, Bellaterra, Spain

The aim of the present work was to study sow voluntary feed intake (VFI) and productive performance in two seasons: spring (SP) and summer (SU). The experiment involved a total of 160 (80 each season) sows and was carried out in a commercial sow farm provided only with natural ventilation in the lactation barns. Ultrasonic back fat at P2 position (BF, mm) one week before farrowing and on day 18 of lactation, total VFI and piglets weight gain during lactation (18 d) were recorded. Temperatures in the lactation barns were controlled daily; mean temperature in the SP was 24.5°C and in the SU 29.1°C. Sows were grouped according to their parity into 3 groups (P1: 1 parity; P2: 2 and 3 parities; P3: > 3 parities) and into 3 groups according to their BF (THIN: <16mm; NORMAL: 16 to 19mm; FAT: >19mm). Average piglet weight on day 18 of lactation was higher in SP compared to that found in SU (5.60 *vs* 4.71kg). Sow VFI (kg/d) was lower in the SU compared to that found in SP (4.84 *vs* 3.69). Voluntary feed intake was also affected by parity and BF being lower in P1 compared to P3 sows (4.48 *vs* 4.05), and higher in the THIN compared to the FAT sows (4.59 *vs* 3.94). The Ancova of the data showed that VFI decreased by 252g per each °C that temperature increase between 23.5°C and 30.3°C, reflecting the importance of environmental temperature on the VFI and performance of sows

Session 32 Theatre 2

Manipulation of liquid feed curves during lactation to increase sow feed intake and its impact on sow weight and piglet performance to weaning

P.G. Lawlor[1], P.B. Lynch[1], K. O'Connell[1], C. Hiet[1,2] and D. Mattras[1,2], [1]Teagasc, Pig Production Development Unit, Moorepark Research Centre, Fermoy, Co. Cork, Ireland, [2]ENITA de Bordeaux, 1 Cours du Général de Gaulle, CS 40201, 33175 Gradignan Cedex, France

The objective was to determine the effect of three liquid feeding regimes on sow lactation feed intake, sow weight change and piglet performance to weaning. At day 109 of gestation sows (N=75) were blocked on parity grouping (gilts, litter 2 to 3 and litter 4 plus) and weight and allocated to treatment: A. Curve 1 (25 MJ DE/day at farrowing to 98 MJ DE/day by day 21 of lactation), B. Curve 1 plus 14.2 MJ DE/day of dry feed from day 4 post-partum, C. Curve 2 (Curve 1 x 1.15). Sows on curves 1 and 2 were fed twice daily a 4.1:1 mixture of feed (dry matter) to water by a computerised liquid feeding system. The lactation diet contained 14.2 MJ DE/kg and 9.1g lysine/kg fresh-weight. The experimental curves were fed between farrowing and weaning (ca.28 days). Mean lactation feed intake was 77.5, 87.6 and 86.6 (s.e. 1.13MJ DE/day; $P < 0.001$) for Treatments A, B and C respectively. Sow weight loss during lactation was 11.8, 8.8 and 13.7 (s.e. 3.57Kg; $P > 0.05$) for Treatments A, B and C respectively. Treatment B tended to have a higher number of pre-weaning deaths per litter (0.7 pigs) than treatment A ($P=0.08$). Treatment had no effect on piglet weaning weight ($P > 0.05$), piglet daily gain ($P > 0.05$), within litter CV for weaning weight ($P > 0.05$) and within litter CV for piglet daily gain ($P > 0.05$).

Session 32 Theatre 3

Periparturient lameness and lactation feed intake predicts sow longevity

J. Deen, L. Anil and S.S. Anil, University of Minnesota, Veteterinary Population Medicine, 1988 Fitch Ave, St Paul, MN 55108, USA

A study involving data from 1357 sows was conducted to analyze the association of of lameness during the periparturient period, lactation feed intake (LFI), parity (1&2, 3-5 or ≥ 6), stillbirths, mummies and preweaning piglet deaths and lactation length with the likelihood of sow removal within 35d post-farrowing or before the next farrowing. Data were analyzed using multivariate logistic regression models (Proc logistic, SAS v 9.1). The likelihood of removal from the herd within 35d post-farrowing decreased ($P \leq 0.05$) by 19% with every additional piglet born alive. The removal risk before 35d post-farrowing decreased ($P \leq 0.05$) by 34 % with every additional kg increase in average lactation feed intake. Sows that did not have lameness during periparturient period had 74% lower ($P \leq 0.05$) likelihood of removal from the herd before 35d post-farrowing compared to the other sows. The likelihood of removal from the herd before next farrowing decreased by 8% with every additional piglet born alive. Incidence of lameness (Odds ratio 0.626) and parity (Odds ratio 0.548 and 0.558 respectively for parities 1&2 and parities 3-5 respectively) influenced the likelihood of removal of sows from the herd before next parity as well ($P \leq 0.05$ for all). The study indicated that lameness, LFI, number of liveborn piglets and parity were associated with sow removal.

Session 32 Theatre 4

Factors affecting piglet weight at weaning in first parity and multiparous sows

A. Cerisuelo, M. Anguita, M.D. Baucells and J. Gasa, Universitat Autònoma de Barcelona, Animal Nutrition, Management and Welfare Research Group, Facultat de Veterinària, 08193, Bellaterra, Spain

This work aimed to study the main factors involved in piglet weight at weaning (PWW), in order to identify the best strategy for cross-fostering. A total of 462 sows from two herds and from 1 to 12 parities were studied. Sows were grouped in 3 parity groups (PG1: parity 1; PG2: parity 2 and 3; PG3: >3 parities). Backfat (BF, mm) was measured at farrowing and sows were grouped in 3 units (THIN: <17mm; MEDIUM: 17-20mm and FAT: >20mm). After farrowing, cross-fostering was carried out according to the strategy followed in each herd. Number of pigs (NPi) and piglet weight (PWi) were recorded after cross-fostering and, again, on day 18 of lactation. Results showed that PWW was lower in PG1 compared to multiparous sows (PG1: 4.8kg; PG2: 5.5kg; PG3: 5.4kg, $P < 0.001$); PWW was also lower in the THIN (5.15kg) compared to MEDIUM (5.39kg) and FAT (5.37kg) sows ($P=0.029$). The impact of NPi and PWi and sow BF at farrowing on PWW was studied through a covariance test. In PG1, PWW was only affected by PWi (+2g per additional g of PWi). However, in multiparous sows, PWW was affected by the NPi (-106.3g per additional NPi), PWi (+1.5g per additional g of PWi) and sow BF at farrowing (+17.2g per additional mm of BF). Thus, in order to maximize PWW, when cross-fostering attention must be paid to the initial piglet weight in first parity sows, but also to the initial number of piglets in multiparous sows.

Session 32 Theatre 5

The effect of biotin in the lactation diet of sows on litter weight gain

M.J. Van Oeckel, S. Millet, M. De Paepe and D.L. De Brabander, ILVO, Animal Sciences Unit, Scheldeweg 68, 9090 Melle, Belgium

The effect of supplementary addition of biotin to the lactation diet of sows on litter weight gain was studied. Four series of on average 17 hybrid sows (in total 68 sows) were allocated to a control diet or a control diet plus 2 ppm biotin when entering the farrowing unit. Piglets were exchanged between sows during the first days of life to standardise the litters as much as possible to 10 to 12 piglets. The average number of piglets per sow was after exchangement 10.9 and 11.2 and at weaning 9.6 and 10.1 for respectively the control and the biotin group. Piglets received creep feed from one week old and consumed on average 5 g/piglet/day for both treatments. Piglets who lost weight or which were outliers for weight gain (average ± 2 x standard deviation) within a treatment were excluded from the results. Litter weight gain was corrected to a standard 28 suckling days. No significant effects of extra biotin addition were found on litter or piglet weight gain. However, litter weight gain (66 kg/control sow *vs* 69 kg/biotin sow), mortality rate of the piglets (12% for control *vs* 10% for biotin sows) and diarrhoea incidence during the third week of life were more favourable for the biotin *vs* the control sows. The improvement (not significant) of litter weight gain with biotin supplementation ranged from 1.0 to 10.3% for the four series and was on average 4.5%. The favourable effect of biotin on litter weight gain was more expressed in unipareous sows versus multipareous sows.

Session 32 Theatre 6

Influence of feeding level on FSH and LH secretion patterns during lactation, on uterus and follicle development after weaning in sows

M. Wähner[1] and J. Kauffold[2], [1]University of Applied Sciences, Strenzfelder Allee 28, 06406 Bernburg, Germany, [2]University of Leipzig, An den Tierkliniken 29, 04103 Leipzig, Germany

Feeding during lactation has been shown to effect reproduction physiology in sows. It was conducted by define the hormonal patterns of FSH and LH in lactation (21 days), the development of uterus and follicles after weaning in primiparous sows fed restricted (RE) or *ad libitum* (AL). 14 sows were used; 7 each for AL or RE (»70% of the diet of AL-sows). An intravenous catheter was inserted on day 6/7 post partum (pp). Blood samples were collected on days 12, 15 and 18 pp. FSH and LH were analysed by RIA. Daily ultrasonography of the ovaries between day 8 and 20/21 pp was used on a few animals per group to monitor follicular growth. All sows were slaughtered either 1 or 7 days after weaning. Mean daily feed intake during lactation was 3.9 kg in AL- and 2.7 kg in RE-sows. FSH concentration were similar in AL- and RE-sows on day 12 pp, but higher in AL-sows on day 15 and day 18 pp ($P < 0,05$). Generally LH concentration were higher in AL- than RE-sows ($P < 0.05$). The results demonstrate that lactating AL-sows both higher LH and FSH concentrations during lactation than RE-sows. An association to simultaneously observed differences in follicular growth is assumed.

Session 32 Theatre 7

Factors affecting voluntary feed intake in the lactating sow

P.B. Lynch and P.G. Lawlor, Teagasc Moorepark Research Centre, Fermoy, Co. Cork, Ireland

The present day commercial sow is leaner with lower fat reserves at first breeding, has fewer empty days each year and has a higher annual production of pigs and weight of pigs at weaning than the sows on which current nutritional recommendations are based. Typically a modern high-output sow will be in negative energy balance in lactation even where feeding is *ad libitum*. Maximising feed intake and minimising weight and tissue loss is essential if fertility in the next cycle is not to be depressed. Voluntary feed intake in lactation is affected by factors of the sow (genotype, body size, age/maturity, litter size, pregnancy feed level, fatness at farrowing, health), diet (nutrient concentration and balance, ingredient composition, wholesomeness, diet form), environment (temperature, light, humidity) and management (feeding method, feeding frequency, water supply). This presentaion will review recent information on factors affecting voluntary feed intake in the lactating sow.

Session 33 Theatre 1

Effects of concentrate supplementation on the expression of the delta 9 desaturase gene in muscle, adipose tissue and liver of grazing beef heifers

S.A. Mcgettrick[1], A.P. Maloney[2], T. Sweeney[1], F.J. Monahan[1] and F.J. Mulligan[1], [1]University College Dublin, School of Agriculture, Food Science and Veterinary Medicine, Belfield, Dublin 4, Ireland, [2]Teagasc, Grange beef research centre, Dunsany, Co. Meath, Ireland

Forty Charolais or Limousin crossbred heifers were randomly assigned to two outdoor and two indoor groups (n=10) for a period of 150 days to examine the effects of feeding grass either alone or in combination with sunflower seed (SFS) and fish oil (FO) on the expression of the delta 9 desaturase gene that is responsible for formation of conjugated linoleic acid in ruminant tissues. The outdoor animals were either offered unsupplemented pasture, or restricted pasture supplemented with 2.5kg of concentrate containing SFS (29%) and FO (6%), while indoor groups were fed a basal concentrate or restricted basal concentrate with 2.5kg of the SFS and FO based concentrate supplement. Delta 9 desaturase mRNA levels from muscle, liver and subcutaneous adipose were quantified using real time RT PCR. Delta 9 desaturase mRNA levels were significantly lower ($P < 0.05$) in muscle and subcutaneous adipose of grass-fed outdoor animals compared to concentrate-fed animals but were unchanged in liver ($P > 0.05$). Supplementation of the diet with SFS and FO had no effect on delta 9 desaturase gene expression in any tissue examined. These results indicate that grass-based diets result in lower delta 9 desaturase gene expression in muscle and adipose tissue of beef animals.

Session 33 Theatre 2

Ruminal degradation of inositol bound P is affected by feed composition and external phytase

J. Sehested and P. Lund, University of Aarhus, Faculty of Agricultural Sciences, P.O. Box 50, DK-8830 Tjele, Denmark

The objective was to study the effect of diet carbohydrate source (fiber *vs* starch) and external phytase (PHYT) on ruminal degradation of inositol bound phosphorous (IP). A 4*4 Latin square experiment with 4 multi-fistulated dairy cows (3 lactating, 1 dry), 4 periods of 21 days and 4 dietary treatments was conducted. Two iso-energetic total mixed rations based on grass-clover silage and rape seed meal was formulated: a high starch TMR (HS) was formulated by adding dehulled oats and maize flour, and a high fiber TMR (HF) by adding soya hulls. Each TMR was mixed with and without external phytase (1900 FTU per kg DM). Daily feed intake was significantly higher on HF (19.4 *vs* 17.5 kg DM), whereas daily intake of IP was significantly higher on HS (29 *vs* 24 g). Daily intake of P (96 g) was not influenced by treatments. Ruminal degradation of IP was significantly increased by HS and by PHYT, and the effects were additive. On the HF diet PHYT increased ruminal degradation of IP significantly from 75% to 80%, whereas on the HS diet PHYT increased ruminal degradation of IP significantly from 83% to 88%. Degradation of IP in the small intestine was low and not influenced by treatments. The results show that ruminal degradation of IP can be increased by diet composition and external phytase. These results show for the first time that dietary IP can be limiting for P utilization in ruminants.

Session 33 — Theatre 3

Feeding ruminally protected sunflower oilcake to sheep

F.K. Siebrits and A. Makgekgenene, Tshwane University of Technology, Animal Sciences, Private Bag X680, Pretoria 0001, South Africa

Sheep fat generally contains relatively high levels of saturated fatty acids (Okeudo & Moss, 2007). It would therefore be advantageous to manipulate the fatty acid composition of lamb to contain more poly-unsaturated fatty acids (PUFA) for consumption in the health conscious market. Sunflower oilcake SFOC has a high rumen protein degradability leading to a loss of essential amino acids such as methionine. It would be advantageous if the PUFA content can be elevated while the rumen bypass protein fraction can be increased at the same time. Converting the residual oil to a calcium soap may achieve this. Mechanically extracted sunflower oilcake (MSFOC) containing 16% residual oil was fed either as–is (n=10), or saponified (n=10) in a complete feedlot diet to SA Mutton Merino weaner lambs(initial weight ca. 23 kg) kept in single pens for 9 weeks. A control diet containing extracted low fat (1%) sunflower oilcake as protein source was fed to a further 10 lambs. Feedlot diets containing about 10% crude protein were fed to the experimental groups while the control diet contained 12%. Rumen degradation was determined by incubating sunflower oilcake samples in 3 canulated oxen. Fatty acid composition was determined on back fat samples. The lambs on the saponified MSFOC diet had a higher intake and growth rate than the group on the untreated MSFOC. The group on the control diet performed similar to the group on the saponified MSFOC. The total PUFA content of lamb fat was not affected by treatment.

Session 33 — Theatre 4

Lactation performance of dairy cows fed a ruminally protected B-vitamin blend

F. Sacadura[1], P.H. Robinson[2], M. Lordelo[1], E. Evans[3] and R.L. Cerri[2], [1]Instituto Superior de Agronomia - Univ. Técnica de Lisboa, Tapada da Ajuda, 1349 Lisboa, Portugal, [2]University of California, Davis, Dpt Animal Science, One Shields Avenue, CA 95616, USA, [3]Essi Evans Technical Advisory Services, 64 Scugog St, Bowmanville, Ontario, Canada

It is unknown whether B-vitamins are available in sufficient levels to support milk synthesis in ruminants. The objective of this study was to determine effects of feeding a ruminally protected B-vitamin complex (BVBlend), containing Biotin, Folic Acid, Pantothenic Acid and Pyridoxine, to dairy cows on milk production efficiency. Two Californian dairy facilities were used, one with mid-lactation primiparous and multiparous cows (trial 1) and another with early-lactation multiparous cows (trial 2). In each trial 247 cows were randomly assigned to treatment in a 2x2 Latin Square design. In trial 1, milk yield and milk fat yield were unaffected by treatment, whereas milk protein yield increased (1.21 *vs* 1.24 kg/d; P=0.02) in cows fed BVBlend. In trial 2, milk yield (39.60 *vs* 40.46 kg/d), milk fat yield (1.40 *vs* 1.47 kg/d) and milk protein yield (1.10 *vs* 1.16 kg/d) increased ($P < 0.02$) with addition of BVBlend. The overall conclusion was that cows responded positively to the BVBlend, with an increase in milk components and milk yield. However, early-lactation multiparous cows were more responsive than mid-lactation and/or primiparous cows.

Rumen fermentation and plasma metabolites in steers offered concentrates differing in energy source either as a supplement to grass silage or *ad libitum*

M. Mcgee and P. O'Kiely, Teagasc, Grange Beef Research Centre, Dunsany, Co. Meath, Ireland

Feeding level and ingredient composition of concentrates offered to beef cattle vary widely. Rumen fermentation was determined using 4 rumen-fistulated steers (661 kg) in two consecutive (i. supplemented (SUP) and ii. *ad-libitum* (AL)concentrate (C) feeding) 4 (diets) × 4 (14 d periods) Latin square design experiments. For SUP, grass silage (GS) was offered *ad libitum* plus 6.0 kg of C per head once daily. For AL, C was offered *ad libitum* plus 1.2 kg DM of GS daily. The 4 C were: rapidly fermentable starch (barley)-based (RFS), slowly FS (maize)-based (SFS), RFS + fibre-based (RFS+F) and fibre (pulp)-based (F). On d 11, rumen fluid samples were obtained at 0, 1 (SC only), 2, 4, 6, 8, 12, 16 (SC only) and 24 h post-feeding. On d 14, blood samples were obtained at 0, 3 and 6 h post-feeding. When offered SUP, there was no effect ($P > 0.05$) of C type on rumen pH, ammonia, lactic acid or total volatile fatty acid (VFA) concentrations or molar proportions of acetate, propionate and butyrate. When offered AL, rumen pH or total VFA concentrations and molar proportion of butyrate did not differ between C but the molar proportion of acetate was lower ($P < 0.05$) for RFS and SFS than RFS+F and F and the molar proportion of propionate was higher ($P < 0.05$) for RFS than RFS+F and F, with SFS being intermediate. Plasma beta-hydroxybutyrate, urea and glucose did not differ ($P > 0.05$) between the C.

Effect of two levels of ground wheat as concentrate replacers on the performance of lactating Holstein cows

S. De Campeneere, D.L. De Brabander and J.M. Vanacker, ILVO, Animal Science Unit, Scheldeweg 68, 9090 Melle, Belgium

The use of home-grown concentrate replacers has gained interest to reduce the feeding costs on dairy farms. To evaluate the effect of two levels (2 and 4 kg) of ground wheat as concentrate replacers, three diets were compared using 18 lactating Holstein cows in a Latin square design with 3 periods of 4 weeks. At the start of the trial, the cows were on average 124 days in milk, producing 29.5 kg milk with a fat and protein content of 4.45 and 3.11%. The basal diet for the control and both wheat diets consisted of maize silage and prewilted grass silage (60/40 on DM base) fed *ad libitum* and 10 kg pressed sugar beet pulp and was further completed with soybean meal, concentrates and urea. In both wheat diets, concentrate and soybean meal were partly replaced (on protein and energy basis) by 2.2, 2.0 and 1.8 kg wheat (treatment 1) and by 4.4, 4.0 and 3.6 kg wheat (treatment 2) in period 1, 2 and 3, respectively. At the start of the trial concentrate level was fixed to supply 105% of the net energy and digestible protein requirements and decreased weekly to correct for the change in lactation stage. Preliminary results indicate that replacing concentrate with 2 or 4 kg of ground wheat did not influence milk production nor milk fat or protein content. By the time of the congress more and final results will be available.

Determination of feed particle size reduction by chewing using an image analysis procedure

I. Schadt[1], M. Caccamo[1], J.D. Ferguson[2], G. Azzaro[1], R. Petriglieri[1], P. Van Soest[3] and G. Licitra[1,4], [1]CoRFiLaC, S.P. 25 Km 5 Ragusa Mare, 97100, Italy, [2]University of Pennsylvania, School of Veterinary Medicine, 382 West Street Road, 19348, Kennett Square, PA, USA, [3]Cornell University, Department of Animal Science, 149 Morrison Hall, 14853, Ithaca, NY, USA, [4]University of Catania, D.A.C.P.A., Via Valdisavoia, 5, 95123, Catania, Italy

Physical effective NDF is defined as feed material that stimulates chewing. This study aimed to examine the particle distribution in swallowed boluses from hay of variable chop lengths using an image analysis procedure described by Licitra *et al.* (J. Anim. Sci. 83:suppl. 1, 252). Three nonlactating, rumen fistulated cows, adapted to Perennial ryegrass hay, were held off feed for 12 hours, rumens evacuated, and offered 0.25 kg of long or chopped hay. Swallowed boli were retrieved from the reticulo-rumen. Treatments were: 1) long hay, 2) hay cut to 5 cm, 3) chopped hay retained on a 1.91 cm sieve, 4) chopped hay passing through a 1.91 cm sieve but retained on a.787 cm sieve, and 5) chopped hay passing the.787 cm sieve and retained on a.127 cm sieve. Long particles in hay treatments and boli were defined as those retained on a 1.6 mm sieve. Mean long particle bolus sizes (mm) per treatment were as follows (superscripts differ by $p<.05$): 1) 9.1^{ab}, 2) 9.1^{ab}, 3) 9.5^{a}, 4) 9.0^{b}, and 5) 7.8^{c}. Hay particles retained above a.787 cm screen were similar in mean size and distribution when swallowed. Particles smaller than.787 cm were smaller in the bolus.

Determination of optimum level of hydrolyzed pith bagasse by steam pressure in feeding of crossbred calves

M. Sabbaghzadeh, N. Dabiri, J. Fayazi and M.R. Fathabadi, Ramin Agri. University, Animal Sci., Ramin University. Mollasani. Ahvaz., 6341773637, Iran

A large amount of agriculture's by-product is produced by food industries of Iran. One of these by-product is pith bagasse.The aim of this experiment was to examine the effect of dietary containing different levels of this by-product.Twenty four crossbred Holstein-Najdi calves were used in this experiment. The calves were divided into 4 groups within each block according their live weight based a randomized block design.4 diets were formulated: diets 1-4.The diet1(control group) formulated according NRC (2000).In diet 2-4, we utilized hydrolyzed pith bagasse by different ratio. The total hydrolyzed pith bagasse content of diets 1-4 were respectively 0, 11, 22 and 33%. The diets were offered adlibitum to all groups. No significant difference was seen in dry matter intake of control group (diet1) and other diets. Diet 4 (containing 33% pith) had significantly lower average daily gain (ADG) than control group ($p < 0.05$). The same trend with ADG was found for feed conversion ratio (FCR). The differences between diets 1-4 for carcass percentage (CP) were not significant. This study is suggests that calves fed diet 3 had the best performance and net return. Moreover we can utilize 22% hydrolyzed pith bagasse in diet without any negative effects, but utilization over than this level, lead to decrease in FCR and ADG of cattle.

Session 33 Theatre 9

Utilization of different levels of sugarcane tops silage supplemented with urea and molasses in fattening buffalo male calves feeding

M. Bojarpour, J. Forozesh, M. Mamoei and M. Sabbaghzadeh, Ramin Universuty, Aimal Science, Mollasani. Ahvaz., 6341773637, Iran

This study was carried out to determine the best level using of sugarcane tops silage supplemented with urea and molasses in fattening buffalo male calves. 16 buffalo male calves with 11 months age and the average live weight 166.96 Kg are used in complete randomized design. 4 experimental treatments with 4 replication per treatment were applied in over 10 weeks. The rations were included as 0, 33.33, 66.66 and 100 percent of sugarcane tops silage which replaced with corn silage. All forage was fed *ad libitum* to buffalo calves. The averages of Dry Matter Intake (DMI) in diet 1 to 4 were 7, 6.55, 7.81 and 4.5 kg respectively. There was significant difference for DMI between diets 1 and 3 with 4 ($p < 0.01$).There was no significant difference for Average Daily Gain between diets. There was significant difference for Feed Conversion Ratio between diet 1 and 3 with 4 and diet 2 with 3 ($p < 0.01$). Final weight have a significant difference between diet 1 and 2 with 4 ($p < 0.01$). The differences between diets 1-4 for carcass weight and dressing were not significant.

Session 33 Theatre 10

Determination of chemical compositions and *in vitro* gas production characteristics of chickpea processing by-products

N. Maheri Sis[1], M. Chamani[2], A.A. Sadeghi[2], A.M. Aghazadeh[3] and A.R. Safaei[4], [1]Islamic Azad University, Department of animal science, Shabestar Branch, Iran, [2]Islamic Azad University, Department of animal science, Science and Research Campus, Tehran, Iran, [3]Urmia University, Department of animal science, Urmia, Iran, [4]Animal Science Research Institute, Animal Nutrition, Karaj, Iran

The aim of present study was to determine the chemical composition and estimation of nutritional value of two types of chickpea wastes including Chickpea Culls (CPC) and Chickpea Dehulling by-Products (CDP) using gas production technique in sheep. The samples were collected from ten pea packaging and processing factories. The results showed that Organic Matter (OM), Non Fibrous Carbohydrates (NFC), starch and Total Phenolic Compounds (TPC) were significantly greater in CPC than that of CDP, but ether extract (EE) and Neutral Detergent Fiber (NDF) were higher in CDP($p < 0.05$). The Crud Protein(CP) and tannins of two feed samples were similar. There were significant differences($p < 0.05$) in 'b' and calculated Dry Matter Digestibility (DMD), Organic Matter Digestibility (OMD) and Metabolizable Energy (ME) but no significant differences were obtained in 'a' and 'c' values between feed samples. Gas productions for 24 hours were significantly higher in CPC than CDP (75.6ml *vs* 60.6ml). The gas productions constants values (a, b and c) for CPC were 4.6, 85.3 and 0.05 while for CDP were 4.9, 78.6 and 0.05 respectively. Overall the nutritive value of CPC was higher than that of CDP.

Session 33 Theatre 11

The interaction between breed and diet on CLA and fatty acids content of milk fat of four sheep breeds kept indoors or at grass

E. Tsiplakou and G. Zervas, Agricultural University of Athens, Animal Nutrition, Iera odos 75, GR-11855, Greece

An experiment was conducted with the objective to study the interaction between breed and diet on conjugated linoleic acid (CLA) and fatty acids profile (FAs) content of milk fat of four pure of dairy sheep breeds (Awassi, Lacaune, Friesland and Chios). All sheep used in this study, were kept in one flock under the same management. During the winter months all sheep were kept indoors and fed with alfalfa hay and concentrates and from April onwards were grazing native pastures without any supplementary feed. A total of 160 individual milk samples were taken from the four breeds at two sampling times: in January (n=80) and in May (n=80) and analyzed FAs. The results of this experiment showed that the sheep breed had no effect on milk FA profile or on atherogenichity index (AI). The only significant effect was on Δ^{-9} desaturase activity index expressed as $C_{18:1}/C_{18:0}$. On the contrary, the diet affected significantly the FA profile of all ewes milk fat, with pasture to cause lower proportions of saturated and higher proportions of unsaturated FAs, compared with those of sheep kept and fed indoors. The interaction between breed and diet was significant for the FAs groups SCFA, MCFA, LCFA, MUFA and for CLA.The S/U FAs ratio and the Δ^{-9} desaturase indexes ($C_{14:1}/C_{14:0}$ and $C_{18:1}/C_{18:0}$) were also affected by the breed× diet interaction. In conclusion, pasture feeding when compared with supplementary diet, induce large variation in the milk FAs profile.

Session 33 Theatre 12

In sacco degradability of crude protein in lupine and soybean

P. Homolka, V. Koukolová and F. Jančík, Institute of Animal Science, Department of Animal Nutritiuon, Přátelství 815, 104 00 Prague Uhříněves, Czech Republic

In this study, nutritive value of three varieties of white lupine *(Lupinus albus)* Amiga, Butan, Dieta and two varieties of blue lupine *(Lupinus angustifolius)* Prima, APR82 were compared with soybean Korada. Chemical compositions were determined according to AOAC. The nitrogen degradability experiments were performed using *in sacco* method in three dry cows (Black Pied) equiped with large ruminal cannulas (120 mm internal diameter). The cows were fed twice a day (at 6 a.m. and 4 p.m.) and their daily rations consisted of 4 kg of alfalfa hay, 10 kg of maize silage and 1 kg of barley meal with a vitamin and mineral supplement. The nylon bags (pore size 42 microns - Uhelon 130 T, Silk and Progress Moravská Chrastová) containing feed samples were attached to a cylindrical carrier. Protein effective rumen degradabilities (rumen outflow rate 6 %/h) were determined in Amiga 76 %, Butan 73 %, Dieta 73 %, Prima 71 %, APR82 79 % and Korada 83 %. Significant differences were registered among soybean Korada and lupine Butan, Dieta and Prima. In lupine, there were significant differences between Prima and APR82. This research was supported by the Ministry of Agriculture of the Czech Republic (MZE 0002701403 and NAZV, project No. QG60142).

Fungal contamination of eight precision-chop grass silages on Irish farms

J. Hassett[1,2], M. O Brien[1,3], J. Mc Eniry[1,3] and P. O Kiely[1], [1]Teagasc, Grange Beef Research Centre, Dunsany, Co. Meath, Ireland, [2]School of Food Science & Environmental Health, Dublin Institute of Technology, Cathal Brugha Street, Dublin 1, Ireland, [3]UCD School of Biology & Environmental Science, Belfield, Dublin 4, Ireland

This small-scale experiment followed a comprehensive assessment of the mycology of baled grass silage on Irish farms, and aimed to describe the fungal contamination on a sample of eight precision-chop grass silages. Horizontal bunker or clamp silos were selected on six farms during February and March 2005. Separate representative samples were obtained from pre-designated sections of each silage (top of feed-face, side/corner of feed-face, centre of feed-face and core from 1 m behind feed-face). Every sample was enumerated for yeasts, total moulds and *Penicillium roqueforti*, and with the latter was subjected to macro- and micro-morphology description. Across the eight silages, mean yeast, mould and *P. roqueforti* counts ranged from 7.9×10^2 to 1.6×10^5, <10 to 3.2×10^5, and 0 to 10^4 colony forming units/g silage, respectively. There was no significant ($P > 0.05$) effect of location within the silo on the counts recorded. Thus, yeasts and moulds were located within each section of every silage, and *P. roqueforti* was found in 7/8 silages and predominantly towards the corners of the top surface of the silage.

Ruminal fermentation and in sacco NDF degradability in growing bull calves fed different starch levels and two types of roughage

K.F. Jørgensen[1], N.B. Kristensen[1], M.R. Weisbjerg[1], O. Højbjerg[1], P. Nørgaard[2] and M. Vestergaard[1], [1]University of Aarhus, Animal Health, Welfare and Nutrition, Foulum, DK-8830 Tjele, Denmark, [2]University of Copenhagen, Basic Animal and Veterinary Sciences, Bülowsvej, DK-1870 Frederiksberg C, Denmark

To study the risk of subacute acidosis in intensively-fed bull calves, 6 ruminally fistulated calves (initial BW 120 kg) were randomly assigned to a 2x2 factorial experiment and fed *ad libitum* with either low starch (LS) or high starch (HS) concentrate and either barley straw (BS) or grass hay (GH) as roughage. DMI of concentrate was 5.2 kg with no difference between treatments. DMI of roughage was 211 and 373 g/d for BS and GH, respectively ($P=0.02$). NDF intake was 1,221 and 912 g/d ($P=0.001$) and starch intake was 1,430 and 2,135 g/d ($P=0.001$) for the LS and HS concentrate, respectively, whereas the NDF and starch intake did not differ due to roughage type. Time spent ruminating was 235 min/d with no effect of treatment. GH compared with BS increased ruminal pH and decreased h/d with pH lower than 5.8. Total and individual VFA concentrations were unaffected by roughage type. LS *vs* HS concentrate only tended to increase ruminal pH and decrease total VFA. Feeding GH compared with BS improved in sacco NDF degradation of barley straw and grass hay, whereas feeding LS compared with HS improved the NDF degradation of barley straw. Inclusion of grass hay in the diet will improve the ruminal environment of intensively-fed bull calves.

Session 33 Poster 15

Variation in the nutritive value of cold-pressed rapeseed cake for ruminants

S. De Campeneere[1], A. Schellekens[2], J.L. De Boever[1] and D.L. De Brabander[1], [1]ILVO, Animal Science Unit, Scheldeweg 68, 9090 Melle, Belgium, [2]LCV, Hooibeeksedijk 1, 2440 Geel, Belgium

With the increasing demand for bio-fuel, the area of rapeseed grown in Flanders has increased in recent years. Most of the rapeseed is pressed in a cold-pressing procedure by farmers themselves (or colleagues). The rapeseed cake is a valuable feedstuff for different animal species. In ruminants however, attention has to be paid for the total fat load in the rumen to avoid a reduced fibre digestion. To be informed about the variation in composition of the cold-pressed rapeseed cake, 28 samples were collected from different farms in Flanders, representing different rapeseed varieties, pressed on different presses and with different drying conditions before pressing. Composition of the rapeseed cake averaged (standard deviation): 93 (15) g moisture/kg product, 299 (27) g crude protein/kg DM, 252 (56) g crude fat/kg DM, 104 (13) g crude fibre/kg DM. Digestibility of the organic matter averaged 83.4 (2.2)% while NEL value amounted to 9.90 (0.90) MJ/kg DM. In sacco incubations indicated that rumen degradability of the protein was very high: 88.5 (1.2)% with very little variation. These results demonstrate a large variation in the fat content of cold-pressed rapeseed cake (range: 153 - 335 g/kg DM) which should be taken into account when formulating diets for ruminants. On the other hand, rumen protein degradability was very high and showed very little variation.

Session 33 Poster 16

Effects of chitosan extracts on *in vitro* ruminal metabolism of maize silage: 2.- Fermentation kinetics

I. Goiri, A. Garcia-Rodriguez and L.M. Oregui, Neiker-Tecnalia, Health and Animal Production, Granja Modelo Arkaute, E-01080 Vitoria-Gasteiz, Spain

The *in vitro* gas production technique was used to study effects of chitosans, with different molecular weights and acetylation degrees, on ruminal fermentation kinetics. Chitosan is a natural biopolymer with demonstrated antimicrobial action. 750 mg/L of culture fluid, of 6 chitosans (CHI1, CHI2, CHI3, CHI4, CHI5, CHI6) was incubated for 144 h recording 59 data points in diluted ruminal fluid with maize silage. The ionophore antibiotic monensin (MON) was used as a positive control and a negative control with no chitosan (CTR) was also included. Each treatment was tested in triplicate and in two periods. Gas production data was adjusted to a biphasic exponential model using the PROC NLIN procedure of SAS using the DUD algorithm. Model parameter mean values were analysed using the PROC GLM procedure of SAS. Non significant differences were found for those model parameters representing the soluble and rapidly fermentable components of maize silage. However, chitosans (CHI1, CHI4 and CHI6) significantly affected the fermentation kinetics of the insoluble but fermentable fraction of maize silage. In conclusion, chitosans have an antimicrobial action on ruminal ecosystem modifying its metabolism, affecting in a different way the fermentability of the feedstuff fractions. However, further research is required to determine the effect of chitosans on ruminal fermentation parameters using commercial diets.

Effects of chitosan extracts on *in vitro* ruminal metabolism of maize silage: 1.- Digestion and fermentation

I. Goiri, A. Garcia-Rodriguez and L.M. Oregui, Neiker-Tecnalia, Health and Animal Production, Granja Modelo Arkaute, E-01080 Vitoria-Gasteiz, Spain

The *in vitro* gas production technique was used to study effects of chitosans with different characteristics on rumen microbial metabolism. 750 mg/L of culture fluid of 6 chitosans (CHI1, CHI2, CHI3, CHI4, CHI5, CHI6) was incubated for 24h in diluted ruminal fluid with maize silage. Each treatment was tested in triplicate and in three periods. Samples were collected for volatile fatty acid (VFA) concentration, pH and gas production values were also recorded. Methane concentration was estimated stoichiometrically. *In vitro* true organic matter digestibility (IVOMD) and partitioning factor (PF) were also calculated. Data were analysed using the PROC MIXED procedure of SAS with day as random effect. All chitosans decreased the IVOMD and the PF values. However, only CHI2, CHI5 and CHI6 decreased total VFA concentration. CHI2 and CHI6 decreased the molar proportion of acetate and increased the molar proportion of propionate, thus increasing c3: c2 ratio. Chitosans did not affect butyrate molar proportions. Except for CHI4 branched chain volatile fatty acids molar proportion was lowered by all the chitosans. Most treatments decreased methane production, but CHI4 had no effect. In conclusion, careful selection and combination of these extracts may allow the manipulation of rumen microbial fermentation, but further research is required to determine the effect of chitosans on ruminal fermentation parameters using commercial diets.

Effect of correcting for a blank on gas production dynamics and prediction of organic matter digestibility of grass silage

A. Garcia-Rodriguez, I. Goiri and L.M. Oregui, Neiker-Tecnalia, Health and Animal Production, Granja Modelo Arkaute, E 01080 Vitoria-Gasteiz, Spain

Effect of correcting for a blank on *in vitro* gas production (IVGP) dynamics and prediction of *in vivo* organic matter digestibility (IVOMD) of grass silage was studied. 72 samples were incubated in buffered rumen fluid for 96 h. Fermentation kinetics of samples corrected (C) or not corrected (NC) for blanks were described according to a generalized Michaelis-Menten model. Pearson correlation test was used to assess repeatability of model parameters and blank gas production between different incubation series. Correlation coefficients were calculated using mean values of A, B and c, corrected and not by sample obtained in each series. The same was done for blank gas production of each series. Stepwise multiple regression analysis was used to obtain prediction equations for IVOMD using IVGP parameters. Blank gas production showed a high repeatability between series (r=0.91). A higher repeatability was found for those parameters adjusted to the non corrected data (r= 0.73 *vs* 0.87; 0.59 *vs* 0.91; 0.57 *vs* 0.90 for A, B and c, respectively). Significant differences were found between NC and C parameters. While the A_{NC} was significantly higher, B_{NC} and c_{NC} were significantly lower. The R^2 was 0.732 and 0.735 for the IVOMD predictive equations using the C or NC parameters, respectively. Not correcting for a blank did not improve IVOMD prediction but improved the methodological repeatability.

Session 33 Poster 19

Effect of addition of malic acid salts on growth performance and ruminal functionality of beef cattle

C. Carrasco[1], V. Dell'orto[2], M. Innocenti[2], S. Vandoni[2], C.A. Rossi[2], M. Puyalto[3], A. Fuentetaja[4] and M.I. Gracia[1], [1]Imasde Agropecuaria, 28224, Madrid, Spain, [2]VSA, U. Studi Milano, 20133, Milan, Italy, [3]Norel, 28007, Madrid, Spain, [4]COPESE, 40480, Segovia, Spain

Three studies were performed to study the efficacy of dietary malic acid salts (MAS) in beef cattle. In Study 1, 72 beefs (Charolaise or Limousine x Spanish crossbreed) were raised from 214 to 337 kg BW. Feeding regime was based on concentrate (0.88UFC) and straw *ad libitum*. Two concentrate diets were used: i) C:control, and ii) M:C+4kg/t MAS. There were 6 replicates of 6 animals per treatment. Only numerical differences were observed on performance & ruminal parameters between treatments. In Studies 2&3, 38 or 40 Charolaise beefs were raised from 451 to 680 & 400 to 680 kg BW, respectively. There were two treatments: i)TMR diet based on concentrate, corn, beet pulp, straw and molasses, and corn silage only in Study 3; providing 0.91-1.05UFC according to animal age, and ii)TMRM (TMR diet+20g/head/d MAS). Beefs fed TMRM tended to grow 10% more ($P < 0.10$), and showed higher propionic acid, ammonia nitrogen ($P < 0.01$) and pH values ($P < 0.10$) than beefs fed TMR. It is concluded that MAS could act as ruminal pH stabiliser and growth promoter when beefs are fed TMR high energy diets. However, no differences were observed due to MAS addition when beefs were fed low energy diets

---

Session 33 Poster 20

Effect of live yeast supplementation on methane and VFA production *in vitro*

A. Fitie and W. Smink, Feed Innovation Services, bv, Wageningen, Netherlands

Methane is the 2nd main greenhouse gas and its production, mainly due to the rumen fermentation, is estimated to account for up to 22% of the anthropogenic sources. Moreover, methane production reduces the energy available for the animal. These losses have been estimated to up to 10% of gross energy ingested. An *in vitro* study with rumen fluid of lactating cows in batch culture was carried out to determine the effect of live yeast (Yea-Sacc^{1026}) on fermentation rate, VFA, ammonia and methane production using the cumulative gas technique. 32 bottles were used for the measurements and two treatments were applied: Control substrate (C), consisting of 50% roughage (grass silage 75%, corn silage 25 %) and 50% concentrate - Yea-Sacc^{1026} (YS) treatment consisting of control diet plus Yea-Sacc^{1026} (8 mg in bottle of 100 ml). The substrates were inoculated during 24 hours. At four different time points (4, 8, 12 and 24 hours) rumen fluid samples were taken for the determination of VFA and methane concentrations. Ammonia was measured after 8 and 24 hours only. YS resulted in a significant increase in VFA (respectively 4.73 and 5.00 mmol/g for C and YS, $p < 0.01$), acetic acid (2.79 and 2.92 mmol/g for C and YS, $p < 0.05$) and propionic acid (1.07 and 1.13 mmol/g for C and YS, $p < 0.01$) production. Acetic / propionic acid ratio and ammonia concentration did not differ between treatments. The concentration of methane was significantly lower for the Yea-Sacc^{1026} supplemented substrate (9.28 *vs* 8.73 % for C and YS, $p < 0.05$).

Session 33 Poster 21

Composition and *in vitro* digestibility of some Israeli agricultural lignocelluloses

E. Yosef[1], E. Zukermann[2], J. Miron[1], M. Nikbahat[1] and D. Ben-Ghedalia[1], [1]Agricultural Research Organization, Volcani Institute, Bet Dagan, 50250, Israel, [2]Extension Service, Bet Dagan, 50250, Israel

The semi-arid climate of Israel is the major impetus for considering agricultural lignocelluloses as potential ruminant feeds. Straws and other agricultural by-products are included occasionally in rations of calves and dry cows, particularly when common forages are scarce. Some relevant materials of recent interest were analyzed in this study. The monocotyledonous straws examined were: wheat, corn, barley, oats, sorghum, pensylaria straws; and the dicotyledonous ones: sunflower, canola, hummus, clover, peanut and cotton straws. The straws were sampled from different regions of Israel. The monocotyledonous straws showed very little compositional variation, containing around 70% NDF, 6-8.8% lignin, ~12% ash. The values of *in vitro* DM digestibility of the straws were, in a decreasing order: corn>sorghum>pensylaria>oats>barley>wheat. NDF-digestibility of corn, sorghum, pensylaria and oats straws were higher than that of wheat and barley straw. The dicotyledonous straw varieties were much more variable compositionally: 9-20% ash, 50-70% NDF, 3.3-10% protein, 3.3-14% lignin. The values of *in vitro* DM digestibility of the straws, in a decreasing order were: sunflower>peanut>hummus>clover>cotton>canola.

Session 33 Poster 22

Influence of iodine supplementation on the iodine concentration of tissues and organs in beef cattle

U. Meyer[1], K. Weigel[1], F. Schoene[2], U. Baulain[3] and G. Flachowsky[1], [1]Institute of Animal Nutrition, Federal Agricultural Research Centre (FAL), Bundesallee 50, 38116 Braunschweig, Germany, [2]Agricultural Research Centre of Thuringia (TTL), Ricarda-Huch-Weg 20, 07743 Jena, Germany, [3]Institute for Animal Breeding, Federal Agricultural Research Centre (FAL), Hoeltystr. 10, 31535 Neustadt, Germany

EU has reduced the upper level for iodine supplementation in dairy cattle to 5 mg/kg feed in order to lower the total iodine intake for consumers. The aim of this study was to test various iodine dosages in the diet of fattening bulls in order to measure the iodine concentration in organs, body liquids and muscle tissue (meat). 34 Holstein bulls (three groups) started the trial at an initial live weight of 222 kg. 0.5, 4 and 10 mg iodine per kg feed dry matter (DM) was added to a maize silage/concentrate diet. The bulls were slaughtered at 550 kg live weight. Fattening results were not influenced significantly. According to the applied dosages daily DM intake was 7.64, 7.50 and 7.54 kg with a daily gain of 1453, 1419 and 1343 g, respectively. Weight of the thyroid gland was 31, 26 and 42 g/animal. As expected the increasing dietary iodine dosage raised the concentration in blood serum (153, 430 and 806 µg/l), M. long. dorsi (16, 45 and 80 µg/kg) and other organs, but the increase is moderate compared to findings in milk and eggs. There is no reduction of iodine upper level required in beef cattle feeding (presently 10 mg/kg in the EU).

Session 33 Poster 23

Intake, digestibility and growth in steers offered grass silage supplemented with sucrose

P. O Kiely and A.P. Moloney, Teagasc, Grange Beef Research Centre, Dunsany, Co. Meath, Ireland

The impacts of fortifying unwilted grass silage with sucrose prior to offering *ad libitum* to cattle were determined. Continental crossbred steers (n = 96; mean (s.d.) starting liveweight 376 (25.4) kg) were individually offered silage (176 (12.3) g dry matter (DM)/kg, pH 4.0 (0.05), *in vitro* DM digestibility 630 (22.2) g/kg) with 0, 30, 60 or 90 g added sucrose (immediately prior to feeding) per kg DM for 109 days. Supplementary concentrates were offered at 0 or 3 kg per head daily. Twelve Continental-crossbred steers (mean (s.d.) liveweight 343 (23.7) kg) was used on two occasions to determine the digestibility of silage fortified with 0, 45 or 90 g sucrose/kg silage DM, alone or with supplementary concentrates. Supplementation with concentrates increased ($P < 0.001$) total DM intake, liveweight gain and the liveweight gain per kg total DM intake, and decreased ($P < 0.001$) silage DM intake. In contrast, fortifying silage with sucrose did not alter ($P > 0.05$) these variables. *In vivo* dietary DM digestibility was increased ($P < 0.01$) by supplementation with concentrates but was not altered ($P > 0.05$) by fortifying silage with sucrose. No interaction was evident ($P > 0.05$) between supplementation with concentrates and fortification with sucrose. It is concluded that fortifying unwilted grass silage offered to growing steers with sucrose did not alter any of the intake, digestion, growth or feed efficiency variables examined, nor did it interact with concentrate supplementation.

Session 33 Poster 24

Evaluation of different models of gas production with grass silage: fitting and biological suitability

A. Garcia-Rodriguez, I. Goiri and L.M. Oregui, Neiker-Tecnalia, Health and Animal Production, Granja Modelo Arkaute, E 01080 Vitoria-Gasteiz, Spain

Statistical and biological criteria were used to evaluate the relevance of an exponential (EXP), logistic (LOG), generalized Mischerlich (MIS) and generalized Michaelis-Menten (MM) model. 72 samples of grass silage of known *in vivo* organic matter digestibility (IVOMD) were analyzed. The goodness of fit of each model was evaluated from the residual variance (RMS) and the Akaike's Information Criterion (AIC_c). Forward stepwise multiple regressions were completed between IVOMD and gas production parameters. Biological suitability was assessed by the coefficient of determination (R^2) and the residual standard deviation (RMS) obtained in this regression analysis. Statistically, all models were considered valid to describe gas production profiles. Non significant differences were found between models for the RMS values. As assessed by the AIC_c values, MM showed the highest likelihood to describe better gas production profiles followed by EXP with MIS and LOG models showing the lowest probabilities. Linear regression showed that models were highly correlated with IVOMD. MIS model accounted better for the variation of IVOMD and showed the best accuracy prediction followed by the MM and LOG with the EXP model being the one with the poorest R^2 and RMS. It is concluded that MM model shows the highest probability to be representing the gas production properly showing at the same time a high sensitivity to variations in IVOMD.

Session 33 Poster 25

Influence of dietary conjugated linoleic acid on ruminant milk fat: a review

É. Cenkvári, Szent István University, Faculty of Veterinary Sciences, Institute of Animal Breeding, Animal Nutrition and Laboratory Animal Sciences, Str. István 2, Bdg. J, H-1078, Hungary

There is a considerable research effort directed to increase the conjugated linoleic acid (CLA) content of foods of ruminant origin, because of the potential health benefits which arise from CLA consumption. Biohydrogenation processes by rumen bacteria make difficult to modify the unsaturated fatty acid content of milk, and the content of n-3 and n-6 fatty acids in dairy milk can only be increased slightly. Some findings show that a high linoleic acid oil content in sunflower increased CLA concentrations to 24.4 mg/g of milk fat compared to the values of 13.3 and 16.7 mg/g of fat for high oleic and high linoleic acid oils, respectively. Some researchers found that grazing cows could attain a CLA concentration of 22.7 mg/g of fat, much higher than the values from cows fed conserved forages. Other data of high interest show that CLA content could be influenced by the ratio of forage and concentrate with a constant supply of linoleic acid. These studies suggest that dietary constituents can provide substances for the optimal bacterial growth in the rumen inducing the production of linoleic acid isomerase, which can maximize CLA output. Such feeding strategies should be developed, with a moderate increase in the secretion of trans-C18:1 compared to CLA. Some results reflect also significant individual variations in cows in the same lactation period having the same daily ration and kept under the same management conditions.

Session 33 Poster 26

Nutritional evaluations of barley malt sprout in the ration of fattening male lambs

M. Pasandi[1], A. Toghdory[2] and A. Kavian[1], [1]Member of scientific Board of agriculture and natural recourse research center, Gorgan, 4915677555, Iran, [2]Member of young researchers club, Islamic Azad university, Gorgan branch, Gorgan, 4914739975, Iran

Two experiments were conducted to nutritional evaluations of barley malt sprout (BMS) in fattening male lamb diet. In experiment 1, four mature wethers, 45-48 kg live weight, were used to determine the apparent nutrient digestibility of the BMS in an *in vivo* method. Mature wethers were initially fed by BMS. The average digestibility coefficient of dry matter, crude fiber, acid detergent fiber and organic matter were 67.90, 69.25, 57.2 and 71.10 percent, respectively. In experiment 2, thirty male lambs with average body weight of 33.01±2.7 kg were used in a completely randomized designed to investigation of the effects of BMS on growth performance of fattening lambs during three month experiment. Treatments include substituting sugar beet pulp and barley grain with BMS at 0, 15 and 30 percent by 10 replicate for each treatment. Substituting sugar beet pulp and barley grain with BMS did not significant affect the mean daily weight gain, feed consumption and feed conversion ($P > 0.05$). However, daily weight gain BMS groups were higher then control group but there were no significant difference between them.

Session 33 Poster 27

Determination of nutritive value of dried and ensiled apple pomace and their effects on performance of finishing lambs

F. Kafilzadeh and G. Taasoli, Razi University, Animal Science, Razi, 6719685416, Iran

This experiment was conducted to study the effect of ensiled and dried apple pomace (AP) obtained from puree making on finishing performance of lambs. Digestibility of both ensiled and dried AP were determined using 4 mature sheep. In finishing experiment, 18 Sanjabi male lambs were used (9 per treatment) in a 120 day which was divided into two periods (60 day each). In the first period the ensiled AP and in the second period the dried AP were fed in an iso caloric, iso nitrogenous total mix ration. The dry matter, CP and NDF of ensiled and dried AP were 247, 63.85, 386.4 and 888.4, 51.2, 385.6 g/kgDM respectively. No significant difference was observed in the apparent digestibility of ensiled and dried AP ($p < 0.05$). The apparent DM digestibility of ensiled and dried AP were 704.3 and 668.2 g/kg DM respectively. Feeding ensiled AP significantly increased Dry Matter Intake (DMI) (0.938 *vs* 0.803 kg/day), Average Daily Gain (ADG) (199.8 *vs* 155.56g) and Feed Conversion Ratio (FCR) (4.69 *vs* 5.16). Use of dried AP had no significant effect on DMI (1.030 *vs* 0.932 kg/day) but significantly improved the ADG (192.3 *vs* 123.82 g) and FCR (5.36 *vs* 7.52). Apple pomace increased carcass dressing percentage ($p < 0.05$) but had no effect on the percentage of different cuts. Results of this study suggest that AP, in both ensiled and dried forms, can improve the performance of finishing lambs.

Session 33 Poster 28

Study on the effects of Ca-LCFA and replacing cottonseed meal with Canola meal on the milk production and composition in Holstein dairy cows

A. Jamshidi, T. Ghorchi, N. Torbatinejad and S. Hasani, Gorgan university of Agricultural sciences and natural rsources, Animal science, Gorgan university of Agricultural sciences and natural rsources(pardis), Basij square, Gorgan,Iran, 49138-15739, Iran

An experiment was done with change over design (with a 2*2 factorial design) inclouding 4 ration diet, 4 period, each period was 21 days and 8 dairy cow after calving. Treatment were 1) 15% cottonseed meal and 3% Ca- LCFA 2) 15% cottonseed meal and 6% Ca- LCFA 3) 15% canola meal and 3% Ca- LCFA 4) 15% canola meal and 6% Ca- LCFA. Results of experiment showed Milk production, milk fat percentage, Solid Non Fat(SNF), Body Weight(BW), were not altered greatly by inclusion of 15% canola meal or cottonseed meal($p > 0.05$), But milk protein percentage increased when cow fed with canola meal($p < 0.05$). Concentration of Ca, TG, blood Glocose were not altered by feeding 15% canola meal versus cottonseed meal ($p > 0.05$). Canola meal decreased concentration of blood Urea($p < 0.05$). Digestibility of DM,CP,CF were not affected($p > 0.05$), but digestibility of ADF decreased by feeding canola meal($p < 0.05$). Milk production, milk protein percentage, milk fat percentage and SNF were not affected by feeding ration with 6% Ca- LCFA($p > 0.05$). Digestibility of ADF, CP, EE significantly decreased when cow fed with 6% Ca- LCFA($p < 0.05$). As a result, ration with 3% Ca-LCFA and replacing cottonseed meal with canola meal in dairy cow ration are suggested.

Session 33 Poster 29

Nutritional value of soybean and sunflower straw in sheep

M. Pasandi[1], A. Toghdory[2] and A. Kavian[1], [1]Member of scientific board of agriculture and natural recourse research center of gorgan, Gorgan, 4915677555, Iran, [2]Member of young researchers club, Islamic Azad university, Gorgan branch, Gorgan, 4914739975, Iran

An experiment was conducted to determine the nutritional value of soybean (SBS) and sunflower straw (SFS) in sheep. In the first stage chemical composition of SBS and SFS were determined by standard methods. In the second stage four mature weathers 57-60 Kg live weight, were used to determine the apparent nutrient digestibility of SBS and SFS. After 15 day adaptation to dietary treatments, a digestion study of 10 days duration, involving quantitative collection of feed, refusals and faeces was conducted to determine the apparent digestibility. The crude protein, crude fiber, acid detergent fiber, nitrogen free extract, calcium, phosphorous, potassium and magnesium content of SBS and SFS were 4.81, 47.09, 52.16, 42.29, 1.63, 0.32, 0.53, 0.95 and 5.64, 25.99, 50.40, 44.58, 0.18, 0.06, 2.45 and 0.06 percent respectively. The average digestibility coefficient of dry matter, organic matter, crude protein, crude fiber, acid detergent fiber in SBS and SFS were 41.6, 45.2, 17.4, 35.1, 27.1 and 49.6, 42.1, 28.3, 44 and 29.9 percent respectively. Total digestible content of SBS and SFS were 38.2 and 45.31 percent.

Session 33 Poster 30

Nutritional evaluation of two forage species for ruminants using *in vitro* gas production technique

A.R. Safaei[1], N. Maheri Sis[2] and A. Mirzaei Aghsaghali[1], [1]Animal Science Research Institute, Animal Nutrtion, Karaj, Iran, [2]Islamic Azad University, Department of animal science, Shabestar Branch, Iran

The objective of this study was to assess the nutritive value of two forage species include Alfalfa (HAM i.e. Hamedani cultivar) and Quackgrass (QCK) grown in Iran by using the chemical composition, *in vitro* gas production kinetics, Organic Matter Digestibility (OMD) and Metabolisable Energy (ME) contents. The rumen mixed microbe inoculums were taken from two fistulated Shall rams. Samples of forages were incubated with rumen fluid to determine gas production. Gas productions were measured at 2, 4, 6, 8, 12, 24, 48, 72 and 96 h. No significant differences found between OM, EE, Ash and ADL contents of experimental forages, although the differences for DM, CP, CF, NDF, ADF and GE were significant ($P < 0.01$). Gas production at all incubation times and gas production constants (a, b, c and a + b) were significantly ($P < 0.01$) higher in HAM hay. The OMD and ME for HAM and QCK hays were 71.2, 43.45 % and 10.96, 6.58 MJ per Kg DM, respectively. Under the climatic conditions of the 2005 growing season, the nutritive value of HAM hay was higher to that of QCK hay, because of lower NDF, greater cell content, OMD and ME.

Session 33 Poster 31

Effects of different protein and energy levels on reproductive efficiency of Atabay ewes during flushing

M. Mohajer[1], H. Fazaeli[2], A.H. Toghdory[3] and R. Kamali[1], [1]Member of scientific board of agriculture and natural recourse research center, Gorgan, 4915677555, Iran, [2]Research institute of animal science, Karaj, 31585, Iran, [3]Member of young researchers club, Islamic Azad university, Gorgan branch, Gorgan, 4914739975, Iran

Four hundred fifty 4-5 years old ewes with body weight 64±2.38 kg were randomly allocated into nine groups. They were grazed in lowland pastures of Golestan province daily and their diets were supplemented with 600 gr of mixed concentrate ration per day in order to determine the effects of protein and energy supply to initiate flushing on ewe productivity. Supplemented concentrate containing barley, sugar beet pulp, wheat bran and cotton seed meal that given for 31 days prior to tupping. A complete randomized design in a factorial arrangement with two factors, crude protein (10, 13 and 16 %) and metabolizable energy (2.08, 2.38 and 2.65 Mcal/kg) was used. For each group of ewes two teaser rams were used to detect heat. The results showed that different levels of protein and energy did not have significant effect on estrus, gestation, lambing and twinning percentage ($P > 0.05$). Litter weight at birth and live weight gain from birth to weaning were not affected by treatments ($P > 0.05$). High level of energy (2.65 Mcal/kg) tendency increased these parameters, but was not significant. High level of energy and protein increased the live weight of ewes during flushing period significantly ($P < 0.05$).

Session 33 Poster 32

Effects of substituting barley with spaghetti wastes in the ration of fattening lambs

R. Kamali[1], A. Toghdory[1], A. Godratnama[2], A. Mirhadi[3] and M. Mohajer[1], [1]Agriculture and natural recourse research, Gorgan, 4915677555, Iran, [2]Agriculture and natural recourse research, Mashhad, 32142, Iran, [3]Research institute of animal science, karaj, 31585, Iran

Sixteen zel lambs with average live weight of 26.24 kg were used in a completely randomized design to investigate the effects on feed intake, animal performance and ruminal parameters of the substitution of barley grain (0, 7, 14 and 21%) with spaghetti wastes during the fattening period. Lambs were housed individually in 1.5 × 1.5 m pens and fed a basal diet consisting of 32% forage and 68% concentrate that were offered *ad libitum*. The lambs were fattened for 90 days and weighted every 15 days. On the last day of experiment ruminal fluid samples were taken with stomach tube before feeding (0), 0.5, 1, 3 and 5 hours post feeding. The results showed that NH_3–N concentrations were not different between treatments in the 0 and 1 h samples ($P > 0.05$), but in the 3 and 5 h samples spaghetti wastes significantly decreased NH_3 –N concentration ($P < 0.01$). Total VFA concentrations were not affected by feeding spaghetti wastes in the 0, 0.5, 3 and 5 h samples ($P > 0.05$), but in 1 h samples were significantly greater ($P < 0.01$). Average daily gain, dry matter intake and feed conversion ratio were not affected by treatments ($P > 0.05$). In conclusion spaghetti wastes in diet for fattening lambs seem to enhance the ruminal environment and reduce the costs of fattening ration.

Session 34 Theatre 1

Immunophysiology of the mammary gland of dairy cows

R.M. Bruckmaier, University of Bern, Veterinary Physiology, Bremgartenstr. 109a, 3001 Bern, Switzerland

Intramammary infection (IMI) is a major problem in dairy farming and both the legal requirements for milk quality and the risk of IMI are increasing. IMI cause a reduction of milk quality and milk production. Because antibiotic treatment in food producing animals is progressively loosing acceptance, a profound knowledge of the immune function of the mammary gland is crucial to adapt the strategies both of IMI treatment and dairy cow breeding. Besides the teat canal anatomy and antibacterial proteins and enzymes, the most important line of defense is the immunological activity which is mainly represented by the inborn unspecific cellular immune response of phagocytes, i.e. macrophages and polymorphonuclear neutrophils (PMN). They are present in the milk of healthy quarters, and while trying to eliminate the pathogens they produce together with the mammary epithelial cells inflammatory factors such as cytokines and lipid mediators after IMI. The chemotactic effect of these factors causes a sharp increase of the transfer of phagocytes, mainly PMN, from blood into milk which can be measured as an increase of milk somatic cell counts (SCC). Thus, high SCC are an indicator of IMI, and bulk tank SCC is the measure of milk quality for the dairy industry. High bulk tank SCC leads to milk price reduction for the farmer. However, extremely low SCC indicate an impaired immune response. Therefore, besides upper limits of SCC also a lower limit needs to be defined to ascertain the immunological function of the mammary gland.

Session 34 Theatre 2

Genetics of udder health in dairy ruminants

R. Rupp, INRA, Animal Genetics, B.P. 52627, 31326 Castanet Tolosan Cedex, France

Genetic variability of udder health has been widely studied in dairy ruminants. Many studies focused on polygenic variation of traits related to udder health, such as mastitis or somatic cell counts, by estimating heritabilities and genetic correlations among phenotypic traits. Some data on the role of Major Histocompatibility Complex, and of the host's immune response are also available. Development from molecular genome mapping led to accumulating information of quantitative trait loci related to udder health. Newly seeded techniques such as transcriptomics and proteomics offer further promise as tools to address the determinism of resistance to udder diseases. Following economic and genetic analyses, and according to welfare and food safety considerations and to breeders and consumer's concern, inclusion of udder health in breeding objective of dairy ruminants is considered worldwide. This paper gives a review of genetic aspects of udder health in ruminants. It highlights challenges for the future including biological questioning on selection strategies, investigation of genes and mechanisms involved in mammary gland's defence and improvement of selection accuracy.

Session 34 Theatre 3

Alternative use of somatic cell counts in genetic selection for clinical mastitis

Y. de Haas[1], J. ten Napel[2], W. Ouweltjes[2] and G. de Jong[1], [1]NRS, P.O. Box 454, 6800 AL Arnhem, Netherlands, [2]ASG, P.O. Box 65, 8200 AB Lelystad, Netherlands

Genetic selection for clinical mastitis (CM) is currently based on indirect traits, of which lactation-average somatic cell count (SCC) is the most important one. The advantage of SCC is data availability. However, the genetic correlation with CM is not perfect and averages 0.7, and a lactation average does not take into account the dynamics of SCC in response to infection. In this study various ways of using SCC are defined, and genetic parameters were estimated. Data was available from 274 dairy herds recording CM with a Management Information System over an 8 year period. The dataset contained 56,709 lactations of 30,145 cows. Three groups of alternative SCC traits were defined. Firstly, by averaging SCC over shorter periods than the classical 305 days in milk. Secondly, by putting weights on outliers in test-day SCC, and summarize these weights on lactation-level. Thirdly, by indicating the presence of patterns of peaks in SCC. Heritabilities were highest for lactation-average SCC (~10%), and lowest for the patterns of peaks (~4%). Of all traits, the patterns of peaks were genetically strongest correlated to CM (~0.9), and SCC averaged over 150d resulted in a slightly stronger genetic correlation with CM (0.73) than SCC averaged over 305d (0.67). It is concluded that SCC can be made a better indicator trait than the current 305d average, by applying one or more SCC-trait definitions from this research.

Session 34 Theatre 4

A hidden Markov model for analyzing somatic cell scores

J. Detilleux, Faculty of Veterinary Medicine, University of Liege, Liege, Belgium, Bd de Colonster n° 20, 4000 Liege, Belgium

A hidden Markov model (HMM) is proposed to compute mastitis-related parameters based on monthly somatic cell scores (SCS). These hierarchical models have been increasingly used to model natural history of many diseases and to characterize the follow–up of patients under varied clinical protocols. Indeed, they allow describing disease evolution, estimating transition rates between health state, and evaluating the therapy effects on progression. Under the HMM proposed here to analyze SCS, two interrelated stochastic processes were defined: The first process is the set of SCS measured each month in milk (MIM) on a cow; the second process is the set of unknown health states of the cow, with (IMI+) and without (IMI-) intra-mammary infection, at the MIM on which the SCS is recorded. Each monthly health state was assumed dependent only on the preceding state (Markov property) with 4 probabilities of transition: IMI- to IMI-, IMI- to IMI+, IMI+ to IMI-, and IMI+ to IMI+. Given the health state, the SCS was modelled with a heteroscedastic two-component normal mixture model with random additive effects. Estimates of the parameters (SCS means and variances and IMI transition probabilities) were obtained via Gibbs sampling. The model was evaluated on simulated SCS data sets using parameters obtained on cows with known IMI states. Biases between Bayesian estimates and values used to simulate the data were computed to evaluate the accuracy of the model. Results are shown and the pro and cons of the proposed method are discussed.

Session 34 Theatre 5

A pathogen-specific analysis of udder health in dairy cows: relationships with cell counts and milking speed

R. Schafberg[1], S. Baumgart[2], L. Döring[2] and H.H. Swalve[1], [1]Institute of Agricultural and Nutritional Sciences, Animal Breeding, Adam-Kuckhoff-Str. 35, 06108 Halle(Saale), Germany, [2]LKV, Angerstr. 6, 06118 Halle(Saale), Germany

Aim of the present study was to assess the udder health status, i.e. the prevalence of intra-mammary infections (IMI), based on bacteriological analyses of milk samples, and its relationship with somatic cell counts as well as milking speed. Data originated from 27 large farms (ave. herd size 347 cows) and comprised a total of 9357 cows. Milk samples (12,820 quarter samples) were collected from 3195 cows and for 5912 cows milk flow parameters were measured using lactocorders. The results showed a prevalence of IMI of 21.7 % (positive samples). Most important pathogens (% of positive samples) were Coagulase negative staphylococci (CNS, 40.3 %), Str. uberis (20.3 %), Alpha-hemolytic streptococci (18.9 %) and S. aureus (12.7 %). Across all pathogens, somatic cell score (SCS) increased significantly with the number of infected quarters. Within pathogens, highest SCS was observed for S. aureus and CNS whereas SCS was only slightly elevated for IMI caused by Alpha-hemolytic streptococci and was equal to negative (healthy) samples for Str. uberis. With the exception of CNS infected cows, milking speeds were higher for cows with IMI as compared to healthy cows. This latter result could be interpreted in two ways: Either IMI is affecting milking speed or cows with increased milking speeds have a greater risk for specific forms of IMI.

Session 34 Theatre 6

Genetic parameters for automatically recorded milk flow rates in Danish Cattle

E. Norberg[1] and M.D. Rasmussen[2], [1]Århus University, Department of Genetics and Biotechnology, P.O. Box 50, 8830 Tjele, Denmark, [2]Århus University, Department of Animal Health, Welfare and Nutrition, P.O. Box 50, 8830 Tjele, Denmark

Milking speed (MS) is included in the Danish total merit index and an intermediate MS is the goal due to problems with very fast milkers (increased risk of mastitis) and very slow milkers (disruption of flow of cows through the milking parlour). Routinely, the trait is recorded as a subjective measure by the farmer, but it is suggested that automatically recording of the trait may improve the accuracy of the evaluation of MS. The objective of this study was to estimate genetic parameters for automatically recorded MS. Data on 1868 cows in 1st, 2nd and 3rd parity from 11 herds was included in this study. Data from about 1 million milkings were analysed for the influence of parity and stage of lactation on average and peak milk flow rates. Flow rates were adjusted to 2nd lactation and 150 DIM and averaged per cow. An animal model with herd and parity as fixed effects and an additive genetic and a permanent environmental effect was fitted. The estimated heritability and repeatability were 0.39 and 0.68 of adjusted automatically recorded average milk flow and 0.49 and 0.87 for peak flow, respectively. Heritabilities and repeatabilities are larger than those found in studies on subjectively recorded MS, and we conclude that the accuracy is larger on automatically recorded MS. The correlation between automatically recorded MS and the MS index of the sires will be investigated.

Session 34 Theatre 7

Mastitis susceptibility in Valle del Belice ewes related to weather conditions

R. Finocchiaro[1], J.B.C.H.M. van Kaam[2] and B. Portolano[1], [1]University of Palermo, Department S.En.Fi.Mi.Zo., Viale della Scienze, 90128 Palermo, Italy, [2]Istituto Zooprofilattico Sperimentale della Sicilia "A. Mirri", Via Gino Marinuzzi 3, 90129 Palermo, Italy

Weather conditions are known to have effects on performance and well-being of livestock animals. Mastitis susceptibility of Valle del Belice ewes to precipitation, solar radiation, sun hours, air pressure, wind speed, wind direction and two temperature-humidity indexes (day and night) was investigated. Production data collected in the period 1998-2006 consisted of 83898 test-day somatic cell score (SCS) records belonging to 8638 lactations of 5562 ewes in 17 flocks. Analyses were based on mixed models that included as fixed effects of flock * year of test-day * season of test day interaction, days in milk * lactation number * litter size interaction and a weather parameter effect, and as random effects the permanent environmental effects within and across lactations. Weather information from the day before the test-day was used. Precipitation, solar radiation, sun hours, wind direction and the temperature-humidity indexes had highly significant effects on SCS. Air pressure and wind speed had no significant effect on SCS. Higher precipitation and sun hours resulted in increased SCS. Extreme values of solar radiation, and day and night temperature-humidity indices resulted in increased SCS. These results show that it is important to include weather information in genetic evaluation models for mastitis resistance.

Session 34 Theatre 8

Genes involved in development and functional differentiation of caprine mammary secretory epithelium

F. Faucon[1,2], E. Zalachas[2], S. Robin[3] and P. Martin[2], [1]Institut Elevage, Paris, 75012, France, [2]Unité Génomique et Physiologie de la Lactation, INRA, Jouy en Josas, 78352, France, [3]Unité Mixte de Recherche 518, AgroParisTech, Paris, 75005, France

Development of mammary secretory epithelium and its functional differentiation occur during pregnancy under combined actions of ovarian steroids, pituitary hormones and growth factors. If the effect of these molecules is relatively well known, effect of differentiation factors expressed locally is not sufficiently characterized. To understand the local regulation of mammary tissue development and differentiation, we realized transcriptional analysis on 5 physiological stages (4 during pregnancy and 1 during lactation). An appropriate experimental design was drawn to follow gene expression profiles during differentiation of mammary tissue. We first validated experimental feasibility by comparing expression pattern of mammary tissue from the first pregnancy (45 days) and from lactation stages. Using 3 goats per stage, this comparison was done on 3 dye-swaps, utilizing oligoarrays. Statistical analysis revealed that among the 8 379 genes spotted, 299 genes were differentially expressed between pregnancy and lactation. Data base examination confirmed that genes controlling cellular activity of conjunctive tissue were over-expressed during pregnancy while genes implicated in milk protein and fatty acid biosynthesis were over-expressed during lactation.

Session 34 — Theatre 9

Mapping of a central ligament defect in Finnish Ayrshire dairy cattle

S. Viitala, V. Ahola, N.F. Schulman and J. Vilkki, MTT Agrifood Research Finland, Biotechnology and Food Research, H2-talo, FIN-31600 Jokioinen, Finland

Weakness of median suspensory ligament can cause severe udder problems in dairy cattle. The incidence of weak central ligament defect among heifers after first parturition has increased recently in Finnish Ayrshire. In this condition, the central ligament stretches badly or breaks down completely causing a pendulous udder that is difficult or impossible to milk and more prone to physical injury and clinical mastitis. The defect is not associated with metabolic symptoms like udder edema. Bad udder conformation is the cause of 6.4% of premature cullings in Finland. The central ligament defect in Finnish Ayrshire can be traced back to two common ancestors. Based on the assumption of the same origin of the mutation causing the defect we performed genome wide homozygosity mapping of affected and unaffected individuals. In addition, a database search of genes relating to connective tissues revealed 34 candidate genes from which 33 were located to the bovine genome. For genotyping, flanking markers for candidate genes were also used. We have identified chromosomal regions that showed skewed allele distribution between the two groups.

Session 34 — Theatre 10

Verification of chromosomal regions affecting the inverted teat development and their derivable candidate genes in pigs

E. Jonas[1], S. Chomdej[1], S. Yammuen-Art[1], C. Phatsara[1], H.-J. Schreinemachers[1], D. Jennen[1], D. Tesfaye[1], S. Ponsuksili[2], K. Wimmers[2], E. Tholen[1] and K. Schellander[1], [1]Institute of Animal Science, University of Bonn, Endenicher Allee 15, 53115 Bonn, Germany, [2]Research Institute for the Biology of Farm Animals, Wilhelm-Stahl-Allee 2, 18196 Dummerstorf, Germany

A genome scan over 18 autosomes was performed to detect QTL for the inverted teat defect in pigs. In the study, animals of dam breeds and of an experimental population showed a high incidence of the inverted teat defect. Different QTL were detected, with a good agreement of the regions linked to the trait. Candidate genes on the verified chromosomes were investigated and association analyses with the inverted teat defect were performed. Additional association analyses of the genes with number of teats were considered. According to linkage mapping, we found that the functional candidate genes *RLN*, *TGFB1*, *LGR7*, and *GH* are located in the QTL regions. Hence, these genes are also positional candidates for traits of udder quality in pigs. Association studies indicate the significant association of *RLN*, *TGFB1* and *GH* with the inverted teat defect in pigs. An 2414 bp fragment of porcine *LGR7* cDNA was sequenced, different alternative splicing forms were detected in different tissues. Expression analysis showed high expression of *LGR7* in teat and inverted teat of pigs. *LGR7* was not differentially expressed in normal compared to inverted teats, splicing variants were also identical among the two types of teat.

Session 34 Theatre 11

Yield losses associated with clinical mastitis occurring in different weeks of lactation

C. Hagnestam[1], U. Emanuelson[2] and B. Berglund[1], [1]Swedish University of Agricultural Sciences, Dept. of Animal Breeding and Genetics, SE-750 07 Uppsala, Sweden, [2]Swedish University of Agricultural Sciences, Dept. of Clinical Sciences, SE-750 07 Uppsala, Sweden

The effects of clinical mastitis (CM) on production were studied in the Swedish Red and Swedish Holstein breeds. The data consisted of 38,535 weekly production records from 1,192 lactations (506 cows), sampled from 1987 to 2004 in a research herd. Daily yields were analyzed using a repeated-measures mixed model with an interaction between a mastitis index and lactational stage among the independent variables. The index was used to distinguish between cows with and without CM, as well as to indicate time (test day) in relation to day of diagnosis. Inclusion of the interaction made it possible to study the effects of CM occurring in different weeks of lactation. Daily milk yield started to decline 2-4 weeks prior to diagnosis. On the day of clinical onset, the milk yield of mastitic cows was reduced by 1-8 kg. After a case of CM, milk yield was suppressed throughout lactation. The magnitude of the yield losses was determined by the week of lactation at clinical onset. The greatest losses occurred when primiparous cows developed CM in week 6, whereas multiparous cows experienced the greatest losses when diseased in week 3. The 305-day milk, fat and protein production in mastitic primiparous cows were reduced by 0-9, 0-8 and 0-7%, respectively. The corresponding reductions in mastitic multiparous cows were 0-11, 0-12 and 0-11%.

Session 34 Poster 12

Association of BoLA-DRB3 alleles with occurrence of mastitis caused by different bactertial species

G. Sender, K.G.A. Hameid and A. Korwin-kossakowska, Institute of Genetics and Animal Breeding, Jastrzebiec, 05-552 Wolka Kosowska, Poland

The objective of the study was to find out association between mastitis caused by different bacterial species and genotypes for *BoLA-DRB3 locus* (*BoLA-DRB3.2*16* and *BoLA-DRB3.2*23*) in Polish dairy cattle. During the period of the study, a total of 722 composite milk samples were collected from 275 lactating cows two or three times during lactation. Bacterial species were identified according to methodology of National Mastitis Council. Cows were genotyped for detection of *BoLA-DRB3.2*23* and *BoLA-DRB3.2*16* alleles using MPT-PCR method. Evaluation was done using analysis of variance (GLM process SAS) to estimate the associations between BoLA genotype and latent or sub-clinical mastitis caused by different bacterial species. The results showed that there was no association between cows carrying allele*BoLA-DRB3.2*16* and *BoLA-DRB3.2*23* with susceptibility/resistance to all typesof *Staphylococus aureus* mastitis or CNS (coagulase negative staphylococci) sub-clinical mastitis. However, *BoLA-DRB3.2*23* - genotype was associated significantly ($p < 0.01$) with increased susceptibility to sub-clinical mastitis caused by *Streptococcus dysgalactiae*. Additionally, *BoLA-DRB3.2*16* - genotype was significantly associated ($p < 0.01$) with increased susceptibility to latent CNS mastitis.

Session 34 Poster 13

Relationship between kind and number of pathogens and SCC and their structures in goat milk

E. Bagnicka[1], A. Winnicka[2], A. Jóźwik[1], M. Rzewuska[2], N. Strzałkowska[1], B. Prusak[1], J. Kaba[2] and J. Krzyżewski[1], [1]Institute of Genetics and Animal Breeding, Department of Animal Science, Postepu 1, 05-552 Jastrzębiec, Poland, [2]University of Life Science in Warsaw, Faculty of Veterinary Medicine, Nowoursynowska 159, 02-776 Warsaw, Poland

The aim of this study was to establish the relationship between kind and number of pathogens and somatic cell count (SCC) and fraction of somatic cells in goat milk. The milk samples were taken from each half of udder of 40 goats two times in lactation (1st – in pick of lactation; and 2nd – between 60th and 120th days of lactation). There were established milk yield and its chemical composition, SCC and their structures and microbiology status of milk samples. About 65% of samples did not contain any pathogens although in $^1/_4$ of them SCC ranged between $1x10^6$ and $1.4x10^6$. About 55% of the samples with pathogens contained coagulase-negative staphylococci, while SCC ranged between $4x10^5$ and $2.7x10^6$. Seven samples contained *Enterococcus* ($2x10^5$ – $1.8x10^6$ of SCC), four samples contained *Staphylococcus aureus* ($5x10^5$ – 2 $x10^6$ of SCC), three samples contained *Streptococcus agalactie* (above $10x10^6$ of SCC) and three samples contained *Streptococcus dysgalactie* ($3.5x10^6$ – 12 $x10^6$ of SCC). There were moderate positive correlation (0.24-0.46) between number of bacteria and: SCC, percent of leucocytes, neutrophils, monocytes in SCC, as well as between number of bacteria and number of eozynophiles, neutrophils, monocytes and lymphocytes in milk.

Session 34 Poster 14

Characterization of Relaxin-3/INSL7 and its receptor being candidate genes for the inverted teat defect in pigs

S. Yammuen-Art[1], D. Jennen[1], E. Jonas[1], C. Phatsara[1], S. Ponsuksili[2], D. Tesfaye[1], E. Tholen[1], K. Wimmers[2] and K. Schellander[1], [1]Institute of Animal Science, Animal Breeding and Husbandry Group, University of Bonn, Endenicher Allee 15, 53115, Germany, [2]Research Institute for the Biology of Farm Animals, Wilhelm-Stahl-Allee 2, Dummerstorf, 18196, Germany

Relaxin is important for the function of reproductive tissues. The central nervous system expression patterns of *RLN3* gives evidence that the G protein coupled receptor 135 (*GPCR135*) coupled with the high affinity interaction to *RLN3*. The objectives of this study were to characterize the *RLN3* and *GPCR135* gene in pig, to determine the chromosomal location of *RLN3* and to test variants in *RLN3* and *GPCR135* for association with the inverted teat defect in a reciprocal cross of Berlin Miniature Pig and Duroc and in animals of the dam breeds German Landrace (DL) and German Large White (DE). Expression study using semi-quantitative RT-PCR showed that *RLN3* is highly expressed in lung, testis and uterus, whereas *GPCR135* is moderately expressed in muscle, heart, spleen, kidney and uterus. The association test between *RLN3* and *GPCR135* genotypes with affection status of teats was not significant. Association analysis using haplotypes revealed only a significant association between *RLN3* and inverted teats. This study was the first to investigate the role of *RLN3* and its receptor for traits of the udder quality in pigs.

Session 34 Poster 15

Relationship between somatic cell score and udder conformation traits in Polish Holstein-Friesian cows

E. Ptak[1], W. Jagusiak[1] and A. Zarnecki[2], [1]Agricultural University, al. Mickiewicza 24/28, 31-059 Krakow, Poland, [2]National Research Institute of Animal Production, Balice k. Krakowa, 32-083, Poland

Somatic cell score (SCS) has been recognized as an indicator of the presence of mastitis. Udder conformation traits also have been used for indirect selection to reduce mastitis. The objective of this research was to evaluate genetic relationships between SCS and udder traits. Data for analysis were first lactation SCS (LSCS) calculated as the average of a minimum 5 test day SCS per lactation of 26,727 cows, sired by 295 bulls. The following udder type traits were included: udder score, fore udder attachment, rear udder height, central ligament, udder depth, rear udder width, fore teat placement and teat length. The multi-trait REML model included fixed effects of herd-year-season-classifier, stage of lactation and linear regression on calving age, and random animal and residual effects. Heritabilities estimated for LSCS and udder traits were low to moderate: LSCS 0.14, udder score 0.12, rear udder width 0.15, central ligament 0.16, fore udder height 0.18, teat placement 0.20, rear udder height 0.21 and teat length 0.31. Genetic correlations between LSCS and udder traits were very low in most cases. The exceptions were genetic correlations between LSCS and udder depth (-0.17) and between LSCS and rear udder width (0.20), which suggested that shallower and narrower udders are associated with lower SCS.

Session 35 Theatre 1

Probiotics: do they have a role in the pig industry?

M. Kenny[1] and E. Mengheri[2], [1]University of Bristol, Clinical Veterinary Sciences, Lower Langford, Somerset, BS40 5DU, United Kingdom, [2]INRAN, Via Ardeatina 546, 00178 Roma, Italy

Probiotics are defined as "live micro-organisms, which when consumed in adequate amounts, confer a health (or production) benefit on the host". In this review the modes of action of probiotics and the hard evidence for positive effects, in properly controlled and analysed trials, will be discussed. In the pig industry both bacteria and yeast have been used in all production stages; to address specific clinical conditions and as non-specific "growth promoters". Critical appraisal is complicated by commercial confidentiality issues. In addition, there are concerns about the viability of organisms in commercial formulations following sub-optimal storage. In pigs, bacteria used as probiotics include *Enterococcus faecium*, various species of *Lactobacillus* and *Bifidobacterium*, non-pathogenic *Escherichia coli* and *Bacillus cereus*. The yeast *Saccharomyces cerevisiae* has also been used. Perhaps the most widely applied "probiotics" are those which ferment liquid feed; these are lactic acid bacteria, usually *Lactobabacillus* or *Pediococcus species.* Fermented feeds can result in significant health improvements and enhanced production parameters, although the effects seen may be due to the presence of lactic acid in the feed as well as purely microbiological effects. It is clear that giving supplementary gut bacteria can have positive health implications but caution must be applied when extrapolating between bacterial strains and different pig life stages.

Session 35 Theatre 2

Alternatives to in-feed antibiotics: use of prebiotics in the nutrition of weaning piglets

W.F. Pellikaan[1], P. Bikker[2], M.W.A. Verstegen[1] and H. Smidt[3], [1]Wageningen University, Animal Nutrition Group, Marijkeweg 40, 6709 PG Wageningen, Netherlands, [2]Schothorst Feed Research, Meerkoetenweg 26, 8218 NA Lelystad, Netherlands, [3]Wageningen University, Laboratory of Microbiology, Hesselink van Suchtelenweg 4, 6703 CT Wageningen, Netherlands

Weaning piglets at an early age increases the risk of gastrointestinal disturbances and increases susceptibility to infections, resulting in decreased gut health and animal performance. In 2004 an EU-funded research consortium was initiated with the aim to provide the European pig industry with natural alternatives to in-feed antimicrobials. The effects of plant extracts and other natural substances (PENS) were studied on different animal parameters, including gut physiology, immunology, gut ecology and overall animal performance. One approach used to manipulate gut health was to make use of feed ingredients with a prebiotic effect. Different *in vitro* techniques were used to assess the influence of various PENS on several gut health parameters. These *in vitro* results were compared with *in vivo* experiments, in which PENS-effects on gut health and growth performance were studied. In this presentation the latest findings of the consortium members will be discussed. Moreover we will reflect these against the currently available literature on prebiotic use in the nutrition of the newly weaned pig. Special attention will be given to reflect the literature to current date and the progress of the consortium in a relevant manner for the pig industry.

Session 35 Theatre 3

Gut immunology: or what keeps the outside world out of the organism

I.P. Oswald[1], H.J. Rothkoetter[2], M. Bailey[3] and C.R. Stokes[3], [1]INRA, BP3, 31931 Toulouse, France, [2]Universität Magdeburg, Leipziger Str. 44, 39120 Magdeburg, Germany, [3]University of Bristol, Langford House, Bristol BS40 5DU, United Kingdom

The immune function of the intestinal mucosa results in either defense against pathogens or in tolerance to food and commensal bacteria. Various mechanisms involving several cell types are responsible for these complementary functions. The piglet is profoundly immunodeficient at birth and is only able to generate limited T and B cell responses. The stages in the development of the mucosal immune system can impact on the piglets ability to mount appropriate responses following weaning. Various "immune-stimulants" can influence the rate of development of the mucosal immune system. The intestinal epithelium can signal the onset of the host innate and acquired immune response through the production of cytokines, crucial for the recruitment/activation of other immune cells. Intestinal inflammation occurs at weaning and its modulation is important when selecting PENS for feeding trials. Mucosal dendritic cells are localised below the intestinal epithelium and are able to extend their processes towards the intestinal lumen. Antigen uptake, processing and presentation by dendritic cells are require to develop specific immune reactions. Co-culture of intestinal epithelial cell lines and dendritic cells allows to follow antigen uptake and intracellular processing in order to understand the impact of selected microbial and nutritional antigens as well as PENS on the intestinal epithelium.

Session 35 Theatre 4

Impact of bioactive substances on the gastrointestinal tract and performance of weaned piglets: a review

J.P. Lalles[1], P. Bosi[2], P. Janczyk[3], S.J. Koopmans[4] and D. Torrallardona[5], [1]INRA, SENAH, Domaine de la Prise, 35590 Saint-Gilles, France, [2]University of Bologna, Via Rosselli 107, 42100 Reggio Emilia, Italy, [3]FBN Research Institute for Biology of Farm Animals, Wilhelm Stahl Allee 2, 18196 Dummerstorf, Germany, [4]Wageningen UR, Edelhertweg 15, 8219 PH Lelystad, Netherlands, [5]IRTA, Animal Nutrition, Ctra. Reus-El Morell, km. 3.8, 43120 Constantí (Tarragona), Spain

The EU ban on in-feed antibiotics has stimulated research on weaning diets for overcoming post-weaning gut disorders in pigs. Many bioactive components including organic acids, animal and plant proteins, amino acids, fatty acids and plant extracts have been studied but only some of them have proven effective. Various organic acids have positive effects lowering gastrointestinal pH, increasing gut and pancreas enzyme secretion and improving gut wall morphology. Particular amino acids are supportive in improving digestion, absorption and retention of nutrients by affecting tissue anabolism, stress and immunology. Spray dried plasma affects positively gut morphology and inflammation through specific and a-specific influences of immunoglobulins and other bioactive components. Bovine colostrums may act similarly to spray dried plasma. *In vivo* effects of plant extracts and other natural substances with anti-bacterial activity vary from large to minimal, depending on the products and doses. Often, bioactive substances protective to the gut also stimulate feed intake and growth performance. New insights on these aspects will be highlighted in this review.

Session 35 Poster 5

The potential effect of plant extracts and other bioactive natural substances (PENS) on GIT microbial activity measured by the *in vitro* gas production technique

W.F. Pellikaan[1], O. Perez[2], L.J.G.M. Bongers[1], S. van Laar-van Schuppen[1], M.W.A. Verstegen[1] and H. Smidt[2], [1]Wageningen University, Animal Nutrition Group, Marijkeweg 40, 6709 PG Wageningen, Netherlands, [2]Wageningen University, Laboratory of Microbiology, Hesselink van Suchtelenweg 4, 6703 CT Wageningen, Netherlands

The *in vitro* cumulative gas production technique was used to investigate how complex microbial communities respond to plant extracts and other natural substances (PENS). In a series of experiments, 42 PENS were tested using faecal inoculum from piglets (two wk post-weaning). The cumulative gas production was measured and fermentation fluids were analysed for pH, volatile fatty acids, NH_3, lactate and microbiological fingerprints. Six inulin products tested were of high fermentability. Seaweed components fucan and ulvan were non-fermentable whilst oligo-mannuronic was difficult to moderately fermentable. Other seaweed components (laminaran, xylan, etc.) and products like carob pulp, citrus pulp, and guar were highly fermentable. A range of PENS (benzoic acid, cinnamaldehyde, naringine, etc.) were tested as additives. None of these PENS affected the gas production kinetics distinctly, but organic oils, sanguinarine and a seaweed derived carotenoid appeared to give the most promising results. It was concluded that different types of inulins may act at different levels of the intestinal tract. The dose-response treatments applied in this study did not seem to affect the microbial activity *in vitro*.

Use of bioassays to assess functional properties of fermented milks

L. Lignitto[1], D. Regazzo[2], G. Gabai[2], L. Lombardi[3], C. Andrighetto[3] and S. Segato[1], [1]University of Padova, Scienze Animali, viale dell Università 16, 35020 Legnaro, Italy, [2]University of Padova, Scienze Sperimentali Veterinarie, viale dell Università 16, 35020 Legnaro, Italy, [3]Veneto Agricoltura, Istituto per la Qualità e le Tecnologie Agroalimentari, via San Gaetano, 36016 Thiene, Italy

The fermentation of milk with proteolytic starter cultures, such as *Enterococcus faecalis* (FAIR E-63) and *Lactobacillus delb. bulgaricus* (LA-2), can generate peptidic compounds from milk proteins with potential immunomodulating and ACE-inhibitory activity implicated in the regulation of blood pressure. The fractions obtained from centrifugation of fermented milks with Centricon Amicon Ultra 15 filters (cut off 5000Da) were used to assess the ACE-inhibitory activity using a rapid enzymatic test. In addition, fraction aliquots were freeze-dried, re-suspended in culture medium and used at different concentrations to assess the immunomodulatory activity by a simple method based on *in vitro* bovine lymphocyte proliferation. The proliferation of lymphocytes obtained from non-pregnant dry cows was measured by the MTT assay after 48 h of incubation. The immunomodulating effects of the two different strains were analysed by ANOVA (SPSS 14.0). FAIR E-63 showed a greater ACE-inhibitory potency (72%) than LA-2 (59%). Both fermented milks inhibited concanavaline-A lymphocyte proliferation and FAIR E-63 was more potent than LA-2. This effect was statistically significant ($P < 0.05$) at low concentrations (5-25 µg/ml).

Effect of supplement of marine algae, β-glucan, yeast on immunity traits of growing pigs

K. Suzuki[1], Y. Kumagai[1], W. Onodera[1], Y. Shimizu[2], Y. Suda[3] and J. Kobayashi[3], [1]Tohoku University, Sendai, 981-8555, Japan, [2]Miyagi Prefec. Anim. Exp. Sta., Miyagi, 989-6445, Japan, [3]Miyagi University, Sendai, 982-0215, Japan

The effect of feed additives on growth and immune traits of pigs was examined between 9 to 17w of age. Sixteen Duroc piglets from three litters were weaned at 3w of age. They were divided into control (4), marine algae (4), β-glucan (4) and yeast (4) groups at 7w of age. The marine algae, β-glucan and yeast were added to feed without antimicrobial agent addition at the rate of 0.8%, 0.1% and 0.5%, respectively. Blood was collected from the 4 groups of pigs at 9, 11, 13, 15 and 17w of age. Moreover, to compare the amount of antibody production to a specific antigen, every group of pigs was injected sheep red blood cell (SRBC) at 15 and 16w of age. Phagocyte capacity (PC), complement alternative pathway activity (CAPA), total leukocyte (WBC), ratio of granular leukocytes to lymph cells (RGL) and SRBC specific IgG (sIgG) were measured. There was no significant difference in body weight among groups. At 17w of age, sIgG and WBC of the marine algae group were significantly higher than the control group. Furthermore, RGL of the marine algae and β-glucan groups were significantly higher than the control group at 17w of age. Though PC of the β-glucan group was higher than that of the control group at several weeks, there were no significant differences among groups. These results suggest that the marine algae influence immune traits in pigs.

Session 35 Poster 8

Effects of cinnamaldehyde, benzoic acid and grapefruit extract on gut health of weanling pigs

N. Andrés Elias[1], I. Badiola[2], J. Pujols[2] and D. Torrallardona[1], [1]IRTA, Animal Nutrition, Ctra. Reus-El Morell, km 3.8, 43120 Constantí (Tarragona), Spain, [2]CReSA Foundation, Campus de Bellaterra, Edifici V, 08193 Bellaterra, Spain

The aim of this study was to evaluate possible alternatives to antimicrobials (GPA) in the diets of weanling piglets. Seventy two animals (24 pens) were offered one of four diets consisting of: control (C), cinnamaldehyde at 175 ppm (CI), benzoic acid at 0.5% (BA) and grapefruit extract at 125ppm (GE) to measure performance over a 28d trial. Another sixty piglets (12 pens) fed the same diets were slaughtered on days 0, 1, 2, 6 and 14 post-weaning to perform histology on the mucosa of the jejunum and to study ileal and caecal microbiota by RFLP. In the performance trial (0-28d), pigs on BA treatment grew faster ($P=0.05$) than those on CI and GE, and had a better feed efficiency ($P=0.1$) than those on GE. BA resulted in higher villus height than C ($P < 0.1$) and CI and GE ($P < 0.05$). Microbiota biodiversity of the BA group was higher ($P < 0.05$) than CI in the ileum and caecum, and C in the ileum. The similarity (intra-treatment) of the ileal microbiota in the BA group was higher ($P < 0.05$) than in the other treatments. That of the caecal microbiota of the GE group was lower ($P < 0.001$) than that of all the other treatments. It is concluded that BA is able to modulate digestive microbiota and can be a valid alternative to GPA for weanling pigs.

Session 35 Poster 9

The effect of protein sources on growth performance, blood metabolism and intestinal microflora in weaned piglets

D.W. Kim, S.B. Cho, S.J. Lee, H.Y. Jeong, H.J. Lee, W.T. Chung, J. Hwangbo and I.B. Chung, National Livestock Research Institute, Division of Nutrition & Physiology, 564 Omokchun-dong, 441-706 Kwonsn-Gu, Suwon, Korea, South

This experiment was conducted to study the effect of dietary supplementation with fishmeal, fishmeal hydrolysate, SPC (soy protein concentrate) and SDPP (spray dried porcine protein) as protein sources on the growth performance of weaned pigs. Landrace pigs (average body weight, 6.55kg) were used to investigate performance, intestinal enzyme activity, fecal microbes and blood characteristics for 28 days. Feed efficiency was significantly different among treatments in the following order: SPC > SDPP > Fishmeal > fishmeal hydrolysate treatment (0.52>0.49>0.45>0.42), respectively. Sucrase, lactase and maltase in duodenum and jejunum tended to be higher in fishmeal hydrolysate treatment than in the other treatments. The fecal *Lactobacillus*, *Enterococcus*, and *Enterobacteriaceae* numbers were 7.7~8.9, 4.7~5.7, and 5.8~7.4 (Log CFU/g feces), respectively. The serum concentrations of glucose, total protein, albumin, GOT and GPT were 95~125 mg/dl, 4.50~6.13 g/dl, 2.77~4.5g/dl, 22.53~39.93 U/L, and 25.90~37.07 U/L, respectively. No significant differences among treatments were found for fecal microbes or serum characteristics. It can be summarized that dietary supplementation with SPC and SDPP as protein sources improved animal performance. However, intestinal enzyme activity, fecal microbes and blood characteristics were similar among treatments.

Session 36 Theatre 1

Irish Thoroughbred industry

M. O'Hagan, Irish Thoroughbred Marketing, The Curragh, Co. Kildare, Ireland

The presentation will cover the area of thoroughbred foal production numbers in Ireland over the past ten years, highlighting the growth in numbers placing Ireland as the third largest producers of thoroughbreds in the world after the USA and Australia. This represents 42% of the European output. It will also show the value and trends of output of equine and other livestock 1998 -2002 in € millions. We will show the number of broodmares at stud in Ireland going back to the mid 1980's. We will show the value of Thoroughbred sales at public auction in recent years. We will outline the commitment of the Irish Government to this industry and the vision of Horse Racing Ireland in promoting better standards on the race track, high prizemoney and a low tolerance of inferior racehorses. The presentation will also outline how Irish Thoroughbred Marketing has marketed the thoroughbred product and how the Irish Government has decided to use the Irish Thoroughbred to assist in marketing this small county called Ireland.

Session 36 Theatre 2

An overview of the Irish sport horse industry

K.D. Hennessy[1] and K.M. Quinn[2], [1]University College Dublin, Centre for Sports Studies, Woodview, Belfield, UCD, Dublin 4, Dublin, Ireland, [2]University College Dublin, School of Agriculture, Food Science and Veterinary Medicine, Agriculture Building, Belfield, UCD, Dublin 4, Dublin, Ireland

Ireland has been renowned over the years for its ability to produce horses with its limestone soils, mild climate and the natural affinity that Irish people have for the horse. These include both Thoroughbreds and Sport Horses (riding horses and ponies of all breeds and types intended to be used for recreational and competitive activities other than racing). Whilst the Thoroughbred sector is centrally organised, well documented and strategy driven, the sport horse sector in comparision is inherently fragmented, with limited documentation, profiling or strategic planning. The Irish Horse Board took the initiative to redress the situation by commissioning a comprehensive study to profile the industry and its activities, with a view to feeding into a strategic planning process. A major survey was conducted to obtain detailed information on and views of participants in the sport horse industry, with over 1,550 responses (21% return rate). This was supported by questionnaires completed by industry organisations, analysis of available industry databases and consultations with industry stakeholders. This paper will provide an overview of the profile of the Irish Sport Horse Industry with particular emphasis on the sport, leisure and business sectors of the industry.

Session 36 Theatre 3

Factors affecting the sale price of sport horse foals at auction

K.D. Hennessy[1] and K.M. Quinn[2], [1]University College Dublin, Centre for Sports Studies, Woodview, Belfield, UCD, Dublin 4, Dublin, Ireland, [2]University College Dublin, School of Agriculture, Food Science and Veterinary Medicine, Agriculture Builiding, Belfield, UCD, Dublin 4, Dublin, Ireland

Sport horse breeders who sell their foals at auction will generally seek to maximise the return on their investment. Foals have variable phenotypic and genetic characteristics, which differentiate them and impact on the potential price they may achieve at auction. Previous research on the variable characteristics affecting equines (predominantly yearlings) sold at auction has focused mainly on the thoroughbred sector. Variables found to be significant included month foaled, sire progeny performance, racing performance of dams, sires and their progeny, and gender where male yearlings achieved better prices than did female yearlings. Only limited research on the variables affecting price at auction has been conducted on the sport horse sector. The purpose of this study was to identify the variables, which significantly affect sport horse foal prices at auction, over an eight-year time frame. Data analysed matched sale price achieved with information provided by the vendor in the sales catalogue, additional information on dam and sire breeding lines and breeding values were sourced. A number of variable characteristics that have a significant impact on price achieved were identified along with changing trends within the market place.

Session 36 Theatre 4

Sport horse and pony breeding in Ireland

K.M. Quinn[1,2], K. Hennessy[3], D.E. Machugh[1], D. Feely[2], N. Finnerty[2] and P.O. Brophy[1], [1]University College Dublin, School of Agriculture, Food Science & Veterinary Medicine, Belfield, Dublin 4, Ireland, [2]Irish Horse Board, Block B, Maynooth Business Campus, Maynooth, Co Kildare, Ireland, [3]University College Dublin, Centre for Sports Studies, School of Public Health & Population Science, Belfield, Dublin 4, Ireland

Ireland has a longstanding tradition of breeding both race and sport horses. The number of sport horses in Ireland is estimated to be 110,000. This equates to 27.5 sport horses per thousand people and makes Ireland the most densely sport horse populated country in Europe. There are approximately 27,500 sport horse broodmares in Ireland with 22,000 mares involved in the breeding of studbook registered sport horse foals. The main Irish studbooks involved in registering sport horses and ponies are the Irish Sport Horse (~ 6,500 foals per year), the Connemara Pony (~ 1,800 foals per year), the Irish Draught (~ 600 foals per year), the Irish Piebald & Skewbald (~ 500 foals per year), the Irish Pony Society (~ 300 foals per year) and the Kerry Bog Pony (~ 20 foals per year). Studbook foal registrations in Ireland increased by 80% from 2000 to 2005 and are expected to increase further. Many breeders operate on a small scale and own, on average, 3.75 broodmares. Fifty-eight percent of broodmares were covered in 2004 and only 45% of broodmares foaled in 2005. Irish sport horse breeders aim to breed for a wide variety of markets, with 69% aiming to breed horses suitable for more than one equestrian discipline.

Session 36 Theatre 5

Equestrian federation of Ireland coaching programme

A. Corbally, Equestrian Federation of Ireland, Millennium Park, Naas, Co. Kildare, Ireland

In Ireland equestrian coaches train and qualify through the Equestrian Federation of Ireland (EFI) Coaching Programme. The program was initiated in 1998 to provide a coach education resource of the highest standard for Irish equestrianism. It is overseen by a committee of representatives from the EFI. Coaches qualify with a recognised qualification, at different levels, from Introduction to Coaching, through Levels 1, 2 and 3, to Level 4 (international). Currently, up to Level 3 courses are available in Ireland The courses are delivered by tutors, who include many of Ireland's top riders, and who themselves are experienced coaches. Tutors are selected on specific criteria and receive further training by the NCTC to enhance their communication and presentation style and ensure uniformity in course delivery. The programme gives coaches a thorough understanding of coaching principles and experience of applying these principles. This practical dimension is crucial in the formation of truly effective, well rounded coaches.

Session 37 Theatre 1

A global perspective of sustainability animal production

G. Bengisu[1], Ü Yavuzer[2] and A.,R. Öztürkmen[3], [1]Harran Universitiy, Faculty of Agriculture, Crop Science, Şanlıurfa, 63040, Turkey, [2]Harran University, Faculty of Agriculture, Animal Science, Şanlıurfa, 63040, Turkey, [3]Harran University, Faculty of Agriculture, Soil Science, Şanlıurfa, 63040, Turkey

We live in a world trouble by different aspects and different problems. Economic and social problems, war, violence, hunger, natural disasters are all around us. Sustainability is defined in many ways and sustainable animal production is a system which keeps the basis of life of future generations. Animal production is a complex effort, directly and indirectly, various segments of people in general. Sustainability advises customer on the risks and opportunities associated with corporate responsibility and sustainable development. Agriculture has changed due to new technologies, increased chemical use, mechanization, specialization and policies that supported maximizing production. The food system relationships agricultural production and animal production. Animal production strategies due to agricultural production. Sustainable agriculture is a sustainable animal. Researchers, farmers, farmworkers, consumers are parts of animal production. A more balanced sustainable food system undoubt due to sustainable animal production. A systems succesfull depend on interdisciplinary efforts in research and education. In view of the many interactions sustainability of animal production ethological, physiological, engineering, genetic and environmental effects.

Session 37 Theatre 2

Economically vulnerable cattle enterprises on Irish farms

W. Dunne, U. Shanahan and M. Roche, Teagasc, Rural Economy Research Centre, Malahide Road, Dublin 17, Ireland

Almost all farms in Ireland have cattle, either alone or in combination with other farm enterprises. Post decoupling, a cattle enterprise with a negative Market Based Gross Margin (MBGM) could be economically unsustainable. In 2004, just prior to decoupling, estimates derived from the Teagasc, National Farm Survey (NFS) show 19.4% or 22,000 Irish farms had such a cattle enterprise. In 2005, due to better beef markets the scale of the loss, and the number of farms declined to 15.2% (17, 000 farms). In both years, the main enterprise on these farms was either dairying, cattle, sheep or crops. In 2004, the negative MBGM cattle enterprises were over represented by those purchasing weanlings (by 30%) and male stores (130%) and selling as fat cattle (40%). With the exception of purchased animals and concentrate feed, most other costs, land area, herd sizes and stocking rates were about 15% lower than the overall average. The number of cows, dairy and sucklers, were under represented by about 50%, and good soil types by circa 20%. When other farm enterprises and the value of direct payments were included, the average family farm income was positive for all farm types. This indicates that the cattle enterprise as formulated could be vulnerable but the farm may be economically sustainable. The scale of the differences in 2005 was reduced due to: stronger beef markets, decapitalisation of the value of the previously coupled payments from cattle cohort prices, and some reformulation of farming activities.

Session 37 Theatre 3

Trends and seasonality of reproductive performance of dairy cows in Tunisia

M. Ben Salem[1], R. Bouraoui[2] and I. Chebbi[2], [1]INRA Tunisia, Forage and Animal Production, Rue Hédi Karray, 2049 Ariana, Tunisia, [2]ESA Mateur, Mateur, 7030 Mateur, Tunisia

Over the last two decades, increases in milk production per cow among dairy herds have been associated with significant decreases in the number of replacement heifers and the profitability of dairy herds. Most of the discussion about such decreases has centered on reproductive inefficiency as a major factor affecting herds profitability. Unfortunately, there were no longe time span studies and analyses to determine whether there is a trend for decreased reproductive parameters over time. The objective of this work is to evaluate reproductive performances of dairy cows in Tunisia, and to identify major factors that may affect it. Reproductive data on individual cows over a 13-year period were collected and reproductive indices were calculated. Measures of reproductive performance changed significantly over time. Calving and breeding seasons and parity were associated with major reproductive indices. Days to first service increased from a low of 63 d in 1993 to 113 in 2005. During that period, the calving to conception interval increased from 135 to 184 d. At the same time, average calving interval was about 422 d. It increased about 5.18 days per year. Services per conception increased, from 1.9 in 1994 to 2.5 in 2005. First service conception rate averaged 40%. Cows that calved during the spring had more open days (+43 d) and longer calving intervals than cows that calved during the fall or winter.

Effects of short-term cooling on pregnancy rate of dairy heifers under heat stress conditions

A.A. Moghaddam, M. Kazemi and M. Kamyab Teimouri, Free (Azad) University, Animal Science Department, Agriculture Faculty, Felestin Square, P.O.Box 366, 39187 Saveh, Iran

The influence of temporary cooling on pregnancy rate was evaluated during heat stress. Estrus was synchronized with two i.m. injections of PG, administrated 11 d apart. All heifers were housed in a shaded enclosed structure. After estrus detection, heifers were randomly divided into three groups: Control (C; n=30), Sprinkler (S; n=30) and Sprinkler & Fan (SF; n=30). Rectal temperatures (R.T) were measured at estrus observation, 0.5, 1 and 2 hours after estrus observation. Group C did not receive any treatment. Group S and SF were exposed to short-term cooling by operating sprinkler and sprinkler with fan, respectively, after estrus detection for 0.5 hour and after AI for another 0.5 hour. RT was similar between groups at the time of estrus. RT at half, one and two hours after estrus detection were significantly lower ($P < 0.05$) in SF (39.3 ± 0.2, 38.9 ± 0.2 and 38.7 ± 0.3, respectively) and S (39.5 ± 0.2, 39.4 ± 0.2 and 38.7 ± 0.3, respectively) groups compared to C (39.9 ± 0.2, 39.7 ± 0.1and 39.7 ± 0.2) group and in SF compared to S group. Pregnancy rate was higher ($P < 0.05$) in SF group (83.3%) as compared to S group (70%) and in both S and SF groups was higher ($P < 0.05$) as compared to C group (53.3%). Results indicate that temporary cooling of dairy heifers increased pregnancy rate.

Breeding objectives and strategy for autochthonous cattle breeds in Serbia

V. Bogdanovic, R. Djedovic and P. Perisic, Faculty of Agriculture, University of Belgrade, Animal Science, Nemanjina 6, 11080 Zemun-Belgrade, Serbia

There are two local, autochthonous cattle breeds in Serbia. Busha is small breed originated from *Bos brachyceros* and the other one is large Podolian cattle originated from *Bos primigenius*. These breeds dominated 50-60 years ago, but they were replaced over the time by more productive cattle, e.g. Simmental or Holstein. According to FAO criteria Busha and Podolian belong to seriously endangered cattle breeds. Number of pure-bred Busha and Podolian cattle in Serbia is very small and do not exceed a couple of hundreds. Breeding objectives and strategy are focusing on sustainable improvement of milk and beef traits and avoiding inbreeding problems. Milk production is of primary importance for Busha cattle while beef production traits and robustness are the most important for Podolian. Average milk yield of Busha is about 1000 kg milk with 4.4% milk fat and 3.1% protein. Milk from Busha cattle is usually used for producing traditional milk products. On the other side, Podolian is very suitable for beef production under extensive conditions. Average daily gain is about 700-900 g while adult body weight of bulls and cows are 800-1000 kg and 500 kg, respectively. Average height at withers, body length and circumference are about 124 cm, 154 cm, 188 cm, respectively.

Session 37 Theatre 6

Beef cattle breeding programmes for sub-saharan Africa

T.O. Rewe[1], P. Herold[1], A.K. Kahi[2] and A. Valle Zárate[1], [1]Institute of Animal Production, Hohenheim University, Stuttgart, Germany, [2]Department of Animal Sciences, Egerton University, Njoro, Kenya

The objective of this paper is to review breeding technologies vital for breeding programme development in sub-Saharan Africa while considering indigenous cattle genetic resources for beef production. Livestock breeding programmes are described based on the general definition modified here to refer to a procedural combination of technologies aimed at genetic improvement of performance, conservation or maintenance of breed integrity while considering biological and socio-economic aspects of the production system. Beef cattle are important when considering the growing demand for meat. The growth in beef output over the years in most of sub-Saharan Africa is attributable to expansion in herd sizes rather than genetic improvement, a phenomenon that cannot be sustained with the current increasing constraints on land. Sub-Saharan Africa is home to a large population of indigenous cattle, however, few examples of successful breeding programmes for them exist. In this paper, technological adjustments and consideration of target group involvement in livestock breeding programmes are given special attention as they are regarded important for low- and medium-input livestock production systems. Breeding programmes are suggested in sub-Saharan Africa within the concept of regional genetic improvement programmes under the control of stakeholders comprising of breed societies, government, and national agricultural research systems.

Session 37 Theatre 7

Different treatments of linseed in the fattening of culled cows and young bulls

V. Robaye, O. Dotreppe, J.F. Cabaraux, I. Dufrasne, L. Istasse and J.L. Hornick, Faculty of Veterinary Medicine, Animal Production - Nutrition Unit, Bd de Colonster, B43, 4000 Liege, Belgium

Linseed is commonly used to increase n-3 content in meat as it is an important source of C18:3n-3. However, due to its small size, linseed must be processed before being offered to cattle. The aim of this study was to compare the inclusion of crushed, flaked and extruded linseed in a fattening diet of Belgian Blue double-muscled young bulls (n=20) and culled cows (n=16). The animals were divided in 4 groups and fattened on an *ad libitum* concentrate based diet. There were no linseeds in the control group while the linseeds were crushed, flaked or extruded in the 3 treated groups. At slaughter, 3 muscles (Longissimus thoracis, Semi tendinosus and Rectus abdominis) were sampled for fatty acids (FA) profiles. Linseed increased the C18:3n-3 content in bulls and cows meats in a same range (respectively 12.90 *vs* 17.85 and 13.95 *vs* 21.03 mg/100g fresh meat (FM); $P < 0.001$). No significant differences were observed between the 3 linseed treatments and there were either no significant differences between the 4 groups in terms of total FA content in meat. Although the n-3 content was slightly higher in cow meat (51.89 *vs* 34.15 mg/100g FM; $P < 0.001$), bull meat was much leaner (389.66 *vs* 787.00 mg FA/100g FM; $P < 0.001$). It was concluded that the use of linseed significantly increased the C18:3n-3 content in bull and cow meats with no differences between the 3 treatments.

Session 37 Theatre 8

Feeding of extruded rape cakes and extruded fullfat soya at dairy cows

J. Třináctý[1], M. Richter[1] and P. Homolka[2], [1]Agrovýzkum Rapotín, Ltd, Dep. Pohořelice, Vídeňská 699, 691 23, Czech Republic, [2]Research Institute of Animal Science, Dep. Animal Nutrition, Přátelství 815, 104 00 Praha 10, Czech Republic

The aim of the trial that was carried out on four dairy cows (2x2), was to evaluate the influence of extruded rape cakes (cakes group) and extruded fullfat soya (soya group) on milk performace and nutrient consumption. The design of trial was as follows: four periods of 42 days with rotation between periods. The ration consisted of maize silage, meadow hay and mixture based on either extruded cake or extruded soya. Cows were fed twice a day, roughage was fed *ad libitum*, mixture was dosed according to milk yield. The nutrients consumption was evaluated according to INRA system. Preliminary evaluation showed that consumption of DM, NEL, PDIN and PDIE in cakes and soya group, respectively, was 18.9 and 18.7 kg, 123.7 and 124.2 MJ, 1.47 and 1.50 kg, 1.44 and 1.62 kg per day. Milk yield was 22.6 *vs* 23.6 kg/day. Except of DM and NEL values all differences were significant ($P < 0.05$). We can conclude that feeding of extruded fullfat soya had positive effect on performance of dairy cows in comparison with extruded rape cakes. *Study supported by MSM2678846201.*

Session 37 Theatre 9

The role of coat colour varieties in the preservation of the Hungarian Grey cattle

A. Radácsi, L. Czeglédi, Z. Szendrei, B. Béri and I. Bodó, University of Debrecen, Institute of Animal Science, Böszörményi Street 138., 4032 Debrecen, Hungary

The Hungarian Grey cattle is characterized by a great variety of coat colours, which vary within the given breed character by age and are affected by the season, as well. This great variety of coat colours form part of the world's genetic heritage. In order to maintain genetic diversity, these qualitative traits that have no economic value at the moment should also be conserved. Research has been done on the largest Hungarian Grey cattle stock, at the Hortobágy Society for Nature Conservation and Gene Preservation. Coat colour measurements were carried out by using a Minolta Chromameter CR-410. The Hungarian Grey cattle belongs to the Podolian group of cattle, which is characterised by the fact that calves are born reddish-coloured and turn into grey at the age of 4-6 months. Proportions of the different shades of grey colour and the mean L*a*b* values of the given varieties were determined in both sexes. The colour of the adult bulls is more diversified. Generally, the front part of the body is darker which is proved by the lower mean L* values (L^*_{neck}: 29.47, L^*_{side}: 47.88 and $L^*_{thigh\text{-}croup}$: 46.13; the L* value shows the lightness of the colour on a scale from 0 to 100, the lower values indicating the darker colour). Several genes are involved in the forming and regulation of coat colour. However, the major regulator is the melanocortin-1-receptor (MC1R). MC1R genotypes of the Hungarian Grey cattle breed were determined.

Session 37 Theatre 10

Establishing a health monitoring system for cattle in Austria: first experiences

C. Egger-Danner[1], B. Fuerst-Waltl[2], W. Holzhacker[3], R. Janacek[4], J. Lederer[1], C. Litzllachner[5], C. Mader[4], M. Mayerhofer[1], J. Miesenberger[1], W. Obritzhauser[3], G. Schoder[4] and A. Wagner[1], [1]ZAR-ZuchtData, Dresdner Straße 89/19, 1200, Austria, [2]University of Natural Resources and Applied Life Sciences Vienna, Gregor Mendel Straße 33, 1180, Austria, [3]Chamber of Veterinaries, Biberstraße 22, 1010, Austria, [4]Animal Health Organisation, Landhausplatz, 3109, Austria, [5]Chamber of Agriculture, Schauflergasse 6, 1014, Austria

A project to establish an Austrian wide health monitoring system for cattle is currently being implemented. Within the project diagnostic data, which have to be documented by law are standardised and recorded. All farms under performance recording are free to join. To increase the health status by management measures health reports are elaborated and available for participating farmers and veterinarians. Responding to an increasing interest in health issues, already 8,500 farms have been participating by the beginning of the year 2007 with more farms continuously joining. In some federal states the percentage is already above 70% of farmers under performance recording. At the same time the recording of diagnostic data has started. For their recording in daily work, the motivation and awareness of farmers and veterinarians are essential. First experiences show that the recording of diagnostic data follows an earlier published model about adoption and diffusion of innovations. An overall aim of the project is the development of a genetic evaluation for health traits for the main Austrian cattle breeds.

Session 37 Theatre 11

Analysis of influences on uneven pressure distribution between hind inner and outer claws in dairy cows

C. Bellmer, H. Hamann and O. Distl, Institute for Animal Breeding and Genetics, Buenteweg 17p, 30559 Hannover, Germany

Uneven pressure distribution underneath the bovine claws is one of the major factors causing claw lesions and productivity losses in dairy cows. The objectives of this study were to determine the influences causing the overload of the hind outer claws in dairy cows. We recorded the punctual pressure distribution using an electronic measurement system in 35 female cattle starting in the third month and ending in the 20-24th month of life and in more than 30 dairy cows during their first, second and third lactation. The hind medial claws showed larger area of ground surface, weight load and pressure per cm^2 in calves and young heifers than in cows. The relative area of ground surface and the relative weight shifted from the medial to the lateral claw with increasing age. After first calving, the relative area of the ground surface of the lateral claws was 57% and the lateral claws had to bear 58% of the weight load. Claw trimming could reduce the relative weight load to 48% underneath the lateral claws which is less than the ratio measured in pregnant heifers. The effect of functional claw trimming was removed in cows after four months. Size and shape of the udder as well as conformation of hind limbs and moving patterns were recorded to investigate their influence on this permanent load shift from the hind medial to the lateral claws in dairy cows.

Session 37 Theatre 12

The value of live animal muscularity scores and carcass grades as indicators of carcass composition

M.J. Drennan and M. Mc Gee, Teagasc, Beef Production Department, Grange Beef Reseach Centre, Dunsany, Co. Meath, Ireland

The value of live animal muscularity scores and carcass conformation (CS) and carcass fat (FS) scores as indicators of carcass composition were determined using 48 bulls and 37 heifers slaughtered at 458 and 612 days of age, respectively. Muscularity scores (MS) were assigned at weaning (8 months) and pre-slaughter. Carcasses were mechanically graded according to the EU Carcass Classification Scheme and carcass meat, fat and bone were determined. Carcass weights (kg), CS and FS (scales 1 to 15) were 323 (sd 45.0), 11.7 (sd 1.35) and 6.1 (sd 1.51) for bulls and 268 (sd 26.4), 10.1 (sd 1.22) and 8.9 (sd 1.85) for heifers, respectively. Correlations of carcass meat proportion with MS at weaning and pre-slaughter were 0.10 and 0.39 for bulls, and 0.35 and 0.65 for heifers, respectively. Correlations of carcass meat proportion with CS and FS were 0.57 and -0.73 for bulls and 0.52 and -0.68 for heifers, respectively. Regression of CS and FS on carcass meat and fat proportion for bulls were: meat (g/kg) = 706 + 8.9 (1.78) CS -11.9 (1.60) FS; fat (g/kg) = 60.0 - 4.0 (1.40) CS + 12.4 (1.25) FS. Corresponding equations for heifers were meat = 723 + 8.1 (3.17) CS – 9.7 (2.09) FS; fat = 34 – 2.9 (2.63) CS + 9.9 (1.73) FS. Carcass value of bulls and heifers increased by 3.5 and 4.4 c/kg, respectively per unit increase in CS. The corresponding figures for FS were -5.4 and -3.8. In conclusion, increasing MS and CS increased meat yield.

Session 37 Theatre 13

Digital Image Analysis for prediction of carcass weight of different breeds of slaughtering beef cattle using some carcass measurements

Y. Bozkurt, S. Aktan and S. Ozkaya, Suleyman Demirel University, Faculty of Agriculture, Department of Animal Science, Cunur-Isparta, 32260, Turkey

The objective of this study was to use digital image analysis system in order to predict carcass weight of different breeds of slaughtering beef cattle by using some carcass measurements and to develop prediction models. A total of 50 digital images and carcass measurements were taken such as carcass weight (CW), carcass area (CA), carcass length (CL) and carcass depth (CD) from different breeds of beef cattle namely, Holstein, Brown Swiss and their crosses. For prediction of carcass weight, CA was found to be the best predictor compared to CL and CD. Linear, quadratic and cubic effects of predictors were examined and R^2 values of CA were higher than those of other measurements for all breeds and were 85.9, 72.9 and 84.1% for Holstein, Brown Swiss and crossbreds respectively. When considering correlation between CW and other measurements, correlation values of CA were greater than the rest for all breeds. The correlation coefficients between CW and CA for were 0.93, 0.85 and 0.92 for Holstein, Brown Swiss and crossbreds respectively and found statistically significant ($P < 0.05$). The results indicated that digital image analysis system could be used to predict CW.However, there is still a need for further studies in order to develop better techniques to use it for prediction.

Beef carcass classification in Hungary

T. Cserhidy, S.Z. Simai, P. Marlok and D. Mezőszentgyörgyi, Cental Agricultural Office, Keleti Károly u. 24., 1024 Budapest, Hungary

National ministerial decrees from 2003 have reregulated the carcass classification system in Hungary. These decrees obligated the categorizations of every slaughter animal in every slaughterhouse not depeding its capacity. Previously slaughterhouse – which were exporting to EU member states – used carcass classification systems only. Beef carcass classification is made by 43 slaughterhouses which have licensed classification place by the Central Agricultural Office and licensed person for classification hired by one of the classification agencies. The identifical data of the slaughtered animals are sent on-line to the central database called OVMR. This database is divided to three parts: the slaughterhouse module, the communication module and the central module. By using the modules, the facts of the slaughterings are reported on-line automatically, because the OVMR and the Cattle identifical and registration (I&R) systems were connected to each other previously. Data checking during the data processing every mistake can be filtered, this way mistakes can be corrigated. After joining the EU, the condition of claiming and clearing the subsidy after classified bovines is the registration in OVMR. As a result of this process there were 24843 slaughtered bulls in 2004 that received special beef premium. This number was 47972 in 2005, and 65941 in 2006. The carcass classifications, the collecting and processing of data are supervised by the inspectors of Cental Agricultural Office regarding EU decrees.

Market quality of cow´s meat in conditions of Slovakia

K. Zaujec[1], J. Mojto[1] and P. Polak[2], [1]Slovak Agricultural Research Cetre, Department of Animal Product Quality, Hlohovska 2, 949 92 Nitra, Slovakia (Slovak Republic), [2]Slovak Agricultural Research Cetre, Department of Animal Breeding, Hlohovska 2, 949 92 Nitra, Slovakia (Slovak Republic)

In 2006 cows formed the majority (58%) of cattle offered to slaughter houses in Slovakia, caused by lower number of fattening bulls being slaugtered. These days bulls are exported as live animals abroad. This results in meat of lower quality available for retail sale. As a consequence today cow's meat appears for sale more often in wholesale as well as in retail sale. The lower buying price for cows compared to bulls is partly reflected in the price for meat in the shop. But the quality of meat plays a significant role for the consumer. In an experiment with cows (n=47), that were randomly selected from different slaughter houses, samples of muscle were taken from m. longissimus dorsi to study some qualitative parameters. Age of cows at slaughter was 2 – 8 years; average weight at slaughter was 249.07 kg. The largest coefficient of variation (54.51 %) was observed in content of intramuscular fat, its average value being 4.37 g.100g-1. Resulting pH value was 5.96 (48 hours p.m.); we found only low occurrence (4.25 %) of DCB meat (dark, cutting beef); meat colour (L*-lightness) was 28.87% and degree of marbling in meat 6.82 (USDA scale). During the panel examination of meat samples we found out that cows' meat is less characteristic in taste compared with meat of young bulls. Shear force (Warner-Bratzler) in meat samples was 11.25 kg.

Session 37 Theatre 16

Estimation of demand function for different types of meat in Iran: application of cointegration

J. Azizi and A.R. Seidavi, Islamic Azad University Rasht branch, Rasht, 4185743999, Iran

In this study by using the Almost Ideal Demand System (A.I.D.S), the demand function for different types of meat in urban and rural societies have been estimated and price as well as non-price elasticities of Marshall and Hicks demand function were investigated, Application of cointegration theory for determining the appropriate demand function was also investigated: Results obtained from this study indicated that during the period under study, household budget allocated to red meat have been decreased both in urban and rural societies, whereas during the same period household budget allocated to the purchase of chicken and fish have been increased. Furtheremore, the share of budget allocated to the purchase of fish in the rural societies increased first but later decreased. Price elasticities of different types of meat in urban as well as rural areas showed that during the period of study the use of price variable for the modification of consumption pattern have not been effective. This indicates that for the modification of consumption pattern, price variable has not been an effective factor. Therefore other variable such as population growth rate is of great significance that should be taken into account. Moreover, because of using time series data in this study, first the unit root in the model's variables wasexamined and then long-term relation of data was investigated. The results of study has revaled the existence of cointegration in this regard.

Session 37 Poster 17

Evaluation of growth ability and muscling in breeding bulls of chosen beef breeds

A. Jezkova[1], L. Stadnik[1], J. Dvorakova[1], F. Kolarsky[1] and F. Louda[2], [1]Czech University of Agriculture Prague, Kamycka 129, Prague 6–Suchdol, 165 21, Czech Republic, [2]Research Institute for Cattle Breeding, Vyzkumniku 267, 788 13 Vikyrovice, 788 13, Czech Republic

Basic physical measures, layer of subcutaneous fat and area of musculus longissimus thoracis et lumborum (MLTL) of breeding bulls (21 bulls of Aberdeen Angus - AA, and 19 bulls of Hereford - HE) at the age 300, 330, 360, 390 a 420 days were measured. Basic physical sizes were height in hips and width of lumbar region on 1st and 6th vertebra measured via tape and Wilkinson´s trammel. Layer of subcutaneous fat (height) and MLTL (height and area) on 1st and 6th vertebra were measured by ultrasound. Data were analyzed by the statistical program SAS STAT 8.0-GLM, by the general linear model. Higher physical measures and also height and area of MLTL at any age in AA bulls were detected ($P < 0.001$; $P < 0.01$). Higher height of subcutaneous fat (difference 0.05-0.15 cm) at the age 360, 390 and 420 days in HE bulls were detected ($P < 0.001$; $P < 0.05$). Physical measures are correlated to one another; intermediate correlation coefficients between heights of subcutaneous fat on 1st and 6th vertebra and between heights of MLTL on 1st and 6th vertebra were found out. Differences in evaluated measures between AA and HE bulls were lower than expected by reason of feed ration and good growth ability of breeding bulls in rearing house.

Session 37 Poster 18

Comparison of growing and finishing performance of different breeds of feedlot beef cattle grown under the Mediterranean conditions

Y. Bozkurt, Suleyman Demirel University, Faculty of Agriculture, Department of Animal Science, Cunur-Isparta, 32260, Turkey

This study was aimed to compare the performance of different breeds of beef cattle during both growing (GP) and finishing periods (FP) under the Mediterranean type of climatic zone. Data comprised of total of 106 beef cattle, including Holstein (11), Brown Swiss (27), Simmental (8) cattle as pure bred (PB) and Boz (12) and Gak (48) as local breeds (LB) with initial average weights of 202, 194, 210, 203 and 220 kg respectively. There were statistically significant ($P < 0.05$) differences in daily liveweight gains (DLWG) of cattle at both GP and FP. While there were no statistically ($P > 0.05$) significant differences in performance between Holsteins (0.90 and 0.68 kg/day for GP and FP respectively), Brown Swiss (0.87 and 0.66 kg/day) and Simmental (0.92 and 0.75 kg/day) cattle and between Boz (0.60 and 0.50 kg/day) and Gak (0.56 and 0.45 kg/day) cattle themselves in both feeding periods, Simmentals tended to perform better than the rest for both periods. The performance of PB cattle was greater than LB cattle in both periods. Overall DLWGs of animals in GP (0.70 kg/day) was statistically higher ($P < 0.05$) than those of FP (0.56 kg/day). The results indicated that since growing and finishing performance of PB cattle were greater than LB cattle any of the PB cattle could be recommended to the feedlot beef systems under the Mediterranean conditions.

Session 37 Poster 19

Examination of slaughter results of Hungarian Simmental paternal half-sib bulls

J.P. Polgár[1], I. Füller[2], M. Török[1], S.Z. Bene[1], Á Harmath[2] and B. Huth[2], [1]Pannon University Georgikon Faculty of Agriculture, Keszthely, Deák Str.16, H-8160, Hungary, [2]Association of Simmental Breeders, Bonyhád, Zrínyi Str. 3, H-7150, Hungary

The examination for determine meat breeding value was started due to the interest of The Federation of Hungarian Simmental Breeders. In every generation 12-15 male offspring of 6-12 sires are fattened from the age of 120day to slaughtering. During the fattening period the feeding of animals is based on *ad libitum* maize silage and concentrate on the ration of 100kg live weight / 1 kg concentrate. At the end of fattening the slaughter was done in the slaughter house of Kóborhús Sc. The data of slaughter and carcass weights and boning out results are collected. Carcasses are judged on the base of EUROP. The database is examined by SPSS 9.0 and Harvey's Least Square Maximum Likelihood Computer Program. In the project the data of 352 slaughtered male offspring have already been calculated. Daily gain (referred on the age of slaughter) and daily gain under the fattening period are found to be 1180 g/day and 1250 g/day, respectively. The overall mean value of killing out percentage is 59.11%, the EUROP quality is R^+ and fat score is 2.56. According to these findings, the results of gain, killing percentage and EUROP quality were significantly influenced by sire.

Session 37 Poster 20

Adaptation stress monitoring in Limousine suckler cows through the behaviour score in the cattle crush and blood indices

P. Nowakowski, A. Rzasa, A. Dobicki and A. Zachwieja, University of Environmental and Life Sciences in Wroclaw, Institute of Animal Breeding, Chelmonskiego 38c, 51-630 Wroclaw, Poland

Research was performed on 50 suckler cows with different time of adaptation to new environment (1 week – Group 1; 3 weeks – Group 2; 12 months – Group 3). Behaviour scored in a cattle crush covered nervousness and vocalisation (1 = low to 3 = high). Simultaneously blood was sampled for monitoring of the levels of: glucose (Glu), hematocrit (PCV), fibrinogen (Fb) and serum amyloid-A (SAA). Significant relation between Glu and increased nervousness behaviour in cattle crush ($r = 0.499$) was stated. High vocalization followed high nervousness of cows ($r = 0.361$). The average Glu level in the Group 1 exceeded upper reference level for cattle and was higher ($P < 0.01$) than in two other groups. High PCV was stated in animals of Group 1 and 2. Fb level for all animals in the Group 3 was within the reference values while 15 and 20% of cows from Group 1 and 2 respectively were characterized by values exceeding norms. Group 1 animals was characterized by high SAA values; only 1 cow had SAA below 10µg/ml, with the maximum stated at 105.9 µg/ml. 28% of cows from Group 2 had high SAA values while cows in the Group 3 were characterized by the negligible ones. Simple behavioural scores in conjunction with blood indices may be a useful tool for monitoring level of adaptation of cows to the new environment. *Supported by the Polish Committee for Scientific Research, grant No. 2 P06Z 063 26.*

Session 37 Poster 21

External measures and matering traits of podolian cattle

S. Stojanovic, Ministry of Agriculture, Nemanjina 22-26, 11000 Belgrade, Serbia

The main activities on management and conservation of podolian cattle in the last five years were: organizing and developing planed reproduction, establishing the herd-book, studying on the external measures and matering traits. This paper shows results of exploring 60 cows, 21 heifers of one year old and 4 bulls. For cows the following average value of linear measures have been established: height of wither is 126.02 cm, height of rump 128.64 cm, the body length 160.66 cm, width of breast 41.92 cm, depth of breast 67.19 cm, heart girth 189.08 cm, and cannon circumference 19.25 cm. One year old heifers have the following average linear measures: height of wither 105.15 cm, height of rump 108.24 cm, body length 117.86 cm, width of breast 30.33 cm, depth of breast 49.86 cm, heart girth 141.29 cm and cannon circumference 15.17 cm. Average values of linear measures for bulls are: height of wither 126.5 cm, height of rump 128.8 cm, body length 158.8 cm, width of breast 46.8 cm, depth of breast 68.5 cm, heart girth 194.3 cm and cannon circumference 21.3 cm. An average values of cattle age at the first mattering was 29.3 months, the first calves was 33.2 months, duration of gestation was 9.3 months, as value between calve intervals is 11.87 months and service period is 2.49 months. The percentage of alive born calves is 94.38 %, the miscarriage percentage 2.04 %, and the percentage of death born calves is 3.57 %.

Session 37 Poster 22

Weaning results of Angus calves in Hungary

F. Szabó[1], J. Márton[2] and S. Bene[1], [1]University of Pannonia, Animal Science and Production, Deák F. str. 16., 8360 Keszthely, Hungary, [2]Hung. Hereford, Angus Br. Association, Dénesmajor 2., 7400 Kaposvár, Hungary

Weaning performance of 2451 Angus calves (1283 male and 1168 female) born between 1989 and 2002 from 930 cows mated with 63 sires were analysed in Angus Beef Cattle Breeding and Dealing Ltd's farm in Adony. The aim of the study was to evaluate the effect of environmental factors on weaning traits. Dam's colour (black, red), age of cows, year of birth, season of birth and sex of calves as fixed, while sire as a random effect was treated. Data were analysed with Harvey's (1990) Least Square Maximum Likelihood Computer Program. The overall mean value and standard error of weaning weight, preweaning daily gain and 205-day weight were 186±3.94 kg, 840±18.81 g/day and 212±3.61 kg, respectively. The average age of the analysed calves was 193 days (SD=42 days). The results of the examination show that weaning weight, preweaning daily gain and 205-day weight increased with increasing dam's age as far as the seven year age of cows (the maximum were 192±4.29 kg, 875±20.71 g/day, 222±4.11 kg). As for the season effect the calves born in winter and spring were heavier (205-day weight were 217±3.94 and 219±3.68 kg) than that of born in the other seasons. Male calves were heavier than females significantly (the difference was 16 kg) only in 205-day weight.

Session 37 Poster 23

Selenium dose response in growing beef receiving organic selenium: Sel-Plex®

D.T. Juniper[1], D.I. Givens[1] and G. Bertin[2], [1]University of Reading, Earley Gate, Reading RG6 6AR, United Kingdom, [2]Alltech EU Reg. Dept., 14 place Marie-Jeanne Bassot, 92300 Levallois-Perret, France

A Latin Square trial (5 diets, five 4-wk periods) examining effects of diets with 0.15 (BR1), 0.30 (BR2) or 0.45ppm (BR3) of organic selenium (Sel-Plex®,CNCM I-3060), or 0.30ppm (BR4) of sodium selenite, or a negative control (BRO) on animal performance and blood parameters was done on 30 Holstein males. Cattle were randomly allocated to diets based on LW (initial mean LW 254.8 kg) and LW gain of previous 4 wks. Individual *ad libitum*. TMR was fed (corn silage 55 %; grass silage 13 %; wheat 15%; soybean meal 15%, DM basis). Ingestion was recorded daily in the last wk of each period, LW on 2 consecutive days at start and end of each period. Blood samples were taken the day before 1st period and on last day of each period. DMI, LW gain or feed conversion didn't differ between diets ($P > 0.05$). There was a positive and linear ($P < 0.05$) response in whole blood Se concentrations to the addition of Sel-Plex®, with blood Se values increasing by 3.8µg/mL for each additional 0.1ppm Se from Sel-Plex®. Both Se content as selenomethionine (Se-Met) and proportion of total Se in Se-Met from blood samples of period 1 were increased between BRO and BR1. At higher Sel-Plex® concentrations (BR1-BR3), there was a marginal increase in Se content as Se-Met but the proportion of total Se in Se-Met remained constant. There were no significant differences ($P > 0.05$) between treatments in any other measured blood parameter

Session 37 Poster 24

Effect of forage to concentrate ratio on slaughter traits and fatty acid composition of Hungarian Simmental young bulls

G. Hollo[1], K. Ender[2], K. Nuernberg[2] and I. Hollo[1], [1]University of Kaposvár, Guba S 40., 7400 Kaposvár, Hungary, [2]Research Institute for the Biology of Farm Animals, Wilhelm Stahl Allee 2, 18196 Dummerstorf, Germany

30 Simmental bulls (initial weight 300.07+43.78 kg, age: 274.57+19.73d) were divided into three groups of 10 each according to different maize silage to concentrate ratio (A:67:33; B:75:25; C:80:20 based on DM). The low concentrate groups (B, C) received linseed supplemented concentrate during fattening period. The average daily gain and the slaughter weight showed not significant differences among groups. Carcass conformation of all groups was assessed as R, while the lowest fattening condition was received by group C. Moreover group C had the lowest amount of kidney fat too. Dressing percentage (A: 58.62%; B: 58.82%; C: 58.93%) and meat (A:71.68%; B:71.85%; C:71.81%) as well as fat (A: 8.58%; B: 8.65%; C: 9.23%) were higher, whilst bone (A:18.65%; B:18.41%; C:17.91%) and tendon (A:1.15%; B:1.10%; C:1.08%) proportion were lower in groups B and C. In groups B and C increased the intramuscular fat content in all muscles (longissimus A:2.00; B:2.30; C:2.47; semimembranosus A:1.20 B:1.62; C:1.78, psoas major A:3.16 B:4.10; C:3.49). Groups B and C showed lower palmitic acid content ($P < 0.001$) in semimembranosus, and higher percentage of the linolenic and eicosapentaenoic acids in all muscles. The beef from groups B and C contained more n-3 fatty acids ($P < 0.05$), and thus the n-6 to n-3 fatty acids ratio ($P < 0.001$) changed favourably.

Session 37 Poster 25

How fast are dietary carbon and nitrogen incorporated into bovine muscle?

B. Bahar[1,2], A.P. Moloney[3], F.J. Monahan[1], A. Zazzo[2], S.M. Harrison[1,2], C.M. Scrimgeour[4], I.S. Begley[5] and O. Schmidt[2], [1]University College Dublin, School of Agriculture, Food Science & Veterinary Medicine, Dublin 4, Ireland, [2]UCD, School of Biology & Environmental Science, Dublin 4, Ireland, [3]Teagasc, Grange Beef Research Centre, Dunsany, Co. Meath, Ireland, [4]Scottish Crop Research Institute, Invergowrie, Dundee, DD2 5DA, United Kingdom, [5]Iso-Analytical Ltd., Millbuck Way, Sandbach, Cheshire, CW11 3HT, United Kingdom

Stable isotope analysis (SIA) of animal tissues can provide information on the dietary history of meat animals. We used SIA to investigate the turnover of carbon and nitrogen in bovine *Longissimus dorsi* and *Psoas major* muscles. The diets of five groups (n=10 each) of continental crossbred beef cattle were switched from a control diet containing barley and unlabelled urea to an isotopically distinct diet containing maize, ^{15}N labelled urea for 168, 112, 56, 28 and 14 days pre-slaughter. Ten animals fed the control diet for 168 days served as an experimental control. Samples of *L. dorsi* and *P. major* muscles were collected at 24 h post-mortem and processed (de-fatted) for SIA. Isotopic equilibrium was not reached in either tissue for $d^{13}C$ or $d^{15}N$ after 168 days of feeding the isotopic diet. The slow turnover of C and N was reflected in half-lives of 151 and 157 days for *L. dorsi* and134 and 145 days for *P. major*, respectively. It is concluded that bovine *L. dorsi* and*P. major* tissues have similar slow turnover rates of C and N which has implications for authenticating long-term dietary changes in cattle.

Relationship of conformation and fat scores with carcass traits

S.B. Conroy[1,2], M.J. Drennan[1] and D.A. Kenny[2], [1]Teagasc, Grange, Beef Research Centre, Dunsany, Co Meath, Ireland, [2]School of Agriculture, Food Science and Veterinary Medicine, University College Dublin, Belfield, Dublin 4, Ireland

Carcass grades are important determinants of carcass price. The aim was to quantify the relationship between carcass grades for conformation and fatness (scale 1 to 15) obtained by mechanical grading and killing-out rate (KO), meat, fat and bone proportions, proportion of high value and carcass value. Bulls (n=75) were slaughtered at a mean age of 458 (sd=41) days and an average liveweight of 575kg (sd=85).The right side of each carcass was dissected into meat, fat and bone. Carcass value (c/kg) was then calculated as the sum of the commercial values of each fat trimmed boneless cuts. Positive correlations (0.52 to 0.84) were obtained for carcass conformation score with KO, carcass meat proportion, proportion of high value cuts and carcass value and negative correlations with carcass fat and bone proportion (-0.40 to -0.89). For carcass fat score high positive correlations were obtained with fat proportion (r=0.61) but correlations with all other traits were low and negative. Regression analysis showed that carcass grades explained 0.69 to 0.80 of total variation in KO, carcass meat and bone proportion and carcass value, 0.55 for fat proportion and 0.28 for the proportion of high value cuts. It can be concluded that carcass grades were good predictors of meat and bone yield and carcass value, modest predictors of carcass fat and poor predictors of the proportion of high value cuts.

Session 37 Poster 27

Correlation of ultrasonic measured fat thickness and ribeye area to the certain values measured on slaughtered bulls

M. Török, J.P. Polgár, G. Kocsi and F. Szabó, University of Pannonia, Georgikon Faculty of Agriculture, Department of Animal Science and Husbandry, Deák Ferenc str. 16., H-8360, Hungary

The aim of this study was to test accurancy of measurements done by Falco 100 (Pie Medical) ultrasonic equipment. 10 Angus, 10 Hungarian Simmenthal, 10 Limousin and 10 Charolais fattening bulls were measured at the feedlot just before slaughtering. Fat thickness at the rump (P8) and ribeye area (REA) were realized from each animal. Slaughter and carcass weights were collected and carcasses were scored on the base of EUROP system. SPSS 9.0 for Windows was used for data processing. Average liveweight was 645±41.5 kg at Angus, 676±41.8 kg at Hungarian Simmental, 655±50.8 kg at Limousin and 694±42.3 kg at Charolais group at the measurement. REA measured with ultrasound was 102.9±8.9 cm^2, 102.7±10.4 cm^2, 111.2±9.6 cm^2 and 106.4±9.5 cm^2, respectively. P8 was 1.05±0.28 cm, 0.62±0.13 cm, 0.62±0.09 and 0.61±0.18 cm, respectively. Correlation between ultrasonic and carcass REA was r=0.74, 0.74, 0.94 and 0.8, respectively. Correlation between P8 and EUROP fat score was r=0.51, 0.73, 0.56 and 0.28, respectively. Overall correlation between ultrasonic and carcass REA was r=0.83 ($P \leq 0.01$), and between P8 and EUROP fat score was 0.69 ($P \leq 0.01$).

Session 37 Poster 28

Relationships between tissue thickness measured by ultrasound and beef carcass quality grading

P. Polák[1], L. Bartoň[2], J. Tomka[1], R. Zahrádková[2], E. Krupa[1] and D. Bureš[2], [1]Slovak Agricultural Research Centre, Department of Animal Breeding, Hlohovská 2, 949 92, Slovakia (Slovak Republic), [2]Research Institute for Animal Production, Department of Cattle Breeding, Přátelství 815, Praha – Uhříněves, 104 01, Czech Republic

The aim of investigation was to find out relationship among tissue thickness measured by ultrasound on live animals and carcass quality grading according EUROP system. Thickness of muscle and fat were measured on 5 positions of live bulls and carcass quality grading for carcass conformation and fatness of 6 genotypes were analyzed. Echocamera Aloka SSD 500 was used for ultrasound measurement. The most frequent grade was R2. According carcass conformation 2 carcasses were in E, 40 in grade U, 64 in grade R and 5 in grade P. The muscle thickness increased with increase of carcass grading almost on all scanned positions. On the other side, the fat layer on all positions of ultrasound scanning decreased with increase of grading for carcass composition. When the trend of grading for fatness was investigated, thickness of muscle decreased and fat cover increased with increase of grading for fatness. The highest correlation coefficient (r=0.33) between muscle thickness and carcass conformation classification was found out for musculus longissimus lumborum on the last lumbar vertebra. The highest correlation coefficient (r=0.30) between fat thickness and carcass grading for fatness was found out on the rump above ischium.

Session 37 Poster 29

Assessment of body composition in growing cattle by chemical analysis and Magnetic Resonance Imaging

U. Baulain[1], U. Meyer[2], S. Brauer[2] and H. Janssen[2], [1]Institute for Animal Breeding, Federal Agricultural Research Centre (FAL), Hoeltystr. 10, 31535 Neustadt, Germany, [2]Institute of Animal Nutrition, Federal Agricultural Research Centre (FAL), Bundesallee 50, 38116 Braunschweig, Germany

To improve feeding recommendations it is important to quantify changes in body composition during growth. Labour intensive determination of carcass composition by dissection or chemical analysis is not feasible as an experimental routine. Aim of this study was to evaluate the accuracy of Magnetic Resonance Imaging (MRI) for predicting composition of cattle carcasses at different age or weight, respectively. 32 female Holstein calves were reared on milk replacer, grass silage and concentrates. The calves and heifers were slaughtered at the age of 14 and 24 weeks (n=8+8) or 15 months at a live weight of 460 kg (n=16). The left carcass side was analyzed for water content, ashes, fat and protein. The other side was dissected into primal cuts which were scanned in a whole body tomograph to measure muscle, fat and bone volumes. In the group of 14 and 24 weeks old animals protein, fat and bone mass derived from chemical analysis could be estimated by MRI data with R^2=0.99, 0.94 and 0.99, respectively. Root mean square errors (RMSE) resulted to 136, 190 and 81 g. For 15 months old animals R^2 were 0.98, 0.86 and 0.95 (RMSE 162, 1312 and 330 g). MRI is suitable to determine protein, fat and bone mass of cattle carcasses. To establish more reliable prediction equations the number of observations per age group should be expanded.

Results of X-ray computer tomography (CT) examination of young calves in relation to slaughter traits

G. Hollo[1], I. Hollo[2], K. Ender[2], K. Nuernberg[1], J. Seregi[1] and I. Repa[1], [1]University of Kaposvár, Guba S. 40., 7400 Kaposvár, Hungary, [2]Research Institute for the Biology of Farm Animals, Wilhelm-Stahl-Alle 2., 18196 Dummerstorf, Germany

In this study the alteration of body composition of altogether 15 Hungarian Simmental male calves was performed using *in vivo* CT method at different age levels. The first and second scanning of whole body (from neck to hock) were made at the age of 7 days and 98 days resp. (average live weight was 43 and 85 kg, resp.). In both cases the muscle, bone, and fat distribution as a proportion of total area did not change significantly. At the same time, during the period between the two scanning dates the highest area incensement can be observed in case of the fat tissue, followed by muscle, bone and connective tissue. The average area of longissimus (LD) at 12.th rib increased from 15 to 23 cm^2. The animals were fattened until the average live weight reached 600 kg. After slaughter the right half carcass was dissected. The muscle tissue of whole body at the first scanning correlated positively to weight of cold half carcass (r=0.63) and meat in the right half carcass (r=0.66). The muscle tissue area of LD at 12.th rib showed a positive relationship (r=0.72) with lean meat content of carcass. Findings reveal on the one hand that calves with higher muscle tissue area at the age of 7 days will also have more meat in carcass, on the other hand the cross sectional scan of LD at 12 th rib of calves can be used as a useful reference scan for the evaluation of meat of carcass.

Investigation of factors affecting on beef marbling score in Japanese beef cattle by image analysis

Y. Hamasaki, N. Murasawa and K. Kuchida, Obihiro University of A&VM, Obihiro, 080-8555, Japan

The degree of marbling in M. longissimus thoracis (ribeye) is economically very important in Japan. Generally, marbling score is evaluated in various factors such as coarseness of the marbling by the beef grader. The aim of this study was to investigate factors affecting the marbling score (MS) by image analysis. Digital images of the 6-7th rib cross section from 1,468 Japanese Black (JB) and 971 crossbreds (JB×Holstein) were used. High quality digital images were taken from April to December in 2006 using a special camera for beef carcass. The ratio of marbling to ribeye area (FAR), the coarseness, and the number of marbling particles were calculated by image analysis. Multiple regression analyses were performed to predict the MS by dates and breeds. The number of selected variables was limited to 3. The MS was used as a dependent variable, and 8 image analysis traits were used as candidates of independent variables. The quadratic effect of FAR was also included. The means of the FAR were 33.4±7.1% in JB and 43.4±8.6% in JB×HO. The most contributing variable in predicting the MS was the quadratic effect of the FAR with a positive regression coefficient in JB and JB×HO and on any dates (R^2=0.80-0.93:JB, 0.75-0.92:JB×HO). The variables except for FAR chosen as significant independent predictors were the number of marbling particles(+) and the coarseness of marbling(-), and their appearance frequencies were 33.3% and 12.5% in JB, 31.1% and 17.8% in JB×HO, respectively.

Session 37 Poster 32

Effect of vacuum or modified atmosphere packaging on some beef quality characteristics

C. Russo, M. D'agata and G. Preziuso, University of Pisa, Department of Animal Production, Viale delle Piagge, 2, 56124 Pisa, Italy

Nowadays, beef shelf-life is very important in relation to various marketing methods that tend to prolong the duration of this product. For this reason, it is interesting to study the effect of different packaging methods on several meat quality parameters. Five steaks were taken from each half-carcass derived from six female veal calves, after 7 days of aging. In each group, one steak was analysed unpackaged, two steaks were vacuum-packed and conserved for 7 and 14 days respectively, while the last two steaks were packaged in a modified atmosphere (60% O2, 30% CO2, 10% N2) and conserved for 7 and 14 days respectively. Thus, meat was analysed for pH, colour, measured on the surface and inside, and water-holding capacity was expressed as Meat/Total ratio. Results showed that pH values were normal and tended to be constant independent of packaging method and length of conservation. The colour parameters measured on the surface of the meat confirm that the vacuum-packaged method induces a slight darkening of the meat (significantly lower values of b* and H*) after both 7 and 14 days of storage. Nevertheless, internal meat colour and water-holding capacity are similar for the two packaging systems, independent of the length of storage. These results, along with the findings of a previous study that testify to the higher microbiological stability of vacuum-packaged meat, lead us to conclude that although the appearance is less attractive, this method permits longer meat storage.

Session 37 Poster 33

The problem of variability in meat quality traits of conventionally processed young bulls: the role of breed and environment effects

F. Vincenti, M. Iacurto, F. Saltalamacchia and S. Gigli, Istituto Sperimentale per la Zootecnia, Via Salaria 13, 00016 Monterotondo (RM), Italy

The aim of this study was to evaluate the variability in meat quality traits of conventionally processed young bulls. In particular we analyzed the influence of two fixed factors: breed (Charolais and Limousin) and environment (4 different farms). A sample of *Longissimus Thoracis* muscle was excised from young bulls within 24 h post-mortem, vacuum packaged and stored at 2±0,5°C until 3 and 7 days post-mortem. The analysed quality parameters were: pH, Warner Shear Force, water losses (drip and cooking loss), Myofibrillar Fragmentation Index, colour (Lightness, Chrome, Hue). Results showed that the most important fixed factor was environment. In fact, extent of variation in many physical parameters analysed, depended on farm where the animals were raised. For breed factor significant differences, according to other works (Wulf *et al.*, 1996; Chambaz *et al.*, 2003) were found on all physical parameters with exception of meat colour. This result is very important because the visual appearance of a meat product determines a consumer's response (Wulf and Wise, 1999). Now further research is required in order to better define farm critical points that cause variability in meat quality traits.

Session 38 Theatre 1

Chromosome regions affecting body weight in egg layers

M. Honkatukia, M. Tuiskula-Haavisto and J. Vilkki, MTT Agrifood Research Finland, Biotechnology and Food Research, Animal Genomics, FI-31600 Jokioinen, Finland

We have previously mapped quantitative trait loci affecting egg production and quality traits using a reciprocal cross of two divergent egg-layer lines, Rhode Island Red and White Leghorn. The lines differ also in body weight, and we initially identified genome-wide significant Mendelian QTL for adult body weight at 40 weeks of age and feed intake at 32-36 weeks of age. In addition, QTL with parent-of-origin effects were detected for feed intake and body weight. In the present study, a total of five body weight traits (weight at 16, 20, 24, 40 and 60 weeks of age) have been analysed from the same mapping population with a slightly different marker maps. Our results confirm the earlier findings but also reveal new QTL. New QTL areas were found on chromosomes 1, 4, 5, 6, and 13. Both Mendelian inheritance and loci with parent-of-origin expression were found. Our conclusions are in good agreement with the results of previous studies from different mapping populations. The results elucidate the most important chromosome regions affecting weight in poultry in general and may add to the understanding of such loci throughout the animal kingdom.

Session 38 Theatre 2

Multivariate models applied to QTL detection for carcass composition on SSC7

H. Gilbert[1], P. Le Roy[2], D. Milan[3] and J.P. Bidanel[1], [1]INRA, SGQA, UR33, 78352 Jouy en Josas, France, [2]INRA, GA, UMR598, 35042 Rennes, France, [3]INRA, LGC, UMR444, 31326 Castanet-Tolosan, France

Multivariate QTL detections were carried out in a Large White x Meishan pig cross to disentangle several QTL mapped in the SLA region for carcass composition traits, backfat thickness measurements (BFT1, BFT2), backfat (BFW) and leaf fat (LFW) weights, intramuscular fat content (IMF). We tested multiple trait and linked QTL models using the INRA QTLMAP software based on approximate likelihood ratio tests assuming mixture of full- and half-sib families with interval mapping techniques. First, groups of traits were selected using a backward selection procedure. Traits were selected from their contribution to the linear combination of traits discriminating the putative QTL haplotypes. Three groups could be distinguished from successive discriminant analyses: 1) external fat (BFT1, BFT2); 2) internal fat (LFW, IMF); 3) BFW. Applying 2-QTL models favored a general hypothesis of at least two linked pleiotropic QTL for external fat, one in the SLA region and the second around 140cM. A more precise pattern was distinguished for internal fat, with two linked QTL, one at 0cM only influencing leaf fat and a pleiotropic QTL on IMF and leaf fat in the SLA region. This approach could not distinguish between the loci in the SLA region. Meishan alleles decreased the values for all traits but IMF, which may be of high interest to improve IMF content while maintaining carcass composition.

Session 38 Theatre 3

Investigation of bovine chromosome 20 for QTL affecting milk production traits using a selective milk DNA pooling strategy and individual genotyping in a daughter design

L. Fontanesi[1], E. Scotti[1], M. Dolezal[2], E. Lipkin[3], S. Dall'Olio[1], P. Zambonelli[1], D. Bigi[1], F. Canavesi[4], R. Davoli[1], M. Soller[3] and V. Russo[1], [1]University of Bologna, DIPROVAL, Sezione di Allevamenti Zootecnici, Via F.lli Rosselli 107, 42100 Reggio Emilia, Italy, [2]University of Natural Resources and Appl. Life Sci., Dep. Sustainable Agricultural Systems, Div. Livestock Sci., Gregor-Mendel-Strasse 33, 1180 Vienna, Austria, [3]The Hebrew University of Jerusalem, Dep. of Genetics, Edmond Safra Campus, 91904 Jerusalem, Israel, [4]ANAFI, Via Bergamo 292, 26100 Cremona, Italy

Bovine chromosome 20 (BTA20) has been the objective of a few studies aimed at identifying QTL affecting milk yield and composition. Then, mutations in two genes, *GHR* and *PRLR*, have been suggested to explain some of these QTL. We scanned BTA20 for QTL influencing milk yield and milk protein percentage using a daughter design in eight Holstein Friesian sires. Two approached were used: selective milk DNA pooling combined with approximate interval mapping and sire haplotype analysis; individual genotyping of the daughters and interval mapping. Twenty microsatellites and the *GHR F279Y*, *PRLR S18N* and *PRLR P186L* mutations were genotyped. The results indicated good agreement between the two approaches and showed the presence of QTL that in part can be ascribed to the mutations in the two considered genes. However, these mutations cannot explain all significant effects observed on BTA20 for the investigated traits.

Session 38 Theatre 4

Need for sharp phenotypes in QTL detection for calving traits in dairy cattle

T. Seidenspinner[1], J. Bennewitz[1], F. Reinhardt[2] and G. Thaller[1], [1]Institute of Animal Breeding and Husbandry, Christian-Albrechts-University, Hermann-Rodewald-Str. 6, D-24118 Kiel, Germany, [2]United Datasystems for Animal Production (VIT), Heideweg 1, D-27283 Verden/Aller, Germany

QTL mapping results for the traits stillbirth and dystocia in dairy cattle are only partly consistent, which might be due to the use of different trait definitions and different recording of phenotypes. The aim of this study was to map QTL in German Holsteins for the traits dystocia and stillbirth in first and second parity as direct calf effect and maternal effect of the sires daughter. Phenotypes were daughter yield deviations estimated in a univariate setting separately for first and second parity. QTL mapping was done using multi-marker regressions including permutation test and false discovery techniques. The results were markedly different for first and second parity, as across traits 18 chromosome-wise significant QTL were found for first, 12 for second parity, but only 3 in common. 3 QTL for maternal stillbirth showed a genome-wise significance, located on chromosomes 7, 15 and 23, respectively. A comparison with mapping results using phenotypes obtained from a multivariate breeding value estimation across parities revealed a different set of significant QTL. It can be concluded that for QTL mapping there is a need for a parity-specific trait observation for dystocia and stillbirth and that daughter yield deviations should be obtained from univariate breeding value estimation.

Session 38 Theatre 5

Detection of quantitative trait loci for udder traits and stature in Finnish Ayrshire

N.F. Schulman, S.M. Viitala and J.H. Vilkki, MTT Agrifood Research Finland, Biotechnology and Food Research, H-talo, FIN-31600 Jokioinen, Finland

Udder traits are important due to their correlation with clinical mastitis, which causes major economic losses to dairy farms. Chromosomal areas associated with udder conformation traits, milking speed and leakage could be used in breeding programs to improve both udder traits and mastitis resistance. Quantitative trait loci (QTL) mapping for udder traits and stature was carried out on bovine chromosomes (BTA) 9, 11, 14, 18, 20, 23, and 29, where earlier studies have indicated QTL for mastitis. A grand-daughter design with 12 Ayrshire sire families and 360 sons was used. The sires and sons were typed for 35 markers. The traits analysed were udder depth, fore udder attachment, central ligament, distance from udder to floor, fore teat length, udder balance, rear udder height, milking speed, leakage, and stature. Associations between markers and traits were analysed with multiple marker regression. Five genome wise-significant QTL were detected: stature on BTA14 and 23, udder balance and central ligament on BTA23, and rear udder height on BTA11. In addition, 13 chromosome-wise significant QTL were suggested. Several of the traits mapped on the same positions on BTA11, 14, and 23. On BTA11 and 14 the suggested QTL positions for udder traits overlap with previously detected QTL for mastitis and SCS.

Session 38 Theatre 6

Fine mapping of QTL for mastitis resistance on BTA11 in three Nordic Red cattle breeds

G. Sahana[1], M.S. Lund[1], L. Andersson-Eklund[2], N. Schulman[3], S. Viitala[3], T. Iso-touru[3], S. Värv[4], H. Viinalass[4] and J. Vilkki[3], [1]Aarhus University, Genetics and Biotechnology, Blichers Alle 20, 8830 Tjele, Denmark, [2]Swedish University of Agricultural Sciences, SLU, 75007 Uppsala, Sweden, [3]Agrifood Research Finland, MTT, 31600 Jokioinen, Finland, [4]Institute of Veterinary Medicine and Animal Sciences, Kreutzwaldi 64, 51014 Tartu, Estonia

Combined linkage and linkage disequilibrium analysis with a variance component based approach was used to fine map QTL affecting mastitis resistance on BTA11. A granddaughter design of 14 grandsire families with 524 progeny tested sons from three related Nordic red breeds (Finnish Ayrshire, Swedish Red and White, and Danish Red) was used in this study. The traits analyzed were clinical mastitis (CM) and somatic cell score (SCS). Thirty-seven markers (both microsatellites and single nucleotide polymorphisms) were genotyped along a segment of the chromosome spanning 85.2 cM. We were able to fine map a QTL affecting CM at 17.8 cM. A QTL for SCS was detected at 62.8 cM with the QTL interval spreading over 14 cM between markers BM6445 and BMS2047. The SCS QTL could not be further fine mapped due to lack of linkage disequilibrium in the region.
This work was partly funded by EC FP5 project Mastitis Resistance (QLK5-CT-2002-01186).

Session 38 Theatre 7

QTL detection for male fertility traits in dairy cattle

T. Druet[1], B. Basso[1], E. Sellem[2], L. Salas-Cortes[2], P. Humblot[2], X. Druart[3] and S. Fritz[4], [1]INRA, UR337, Domaine de Vilvert, 78352 Jouy-en-Josas, France, [2]UNCEIA, 13, rue Jouët, 94703 Maisons-Alfort, France, [3]INRA, UMR85, Centre de Tours, 37380 Nouzilly, France, [4]UNCEIA, 149 rue de Bercy, 75595 Paris, France

A QTL detection experiment has been implemented in France to search for QTL related to male fertility in dairy cattle. Ten families, involving in total 515 bulls, were measured for ejaculate volume and spermatozoa concentration, number of spermatozoa, motility, velocity, percentage of motile spermatozoa after thawing and spermatozoa abnormalities. 148 microsatellite markers were used to realize a genome scan. First, genetic parameters were estimated for all the traits. Production traits presented moderate heritabilities (from 0.15 to 0.30) while some of the quality traits such as motility had high heritabilities (close to 0.60). Genetic correlations among traits showed, for instance, strong negative relationships between volume and motility or between velocity and spermatozoa abnormalities. Only three QTL were significant at $P < 0.001$, all related to spermatozoa abnormalities. In addition 11 QTL ($P > 0.01$) and 18 QTL ($P > 0.05$) were detected. However, due to lack of power of the design further analyses are required to confirm these QTL. Multitrait techniques such as Discriminant Analysis were applied to increase the power of detection of these QTL. The LRT test was increased for most QTL and even new QTL were detected.

Session 38 Theatre 8

QTL detection for fatty acid composition on porcine chromosome 12 and analysis of the candidate genes FASN, GIP and ACACA

E. Alves[1], G. Muñoz[1], A. Fernández[1], A.I. Fernández[1], C. Barragán[1], J. Estellé[2], R. Quintanilla[3], L. Silió[1], M.C. Rodríguez[1] and C. Óvilo[1], [1]INIA, Mejora Genética Animal, 28040 Madrid, Spain, [2]UAB, Ciència Animal i dels Aliments, 08193 Barcelona, Spain, [3]IRTA, Genética y Mejora Animal, 25198 Lleida, Spain

We have carried out a QTL mapping and candidate gene analyses for fatty acid composition on SSC12 in an Iberian x Landrace cross. The content of ten fatty acids was measured in backfat samples from 377 F2 animals. Seven markers were genotyped in order to perform the QTL scan. Two significant QTL were detected: QTL1 for C18:3(n-3), C20:1(n-9), C16:0 and C14:0 (10-35 cM), and QTL2 for C16:1(n-9), C18:0 and C18:1(n-7) (70-80 cM). *FASN* gene was studied as candidate gene for QTL1. cDNA characterization and sequence analysis revealed the presence of ten SNPs, linkage mapping showed that *FASN* gene is away from the confidence interval of the QTL1. *GIP* and *ACACA* genes were analysed as candidate genes for QTL2. Both cDNA sequences were characterized and analysed in the animal material. The *GIP* sequence analysis allowed us to detect three SNPs, none could be associated to the QTL effect. *ACACA* sequence analysis allowed us to detect 15 SNPs. Two of them, *ACACA*:c.5634C>T and *ACACA*:c.6681G>T, were genotyped and four haplotypes were identified. SNPs and haplotypes association analyses were performed. Highly significant effects for SNP *ACACA*:c.5634C>T on C16:1(n-9) (0.12), C18:0 (-0.31) and C18:1(n-7) (0.07) acids were detected.

Weight gain of F2-gilts depends on its paternally inherited IGF2-allele

H.C.M. Heuven[1,2], B.T.T.M. van Rens[2], E.M. van Grevenhof[2] and H. Bovenhuis[2], [1]University of Utrecht, Veterinary faculty, Yalelaan 7, 3584 CL Utrecht, Netherlands, [2]Wageningen-UR, Animal Breeding and Genetics Centre, Marijkeweg 40, 6709 PG Wageningen, Netherlands

F2-gilts, originating from a cross of 5 F1-boars and 21 F1-sows (Large White x Meishan), were reared on a restricted feeding regime to maximize their reproductive performance. The goal was to determine the effect of the *IGF2* gene on growth trajectory of F2-gilts from birth until weaning of their first litter. For this purpose, weight measurements at weekly intervals from birth until 10 weeks of age and subsequently on a monthly basis until weaning of their first litter were available on F2-gilts. From birth until weaning F2-gilts were reared by F1-sows; fed ad lib until 10 weeks and from then on fed a restricted diet using commercially available complete feeds. Grand parents, F1 parents and F2-gilts were genotyped for the causative *IGF2*-SNP (A/G) as well as 13 other SNPs in a 6 cM region. F1-parents were heterozygous for *IGF2*-SNP. The surrounding marker information was used to determine parental origin of the *IGF2* alleles in the F2-gilts. Analysis showed a significant effect of the paternally inherited *IGF2*-allele on weight at different ages. When the F2-gilts inherited the *IGF2*-A-allele the weight was increased compared to the gilts that inherited the G-allele. A small difference existed at birth and it increased to approximately 6 kg at weaning of the first litter. The difference in weight gain might be due to increased lean muscle growth at the expense of fat deposition.

Session 38 Theatre 10

A novel approach for estimating allele frequencies of lethal autosomal-recessive genetic disorders

S. Manatrinon[1], C. Egger-Danner[2] and R. Baumung[1], [1]University of Natural Resources and Applied Life Sciences Vienna (BOKU), Gregor-Mendel-Strasse 33, A-1180 Vienna, Austria, [2]Zuchtdata-EDV Dienstleistungs GmbH, Dresdner Str. 89, A-1200 Vienna, Austria

Based on the gene dropping method we developed a new approach to estimate allele frequencies for lethal autosomal-recessive genetic disorders. The advantage of this stochastic approach is that it provides allele frequency distributions. We tested the method in the deep and complex pedigrees of the Austrian Brown Swiss population and compared it with several other approaches that allow the estimation of frequencies of recessive alleles based on known carrier animals. Applying classical pedigree analysis we identified carriers of Arachnomelia (A), Spinal Dysmyelination (SDM), Spinal Muscular Atrophy (SMA) and Weaver (W) 4, 4, 10 and 11, respectively among the 500 genetically most important ancestors contributing to a defined reference population, e.g. animals born 2005/2006. Estimates derived by different methods were quite similar. Arithmetic mean for allele frequencies in % and standard deviation (in brackets) from 1000 gene dropping simulation runs for a cohort of animals born 2005 and 2006 are: 1.94 (1.22), 5.56 (1.89), 6.77 (1.85) and 3.12 (1.13) for the disorders A, SDM, SMA and W, respectively. For a cohort born one generation earlier, 2001, the results were as follows: 1.67 (0.86) for A, 4.34 (1.47) for SDM, 5.48 (1.27) for SMA and 3.33 (1.07) for W, showing an increase in frequencies for all disorders except for W.

Session 38 Theatre 11

Concordance between IBD probabilities and linkage disequilibrium

F. Ytournel, D. Boichard and H. Gilbert, INRA, UR337, Domaine de Vilvert, 78352 Jouy en Josas, France

Studies (Grapes *et al.*, 2006; Zhao *et al*, 2007) have been conducted to estimate the optimal haplotype length to estimate IBD probabilities as defined by Meuwissen and Goddard (2001). We simulated different genetic maps with various lengths, marker densities and markers composing them (SNPs, microsatellites or a mixture of both). The simulated populations included 100 individuals. IBD probabilities were computed either at the QTL position or at the middle point of the marker bracket preceding the QTL. We evaluated:

- the distribution of the IBD probabilities for haplotypes of 4, 6 or 10 markers depending on the real IBD status at the QTL, to evaluate the ability of the probabilities to discriminate IBD from non-IBD QTLs,
- the evolution of the correlations between the real QTL IBD status and IBD probabilities depending on a) the location of marker in highest linkage disequilibrium with the QTL (LD, evaluated with χ^{2}' (Yamazaki, 1977)) and b) the values of LD between the QTL and its closest marker.

It appeared that non-IBD QTLs were better identified than IBD ones with all designs. The discrimination ability improved with the presence of microsatellite markers in the haplotype, with the increase of the haplotype length or the increase of the number of markers defining the haplotype, and when IBD probabilities were computed at the QTL position. The correlation between IBD probabilities and real IBD grew with the intensity of LD and with shortest distances between the QTL and the marker in maximum LD with it.

Session 38 Theatre 12

Benefits of using phenotypic measures of dams for estimating QTL variance components and MA-BLUP EBV

S. Neuner[1], R. Emmerling[1], G. Thaller[2] and K.-U. Götz[1], [1]Bavarian State Research Centre for Agriculture, Institute of Animal Breeding, Prof.-Dürrwaechter-Platz 1, 85586 Poing-Grub, Germany, [2]Christian-Albrechts-University, Institute of Animal Breeding and Husbandry, Olshausenstraße 40, 24098 Kiel, Germany

Reliable estimates for variance components (VC) in QTL-models are important for fine mapping experiments and marker assisted genetic evaluations (MAGE). In cattle populations only a small fraction of the population is genotyped at genetic markers in most cases and only these animals are included in MAGE models. Phenotypic measures in MAGE models are pre-corrected phenotypes (daughter yield deviations for bulls, DYD, yield deviations for cows, YD) estimated in polygenic animal models for the entire population. Since DYD and YD may represent different amounts of information, the problem of weighting arises. To detect the best combination of observations and weighting factors in MAGE models, a stochastic simulation for the trait milk yield was applied. The results show that the use of MAGE DYD models is the most appropriate strategy to estimate QTL VC, but weighting is essential. To ensure that marker assisted selection improves selection efficiency even for moderate QTL effects ($\geq$10%) MAGE models should include DYD and YD. A useful strategy for practical MAGE is to estimate VC in DYD models and EBV in DYD-YD models. Additionally, the benefit of using MA-BLUP EBV for the selection among paternal half sibs inheriting alternative QTL alleles was investigated.

Session 38 Theatre 13

Using bioinformatics to reduce the search for genes to a known 4% of the bovine genome

G.E. Pollott, SAC, Sustainable Livestock Systems Research Group, Bush Estate, Penicuik, Midlothian, EH26 0PH, United Kingdom

The 3rd release of the bovine genome comprises about 3 billion bases. Searching for genes in such a large database is a formidable task. The Neutral Indel Model combines methods from molecular evolution and comparative genomics to identify functional DNA in a target genome. The method uses the gaps between insertions/deletions across the two genomes to identify DNA under purifying selection. The bovine and human genomes, and their alignment, were downloaded and analysed using the Neutral Indel Model. The model identified 3.8% of the bovine genome as being functional DNA and a database of functional DNA segments, known as indel-purified segments, was constructed. In order to test the efficacy of the method, a file of all known bovine genes was downloaded and compared to the database of indel-purified segments. At a false discovery rate of 10%, some 63% of known bovine exonic DNA was located within the indel-purified segments database. This encouraging result implies that the Neutral Indel Model was able to identify a large proportion of bovine functional DNA, with no prior knowledge of its purpose. The implications of this result are that the search for unknown genes, or all genes affecting a specific quantitative trait, only needs to be concentrated in the indel-purified segments. This reduces the search space to 3.8% of the genome. Future research will refine the Neutral Indel Model in order to increase the proportion of exonic DNA found in the indel-purified segments to 90-100%.

Session 39 Theatre 1

Predictive ability of different models for clinical mastitis in joint genetic evaluation for Sweden, Denmark and Finland

K. Johansson[1,2], S. Eriksson[1], J. Pösö[3], U. Sander Nielsen[4] and G. Pedersen Aamand[5], [1]Swedish University of Agricultural Sciences, P.O. Box 7023, SE-750 07 Uppsala, Sweden, [2]Swedish Dairy Association, P.O. Box 7023, SE-75007 Uppsala, Sweden, [3]FABA Breeding, PB 40, FI-01301 Vanta, Finland, [4]Danish Agricultural Advisory Service, Udkaersdej 15, DK-8200 Aarhus, Denmark, [5]Nordic Cattle Genetic Evaluation, Udkaersdej 15, DK-8200 Aarhus, Denmark

Clinical mastitis (CM) and somatic cell count (SCC) in three lactations, and first lactation udder conformation traits (UC), are included in the joint genetic evaluations of Nordic Red and Holstein breeds in Sweden, Denmark and Finland since 2006. The aim of this study was to compare predictive ability of different multi-trait models for udder health on a Nordic level. Linear sire models including different number of udder health traits (CM, SCC, UC) were used to estimate breeding values (EBVs), based on data comprising 2.7 million observations recorded until 2002. Correlations were estimated between these EBVs and daughter group means for clinical mastits recorded from 2003 and onwards. The comparison was made for 654 bulls born between 1997 and 2000. Correlations between daughter group means and EBVs from models with only CM, only SCC and only UC were 83%, 85% and 66%, respectively, of the correlation estimated using the full multi-trait model (CM+SCC+UC).

Session 39 Theatre 2

A bivariate threshold model analysis of calving difficulty and stillbirth in Norwegian Red cows

B. Heringstad[1,2], Y.M. Chang[3], M. Svendsen[2] and D. Gianola[1,4], [1]Department of Animal and Aquacultural Sciences, Norwegian University of Life Sciences, P.O. Box 5003, N-1432 Aas, Norway, [2]Geno Breeding and A. I. Association, Aas, N-1432, Norway, [3]Division of Genetic Epidemiology, Cancer Research UK Clinical Centre, St James s University Hospital, Leeds, United Kingdom, [4]Department of Dairy Science, University of Wisconsin, Madison, Wisconsin, USA

The objectives were to infer genetic parameters of calving difficulty (CD) and stillbirth (SB), and to evaluate phenotypic and genetic change for these traits for Norwegian Red cows. SB is a binary trait and CD has three categories: "easy calving", "slight problems", and "difficult calving". Average stillbirth rate for Norwegian Red has remained unchanged since 1978; at 3 % for first calving and 1.5 % for second and later calvings. The percentage in the category "difficult calving" has not changed over years (2-3 % for heifers and 1 % for cows), but the category "slight problems" increased from 4 % to 7 % for heifers and from 2 % to 3 % for cows. A total of 528,475 first calving records were analyzed with a Bayesian bivariate sire-maternal grandsire threshold liability model. Posterior means of direct and maternal heritability were 0.13 and 0.09 for CD, and 0.07 and 0.08 for SB. Genetic correlations were 0.79 (direct SB and direct CD), 0.63 (maternal SB and maternal CD), and close to 0 between direct and maternal effects. No genetic change for SB was found, while a slight genetic improvement for CD was detected in Norwegian Red.

Session 39 Theatre 3

An improved model for the genetic evaluation for length productive life in Swiss Red & White, Swiss Holstein and Swiss Brown cattle

C. Stricker[1], A.R. Sharifi[2], U. Schnyder[3] and V. Ducrocq[4], [1]applied genetics network, Boertjistrasse 8b, 7260 Davos, Switzerland, [2]Institute of Animal Breeding and Genetics, Albrecht-Thaer-Weg 3, 37075 Goettingen, Germany, [3]Swiss Brown Cattle Breeders Federation, Chamerstrasse 56, 6300 Zug, Switzerland, [4]UR337, INRA, 78352 Jouy-en-Josas, France

Animal tracing data bases allow to determine more precisely the time of culling of an animal. Furthermore, Ducrocq (2003) showed how lactation stage specific baseline hazards can be accounted for in a Weibull model to remove unrealistic genetic trends, a problem that was inherent in the genetic evaluation in Swiss dairy breeds. We analyzed 215'000/1.2Mio/800'00 lactations of daughters of 2455/9685/5942 sires for the Holsteins, Swiss Browns and Red and Whites, respectively. The sire-maternal grand sire Weibull model additionally included the effects for age at first calving, herd*year*season, region*season, deviations of within herd*year performance for milk yield and the sum of fat an protein content. For all three breeds, the model revealed unrealistically high sire variances, due to a huge genetic trend in the sire proofs. Thus, we changed the effects from deviations of herd*year performance to deviations*year of herd*year performance. This resulted in a plausible genetic trend in all breeds and a heritability of 0.08 for length of productive life in Holsteins. However, preliminary results still show unrealistically high sire variances for Red and Whites and Swiss Browns, and need further investigation.

Session 39 Theatre 4

Heritability of lifetime milk yield and productive life and their relationship with production and type traits in the Simmental, Swiss Fleckvieh and Red Holstein populations in Switzerland

M. Gugger[1], F. Ménétrey[2], S. Rieder[1] and M. Schneeberger[1], [1]Swiss College of Agriculture, Länggasse 85, 3052 Zollikofen, Switzerland, [2]Swiss Simmental and Red&White Cattle Breeders' Association, P.O. Box 691, 3052 Zollikofen, Switzerland

Costs of milk production can be reduced through lower replacement rate by increasing productive life of cows. Lifetime production (LP, production to 6th lactation) and productive life (PL, number of completed lactations) of 112,850 daughters of 766 test AI bulls were used to obtain daughter averages and to estimate heritabilities. Bulls belonged to three sections of the Swiss Simmental and Red&White cattle herd book, differing in percentage of Red Holstein genes. Correlations of daughter average LP and PL with sire EBVs for production, functional and type traits and with composite indices differed among herd book sections and, in some instances, changed signs (e.g., for correlations with EBV for linearly scored muscling) due to different breeding objectives. The strongest correlations of LP were found with EBV milk (>0.69), of PL with total merit index (0.44 to 0.52) and the composite fitness index (0.32 to 0.56). Heritabilities were estimated using two sire models (without or with including a fixed effect of herd book section). They were around 0.19 and 0.13 for LP, and 0.111 and 0.097 for PL from the two models. Estimates obtained from the first model may be more appropriate because breeding objectives differ among herd book sections.

Session 39 Theatre 5

Relationship between milk production traits and fertility in Austrian Simmental cattle

B. Gredler[1,2], C. Fuerst[2] and J. Sölkner[1], [1]University of Natural Resources and Applied Life Sciences Vienna, Department of Sustainable Agricultural Systems, Gregor Mendel Str. 33, 1180 Vienna, Austria, [2]ZuchtData EDV-Dienstleisungen GmbH, Dresdner Str. 89/19, 1220 Vienna, Austria

The effects of milk urea nitrogen (MUN), fat-protein-ratio (F:P), milk lactose percentage (MLP) and milk yield (Mkg) on fertility traits days to first service (DFS) and days open (DO) were analysed. In total, records of 15,662 first lactation dual purpose Simmental cows were examined. The test day record closest to the date of first insemination was chosen. The SAS procedure MIXED was used to estimate the effect of milk production on fertility. The model included the fixed effects of herd, year and month of calving, a continuous effect of calving age and separately MUN, F: P, MLP or Mkg. All milk production traits significantly affected DFS and DO ($P < 0.0001$). As a first step for heritability estimation univariate analyses were run for milk production and fertility. Effects accounted for production traits were the fixed effects of herd, year and month of test-day milk recording, AM/PM milking, a continuous effect of days in milk (linear and quadratic) and a random genetic animal effect. For DFS and DO the same model as for statistical analyses (omitting milk production traits) was applied. Heritabilities for MUN, F:P, MLP, Mkg, DFS and DO were 0.12±0.013, 0.11±0.013, 0.36±0.02, 0.30±0.017, 0.045±0.007 and 0.031±0.007, respectively. In a further step multivariate analyses will be run.

Session 39 Theatre 6

Parameter estimation for fertility traits of dairy cattle using a multiple trait model

Z. Liu, J. Jaitner, E. Pasman, S. Rensing, F. Reinhardt and R. Reents, VIT, Genetic Evaluation Department, Heideweg 1, 27283 Verden, Germany

Genetic parameters were estimated using a multiple trait model for six fertility traits of dairy cattle: age at first insemination (AF) and non-return rate 56 days (NRh) of maiden heifers, interval calving to first insemination (CF), non-return rate (NRc), days open (DO) and calving interval (CI) of cows. The statistical model contains three fixed effects, permanent environmental effect of cow, and additive genetic effect. For NRh and NRc, a permanent effect of service sire was fitted additionally. A total of 20 trait-wise sub-analyses were performed using a very large dataset to estimate the parameters via VCE 5, followed by a bending procedure, based on eigen- and eigenvectors, to make all (co)variance matrices positive definite. Heritability estimates of the fertility traits were low: 1.2% (NRh), 5.2% (CF), 1.5% (NRc), 5.2% (DO), and 4.3% (CI), with exception of the growth trait 22.4% (AF). Genetic correlations among the time interval traits CF, DO and CI were high, ranging from 0.82 to 0.95. The non-return rates had a genetic correlation of 0.63 between maiden heifers and cows. Low genetic correlations were found between the non-return and the other traits of cows. Service sire permanent effect accounted only for 0.5% of phenotypic variance, but reasonably large variation existed among service sires in both non-return rate traits.

Session 39 Theatre 7

Claw disorders diagnosed at hoof trimming: relationships with fertility

H.H. Swalve[1], H. Alkhoder[1] and R. Pijl[2], [1]Institute of Agricultural and Nutritional Sciences, Animal Breeding, Adam-Kuckhoff-Str. 35, 06108 Halle, Germany, [2]Pijl-Hooftrimming, Fischershäuser 1, 26441 Jever, Germany

Claw disorders can be diagnosed at the time of hoof trimming. Findings were collected at hoof trimming using a personal digital assistant (PDA) with an interface to a data base on a PC and an interface to herd data stemming from the central milk recording computer. A total of around 40,000 records from 14,000 cows were collected over a period of six years. Data comprised the pathological findings (sub-clinical and clinical), herd environment information, milk yields, pedigree information as well as records on all inseminations from which fertility parameters were derived. The most prominent disease of the claw was laminitis, found in around 33% of all cows. Other diseases were of lesser importance but summarizing all findings for all 16 diseases recorded revealed that only around 39 % of all records showed no disease at all. Relationships with fertility parameters were not as strong as anticipated. The number of inseminations per pregnancy was unaffected by claw diseases while days open showed tendencies for an increase for most diseases and significant differences (+ 14 days) for dermatitis interdigitalis. It may be concluded that only severe cases of claw diseases show a pronounced effect on fertility. The present study also is an example for general problems which arise when using field data on fertility, as strategies for data edits become more important than statistical modelling.

Environmental and genetic effects on claw disorders in Finnish dairy cattle

A.-E. Liinamo, M. Laakso and M. Ojala, Helsinki University, Department of Animal Science, P. O. Box 28, FIN-00014 Helsinki, Finland

The aim was to study the environmental and genetic effects on the most common claw disorders in the Finnish dairy cattle population. The data were obtained through a nationwide claw health programme during 2003 and 2004. Altogether, 74 410 observations on 41 087 cows and heifers originating from 1 462 dairy farms were included in the data, representing 15% of the milk recorded dairy cows. Claw disorder information had been collected by claw trimmers during their routine visits to the dairy farms participating in the programme. Studied traits included sole hemorrhages, white line disease, heel erosion, screw claws, and all claw disorders combined. Fixed effects were analysed both with linear and non-linear models, and genetic parameters were estimated with a linear approximation of the binary data using repeatability animal model and REML method. Breed, parity, lactation stage, 305-d milk production, claw trimming frequency, claw trimming season, feed type, barn type, bedding type, manure removal method, herd size, herd, and claw trimmer all had a statistically significant effect on most of the studied claw disorders. Heritability estimates of the claw disorders varied between 0.01 and 0.07, while corresponding repeatability estimates varied between 0.05 and 0.33. Genetic and phenotypic correlations between different claw disorders were positive and low to moderate.

Relationships between claw disorders and milk yield in Holstein cows estimated from recursive linear and threshold models

S. König[1,2], X.L. Wu[3], D. Gianola[3] and H. Simianer[2], [1]University of Guelph, Department of Animal and Poultry Science, N1G2W1 Guelph, Canada, [2]University of Göttingen, Institute of Animal Breeding and Genetics, 37075 Göttingen, Germany, [3]University of Wisconsin, Department of Dairy Science, Madison WI 53706, USA

Relationships between 4 different claw disorders and test day milk yield were analyzed in a Bayesian framework by fitting standard linear, threshold, recursive linear, and recursive threshold models to data from 5360 Holstein cows. A two-way causal path was postulated describing first the influence of test day milk yield on claw disorders and, secondly, the effect of the disorder on milk yield at the following test day. Heritabilities of disorders were slightly larger when applying threshold or recursive threshold models relative to linear or linear recursive models. Posterior means of genetic correlations between level of milk yield at individual test days and disorders ranged between 0.16 and 0.43 suggesting that breeding strategies focussing on increased milk yield increase susceptibility to claw disorders. Structural coefficients in the model described recursiveness at the phenotypic level. The increase of disease probability per 1 kg increase of test day milk yield was between 0.003 and 0.024. Structural coefficients ranging between –0.121 and –0.670 predict that one unit increase in the incidence of any disorder results in a reduction of milk yield at the following test day by up to 0.67 kg.

Uniform breeding goals increase the possibilities of co-operation across dairy cattle populations

L.H. Buch, M.K. Sørensen, A.C. Sørensen, J. Lassen and P. Berg, Department of Genetics and Biotechnology, University of Aarhus, 8830 Tjele, Denmark

Differences in trait definitions and relative economic weighting of breeding goal (BG) traits reduce similarity of BG and consequently possibilities of co-operation between populations. This study had two objectives: first, to quantify the effect of a uniform definition of the traits included in the fertility indices in two populations on genetic gain, and second to quantify the effect of a more similar relative weighting of BG traits across populations on genetic gain. The objectives were investigated using stochastic simulation. Two dairy cattle populations were simulated in a 25 years period doing 15 replicates. One population mimics the Nordic Holstein population and the other mimics the US Holstein population. Both BG consisted of production, udder health, and fertility. Three scenarios were simulated: (1) the current situation where trait definitions and relative economic weighting differ, (2) the traits included in the US fertility index were changed into the traits registered in the Nordic countries, and (3) lower economic weight was placed on production in the US BG which made the relative weighting of the traits in the two BG more similar. Genetic correlations between BG were 0.81, 0.81, and 0.84 for scenario 1, 2, and 3, which is about the supposed break-even genetic correlation for profitable co-operation. Genetic correlations between fertility indices were 0.82, 0.85, and 0.85 for scenario 1, 2, and 3.

Authors index

C

D

G

I

J

K

M

Q

R

S

T

U

V